William Jerome Harrison

Geology of the Counties of England and of North and South Wales

William Jerome Harrison

Geology of the Counties of England and of North and South Wales

ISBN/EAN: 9783337002817

Printed in Europe, USA, Canada, Australia, Japan

Cover: Foto ©berggeist007 / pixelio.de

More available books at **www.hansebooks.com**

GEOLOGY

OF THE

COUNTIES OF ENGLAND

AND OF

NORTH AND SOUTH WALES.

BY

W. JEROME HARRISON, F.G.S.,

SCIENCE DEMONSTRATOR FOR THE BIRMINGHAM SCHOOL-BOARD; LATE CURATOR
LEICESTER TOWN MUSEUM;

AUTHOR OF "A MANUAL OF PRACTICAL GEOLOGY," ETC., ETC.

LONDON:
KELLY & CO., 51, GREAT QUEEN STREET, LINCOLN'S INN FIELDS, W.C.;
AND
SIMPKIN, MARSHALL & CO., STATIONERS' HALL COURT, E.C.

MDCCCLXXXII.

PREFACE.

In this book I have endeavoured to give, with as much detail as was consistent with clearness, a description of the geological structure of England and Wales.

The main object at which I have aimed, has been to enable any person, albeit little skilled in the science of geology, to readily ascertain the nature of the rocks of any part of the country in which he may dwell, through which he may be travelling, or respecting which he may need information.

I have, accordingly, made this book a *geographical* geology. Considering that every Englishman knows fairly well the position of the counties generally, while with his own county he is usually very familiar, I have written a separate account of the geology of each county.

Taking the principal rock-masses, or geological formations in turn, I have traced the strata more or less minutely through each county, indicating the position and direction of the beds by naming the towns and villages situated upon the respective divisions of the rocks. An account of the nature, thickness, and characteristic fossils of each stratum is given, together with its economic products, its influence on the scenery of the district, and its value (if any) as a water-bearing bed.

By the use of the Index each geological formation may be tracked throughout England, and its development and nature in one county compared with the conditions under which it occurs in any other: as a rule, each rock-bed has been described most fully in the county in which it occupies the largest area and is best displayed.

PREFACE.

I have prefixed to the geological description of each county lists of (1) its scientific societies, (2) its museums, (3) the maps and books of the Geological Survey referring to the county, (4) the more important books and papers written respecting its rocks by private workers. Altogether, I have given in this way the names of about 190 societies, 140 museums, and about 700 titles of books, papers, &c. It will give some idea of the *quantity* of geological work which has been done by private individuals in England, when I state that I have consulted in the preparation of this book more than 4,000 papers, memoirs and pamphlets, mostly published in the transactions, journals, &c., of the various learned and local societies during the present century.

Although, as (late) Curator of the Leicester Town Museum, and as Secretary to the Midland Union of Scientific Societies, I have had special facilities for the compilation of the above lists, I feel that they must necessarily be incomplete, and I would earnestly ask for correction and assistance from all who are interested in the progress of geology.

Many of the articles in this book have already appeared in print as part of the local topographies in the Post Office Directories of the Counties, published by Messrs. Kelly & Co., who are also the publishers of this work.

It was, indeed, the interest excited by the articles in this form, as shown by the very numerous letters of enquiry on geological subjects, which were received after the publication of the Directory for any single county, that led to the completion of the work, and its issue in its present form.

To Mr. E. R. Kelly I would tender my sincere thanks for the personal interest which he has shown in this book, and for the great pains which he has taken to help me in its preparation.

A preliminary list of text-books, &c., on geology will be found at p. xxv., including such as should, at least, be possessed by every public library; few provincial libraries, however, contain even those here enumerated, and the want of books of reference is a great hindrance to country workers.

I have received help from many quarters, which it is my duty gratefully to acknowledge. Having had occasion to consult every map which has been drawn, and every line that has been written by the officers of the Geological Survey of England, I wish to bear special testimony to the high value, thorough accuracy, and the immense painstaking, evinced by their work; to the Director-General of the Survey—Professor Ramsay—I am indebted for the kind permission to copy from the published memoirs of the Survey nearly all the sections which illustrate this book; sections which are especially valuable since they have in all cases been drawn and measured on the spot, and are records of facts and not of theories. The cuts of flint implements are copied from Mr. John Evans's great work—"The Ancient Stone Implements of Great Britain," and those of fossils from Professor Nicholson's equally unique book on Palæontology.

Many geologists have personally aided me, and I would heartily thank Messrs. W. Whitaker, H. B. Woodward, T. Davidson, W. Adams (of Cardiff), E. B. Marten (of Stourbridge), C. Moore, S. Allport, John Evans, Grenville Cole, E. Wilson (of Nottingham), Professors Judd, Rupert Jones and Lebour, the Rev. P. B. Brodie, and the Rev. H. W. Crosskey, for the readiness with which they have at all times afforded me help and information. I would also thank the hundreds of local geologists whose papers I have read, and whose labours I have utilized; their individual work is, I trust, in each case recognized in connection with the county to which their discoveries refer.

To the Council of the Geological Society I am indebted for the loan of the blocks for Figs. 55 and 56, originally drawn to illustrate my paper on the Rhætic Beds in their "Quarterly Journal."

The compilation of this book has occupied my leisure time for a long period, but if it be found a handy work of geological reference by the general enquirer, and also prove useful to the local student by showing him what is known of his district, and where to go for further information, the objects which I had in view in writing it will have been fully attained.

W. JEROME HARRISON.

Birmingham, Nov., 1881.

CONTENTS.

INTRODUCTION	i
LEARNED SOCIETIES	xxii
GEOLOGICAL SURVEY OF THE UNITED KINGDOM	xxiv
GEOLOGICAL BOOKS AND PAPERS (GENERAL LIST)	xxv
GEOLOGY OF—	
BEDFORDSHIRE	1
BERKSHIRE	7
BUCKINGHAMSHIRE	14
CAMBRIDGESHIRE	20
CHESHIRE	29
CORNWALL	37
CUMBERLAND	45
DERBYSHIRE	55
DEVONSHIRE	61
DORSETSHIRE	71
DURHAM	80
ESSEX	86
GLOUCESTERSHIRE	92
HAMPSHIRE	99
ISLE OF WIGHT	105
HEREFORDSHIRE	114
HERTFORDSHIRE	120
HUNTINGDONSHIRE	125

CONTENTS.

KENT	129
LANCASHIRE	139
LEICESTERSHIRE	151
LINCOLNSHIRE	163
MIDDLESEX	171
MONMOUTHSHIRE	178
NORFOLK	183
NORTHAMPTONSHIRE	193
NORTHUMBERLAND	199
NOTTINGHAMSHIRE	207
OXFORDSHIRE	212
RUTLAND	217
SHROPSHIRE	221
SOMERSETSHIRE	229
STAFFORDSHIRE	239
SUFFOLK	249
SURREY	258
SUSSEX	263
WARWICKSHIRE	269
WESTMORLAND	275
WILTSHIRE	282
WORCESTERSHIRE	292
YORKSHIRE—EAST RIDING	298
,, NORTH RIDING	304
,, WEST RIDING	310
NORTH WALES	322
SOUTH WALES	334
APPENDIX (GLOSSARY OF GEOLOGICAL TERMS)	343

INDEX.

Adams, W., 340.
Agassiz, Prof., 331.
Aitken, J., 150.
Allport, S., 37, 151, 221, 222, 270.
Alluvial deposits, 182.
Alluvium, 118, 174, 211, 216.
Alum shales, 307.
Ammonites. 73.
Ammonite zones, 300.
Andrews, W., 270.
Andrews, Rev. W. R., 282.
Annelid burrows, 336.
Anning, Miss M., 73.
Anstie, John, 93, 228.
Anthracite, 65, 340.
Anticlinal curve, x.
Aqueous rocks, iv.
Arenaceous rocks, iv.
Arenig slates, 46.
Arenig beds, 223. 326, 337.
Argillaceous rocks, iv.
Arvonian formation. 324. 336.
Ash-beds, 153. 326.
Atherfield clay, 100, 107, 132, 259.
   '265.
Atthey, T. 205.
Austen, R. A. C. Godwin, 62, 175, 264.
Austin, T., 92.
Aveline, W. T., 45, 139, 151, 193, 207,
   209. 275.
Aymestry limestone, 117, 179, 225,
   241, 294.

Baggy and Marwood beds, 63.
Bagshot beds, 11, 78, 88, 103. 110.
   137, 173, 261, 289.
Baker, J. G., 304.
Bala beds, 224, 326, 337.
Bala limestone, 278.
Banks, R. W., 114.
Bannisdale slates, 141, 278, 312.
Barkas, T. P., 205.
Barr limestone, 241.
Barrett, L., 21.
Barrois, C., 109.
Barton beds, 103, 111.
Basalt, 201, 226, 246.
Bate, C. S., 80.
Bateman, T., 248. 309.
Bath oolite, 214, 236, 284.

Bauerman, H. 1, 14.
Bayfield. T. G., 188.
Beale, C. 239.
Beche, Sir H. de la, 38, 61, 217, 228,
   229, 334.
Beckles. S. H., 76, 264.
BEDFORDSHIRE, 1.
Beesley, T., 213.
Belt, T., 38, 184.
Belemnites, 197.
Bellamy, J. C., 322.
Bembridge beds, 111.
Bensted. W. H., 132.
BERKSHIRE, 7.
Bernician beds, 81, 202.
Bevan, Dr., 340.
Beyrich, Prof., 112.
Binney. E. W., 30, 45, 140, 311, 315.
Bird, C., 45, 311.
Blackmore, Dr. H. P., 282.
Blake, Prof. J. F., xxviii.; 21, 71, 163,
   283, 300, 304, 306, 338.
Blue John (fluor spar), 56.
Blown sand, 190.
Bognor beds, 78, 103, 110, 267.
Bolckow & Vaughan, Messrs., 306.
Bolton, J., 45.
Bone beds, 95, 96, 117, 225. 273, 316.
Bonney, Prof. T. G., xxviii., 20. 23,
   151, 239, 276, 323.
Books, list of geological, xxv.
Boot, J. T., 207.
Borings, 90, 137, 147, 154, 155, 166,
   169, 173. 175, 187, 211, 213, 233,
   250, 264, 273, 306.
Borings in Herts, 123.
Borrowdale series, 47, 141, 276.
Boulder clay, 36. 89, 118, 123, 127,
   148, 160, 169, 174. 190, 197, 205,
   220, 227, 255. 297, 303, 319.
Box stones, 253.
Bradford clay, 236, 285.
Brady, H. B., 81.
Bracklesham beds, 11, 103, 111, 267.
Brandon beds, 255.
Breccia, 295.
Breccias, 226.
Brick-earth, 90, 137.
Brine, 296.

Bristow, H. W., 72, 105, 232.
British Association, xxii.
British fossils, xiv.
Brockenhurst series, 103.
Brockram, 52, 280.
Bronze age, 28.
Brodie, Rev. P. B., 14, 29, 114, 151, 269, 270.
Brown, R., xxviii.
Brown, J., 86.
Brown, T. F., 334.
Browne, A., 9, 105.
Browne, A. J. Jukes, xxviii, 3, 20, 25, 163.
BUCKINGHAMSHIRE, 14.
Buckman, Prof. J., 71, 74, 228, 238, 285.
Buckland, Dr., 188, 212, 309.
Budleigh Salterton pebbles, 66.
Bunbury, C. J., 184.
Bunbury, Sir C., 192.
Bunter sandstone, 33, 59, 66, 96, 147, 155, 209, 226, 234, 247, 272, 295, 306, 319, 330.
Bure Valley beds, 189, 255.
Buried forests, 27.
Burns, D., 206.
Burtle beds, 237.
Burton, F. M., 163.
Bury, W., 285.
Busk, Prof., 321
Cainozoic rocks, xvii.
Calcareous rocks, iv.
Calciferous sandstone series, 81.
Calcite, iii.
Callaway, Dr. C., 114, 221, 222, 223, 323.
Cambrian formation, 115, 223, 293, 324, 336.
Cambridge greensand, 24.
CAMBRIDGESHIRE, 20.
Cannel coal, 145.
Caradoc beds (see Bala beds).
Carruthers, W., 286.
Carstone, 23, 100, 186, 266.
Carboniferous formation, 40, 51, 65, 81, 95, 118, 142, 153, 180, 202, 225, 231, 241, 270, 279, 294, 305, 312, 329, 339.
Carboniferous limestone, 30, 51, 56, 65, 95, 118, 142, 153, 180, 225, 231, 243, 279, 313, 330, 339.
Caverns, 69, 231, 309, 321, 332, 341.
Cephalopoda bed, 236.
Chalk, 3, 9, 18, 77, 87, 101, 108, 121, 132, 169, 173, 187, 215, 237, 250, 260, 266, 288, 301.
Chalk marl, 3, 24, 67, 77, 101, 109, 121, 187, 266, 288.

Chalk rock, 4, 9, 18, 25, 77, 109, 121, 215, 251, 288.
Chalky boulder clay, 26.
Champernowne, A., 62, 64, 228, 232.
Charlesworth, E., 249.
Charnwood Forest rocks, 152.
Chert, 306.
CHESHIRE, 29.
Chillesford beds, 89, 189, 255.
Chines, 107.
Chloritic marl, 77, 101, 108, 237, 266, 288.
Clarke, Rev. W. B., 183.
Clough, C. T., 80.
Clunch, 214.
Clutterbuck, Rev. J. C., 120, 172, 177.
Coal (how formed), 144, 204.
Coal measures, 32, 52, 58, 65, 81, 82, 95, 144, 154, 181, 204, 208, 225, 231, 232, 243, 270, 294, 315, 316, 330, 340.
Coal mines (deep), 145.
Codrington, T., 99, 105.
Colchester, W., 253.
Collins, J. H., 37, 38.
Collyweston slates, 167, 195, 219.
Condamine, Rev. H. M. de la, 125.
Conglomerates, 279.
Coniston flags, 278.
Coniston grits, 141.
Coniston limestone, 141, 277, 312.
Copper, 42.
Copper ore, 327.
Coprolites, 3, 17, 18, 23, 24, 89, 121, 254, 260.
Corder, H., 86.
Coralline crag, 89, 254.
Coral rag, 2, 8, 16, 22, 75, 127, 215, 237, 286, 301, 309.
Cornbrash, 2, 15, 74, 97, 126, 167, 196, 214, 220, 237, 285.
Cornstones, 118, 180, 225.
Cotham marble, 96, 234.
Crag beds, 189.
Cretaceous formation, 2, 67, 76, 87, 101, 108, 120, 168, 187, 215, 237, 250, 260, 266, 287, 301.
Crofton, Rev. A., 276.
Cromlechs, 333.
Cross, Rev. J. E., 163.
Crust of the Earth, vii.
Cucullea zone, 63.
Culm, 40.
Culm measures, 65.
CUMBERLAND, 46.
Curley, T., 118
Dakyns, J. R., 310.
Dalton, W. H., 86, 314.
Damon, R., 71.

INDEX.

Darbishire, R. D., 54.
Davidson, T., xxviii., 62, 221.
Davies, D. C., 226, 322.
Davies, W., 129, 287.
Davis, J. W., 311, 316.
Davis, L., 325.
Dawkins, Prof. W. B., 30, 55, 184, 234, 238, 249, 264.
Day, E. C. H., 71.
Delabeche, Sir H. (see Beche, Sir H. de la).
Denbighshire grits, 329.
Denudation, xi., 191, 331.
De Rance, C. E. (see Rance, C. E. de).
DERBYSHIRE, 55.
Devonian formation, 39, 63, 117, 230, 231.
DEVONSHIRE, 61.
Dewstone, 295.
Diabase, 291.
Diatoms, 137.
Dickenson, P., 32.
Dickinson, J., 140, 145.
Dimetian formation, 324, 336.
Diorite, 270, 271.
Dip, ix.
Dogger, 308.
Dolerite, 154, 246.
Dolomite, 153, 318.
Dolomitic conglomerate, 96, 181, 234, 340.
Downes, Rev. W., 70.
Downton sandstones, 94, 117, 225.
Drew, F., 129.
Drift, 4, 19, 25, 36, 43, 53, 60, 65, 85, 89, 98, 112, 118, 123, 127, 148, 160, 169, 174, 190, 197, 205, 211, 216, 217, 220, 227, 247, 255, 274, 280, 289, 297, 309.
Druid stones, 289.
Duncan, Prof. P. M., xxviii., 62.
Dunes, 44, 85, 149.
D'Urban, W. S. M., 62.
DURHAM, 80.
Dykes, 309.
EAST RIDING, 298.
Eccles, J., 142.
Egerton, Sir P. G. M., 29, 71, 221.
Elements, ii.
Elvans, 41, 68.
Eocene formation, 10, 19, 78, 88, 102, 110, 122, 133, 188, 216, 251, 261, 266, 289.
Erratics, 148, 160, 162, 169, 220, 247, 267, 281.
Eskers, 281.
Etheridge, R., xxviii., 61, 64, 65, 92, 282.
Ettingshausen, Baron von, 129.
Evans, C., 99, 258.

Evans, J., 5, 19, 124, 128, 150, 176, 198, 216, 238, 297.
Evans, J. F., 322.
Eyton, Miss, 227.
Falconer, Dr. H., 71, 341.
Farewell rock, 231.
Farrer, J., 314.
Faults, vii., 84, 146, 154, 205, 313.
Feather-edge coal, 31.
Felspar, iii.
Fen beds, 26, 128, 169, 190, 256.
Fergusson, Jas., 291.
Firestone, 260.
Fisher, Rev. O., 20, 71, 105, 111, 184.
Fish & insect limestones, 159, 166, 235.
Fitton, Dr. W. H., 217
Fitch, R., 184.
Flint workings, 192.
Flint implements (see Prehistoric Man).
Flower, J. W., 184, 249.
Flower, Prof. W. H., 249.
Fluvio marine crag, 189.
Fluor spar, iii.
Folkestone beds, 100, 132, 260, 265.
Foraminifera, 289.
Forbes, D., 187.
Forbes, E., 105.
Foreland sandstones, 63, 230.
Forest bed, 189, 255.
Forest marble, 15, 74, 97, 196, 214, 236, 285.
Forster, W., 202.
Foster, C. Le Neve, 38, 62.
Fossils, study of, xiii.
Fossils, classification of, xv.
Fossils, use of, xvi.
Fossil plants, xvi.
Fossils, how to name, xvii.
Fox, Col. Lane, 172, 176, 216, 267.
Fox, Rev. W., 106.
Fuller's Earth, 74, 97, 236, 260, 284.
Gabbro, 41.
Ganister beds, 58, 81, 144, 203, 316.
Ganoid fishes, 117.
Gardner, J. S., 67, 99.
Gault, 3, 8, 18, 23, 67, 76, 101, 108, 121, 132, 186, 215, 237, 260, 266, 287.
Geikie, Prof. A., 322.
Geikie, Dr. J., 200.
Geological Society, xxii.
Geological Survey, xviii., xxvi.
Geologist's Association, xxii., 172.
Geology, objects of, ii.
George, E. S., 311
Glacial deposits, 174, 190, 341.
Glacial Period, 53, 118, 123, 148, 169, 205, 280, 297, 319, 331.
Glass, Rev. N., 178.

Glauconite, 24, 108.
Gold, 325, 338.
Goodchild, J. G., 275.
Granite, 40, 68, 152.
Grantham, R. B., 184.
Graptolites, 326.
Grauwacke, 39, 323.
Great Oolite, 2, 15, 74, 97, 126, 167, 196, 214, 220, 236, 284.
Green, Prof. A. H., 1, 14, 46, 55, 139, 143, 212, 239, 243, 310, 316.
Greenstone, 40.
Greenwell, Canon, 192, 206, 309.
Grey-wethers, 12, 79, 104, 261, 289.
Grindrod, Dr., 114, 294.
Gunn, Rev. J., 183, 188, 190, 249.
Gunn, J., 249.
Gunn, W., 80.
Gurney, Miss, 190
Gypsum, iii, 60, 155, 209, 264.
Hall, T. M., 61, 62.
Hampshire Basin, 289.
Hangman grits, 63, 230
Harbottle grits, 203.
Harkness, Prof., 45, 275, 334.
Harlech beds, 325
Harmer, F. W., xxviii, 184, 249.
Harris, W. H., 335.
Harrison, P., 268.
Harrison, W. J., 151, 172, 217, 293.
Haslingden flags, 31.
Hastings beds, 76, 131, 259, 265, 287.
Hawkes, W., 239.
Hay Fell flags, 278.
Headon beds, 103, 111.
Heer, O., 62.
Hempstead beds, 111.
Henwood, W. J., 37.
Henslow, Prof., 255.
Herman, W. D., 193.
Herefordshire, 114.
Herries, W. H., 7.
Hertfordshire, 120.
Hessle beds, 303.
Hicks, Dr. H., 322, 323, 324, 325, 335, 336.
High-level gravels, 112.
Hill gravels, 190.
Hill, Rev. E., 151.
Hoare, Sir R. C., 289.
Hodgson, Miss E., 46
Holloway, W. H. 164.
Holmes, T. V., 46, 311.
Holl, H. B., 37, 61, 115, 292.
Homfray, 326.
Hollybush sandstone, 115, 223, 293.
Holstone (boulder), 162.
Hopkinson, J., 120, 334.
Horses, 95.

Howell, H. H., 1, 80, 151.
Houghton, F. T. S., 276.
Howorth, H. H., xxviii.
Hudleston, W. H., xxviii., 14, 21, 72, 283, 304, 308.
Hughes, T. McK., 120, 275, 311, 321, 323.
Hulke, J. W., 71.
Hull, Prof. E., 1, 14, 29, 92, 139, 140, 145, 151, 179, 212, 239, 243, 247.
Hunstanton limestone, 168.
Hutton, W., 200.
Hunt, R., 42.
HUNTINGDONSHIRE, 125.
Huxley, Prof. T. H., 269
Hyperodapedon, 66.
Hythe beds, 100, 132, 259, 265.
Ibbetson, Capt., 106.
Ichthyosaurus, 73.
Igneous rocks, iv, 40, 50, 67, 84, 200, 246, 279.
Ilfracombe slates, 63, 230.
Inferior oolite, 73, 97, 125, 195, 214, 236, 273, 284, 297.
Inghoe grits, 203.
Inliers, 101, 122, 179, 214, 243.
Ingram, Rev. A. H. W., 270
Insect limestone, 297.
Iron ore, 52, 131, 142, 165, 168, 180, 195, 213, 265, 286.
Ironstone, 307.
Irving, Rev. A., 207, 211.
ISLE OF WIGHT, 105.
Jackson, 321.
Jenkinson, H. I., 46.
Jenkinson, Rev. E., 203.
Jet, 307.
Jones, Prof. Rupert, 7, 55, 221, 263.
Jordan, T. B., 172.
Judd, Prof. J. W., 99, 103, 105, 112, 125, 151, 159, 163, 196, 202, 217, 273, 299, 301.
Jukes, Prof. J. B., 61, 64, 151, 231.
Kaims, 205.
Kaolin, 43
Keeping, H., 105.
Keeping, W., 17, 23, 286, 335.
Kellaway's rock, 2, 16, 75, 126, 168, 285, 309.
Ketton freestone, 220.
Kendall, J. D., 45, 46.
Kentish rag, 132, 259.
KENT, 129.
Keuper beds, 33, 34, 54, 60, 66, 96, 147, 155, 164, 181, 209, 226, 234, 247, 272, 296, 299, 306.
Killas, 40, 69.
Kimmeridge clay, 2, 8, 16, 22, 75, 127, 168, 186, 215, 286, 301, 309.

Kinahan, G. H., 71.
Kinderscout grit, 31, 143.
King, C. C., 7.
King, Prof. W., 81.
Kirkby, J. W., 80.
Kirkby Moor flags, 278.
Kirshaw, — 296.
Kirwan, Rev. R., 70.
Labyrinthodon, 210, 272.
Lamplugh, G. W., 298.
LANCASHIRE, 139.
Lancaster, J., 207.
Landscape limestone, 96, 234.
Landslips, 70, 78.
Langley, A. A., 322.
Lankester, E. R., 249.
Lapworth, C., 334.
Laurentian formation, 115.
Laurentian Period, 293.
Lavis, H. J. J., 62.
Layton, J., 189.
Lead, 42.
Lead Ore, 52, 56, 81, 82, 203, 277, 305.
Learned Societies, xxii.
Lebour, Prof. G. A., 34, 46, 81, 199, 203, 206, 276.
Lee, J. E., 62, 64, 178, 225.
Lees, F. A., 311.
LEICESTERSHIRE, 151.
Lias, 15, 36, 52, 67, 72, 96, 159, 164, 181, 194, 211, 213, 218, 227, 235, 247, 273, 283, 296, 300, 306, 341.
Lignite, 67.
Limestone shales, 231.
LINCOLNSHIRE, 163.
Lincolnshire limestone, 126, 159, 167, 196, 219, 301.
Linchets, 288.
Lingula flags, 223, 325, 336.
Lister, Rev. W., 239.
Llanberis Beds, 325.
Llandeilo (lower) beds, 223.
Llandeilo (upper) beds, 223.
Llandeilo flags, 326, 337.
Llandovery (upper) beds, 94, 116, 224, 241, 294, 335, 338.
Llandovery (lower) beds, 328.
London Basin, 261, 289.
London clay, 11, 19, 78, 88, 102, 103, 110, 123, 137, 173, 216, 252, 261, 267, 289.
Longe, F. D., 93.
Longmynd rocks, 223.
Lonsdale W., 40.
Lower Cretaceous Formation (*see* Neocomian).
Lower Greensand, 2, 8, 17, 23, 76, 100, 107, 127, 131, 186, 215, 259, 265, 287.

Lubbock. Sir J., 289.
Lucas, J., 129.
Ludlow Beds, 94, 116, 179, 225, 338.
Luxullianite, 41.
Lyell, Sir C., 183, 188, 224, 267.
Lynton slates, 63, 230.
Mackintosh, D., xxviii., 29, 36, 46, 140, 275, 322, 323.
Magnesian limestone, 52, 84, 146, 205, 306, 319.
Malm rock, 101, 260, 288.
Malvern shales, 115, 293.
Mammaliferous crag, 189.
Mammatt. E., 151.
Mantell, G. A., 105, 263.
Maps, list of geological, xxvi.
Marl slate, 84, 205.
Marlstone, 96, 159, 194, 165, 213, 218, 235, 273, 297.
Marr, J. E., 276.
Marshall, J. G., 276.
Martin, R. F., 46.
Maskelyne, Prof. N. S., 291.
Mather & Platt, Messrs., 147, 187.
Maw, G., 29, 221.
May Hill sandstone. 294 (*see* Llandovery (upper) beds).
M'Coy, Prof. F., 39.
Mello, Rev. J. M., 55.
Menevian beds. 325, 336.
Mesozoic rocks, xvii.
Metamorphic rocks, v, 50.
Miall, L. C., 272, 311, 317.
Mica, iii.
Mica trap, 279.
Microlestes, 234.
Midford sands, 97, 235, 284, 308.
Middle Glacial beds, 26, 89, 255.
Middle Glacial gravels. 123, 148, 160, 174, 190.
Middle Glacial sands, 36.
MIDDLESEX, 171.
Millepore bed, 301.
Miller, H., 199.
Millstone grit, 31, 52, 57, 65, 81, 82, 95, 143, 153, 180, 203, 225, 231, 243, 270, 280, 305, 314, 330, 339.
Mineralogical Society, xxii.
Minerals, ii.
Minette, 279.
Miocene formation, 67.
Mitchell, Dr. J., 1, 125, 183.
Mitchell, W., 105.
Model of London, 172.
MONMOUTHSHIRE, 178.
Moore, C., xxvii., 228, 233, 235, 287, 334, 335, 341.
Morris, Prof. J., 163, 221, 292, 295.

INDEX.

Morte slates, 63, 230.
Mortimer, R., 298, 302.
Morton, G. H., 30, 140.
Mountain limestone (*see* carboniferous limestone).
Munford, Rev. G., 184.
Murchison, Sir R. I., 45, 114, 118, 223, 224, 226, 275, 317, 323.
Murchisonite, 66.
McMurtrie, J., 232.
Muschelkalk, 234, 295.
Neocomian formation, 76, 100, 106, 130, 168, 186, 259, 265, 287, 301.
Neolithic age (*see* Prehistoric Man).
Newcastle, Duke of, 208
New Red sandstone (*see* Trias).
Newton, Prof. A., 192.
Newton, E. T., 184.
Nicholson, Prof. H. A., 45.
Nicolls, W. T., 99.
NORFOLK, 183.
Northampton sand, 15, 125, 159, 166, 195, 213, 218, 273, 301.
NORTHAMPTONSHIRE, 193.
NORTH RIDING, 304.
NORTHUMBERLAND, 199.
Norton, H., 184.
NOTTINGHAMSHIRE, 207.
Norwich crag, 189, 255.
Norwood, Rev. T. W., 300.
Oldhaven beds, 88, 133, 261.
Old Red sandstone, 63, 94, 117, 142, 179, 225, 229, 294, 329, 338.
Oligocene formation, 103, 112.
Oolite, 1, 73, 97, 125, 159, 166, 186, 194, 213, 218, 235, 264, 273, 284, 301, 307.
Ooze, 302.
Ormerod, G. W., 29.
Ormerod, W. G., 62.
Osborne beds, 111.
Outcrop, ix.
Outliers, 104, 123, 159, 175, 227, 339.
Owen, Prof. R., 213, 253, 267, 287.
Oxford Clay, 2, 8, 16, 22, 74, 97, 126, 168, 197, 214, 237, 285, 301, 309.
OXFORDSHIRE, 212.
Painter, Rev. W. H., 55.
Palæolithic age (*see* Prehistoric Man).
Palæontology, xiii.
Palæontographical Society, xxii.
Palæozoic rocks, xvii.
Papers, list of, xxvii.
Parker, W. K., 55.
Passage beds, 226.
Pattison, S. R., 37.
Peach, C. W., 37, 39.
Pearce, J. C., 286.
Peat, 27, 128, 105, 169, 206.

Pebidian formation, 324, 336.
Penarth beds, 96, 234
Pengelly, W., 37, 62, 70.
Pennant grit, 185, 232, 340.
Pennine Chain, origin of, 147.
Pennine Fault, 51.
Penning, W. H., xxviii., 20, 86, 124.
Pennington, R., 55.
Perkins, C. H., 335.
Permian formation, 52, 54, 84, 96, 146, 155, 204, 208, 226, 233, 246, 271, 280, 295, 306, 317, 330.
Peyton, J. E. H., 264.
Phillips, J. A., xxviii., 38, 323.
Phillips, Prof. J., 61, 71, 212, 228, 284, 299, 300, 304, 308, 311, 313, 314.
Phosphorite, 327.
Phonolite, 41.
Pickwell Down beds, 63, 230
Pilton beds, 64.
Pine raft, 107.
Pipe-clay, 79.
Plastic clay (*See* Woolwich & Reading beds).
Plateau gravels, 256.
Pleistocene formation, 89, 104, 189.
Plesiosaurus, 73.
Pleydell, J. C. Mansel, 71.
Pliocene formation, 89, 189, 253.
Plutonic rocks, v.
Plumbago, 49.
Pontefract rock, 318.
Porphyrite, 200.
Porter, Dr. H., 20, 125.
Portland beds, 16, 75, 215, 287.
Portland cement, 9, 133.
Portland sand, 8.
Post-glacial beds, 90, 320.
Post-glacial deposits, 149, 174, 190.
Post-glacial period, 169
Post-pliocene formation, 237, 267.
Post-tertiary deposits, 137.
Postlewaite, J., 46.
Potato-stones, 234.
Poulton, E. B., 12.
Pre-Cambrian formation, 115, 152, 222, 325, 335.
Pre-glacial beds, 189.
Prehistoric man, 5, 27, 36, 44, 53, 60, 70, 79, 85, 91, 98, 104, 113, 118, 123, 137, 150, 169, 176, 182, 191, 197, 206, 216, 227, 247, 256, 261, 267, 274, 281, 289, 297, 303, 309, 320, 332, 341.
Prestwich, Prof. J., 7, 71, 86, 99, 120, 133, 172, 175, 184, 188, 249, 264.
Price, F. G. H., 24, 129.
Pudding stone, 123.

INDEX.

Punfield series, 76, 107.
Purbeck beds, 17, 78, 215, 264, 287.
Purbeck marble, 76.
Quartz, iii.
Quartz Felsite, 51.
Quartzite pebbles, 66, 274.
Quartzites, 222.
Quaternary period, 189, 255, 303.
Querns, 150, 274, 332.
Raised beaches, 44, 70, 206, 237.
Ramsay, Prof. A. C., xxvii., xxviii., 92, 148, 178, 226, 246, 271, 279, 289, 295, 313. 318, 322, 324, 331.
Rance, C. E. de, 36, 139, 149, 275.
Randall, J. 221.
Ravines, 232.
Reade, T. M., 29, 36, 140.
Reading Beds (see Woolwich & Reading Beds).
Recent deposits, 256
Red chalk, 168, 186, 302.
Red crag, 89, 254.
Red rock fault, 30.
Reed, W., 306.
Reid, C., 184, 191.
Reports, British Association, xxvii.
Rhaetic beds, 36, 52, 66, 72, 96, 159, 164, 181, 211, 227, 234, 247, 273, 296, 299, 341.
River gravels, 197.
Roaches grit, 31.
Roberts. G. E., 221, 292, 295.
Rocks, iv.
Rock salt, iii. 34, 85, 306.
Rofe, J. 306.
Rome, Rev. J. L., 163, 299.
Rose, C. B., 190.
Rough rock, 31, 143.
Ruddy. T., 323.
Russell, R., 46.
RUTLAND, 217.
Rutley, F., 61, 233.
Salt, 35, 148, 155.
Salter, J. W., 221, 223, 275, 322, 325.
Sandgate beds, 100, 132, 260, 265.
Sand-pipes, 251.
Sands of North Downs, 135.
Sarsen stones, 12, 289, 291.
Saunders, J., 1.
Scar limestone, 81, 142, 313.
Scenery, 174.
Scott, Sir G., 196, 209, 219.
Scudder, S. H., 239.
Sedgwick, Prof. A., 207, 224, 323.
Seeley, Prof. H. G., 1, 2, 20, 22, 23, 99, 184.
Selenite, 253.
Septaria. 74, 88, 137, 173, 253, 261.
Serpentine, 41.

Shap granite, 279.
Sharp, S., 193, 217.
Shelley, — 262.
Shineton shales, 223.
Shipman, J., 207.
Shone, W., 30.
SHROPSHIRE, 221.
Shrubsole, W. H., 129, 137, 173.
Silkstone coal, 316.
Silt, 169.
Silurian formation, 39, 46, 64, 81, 94, 116, 141, 178, 202, 223, 241, 276, 294, 312, 326, 329, 337.
Simonside grits, 203.
Skertchly, S. J. B., 20, 28, 125, 163, 183, 185, 191, 249, 256.
Skiddaw slates, 46, 141.
Slate, 325.
Smith, J., 221
Smith, W., xvii, 83, 229, 302, 299.
Smith, W. G., 124, 172, 176.
Sollas, W. J., 20, 178, 340.
SOMERSETSHIRE, 228.
Sorby, H.C., xxviii., 151, 299, 311, 315.
SOUTH WALES, 334.
Speeton beds, 301.
Spencer, Jas., xxviii.
Spencer, T., 311.
Spirorbis limestone, 225.
Spurrell, F. C. J., 138.
Stanley, Hon. O. 322, 333.
Stephens, D., 74.
Stevens, E. T., 282
Stockdale shales, 141, 278.
Stone-bed, 189.
Stonehenge, 289.
Stone implements, 320.
Stonesfield slate, 214.
Strahan, A., 29, 30.
Strangways, C. Fox., 298, 310.
Strike, ix.
Stukeley, Dr., 36.
Submerged forests, 70.
Sub-wealden boring, 264.
SUFFOLK, 249.
Suffolk bone bed. 253.
Sun-cracks, 296.
SURREY, 258.
Sussex marble, 265.
SUSSEX, 263.
Swallow-holes 11, 102, 306.
Syenite, 51, 152.
Symonds, Rev. W. S., 114, 119. 292, 293, 322, 334.
Symons, J. G., 172.
Synclinal curve, x.
Table of formations, xxi.
Tarannon shales, 329, 338.
Tate, G., 80, 199, 201, 202.

INDEX.

Tate, R., 92, 166, 228, 304.
Taylor, J. E., xxviii., 250, 252.
Taylor, J., 183.
Taylor, R. C., 183, 249.
Taylor, Dr. W., 335.
Tawney, E. B., 105, 232.
Teall, J. J. H., 1.
Ten-yard coal, 245.
Tertiary period, 133, 188, 216.
Tertiary epoch, 251, 261.
Thanet beds, 88, 133, 252, 261.
Thomas, A., 92.
Thomas, T. H., 340.
Thompson, B., 193.
Tiddeman, R. H., 140, 275, 311, 321.
Till, 53, 190.
Timins, Rev. J. H., 293.
Tomes, R. F., 269.
Topley, W., 46, 129, 258, 263, 276.
Totternhoe stone, 3, 9, 18, 25, 121.
Trap dykes, 279.
Travertin, 126.
Trevelyan, Sir W. C., 201.
Tremadoc beds, 115, 325, 336.
Trias, 52, 59, 66, 85, 96, 147, 155, 164, 181, 209, 226, 234, 247, 272, 295, 299, 306, 319, 330, 340.
Trilobites, 336.
Trimmer, J., 183.
Tuedian beds, 202.
Tute, Rev. J. S., 319.
Twigg, G. H., 239.
Upper Greensand, 8, 18, 67, 76, 101, 108, 121, 132, 187, 215, 237, 260, 266, 288.
Ussher, W. A. E., 38, 62, 228, 232, 234.
Vernon, Rev. W. V., 298.
Voelcker, Dr., 3.
Volcanic ashes, 328.
Volcanic rocks, v.
Volcanic series of Borrowdale, 276.
WALES, SOUTH, 334.
WALES, NORTH, 322.
Walford, E. A., 213.
Walker, H., 86.
Walker, J. F., 20.
Wallace, W., 45.
Ward, J. C., 45, 46, 276.
Ward, T., 29.
Warp, 169, 320.
WARWICKSHIRE, 269.
Waterstones, 34, 155, 272.
Water supply, 175.
Watts, W., 140.
Weald clay, 76, 106, 131, 259, 265.
Wealden beds, 130, 265.

Wenlock beds, 94, 116, 179, 224, 241, 294, 329, 338.
Westleton shingle, 255.
WESTMORLAND, 275.
Wethered, E., 93.
Weybourne sands, 189.
Whin Sill, 51, 81, 84, 201, 280.
Whitaker, W., 1, 6, 7, 14, 20, 29, 37, 62, 71, 86, 99, 105, 120, 129, 133, 140, 172, 212, 217, 228, 249, 251, 255, 258, 263, 267, 269, 282.
Whiteaves, J. F., 213.
White, Gilbert, 100.
White lias, 66, 234, 273, 300, 341.
Whiting, 9.
Wild, G., 140.
Wilkins, E. P., 105.
Willett, H., 264.
Williamson, Prof. W. C., xxviii.
Wilson, E., 55, 140, 163, 207.
WILTSHIRE, 282.
Wiltshire, Rev. Thos., xxii., 184, 299.
Winn, C., 165.
Winwood, Rev. H. H., 341.
Wise, J. R., 99.
Woburn sands, 2, 17.
WORCESTERSHIRE, 262.
Wood, Col., 341.
Wood, E., 306.
Wood, Rev. H. H., 74.
Wood, S. V., 254, 299.
Wood, S. V., jun., xxvii., xxviii., 26, 28, 86, 120, 163, 172, 184, 263.
Woodhouse, J. T., 271.
Woodward, C. J., 56.
Woodward, H., 221, 282.
Woodward, H. B., 62, 92, 183, 184, 188, 228, 229, 232, 285, 335.
Woodward, S., 183, 188.
Woodward, S. P., 81.
Woolhope beds, 241, 329, 338.
Woolhope limestone, 94, 116, 224, 294.
Woolwich and Reading beds, 10, 19, 78, 88, 102, 103, 104, 110, 122, 133, 173, 216, 252, 261, 267, 289.
Wright, C. C., 207.
Wright, Dr. T., xxvii., 93.
Wurzburger, P., 45.
Wyatt, J., 1, 5.
Yeovil marble, 74.
Yoredale beds, 31, 52, 57, 81, 142, 225, 243, 280, 305, 314.
YORKSHIRE, East Riding, 298.
  ,,   North Riding, 304.
  ,,   West Riding, 310.
Zones of the chalk, 25, 109, 133.

INTRODUCTION.

THE PRINCIPLES OF GEOLOGY.

CHAPTER I.

THE object of this book being to describe as simply as possible, yet with a fair amount of detail, the nature, arrangement and history of the rocks of England and Wales, it seems advisable, in a few preliminary pages, to explain briefly the main facts and principles of geological science, such as can be only shortly alluded to, or a knowledge of which must be taken for granted in the body of the work.

Objects of Geology.—The Historic Period dates back some 5,000 years from the present time; here the task of the geologist begins, and his work extends backwards to the time when the world first became fit to be the habitation of living beings. With the origin of the earth, or of the solar system, geology is not strictly concerned; these are questions for the astronomer and the physicist.

What the geologist aims at, is to endeavour to explain, as far as possible, all the changes which our planet has passed through since its surface was first composed of land and water. For this purpose we find it necessary carefully to study all the natural phenomena which are now taking place, such as volcanoes, earthquakes, the action of the sea, of rivers, ice, &c., the forms and habits of organized beings, &c. For we believe that the forces which are now acting on our globe are those which have acted during the past, and so we explain the past by the aid of the present.

If the science of geology could be perfected, it would enable us to reconstruct the physical geography of any given epoch; to say exactly what was then the arrangement of seas and continents, and with what kind of life they were peopled. Although through the imperfection of the geological record we can never hope to attain this complete knowledge, yet the approach we can make towards it is of the highest interest and importance.

The broad principles of geology are very simple and easily comprehended, and a knowledge of them would prove a source of pleasure and profit to everyone : the scenery which surrounds us, the soil beneath our feet, the mineral treasures beneath that soil, the sites of our towns and villages, the occupations of the people, the nature of the water we drink, and countless other facts which meet us in our every-day life, all depend upon the science of geology for their reasonable explanation. By bringing home to every one's door, a description of the geology of the district in which they live, it is hoped that this book will render the acquisition of such knowledge more easy and direct.

THE ELEMENTS.—By chemical analysis it has been found that all the matter which we have been able to examine, and which constitutes all the known part of our globe and its surroundings, is made up of about seventy simple substances, called the elements. Of these, the gas called oxygen is by far the most abundant, constituting about one-half by weight of the earth's crust, while an element called silicon forms one-quarter ; if to these two non-metals we add the six metals, aluminium, iron, calcium, magnesium, sodium and potassium, we shall have ninety-nine hundredths of the weight of the solid exterior of the earth now known to us.

MINERALS.—The elements, with the exception of a few of the metals, as gold, silver, &c., are rarely found in a pure or separate state ; two, three, or more of them generally occur combined chemically with one another, forming what we call minerals. A mineral may be defined as an inorganic homogeneous (*i.e.*, having a definite chemical composition) substance occurring naturally in the crust of the earth. Some thousands of minerals are now known, and as examples we will describe a few of the commonest.

Quartz, composed of oxygen and silicon (Si. O_2), crystallizes in six-sided prisms and pyramids; so hard that it cannot be scratched by a knife; usually colourless or white. Sand is formed of small broken-up particles of quartz. Rock-crystal, amethyst, cairngorm, cat's-eye, and avanturine are all forms of quartz; chalcedony, cornelian, bloodstone, agate, flint, jasper and opal, consist of quartz combined with a small quantity of water, the various colours being mostly due to minute quantities of oxide of iron.

Felspar.—The name of a *group* of minerals, all being silicates of alumina combined with other silicates, usually of potash, soda, lime, or iron; can be scratched with difficulty by a knife; colour usually white or flesh-colour.

Mica.—Also the name of a group, characterised by readily splitting into thin plates.

The above three minerals —quartz, felspar, and mica— are the essential ingredients of granite.

Calcite or Calc-spar.—Crystallized carbonate of lime, (Ca. C. O_3) consisting of calcium, carbon, and oxygen; effervesces with acids (strong vinegar will do), and can be readily scratched by a knife. Colour usually white; also called dog-tooth spar, satin spar, and Iceland spar. The cavern-deposits called stalactites and stalagmite, and the well-known rocks named calcareous tufa, chalk, and limestone, are also composed of carbonate of lime.

Gypsum —Sulphate of lime, (Ca. S. O_4), composed of calcium, sulphur and oxygen; can be scratched by the finger-nail. When crystallized is called *selenite*, or, when pure and compact, *alabaster*.

Fluor-spar.—Fluoride of calcium (Ca. F_2), composed of calcium and fluorine; commonly called Blue John; occurs in beautiful cubical crystals.

Rock-salt.—Composed of sodium and chlorine (Na. Cl.); readily recognised by its taste.

ROCKS.—Of minerals such as we have been describing all rocks are formed. Sometimes a rock, as limestone, is formed of one kind of mineral only, but more generally two or more minerals occur mingled together and forming a rock.

In the geological use of the term rock, the idea of hardness, which we are so apt to associate with it, must be utterly discarded. A rock is any substance which makes a definite bed or layer in the crust of the earth, no matter whether it be soft and yielding like clay or sand, or hard and impenetrable like granite or sandstone.

The diversity of rocks, like that of minerals, is very great, and it is only of late years that their study has made much progress, at least in this country. First of all we classify rocks according to their origin, terming those Igneous which seem to have been melted by heat: we call others Aqueous which show signs of having accumulated in water; while a sub-class, linking these two together, is found in the Metamorphic Rocks, for these bear evidence of having been subjected to great heat (and other agencies also) which, however, was not enough to melt them.

Aqueous Rocks.—These are, with a little care, readily recognised. In the first place they are made up of irregular fragments, such as flakes of mud or grains of sand; then they are stratified or arranged in beds or layers, just as these particles settled down on the bottoms of the old seas or lakes in which the aqueous rocks were formed; and, again, they often contain the remains of animals or plants (*fossils*) which inhabited or were carried into the waters and, dying, became ultimately enclosed in the sedimentary matter which was accumulating then, as now, on the ocean floor.

When clayey matter predominates the rock is termed *argillaceous*; if sandy, *arenaceous*; or if limy, *calcareous*.

Igneous Rocks are crystalline; sometimes the crystals of which they are composed are so minute as to be only visible when a very thin slice of the rock is examined under the microscope. The igneous rocks occur in irregular masses of variable extent, which break through and alter the neighbouring aqueous rocks. Of course we cannot expect to find any fossils in them. If the nebular theory be correct, according to which the earth has cooled down from a molten mass to its present condition, then the rocks which formed the first crust of our planet must all have belonged to this igneous division; but of these no trace now remains, all have been removed by the incessantly operating forces of denudation since that far distant period, whose most probable date is now fixed by scientists at about one hundred millions of years ago. The exact period in years, geologists do not profess to be able to determine; but it is certain that our earth is of exceeding antiquity, and we believe in its great age because the evidence given by the rocks reveals changes, for whose accomplishment periods of time would be required, which we may attempt to estimate in figures, but whose real significance the human mind can scarcely appreciate. Igneous rocks which have cooled down and solidified at great depths are called *Plutonic*. These include granite, syenite, felstone, quartz-porphyry or elvans, &c. When we take blocks of these rocks and melt them, we find that they cool down to stony masses quite different in appearance from the original rocks. The fact is that such rocks as granite have been formed in the earth's interior by processes which we cannot imitate. They have been produced by a moist heat, acting for perhaps many years or even centuries, and aided by the pressure of thousands of feet of superincumbent rocks. The chemical force, too, has come into play, aided by those electric and magnetic currents which are continually traversing the earth's crust, and the result is a crystalline mass which we have not yet been able to imitate in the laboratory.

The division of igneous rocks called *Volcanic*, includes such as have been poured forth in a melted state at the surface, or have cooled down from a liquid condition at no great depth, or under water. Of modern volcanic rocks, lava and volcanic ashes are well known; basalt, pitchstone, and trachyte represent the lava of old times; while the coarse slates of

Wales, Cumberland, Charnwood Forest, &c, are for the most part only volcanic ashes, which have been consolidated and hardened.

Metamorphic Rocks form a sub-class linking together the two great divisions of the aqueous and igneous rocks. They include *gneiss* and *mica-schist*, which are probably altered slate, *slate* itself, *quartzite* or altered sandstone, *marble* or crystalline limestone, and many other varieties. Such rocks generally occur in the immediate neighbourhood of large masses of igneous rocks, and can often be traced passing into ordinary aqueous rocks on the one hand, while they become more highly altered as their proximity to the igneous mass increases. This is often well shown by the thin beds of basalt which occur in the coal measures; the shale for a few inches on either side of the basalt is greatly hardened and converted into *porcellanite;* the next foot or so of rock is slightly hardened; but further off no effects are visible.

CHAPTER II.

INTERNAL HEAT OF THE EARTH.—Very numerous observations have shown, that in mines or bore-holes the heat of the rocks increases at the rate of one degree Fahrenheit for every 50 or 60 feet we go down, (not counting the first hundred feet, the temperature of which varies with the seasons). If the heat continues to increase at this rate, it must be sufficient, at a depth of 30 or 40 miles, to melt any known substance, supposing the experiment were tried at the earth's surface. But in considering what would happen at a great depth, we must take into account the influence of the enormous pressure which would be exerted there, by the weight of all the rocks between the given point and the surface; for we know that the greater the pressure, the greater will be the heat required to melt any substance. In this way the increased pressure tends to counteract the increased heat as we go downwards, and it is now considered probable that the earth is solid to the very centre.

Thus, our earth is still cooling, and the internal heat which it now has, is probably only a fraction of that which it possessed when in the liquid state, or when, still earlier, it formed with the other members of the solar system a mighty gaseous cloud or nebula.

MOVEMENTS OF THE EARTH'S CRUST.—It is an almost universal law, that as substances cool they contract, therefore "a cooling earth is a contracting earth." As matters now stand, the interior or "nucleus" of our earth is cooling more rapidly in proportion than the outer portion or "crust." The latter endeavours to follow and accommodate itself to the shrinking nucleus, and in doing this it is thrown into ridges and hollows which form our continents and ocean basins. This shrinking is continuous in its action, although being so gradual its effects only become visible in the course of centuries. Thus we now know that while the extreme south of the Scandinavian peninsula is sinking, all the northern portion is slowly rising, so that raised beaches occur at North Cape, at an elevation of 600 feet above sea-level: Greenland is oscillating in a similar manner. Mr. Darwin, too, has shown that the occurrence of the Coral Islands, known as Atolls and Barrier Reefs, is only explicable by supposing the coral polyp to build upwards on slowly sinking land.

Where the rocks yield to the pressure, they are thrown into long curves, and every chain of mountains shows this folding, sometimes in a very marked manner; yet we know that these rock-beds were originally deposited in a horizontal position on the sea-floor; they must, therefore, owe their present undulations to the contracting forces to which we have alluded.

When the rocks break along the line of pressure, instead of yielding to it, we have *faults* produced, for the elevation or depression is then confined to one side of the fault, and so beds which were once continuous, become dislocated and parted : thus the Coal-measures of the centre of Scotland owe their preservation to their situation between two great faults, one of which runs across the country from Stonehaven to the Firth of Clyde, and the other from Dunbar to the coast of Ayr ; all the beds between these two faults are, geologically speaking, some two or three miles below their proper position as compared with the rocks which form

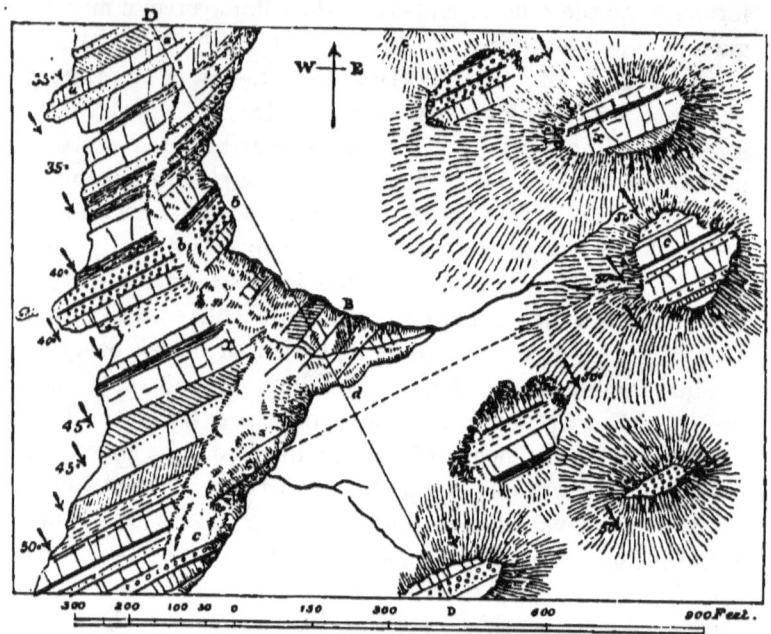

Fig. 1.—This represents a geological map of a rocky coast line. The successive strata are seen, one after another, on the beach ; they form lofty cliffs, and are exposed inland at various openings, quarries, &c., as at x and c. The *strike* of the beds is from north-east to south-west, and their *outcrop* agrees mainly with this. The *dip* is to the south-east, as indicated by the arrows, at angles increasing from 35 to 50 degrees. (*From Brit. Encyc.*)

the Highlands on the one hand, and the Lammermuir Hills and Lead Hills on the other.

ARRANGEMENT OF ROCK MASSES.—When layers of rock (strata) lie in a horizontal position, we judge that they have been elevated very evenly, for we know that they were deposited horizontally on the floors of seas, although we now see them perhaps hundreds or thousands of feet above the level of the ocean.

Scarcely any stratum, however, retains this exact horizontality over a large area; we shall, if we follow it, find it inclining or dipping in one direction, and rising in the opposite until it "crops out," *i.e.*, comes to the surface. The older rocks we shall find, as a rule, much more sharply inclined, sometimes even standing on end or vertical.

DIP.—The amount measured in degrees by which any rock-bed deviates from horizontality is termed its *dip*; this amount, or angle of dip, is determined by an instrument called a clinometer; it is also necessary to find the *direction* of dip, which is of course done by a compass: thus we may

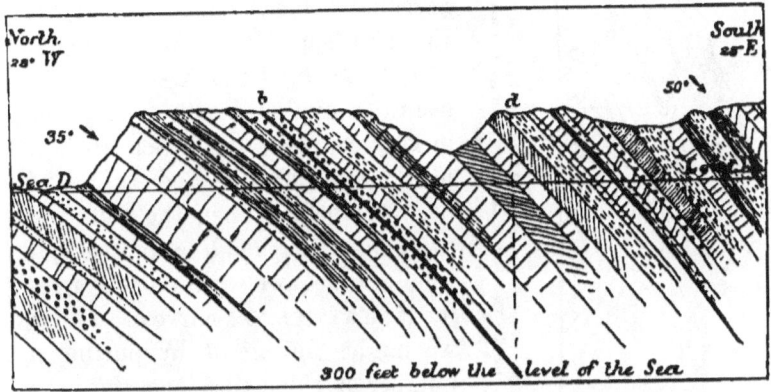

Fig 2. Horizontal section along the line D D on fig. 1. This section is drawn at right angles to the strike, and consequently shows the true *dip* of the strata. From this it is apparent that a boring at *d* would strike at a considerable depth the bed which crops out or comes to the surface at *b*.

find that a bed is dipping 20 degrees to the south, or 35 degrees to the north-east, &c., &c.: the dip cannot, of course, exceed 90 degrees, as the strata are then vertical.

STRIKE.—This word is said to be derived from the German *streichen*, to extend; it is a line drawn at right angles to the dip. The term "strike" is often confounded with the OUTCROP of strata; the two, however, will only coincide (1) when the surface of the ground is perfectly level, or (2) when the beds are vertical. In a hilly country composed of beds dipping at a moderate angle, the outcrop or line along which the strata come to the surface, will be found to zigzag down the valleys and up the hills; while so long as the dip remains constant, the strike is invariable: thus if beds dip to the east their strike must be due north and south; if the dip is to the south-east the strike will run from north-east to south-west. When traced for any distance, it will generally be found that the outcrop agrees with the strike in its general direction. This is well seen in the case of the Secondary Rocks which run across England from Yorkshire to Dorset.

CURVES OF STRATA.—It may be useful to explain briefly, two terms in very common use, which are applied to the folds or undulations of beds of rock. When the beds dip away on either hand from a central line or axis, we call the arrangement an ANTICLINAL curve or saddle; we might imitate it by putting a number of ridge-tiles one upon the other. The reverse of this arrangement is where the beds dip down to meet each other, forming a SYNCLINAL curve or trough; a nest or set of basins fitting one within the other might roughly represent such an arrangement. (See figs. 3, 4.)

At first sight we should expect to find the strata which form the hills and mountains arranged in anticlinal curves, with rocks lying in synclinal curves forming the valleys between. Such might possibly be the result if the rocks

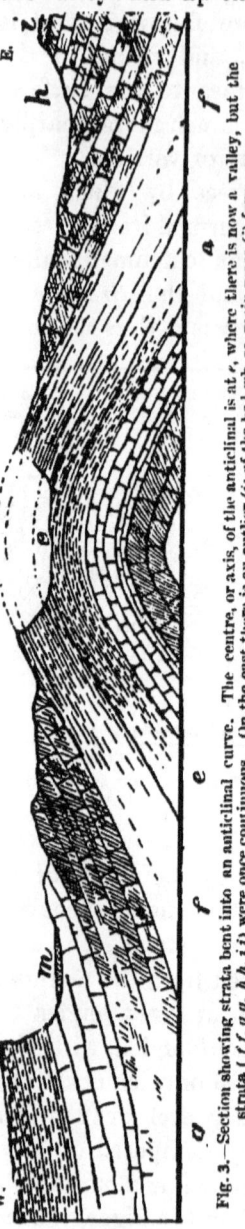

Fig. 3.—Section showing strata bent into an anticlinal curve. The centre, or axis, of the anticlinal is at *e*, where there is now a valley, but the strata (*f f. g g, h h, i i*) were once continuous. On the east there is an outlier (*i*) of the bed whose main mass (*i*) forms an escarpment on the west. The surface deposit at *m*, may represent either glacial drift, or a filled-up lake-basin.

remained unaltered, after they were elevated above the sea; but a very little geological study shows that our mountains have been carved out, by atmospheric forces acting since that upheaval.

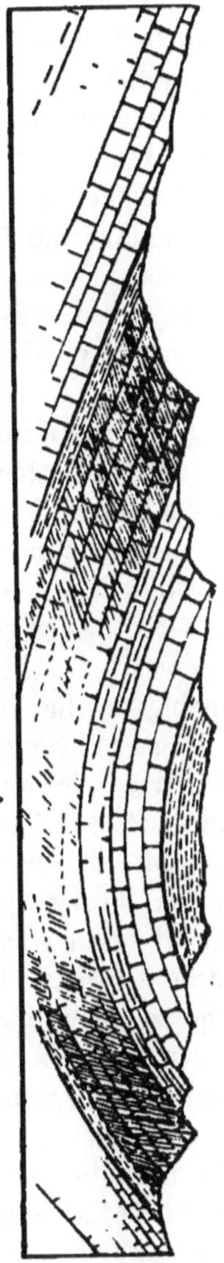

DENUDATION—*The Sea.*—Geological examination shows that nine-tenths of the land is composed of aqueous rocks, *i.e.*, of rocks which have accumulated as grains of sand or flakes of mud on ocean floors. When we consider the effect of the waves upon any mass of matter slowly rising above the sea, we can understand the wonder which has been expressed by able geographers that there should be any land at all! But the sea does not cease its assaults upon the land, although the latter may rise far above it; continually on every coast the waters dash upon the shore; where the cliffs are lofty they are rapidly undermined and worn back, the destruction being mainly effected by the hurling against the cliffs of blocks which have been detached by frost, &c., the rocks thus being made to aid in their own destruction. On the east coast of England, where there are cliffs of clay, the waste has been especially rapid, being at the rate of from one to three feet per annum.

Atmospheric Agencies.—While the wearing action of the sea is confined to the coast line, the denuding agencies which are classed together under the term "atmospheric," have for their field of work the whole area of the land, and

Fig. 4 Section showing strata bent into a synclinal curve, or trough.

the effects they produce are correspondingly more powerful. *Rain* loosens the particles of rocks and carries the sand and mud into rivers; moreover by virtue of the carbonic acid which it contains, it dissolves limestone rocks. *Rivers* wear away their beds and banks, and transport all the detritus onwards towards the sea ; it is estimated that an average river lowers the surface of the country it drains, at the rate of one foot in 6,000 years. Now, as the average elevation of the land but little exceeds 1,000 feet, it is evident that in the course of six millions of years (if there were no counteracting agencies) the whole of the land would have been removed by this powerful agent of denudation and cast into the sea.

Ice in the form of glaciers, bergs, and avalanches, exercises an eroding action ; the conversion of water into ice by cold has a powerful influence in rending or splitting off blocks of rock, for water expands when it freezes, and the force so exerted, being a molecular one, is practically irresistible : this is the origin of the piles of freshly fallen stones, "screes" as they are often called, which we find each spring at the foot of almost every mountain crag.

These are the principal agents of denudation—the sea, rain and rivers, ice and frost ; by their combined action the outlines of continents are being changed, and the scenery altered. The work is done so slowly that a human lifetime is too short to recognise the universality of these changes ; but they are sure though slow. The hills we see are not those which our fathers saw, though the amount of alteration is so small as to be practically unrecognisable.

DENUDATION COUNTERACTED BY UPHEAVAL.—As we have every reason to believe that the denuding forces have been at work for an enormous period of time, in fact, ever since the formation of a solid earth, it may naturally be inquired how it is that any land is left. Why has it not been entirely worn away ? The answer to this point will be found in those movements of the earth's crust to which we have already alluded ; the slow rising of large areas, together with more sudden upheavals in regions where volcanic forces are active, have so counteracted the denuding forces, that probably the relative proportions of land and sea have always remained constant or nearly so.

CHAPTER III.

THE STUDY OF FOSSILS, OR PALÆONTOLOGY.—Lyell defines a fossil as "any body, or the traces of the existence of any body, whether animal or vegetable, which has been buried in the earth by natural causes." The art of finding and recognising fossils requires considerable practice; workmen, as a rule, only notice those which are large and unmistakable; it is necessary to look very closely, very slowly, and very often at the rocks, and to have, moreover, some idea of what to look for; which will be best gained by previous examination of specimens, such as may be seen in museums or in the private collections of our friends.

It is, however, much easier to find fossils than to be able to correctly describe and recognise them when found. The only way to do this is first to acquire some familiarity with the structure and classification of living animals and plants. Thus marine shells constitute the majority of the ordinary fossils we find embedded in rocks, but only their hard parts occur, and these are commonly imperfect; the living forms, however, we can obtain complete and perfect, we can study their habits, modes of life, and places of occurrence; in fact, all true knowledge of the life of the past must be founded upon a study of the animals and plants which now exist, and which are, indeed, the descendants, modified, it is true, by many causes, of those which formerly peopled the earth. Students of geology frequently complain as to the difficulty of obtaining good text-books treating of fossils;

but the fact is these grumblers wish to begin at the wrong end; they should first take up the study of living things, (Zoology or Botany), and then they will find the fossil forms easy of comprehension.

CLASSIFICATION OF FOSSILS.—As we have pointed out above, it is impossible to separate the study of the past life of the globe from the study of the living beings which exist on its surface to-day. The great difficulty in the former task is the incompleteness and imperfection of the materials with which we have to work. Thus in the rocks we find only the scantiest traces of land animals: their bones decayed away on the surface of the land on which they died. Of the multitude of creatures which possessed no hard parts we can evidently scarcely hope to find traces. Then the beds of aqueous rocks which form our earth's crust are only a tithe of those which have been formed and swept away; while the excavations or openings where fossils may be gathered, bear a much smaller proportion to the vast expanse of land, than a pin-prick in the rind of an orange does to its whole exterior. Considering these facts, the progress which has been made in palæontology is surprising, and the results are of the highest interest. Speaking broadly, it may be said, that as we go backwards in time to older and older rocks we find them to contain a greater proportion of simply organized beings. We discover, too, numerous forms for which it is difficult or impossible to find a place in a classification derived only from a study of living beings. These are the extinct species, and they generally serve to fill up gaps in any good system, linking together groups of animals or plants which now seem widely different, but which may nevertheless have a common ancestry. In 1843 Prof. Morris in his Catalogue of British Fossils enumerated 5,300 species; Mr. Etheridge's new catalogue contains 15,000. This fact alone suffices to show the great progress which the science of palæontology has made and is making.

Fossil Vertebrates.—Of the animals possessing a vertebral

column, or backbone, the oldest known to us is a *fish* which occurs in the Upper Silurian strata. *Amphibians* appear in Carboniferous rocks, and *true reptiles* in the Permian. The sandstones of the Trias contain foot-prints which are probably those of *birds*, but no actual remains of birds have yet been found older than those in the Middle Oolite (Solenhofen Stone) of Germany. Small mammals of the lowest type *(marsupials)* appear in the Rhætic Formation, but no remains of *large* mammals have yet been met with in either Palæozoic or Mesozoic deposits.

Fossil Invertebrates.—The creatures without a backbone are now usually classified in five or six sub-kingdoms. First come the PROTOZOA, including microscopic animals, as the *foraminifera* and *infusoria*, and among larger forms the *sponges*. If the *Eozoon* of Canada be a true fossil, then the foraminifera date back to the Laurentian Epoch, and are the oldest known fossils. These minute animals occur in great numbers in almost all later rocks, forming the great mass of the chalk and of some other limestones. Sponges first appear in the Cambrian rocks of Wales and Canada.

THE CŒLENTERATA are divided into two classes—(1) the *Hydrozoa*, and (2) the *Actinozoa*. Of the former, the Graptolites are an important example, characterising the Silurian rocks. Corals belong to the Actinozoa, and first appear in Lower Silurian beds.

The sub-kingdom ANNULOIDA includes the sea-urchins, star-fishes, crinoids, &c.; these date back to Cambrian and Silurian times.

To the sub-kingdom ANNULOSA belong the *worms*, whose tracks and burrows are found in Cambrian and possibly even in Laurentian strata. Here, too, are placed the *crustaceans*, whose first important representatives were the trilobites. which lived from the Cambrian to the Carboniferous age, True *insects* first occur in the Devonian rocks of North America.

The most highly-organized invertebrates are the MOLLUSCA,

which form perhaps the most important class of all fossils. Oldest of all shells are the *Brachiopoda*, which have two representatives in the very lowest group of the Cambrian rocks of St. David's (South Wales). The first common bivalves *(Lamellibranchiata* or *Conchifera)* are found in Upper Cambrian strata *(Lingula Flags),* and here too the first univalves *(Gasteropoda)* come in. The most highly organized mollusks are the *Cephalopoda,* of which the oldest is a straight form called *Orthoceras,* found in the *Tremadoc Slates* (Upper Cambrian): in Lower Silurian rocks that long-lived genus the *Nautilus* is found, whose descendants still swim in tropical seas; but the *Ammonite* and *Belemnite* are confined to Secondary (Mesozoic) strata.

FOSSIL PLANTS.—The oldest plant remains are impressions of sea-weeds in the Cambrian beds, but they are of a doubtful and obscure nature. The earliest known true land-plants come from the Upper Silurian rocks, and are allied to our club-mosses. In Devonian rocks ferns and horsetails appear, and here too the first *trees* are met with: they evidently belonged to the *Conifers* or pine and fir order. Of flowering plants and ordinary trees, such as the oak, &c., we find scarcely a trace before Upper Cretaceous times.

USE OF FOSSILS.—From the examination of a number of fossils, and very often from an inspection of a single specimen only, a skilled geologist is enabled to form an opinion, as to the age of the beds of rock from which the fossils have come. Large sums of money have been spent in this country in boring or sinking for coal, in rocks which lie *underneath* the coal measures, and this fact would have been recognised if the persons who superintended such works had had even an elementary acquaintance with the science of geology. The fact is that *particular species of fossils characterize different strata,* every species having a limit, above or below which it rarely if ever occurs. Thus, suppose we found in a quarry a specimen of *Ammonites Bucklandi,* we should feel certain that the rock from which we extracted it belonged to the Lower Lias, for this fossil has never been found in an other strata. It was, indeed, the great discovery of William Smith, the "Father of English Geology," that *strata could be identified by their fossil remains.*

How to find out the Name of a Fossil.—Of course the easiest plan is to send the specimen to a person who is a skilled palæontologist. If, for instance, we forward it to the editor of that excellent periodical, "Science Gossip" (3, St. Martin's place, Trafalgar square), we shall probably get a satisfactory reply in the next number, and this is a great boon to dwellers in the country.

When we live near a good museum, or have access to the private collections of geologists, we can compare the specimen, whose name we wish to find out, with those exhibited. In addition, the volumes of the Palæontographical Society (established for the purpose of figuring the whole of the British fossils) will be of the greatest use, while Professor Nicholson's Manual of Palæontology will also be very useful. The localities of all fossils collected, should always be carefully marked upon them in ink, in some inconspicuous place at the earliest possible moment.

CHAPTER IV.

Geology of England.—England includes a wonderful variety of rock-beds or strata; but up to the year 1790 no one seems to have noticed that these beds follow one another in a regular and definite order, lying one upon the other, with a general inclination, slant, or dip to the south-east. A civil engineer and surveyor, named William Smith, discovered this fact, and worked it out so far as he was able. In his large geological map, published in 1815, we see bands of colour running across the country from north-east to south-west, each band of colour representing the outcrop, or surface area, occupied by any set of beds or *formation.* By this term *formation*, we mean a group or series of beds having some characters in common, such as similarity of dip, occurrence of the same fossils throughout, &c. In consequence of this arrangement of the strata, it is possible to walk on the edge of one and the same kind of rock, starting, say, from the coast of Dorset, and continuing in a north-east direction through Northampton and Leicester to Yorkshire. On this line we should find the clays and limestones both of the Lias and the Oolite running right across England. A traveller who goes in the opposite direction, or from London to Holyhead, meets with a continual variation of rocks: this is because he is *crossing* the line of strike and outcrop of the strata.

THE GEOLOGICAL SURVEY.—This is a department of Government which is charged with the thorough examination and mapping of the rocks of the United Kingdom. It was established by Sir Henry de la Beche, in 1835, and under him and his able successors, Murchison and Ramsay, the work has been carried on with great accuracy and efficiency. The Maps, Sections and Memoirs, published by the Geological Survey, are simply indispensable to every working geologist. The purchase of such as relate to his own neighbourhood, should be one of the first steps taken by anyone who wants to study the geology of the district in which he dwells.

PALÆOZOIC ROCKS (from *palaios*, ancient, and *zoe*, life: also called Primary). These are the old rocks which form most of Wales, the Lake District and Cornwall, together with the coal-fields of our western and northern counties: their total thickness is enormous, being from 70,000 to 100,000 feet. Owing to their antiquity, they have undergone far greater changes than the newer strata, which rest upon them and form the centre and east of England. The Palæozoic strata are sub-divided into the following formations:—

    6. Permian.
    5. Carboniferous.
    4. Devonian, or Old Red Sandstone.

3. Silurian.
2. Cambrian.
1. Pre-Cambrian, or Laurentian.

The strata below the Carboniferous beds have been so altered, broken, dislocated and disturbed since their formation, that their arrangement and classification was for many years considered a hopeless task by the early geologists. But between 1830 and 1840 they were vigorously attacked and reduced to order by Sir Roderick Murchison and Professor Sedgwick. They contain many fossils, but these are all of extinct forms, differing widely from the animals and plants which now exist. The Palæozoic formations are rich in minerals, indeed our coal, lead, tin, zinc, &c. (and a large proportion of our iron) are exclusively obtained from them.

MESOZOIC ROCKS (*mesos*, middle, and *zoe*, life).—Resting on the Palæozoic Rocks come the strata which form the middle and much of the east of England: they are often spoken of as the Secondary strata.

They include five formations:—

11. Chalk, or Cretaceous.
10. Oolite.
9. Lias.
8. Rhætic.
7. Trias.

These are formed by alternations of beds of clay with limestones and sandstones, and the total thickness may be estimated at from 8,000 to 10,000 feet. The fossils are still all of extinct species, but in their forms they present a far greater similarity to those of the present day, than those met with in the older Palæozoic strata. The Secondary rocks are not rich in useful minerals, except iron-ore, which has of late years been largely extracted from the Middle Lias (in Cleveland), and Inferior Oolite (Northampton).

CAINOZOIC ROCKS (*cainos*, recent and *zoe*, life).—

This division includes the newest or latest formed strata, viz.:

17. Recent or Post-Glacial Deposits } Quaternary
16. Pleistocene or Glacial Beds } Epoch.
15. Pliocene
14. Miocene } Tertiary
13. Oligocene } Epoch.
12. Eocene

The extreme thickness of the Cainozoic rocks of England is about 2,500 feet. The lower beds (Eocene to Pliocene) occur only in two areas or "basins"—(1) the London Basin, and (2) the Hampshire Basin.

In the Eocene strata a few of the shells are identical with living forms, and the number goes on increasing until in the Pleistocene strata all the mollusks found fossil are of living species.

Lastly, in the Recent or Post-Glacial deposits, all the fossils whatever, whether shells or bones, &c., belong to yet living forms.

SOILS, AND THE DRIFT.—As all rocks wear away under the influences of the atmosphere, they become covered with a greater or less thickness of broken and softened matter, or "soil," which is a hindrance to their geological examination. In the south of England it is found useful to use a sort of sharpened steel cylinder, or "geological cheese taster," by which specimens can be brought up from a depth of two or three feet; still, by experience, a geologist soon learns to recognise the kind of soil which is produced by any given rock.

The Drift is a more serious obstacle to the study of the stratified rocks, which it covers over and conceals. It is the term applied to the masses of stony clay and sand left by glaciers and icebergs during the last glacial period; its absence south of the Thames probably indicates that these ice-masses did not then extend much beyond the line of that river.

INTRODUCTION.

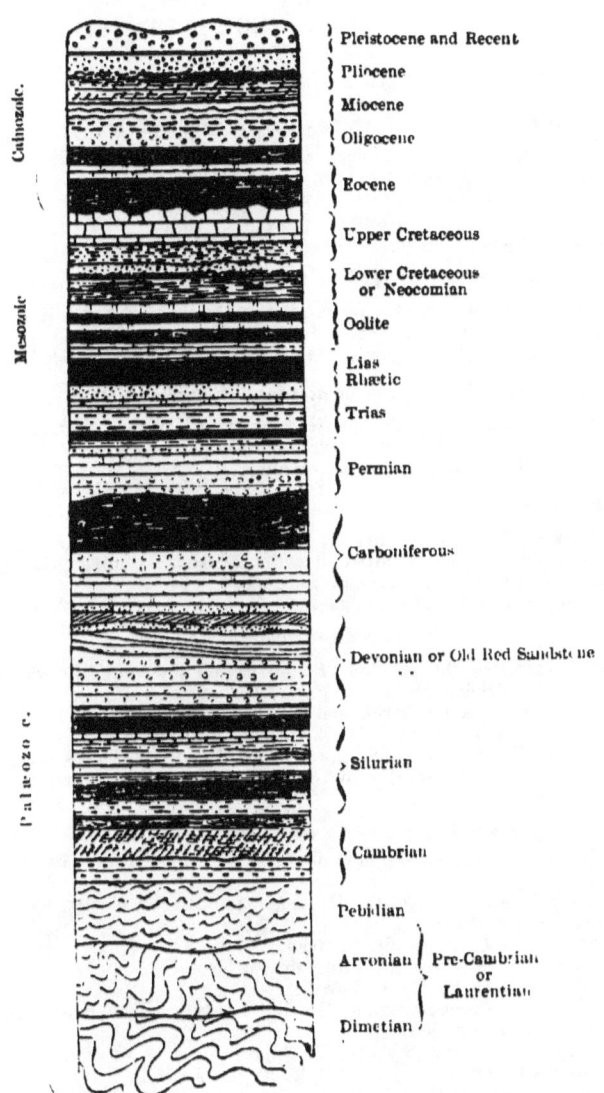

Fig. 5.—Ideal Section, showing the order of succession of the stratified rocks.

LEARNED SOCIETIES OF ENGLAND,

WHICH ARE CONNECTED WITH THE GENERAL STUDY OF GEOLOGY.

GEOLOGICAL SOCIETY OF LONDON. Established 1807.

Head Quarters, Burlington House, Piccadilly, London. Candidates for membership must be proposed by three Fellows, to one of whom they must be known personally. Entrance fee, six guineas; annual subscription, two guineas. Number of Members, 1,400.
Publish *Quarterly Journal*, Longmans, 5s. per number.

THE MINERALOGICAL SOCIETY OF GREAT BRITAIN AND IRELAND.

Entrance fee, one guinea; annual subscription, one guinea. General Secretary, J. H. Collins, F.G.S., Truro. Number of Members, about 200.
Publish the *Mineralogical Magazine*; 2s. quarterly, Simpkin, Marshall & Co.

GEOLOGISTS' ASSOCIATION OF LONDON. Established 1858.

Head Quarters, University College, Gower Street, London. Entrance fee, 10s. 6d. Annual subscription, 10s. Number of Members, 420.
Publish *Proceedings*. Bogue, 1s. 6d. per number.

BRITISH ASSOCIATION FOR THE ADVANCEMENT OF SCIENCE. Established 1831.

Head Quarters, 22, Albemarle Street, London. Meetings held annually at the principal towns in the United Kingdom. Annual subscription, £1. Number of Members and Associates, about 3,000.
Publish *Annual Report*. Murray, 21s.

THE PALÆONTOGRAPHICAL SOCIETY.

Established 1847, for the purpose of figuring and describing the whole of the British Fossils. Subscription, one guinea per annum.

All the back volumes are in stock, and may be obtained from the Hon. Sec., Rev. Thos. Wiltshire, 25, Granville Park, Lewisham, Kent, S.E.

Up to and including the year 1881 the number of species of fossils described in the volumes published annually by the Society was 4,934, illustrated by 1,430 plates.

1. *Monographs which have been Completed and Issued to the Members.*

The Carboniferous and Permian Foraminifera (the genus Fusulina excepted), by Mr. H. B. Brady.
The Tertiary, Cretaceous, Oolitic, Devonian, and Silurian Corals, by MM. Milne Edwards and J. Haime.
The Polyzoa of the Crag, by Mr. G. Busk.
The Tertiary Echinodermata, by Professor Forbes.
The Fossil Cirripedes, by Mr. C. Darwin.
The Post-Tertiary Entomostraca, by Mr. G. S. Brady, the Rev. H. W. Crosskey, and Mr. D. Robertson.
The Tertiary Entomostraca, by Prof. T. Rupert Jones.
The Cretaceous Entomostraca, by Prof. T. Rupert Jones.
The Fossil Estheriæ, by Prof. T. Rupert Jones.
The Fossil Merostomata, by Mr. H. Woodward.
The Tertiary, Cretaceous, Oolitic, Liassic, Permian, Carboniferous, Devonian, and Silurian Brachiopoda, by Mr. T. Davidson.

The Eocene Bivalves, vol. i. and Supplement, by Mr. S. V. Wood.
The Eocene Cephalopoda and Univalves, vol. i., by Mr. F. E. Edwards and Mr. S. V. Wood.
The Mollusca of the Crag, by Mr. S. V. Wood.
Supplement to the Crag Mollusca, by Mr. S. V. Wood.
Second Supplement to the Crag Mollusca, by Mr. S. V. Wood.
The Great Oolite Mollusca, by Professor Morris and Dr. Lycett.
The Trigoniæ, by Dr. Lycett.
The Oolitic { Echinodermata, vol. i., Echinoidea } by Dr. Wright.
            "   ii., Asteroidea
The Cretaceous (Upper) Cephalopoda, by Mr. D. Sharpe.
The Fossils of the Permian Formation, by Professor King.
The Reptilia of the London Clay (and of the Bracklesham and other Tertiary Beds), by Professors Owen and Bell.
The Reptilia of the Cretaceous, Wealden, and Purbeck Formations, by Professor Owen.
The Fossil Mammalia of the Mesozoic Formations, by Professor Owen.
The Fossil Elephants, by Professor Leith Adams.
The Reptilia of the Liassic Formations, by Professor Owen.

2. Monographs in Course of Publication.

The Eocene Flora, by Mr. J. S. Gardner and Baron Ettingshausen.
The Flora of the Carboniferous Formation, by Mr. E. W. Binney.
The Crag Foraminifera, by Messrs. T. Rupert Jones, W. K. Parker, and H. B. Brady.
Supplement to the Fossil Corals, by Dr. Duncan.
The Echinodermata of the Cretaceous Formation, by Dr. Wright.
The Carboniferous Entomostraca, by Messrs. T. Rupert Jones, J. W. Kirkby, and G. S. Brady.
The Trilobites of the Mountain-Limestone, Devonian, and Silurian Formations, by Mr. J. W. Salter (to be completed by Dr. H. Woodward).
The Malacostracous Crustacea, by Professor Bell.
Supplement to the Fossil Brachiopoda, by Mr. T. Davidson.
The Ammonites of the Lias, by Dr. Wright.
The Belemnites, by Professor Phillips (to be completed by Mr. R. Etheridge).
The Sirenoid and Crossopterygian Ganoids, by Professor Miall.
The Fishes of the Carboniferous Formation, by Professor Traquair.
The Fishes of the Old Red Sandstone, by Messrs. J. Powrie and E. Ray Lankester, and Professor Traquair.
The Reptilia of the Wealden Formation (Supplements), by Professor Owen.
The Reptilia of the Kimmeridge Clay, by Professor Owen.
The Reptilia of the Mesozoic Formations, by Professor Owen.
The Pleistocene Mammalia, by Messrs. Boyd Dawkins and W. A. Sanford.
The Cetacea of the Crag, by Professor Owen.

(For Local Societies, see Lists prefixed to each County).

Geological Survey of the United Kingdom.

Head Office, 28, Jermyn Street, London, S.W. Director-General, Prof. A. C. Ramsay, LL.D., F.R.S. The staff for England consists of 36 geologists and assistants; for Ireland, 17; and for Scotland, 12. This is a Government Department, established under Sir Henry De la Beche in 1835, for the purpose of examining, mapping, and describing the whole of the British strata. De la Beche was succeeded by Sir. R. I. Murchison, in 1855; and the latter by Prof. Ramsay, in 1871.

Commencing in the south-west counties and in Wales, the work of the Survey has been continuously advancing eastwards, until now (1881) only portions of the counties which lie on the east coast (Northumberland to Suffolk) remain unexamined. Here nearly the whole strength of the Survey is at present concentrated, and it is considered probable that the "one-inch map" will be finished by the end of 1885.

Maps.—The general geological map is on the scale of 1 inch to 1 mile; it is published partly in sheets, price 8s. 6d. each, and partly in quarter-sheets, price 3s. each. Originally only the solid stratified rocks, or "deep deposits" were mapped; but in 1866 it was resolved to show the superficial or drift deposits of clay, sand, &c., on a separate set of maps; very few of the latter have, however, yet been published. Maps of the coal-fields are also published on a larger scale, viz., 6 inches to 1 mile.

Horizontal Sections.— These illustrate the subterranean relations of the strata to a depth of 1,000 feet, and give the true outline of the ground, and the actual inclination and thickness of the strata: 121 published, price 5s. each. Scale 6 inches to 1 mile.

Vertical Sections. These are arranged in columns on a scale of 40 feet to 1 inch, and show minutely the nature and relations of the different beds of rock. Sections of many colliery shafts are given. 63 have been published, price 3s. 6d. each.

Memoirs, Explanations, &c. Numerous and valuable works, written by the officers of the Survey, describing the strata laid down upon the maps, and the ores, fossils, &c., contained in them, have from time to time been officially published.

Catalogues, containing full particulars of all the publications, maps, &c., of the Geological Survey may be obtained from Stanford, Charing Cross; Longmans, Paternoster Row, &c.

(*For particulars of Maps, &c., see Lists prefixed to each County.*)
(*For Works on Local Geology, see Lists prefixed to each County.*)

LIST OF THE MORE IMPORTANT BOOKS

TREATING GENERALLY OF THE GEOLOGY OF ENGLAND AND WALES.

Students' Elements of Geology, by Sir Charles Lyell. 3rd edition. Murray, 1878, 9s.
Principles of Geology, by Sir C. Lyell. 12th edition. Murray, 2 vols., 32s.
School Manual of Geology, by Prof. Jukes. 3rd edition. A. and C. Black, 1876, 4s. 6d.
Physical Geology, by Prof. A. H. Green. 2nd edition. Rivingtons, 1880, 12s. 6d.
Text-Book of Geology, by Prof. A. Geikie. Macmillan, 1881.

Elements of Chemical and Physical Geology, by G. Bischof (Cavendish Society). 1854. 3 vols., 50s.
Outlines of Field Geology, by Prof. A. Geikie. 2nd edition. Macmillan, 1879, 3s. 6d.
Practical Geology, by W. J. Harrison. Stewart, 1878, 2s.
Field Geology, by Penning and Jukes-Browne. 2nd edition. Baillière, Tindall, and Cox, 1879, 7s. 6d.
Town Geology, by Rev. C. Kingsley. Strahan and Co., 1872, 5s.
Physical Geology and Geography of Great Britain, by Prof. Ramsay. 5th edition. Stanford, 1878, 15s.
Geology of England and Wales, by H. B. Woodward. Longmans, 1876, 14s.
Records of the Rocks (Palæozoic), by Rev. W. S. Symonds. Murray, 1872, 12s.
Siluria (includes all the Palæozoic Rocks), by Sir R. Murchison. 5th edition. Murray, 1872, 18s.
Synopsis of the Classification of the British Palæozoic Rocks, by Prof. Sedgwick. Whareham, 1855, 21s.
Geology of Oxford and the Valley of the Thames, by Prof. Phillips. Macmillan, 1871, 21s.

System of Mineralogy, by J. D. Dana. Trubner, 42s.
Text-book of Mineralogy, by J. D. Dana. Trubner, 18s.
Systematic Mineralogy, by H. Bauerman. Longmans, 1881, 6s.
Glossary of Mineralogy, by H. W. Bristow. Longmans, 1861, 6s.

Study of Rocks (Petrology), by F. Rutley. Longmans, 1879, 4s. 6d.
Rocks, Classified and Described, by Von Cotta (translated by Lawrence). Longmans, 1866, 14s.
A Handy-Book of Rock Names, by G. H. Kinahan. Hardwicke, 1873, 4s.
Building and Ornamental Stones of Great Britain and Foreign Countries, by Prof. E. Hull. Macmillan, 1872, 12s.
Rock-Metamorphism (or an old Chapter of the Geological Record), by Profs. King and Rowney. Van Voorst, 1881.

Mineral Statistics of the United Kingdom (Geological Survey Memoir). Published annually. Stanford, 2s.
Economic Geology, by Dr. D. Page. Blackwood, 1874, 7s. 6d.
The Coal-fields of Great Britain, by Prof. E. Hull. 4th edition. Stanford, 1880, 16s.

Coal and Coal Mining, by W. W. Smyth. Lockwood, 1873, 3s. 6d.
Report of the Royal Commission on Coal. Blue-book, 1871.
Mines and Miners, by L. Simonin (translated by H. W. Bristow). Mackenzie, 1868, 42s.

Manual of the Mollusca, by Dr. S. P. Woodward. Weale's series. 2nd edition. 1868, Lockwood, 6s. 6d.
A History of British Fossil Mammals and Birds, by Prof. R. Owen. London, 1846, 31s. 6d.
The Mineral Conchology of Great Britain, by J. Sowerby. 6 vols. 1812-29 (Quaritch), 16 guineas.
Manual of Palæontology, by Prof. Nicholson. Blackwood and Sons, 1879, 42s.
Ancient Life-History of the Earth, by Prof. Nicholson. Blackwood and Sons, 1877, 10s. 6d.
Characteristic Palæozoic Fossils (British), by W. H. Baily. Van Voorst, 1875, 21s.
Fossil Flora of Great Britain, by Lindley and Hutton. 1831-37 (Quaritch), 5 guineas.
Catalogue of the British Fossils, including about 15,000 species, by R. Etheridge. (In the press.)
Catalogues of the Collection of Fossils in the Museum of Practical Geology, Jermyn Street, London.
Catalogue of British Fossils, by Prof. J. Morris. 2nd edition. 1854.
Monograph of British Graptolites, by Prof. Nicholson. Blackwood and Sons, 1872, Part I., 5s.
Illustrations of Fossil Plants, by Lindley and Hutton. Edited with Catalogue, by Prof. Lebour. Reid, Newcastle-on-Tyne, 1879, 25s.
Chart of British Tertiary Fossils (800 figures), by Lowry and Bone.
Chart of Fossil Crustacea, by Woodward & Salter. Tennant, 1865, 10s. 6d.
Chart of Characteristic British Fossils. S.P.C.K., 1853.

A Monograph of the Gault, by F. G. H. Price. Taylor and Francis, 1880, 3s. 6d.

Volcanoes, by Prof. J. W. Judd. C. K. Paul & Co., 1881, 5s.
Volcanoes, by G. P. Scrope. Longmans, 1872, 15s.

Pre-historic Europe, by Dr. J. Geikie. Stanford, 1881, 25s.
Climate and Time, by Dr. Croll. Bouge, 1875, 24s.
The Great Ice Age, by Dr. James Geikie. 2nd edition. Isbister and Co., 1876, 24s.

Ancient Stone Implements of Great Britain, by John Evans. Longmans, 1872, 28s.
Ancient Bronze Implements, Weapons and Ornaments of Great Britain and Ireland, by John Evans. 509 pp., 540 wood-cuts. Longmans, 1881, 25s.
Geological Evidences of the Antiquity of Man, by Sir C. Lyell. 4th edition. Murray, 14s.
Origin of Civilization and the Primitive Condition of Man, by Sir J. Lubbock. Longmans, 18s.
Transactions of the Congress of Prehistoric Archæology at Norwich. 1868, Longmans.

Geological Map of England and Wales, by Prof. Ramsay. Stanford, 32s.
Greenough's Geological Map of England and Wales. Published by Geological Society, in 6 sheets, 42s.
Geological Map of England and Wales, by H. W. Bristow. Lett's Popular Atlas; Part IV., price 7d.

BRITISH ASSOCIATION ANNUAL REPORTS.

Exploration of Kent's Cavern, Torquay, see vols. from 1865 to 1880.
Increase of Underground Temperature, 1868 to 1881.
Circulation of Underground Waters in the New Red Sandstone, &c., 1875 to 1881.
Thermal Conductivities of Rocks, 1874 to 1879.
Erratic Blocks or Boulders, 1873 to 1881.

The Geologist, a monthly magazine, edited by S. J. Mackie, 1858 to 1864 (June).

The Geological Magazine, monthly, price 1s. 6d., edited by Dr. H. Woodward, 1864 (June) to 1881.

The Geological Record (edited by W. Whitaker), contains titles and short abstracts of all works (British or foreign) on geology published during the year. Commenced in 1874. Published annually. London, Taylor and Francis, 16s. (to subscribers, 10s. 6d.).

Catalogue of the Library of the Museum of Practical Geology and Geological Survey, by White and Newton (about 28,000 vols.). 1878, 15s.
Catalogue of the Library of the Geological Society of London. Taylor and Francis, 1881, 8s.
Wigan Free Library: Catalogue of Books and Papers on Mining, Metallurgy and Manufactures, by H. T. Folkard. 1880.
Catalogue of Scientific Papers (British and Foreign) published in Journals, Transactions, &c., between 1800 and 1863, arranged under authors' names. Published by the Royal Society of London. Six vols., 20s. per vol., 1867-72.
Second Series, containing Scientific Papers published between 1864 and 1873. Two vols., 20s. per vol., 1877-79.

TITLES OF A FEW IMPORTANT PAPERS,

TREATING GENERALLY OF THE GEOLOGY OF ENGLAND AND WALES.

1860. Wright, Dr. T. On the Zone of *Avicula contorta*, and the Lower Lias of the South of England. Q. Journ. Geol. Soc., vol. xvi. p. 374.
1861. Moore, C.—On the Zones of the Lower Lias and the *Avicula contorta* Zone (Rhætic Beds). Q. Journ. Geol. Soc., vol. xvii. p. 483.
1862. Ramsay, Prof. A. C.—The Glacial Origin of Certain Lakes. Q. Journ. Geol. Soc., vol. xviii. p. 185.
1867. Moore, C.—Abnormal Conditions of the Secondary Formations in the South-West of England. Q. Journ. Geol. Soc., vol xxiii. p. 449.
1869. Wood, S. V., Junr.—Relation of the Boulder Clay without Chalk of the North of England to the Great Chalky Boulder Clay of the South. Q. ourJn. Geol. Soc., vol. xxvi. p. 90.

1870. Brown, R.—On the Physics of Arctic Ice. Q. Journ. Geol. Soc., vol. xxvi. p. 671.
1871. Ramsay, Prof. A. C.—Physical Relations of the New Red Marl, Rhætic Beds, and Lower Lias. Q. Journ. Geol. Soc., vol. xxvii. p. 189.
1871. Ramsay, Prof. A. C.—The Red Rocks of England of Older Date than the Trias. Q. Journ. Geol. Soc., vol. xxvii. p. 241.
1872. Ramsay, Prof. A. C. River Courses of England and Wales. Q. Journ. Geol. Soc., vol. xxviii. p. 148.
1875. Blake, Rev. J. F.—Kimmeridge Clay of England. Journ. Geol. Soc., vol. xxxi. p. 196.
1876. Penning, W. H. Physical Geology of East Anglia during the Glacial Period. Journ. Geol. Soc., vol. xxxii. p. 191.
1877. Blake and Hudleston.—Corallian Rocks of England. Journ. Geol. Soc., vol. xxxiii. p. 260.
1877. Davidson, T. What is a Brachiopod? Geol. Mag., pp. 145, 199, 262.
1877. Wood, S. V., Junr., and Harmer, F. W.—The Later Tertiary Geology of East Anglia. Q. Journ. Geol. Soc., vol. xxxiii. p. 74.
1877-8. Wood, S. V., Junr.—American and British Surface Geology. Geol. Mag.
1878-81. Taylor, J. E.—Our Common British Fossils and where to find them. Science Gossip. A Magazine, monthly, price 4d.
1879. Duncan, Prof. P. M. The Carboniferous Formation. Report British Assoc. (Sheffield), p. 326.
1879. Mackintosh, D.—Boulders of West of England and East of Wales. Journ. Geol. Soc., vol. xxxv. p. 425.
1879. Sorby, H. C.—Microscopical Researches on the Structure and Origin of Limestones. Geological Soc., Presidential Address, vol. xxxv.
1880. Blake, Rev. J. F.—Portland Rocks of England. Journ. Geol. Soc., vol. xxxvi. p. 189.
1880. Bonney, Prof. T. G.—Pre-Cambrian Rocks of Great Britain. Proc. Birmingham Philosphical Soc., vol. i., p. 140.
1880. Jukes Browne, A. J. Sub-divisions of the Chalk. Geol. Mag., p. 248.
1880. Sorby, H. C.—Microscopical Researches on the structure and origin of Non-Calcareous Rocks. Geological Soc., Presidential Address, vol. xxxvi.
1880-81. Howorth, H. H.—The Mammoth in Europe. Geol. Mag. pp. 561, 198, 251, 309, 403.
1881. Etheridge R. Analysis and Distribution of the British Palæozoic Fossils (Presidential Address), Geol. Soc. vol. xxxvii.
1881. Phillips, J. A. Grits and Sandstones, their Constitution and History. Q. Journ. Geol. Soc., vol. xxxvii., p. 6.
1881. Spencer, Jas.—Recreations in Fossil Botany. Science Gossip.
1881. Williamson, Prof. W. C. Organization of Fossil Plants of Coal Measures. Part x., Phil. Trans., vol. 171.

(*For Local Publications see Lists Prefixed to each County*).

Fig. 6 (see page 4).—Terraced Lower Escarpment of the Chalk, east of Chalgrave. The terrace-like ledges or "lynchets" are conspicuous on the right.

No. 1.

GEOLOGY OF BEDFORDSHIRE.

NATURAL HISTORY AND SCIENTIFIC SOCIETIES.
Bedfordshire Natural History Society and Field Club. Proceedings.

PUBLICATIONS OF THE GEOLOGICAL SURVEY.
Maps.—Quarter Sheets: 46 N.E. Hitchin; 46 S.E. Hatfield, Luton, Stevenage; 46 N.W. Newport Pagnel, Woburn; 46 S.W. Leighton Buzzard, Aylesbury, Tring.
Books.—Geology of the London Basin, by W. Whitaker, 13s.

IMPORTANT WORKS OR PAPERS ON LOCAL GEOLOGY.

1862. Wyatt, J.—Flint Implements, &c., in the Gravels of the Ouse. Journ. Geol. Soc., vol. xviii. p. 113; xx. p. 183.
1865. Whitaker, W.—On the Totternhoe Stone. Journ. Geol. Soc., vol. xxi. p. 398.
1873. Teall, J. J. H. - On the Potton and Wicken Deposits. Sedgwick Prize Essay: Cambridge.

See also General Lists, p. xxv.

THE rocks of Bedfordshire have been carefully examined by different geologists, to whose thorough and minute work we owe our knowledge of the geological structure of the county. The beautiful maps, coloured so as to show exactly where each stratum reaches the surface, issued by the Geological Survey in 1864 and '65, were chiefly the work of Messrs. Hull, Whitaker, Green, Bauerman and Howell, and it is to be regretted that they are not accompanied by full descriptive memoirs. Mr. Saunders, of Luton, has done good work in the Chalk district, and the coprolitic beds have been well described by Mr. Teall. To the surface geology Mr. J. Wyatt, of Bedford, has made most valuable contributions, and we also owe much to Professor H. G. Seeley, Dr. J. Mitchell, and others. The general structure of the county is plainly owing to the fact of the central line being occupied by a broad band of clay, to the north-west of which we have undulating ridges of limestone, while on the south-east rises the chalk escarpment; thus the town of Bedford is 100 feet above the sea level, whilst Luton is 351 feet, Dunstable 483 feet, and Dunstable Downs 799 feet respectively. As is usual, the oldest beds are found in the north-west of the county, and we shall commence with them.

THE OOLITES. -The lower oolitic strata do not enter the county; the well-known white limestone which we find in the north-west corner is of *Great*

Oolite age, being of the same age as that which at Bath and elsewhere in the West of England furnishes such excellent building stone. We can trace it from Cold Brayfield by Carlton and Harrold, northwards to Puddington and Farndish, and eastwards along the Ouse valley to the western suburbs of the town of Bedford, where it is seen in a stone pit on the north side of the Ouse, towards Kempston; it is a compact stone, got in some places for lime-burning, but good sections of it are rare; it is associated with bands of *Great Oolite Clay*, which form a cold, unkindly, but often well-wooded district. One or two small species of oyster, as *Ostrea subrugulosa* and *O. Sowerbyi* are common in the fossil state.

The Cornbrash. This very continuous band of reddish rubbly-limestone can be traced from Newton Blossomville by Turvey Farm, to just west of Pavingham; here it turns south, the outcrop having been worn back by the river Ouse, but at Wick End, about a mile north of Stagsden, it ceases to be traceable. We find it again on the east side of the Ouse in some brick-pits and lime-kilns, whence it curves round by Clapham to Oakley Hill, where it is cut off by an east and west fault, re-appearing at Milton Wood about a mile eastwards; thence it runs northwards out of the county by Milton Ernest and Souldrop. The width of the outcrop is remarkably even, averaging one-quarter of a mile. The beds appear nearly horizontal. *Ostrea Marshii* is the most frequently occurring fossil.

Oxford Clay. This is a bluish clay, weathering yellow, and several hundred feet thick, which occupies all the central and north-eastern portions of the county. At the base sandy beds locally occur, known as *Kellaway's Rock*, forming a link between the Cornbrash, which was accumulated in shallow water, and the Oxford Clay, which was a deep-sea deposit. The latter forms a stiff retentive soil, which is largely in pasture, but it is often so deeply covered by surface deposits to be noticed further on as not to exercise any influence on the character of the soil; it forms all the substratum of the Fens further north, and Professor H. G. Seeley has proposed the name of Fen-clay for it, a term however which is only locally used; it is largely dug for brick-making, and for claying the soil of the fens. The band of gritty limestone known as the *Coral Rag* is absent in this district, so that there is no well-marked line of division between the great mass of the Oxford Clay below, and the *Kimmeridge Clay* above. The latter has not in consequence been mapped by the Geological Survey in this county, but it is probably represented on the eastern side. Local geologists must determine this by carefully collecting and noting the localities of the fossils found in their neighbourhood. From a good section near Ampthill fossils have been obtained, which show that the beds of clay there are near the transition line from the Lower to the Upper Clays; they are about the horizon of the Coral Rag. In the Oxford Clay proper, *Gryphœa dilatata*, a wide thick-shelled kind of oyster, is very common; bones of reptiles, fish teeth, &c., also occur.

THE CRETACEOUS FORMATION.—As the Upper Oolitic beds the Purbeck and Portland strata are here absent, we pass at once from the thick Oolitic Clays to those rocks of which the Chalk is the most distinguished member.

The Lower Greensand. -This bed presents a complete contrast to those last noticed (the Oxford and Kimmeridge Clays), being a light buff-coloured or reddish-brown sand, often showing false bedding and full of fragments and concretions of brown peroxide of iron; it enters the county at Leighton Buzzard and Linslade, and passes by Little and Great Brickhill and Wavendon Plantations to Ridgmont, Eversholt and Flitwick, constituting here the well-known Woburn Sands. Fullers' earth has been dug at Wavendon, where the beds have a thickness of from 150 to 200 feet; thence by Ampthill and Shefford the sands extend to Biggleswade, Sandy and Potton, where they pass out of the county. In the south, near Little Brickhill, these beds are 250 feet in thickness, and they occupy a belt of country 4 miles wide; at Millbrook, near

Ampthill, the thickness is not more than one-tenth of this amount, or 25 feet, the upper portion having probably been removed by denudation; near Potton the series is about 100 feet thick, and is capped by about 35 feet of Carstone (or Quern stone). At Sandy these Lower Cretaceous beds form a very picturesque escarpment on the right bank of the Ivel; between Sandy and Potton they are well exposed in the cuttings of the Cambridge and Bedford Railway. The land is so barren in some parts as to produce little except plantations of Scotch fir; but where it is mixed with clay, as along the western side of the Great Northern line, the result is a sandy loam of great productiveness, in which onions, potatoes, and market produce generally are grown with great success. Over the greater part of its extent this rock forms a range of undulating hilly ground with beautiful woodland scenery. Economically the Lower Greensand is celebrated for the band of phosphatic nodules which occurs at the base: this varies from 6 inches to 2 feet in thickness, and is full of waterworn fossils, including many saurian and fish remains, and the rounded shapeless masses generally termed coprolites. In the pits at Potton we see under the surface soil about 9 feet of yellowish sands, and then the coprolite bed, about 2 feet thick, under which again is a considerable thickness of sand, a well 50 feet deep not having passed through it; sometimes the nodule bed is as much as 6 feet thick; numerous pebbles of quartz occur, which have to be picked out. The word *coprolite* should properly be applied only to the fossil dung of reptiles, fishes, &c., but most of the masses which go by that name are pieces of wood mineralized by infiltration of phosphatic matter, casts of shells, bones, &c., and from their worn and rolled appearance they seem mostly to have been washed out of earlier deposits, as the Oxford and Kimmeridge Clays. They contain 49 per cent. of phosphate of lime, and 7 per cent. of carbonate of lime, according to an analysis by Dr. Voelcker; at Millbrook there is only 15 feet of sand below the phosphate bed, and at Little Brickhill it rests upon the Oxford Clay. In 1879 there were raised 5,000 tons of coprolites from the Lower Greensand of Bedfordshire; the value being 28s. per ton.

The Gault is a bed of bluish-grey clay, 200 feet thick, which rests upon the Lower Greensand; it decreases in thickness as we follow it north-eastward into Cambridgeshire. From Eggington and Eaton Bray, we can trace it by Milton Bryant, Toddington, Westoning, Shillington, Arlsey, Shefford, Dunton, Wrestlingworth and Cockayne Hatley. It furnishes a soil known as "black land," and forms part of the plain which stretches from the foot of the chalk escarpment; it is largely dug for bricks, and in the excavations at Arlsey a face of 50 or 60 feet in height is exposed; fossils are tolerably numerous, *Avicula* and *Plicatula* being the commonest species.

Upper Coprolite Bed.—At least two well-marked seams of "coprolites" occur in the Gault, south of Barton. North of this place, the upper band forms a line between the gault clay and the lower marly beds of the Chalk; the nodules are dark-coloured and very rich in phosphate; this is the same band which is so largely worked near Cambridge, and it has been supposed to represent the *Upper Greensand*. Mr. Jukes-Browne has shown, however, that it is a result of the denudation of the *Gault*, and agrees with Mr. Whitaker in considering it to be the base of the chalk marl. Rock-fragments occur in it, some of which seem to show signs of ice action.

The Chalk.—The lowest beds are comparatively soft, and are termed the *Chalk Marl.* This bed is about 80 feet thick, and forms undulating rising ground; its top is marked by a hard band about 6 feet thick, generally in two layers, and called the *Totternhoe Stone,* because it is well seen near the village of that name; it is a hard and rather brownish sandy chalk, with dark grains. In south Bedfordshire it has been worked for centuries by means of galleries driven into the escarpment which it forms. Eddlesborough church stands on a conical hillock, all round which this hard band crops out. Passing Maiden Bower, where it is exposed in the railway cutting, we trace the Totternhoe Stone by Houghton Regis, Upper Sundon Sharpenhoe and Barton, out of the county. *Ammonites varians* is common. The bed generally forms a ridge or shelf along the chalk escarpment.

The *Chalk without Flints* comes next; and is, perhaps, 400 feet thick; it rises from above the Totternhoe Stone to within a small distance of the top of the high escarpment, forming the slopes of the combes, and running eastward along the sides of the valleys; good sections may be seen in the railway cuttings near Dunstable and Luton. Here Mr. Saunders obtained many fossils, but of few species, *Terebratula* and *Inoceramus* being the common shells. Near Leagrave station there is a cutting more than a mile long, reaching to the lower escarpment; at its southern end the Chalk is capped by a gravelly soil and hollows full of gravel, but further north, where deeper, it gives a good section of the Lower Chalk. On the road from Streatley to Sharpenhoe, we find an exposure, in the road-cutting, of the hard beds near the top; they contain large but broken fish vertebræ. That rare plant, *Anemone pulsatella* grows plentifully on some of the slopes of the chalk combes in this neighbourhood; it occurs chiefly on the south-west slopes, especially near Barton, where its pretty purple flowers cover the hill-sides.

The *Chalk Rock* is a thin, hard, cream-coloured bed, containing green-coated nodules, and about 4 feet thick; it is regarded as the top bed of the Lower Chalk, and by its superior hardness forms the top of the high escarpment known as the Royston and Luton Downs; it contains numerous fossils, about twenty species having been collected from the railway cutting at St. George's Wood, Chiltern Green station, &c.

The *Upper Chalk with Flints* hardly enters the county: its fossils are mostly *Echinoderms* (sea-urchins), *Sponges*, and *Ventriculites*; it may be seen in many pits where it is dug for lime-burning, whilst the flints are used to repair roads and for glass making. It forms a barren tract of land, covered with short turf, and its swelling rounded outlines, usually destitute of trees, are easily recognisable from a distance. Between Sundon and Pirton, and above Barton-in-the-Clay, the chalk scenery is very fine; Deacon Hill and Lilley Hoo are conspicuous objects, and the "lynchets" or terraces on the sides of some of the combes are very remarkable objects; they are well seen east of Chalgrave. (See Fig. 6.)

The Drift.— The beds of rock we have been describing, especially those in the centre and north of the county, are often concealed by a thick covering of what is commonly termed Boulder Clay; it is a stiff, brownish or bluish tenacious clay, full of subangular fragments of all kinds of rocks, mostly such as are not found in Bedfordshire; north of Bedford it is about 50 feet thick, and covers over the Oxford Clay from Coldington to Eaton Risely and Shelton; it lies thick, too, between Wootton, Carlton and Cranfield. Often by cultivation it has become incorporated with the Oxford Clay beneath, and we can only recognise it by the presence of scattered boulders. Fragments of granite, syenite, quartzite, &c., are common, with lumps of hard chalk and limestone; liassic and oolitic fossils are of frequent occurrence. All this heterogeneous mass is probably the deposit of a great glacier which pushed southwards down Lincolnshire from the Cumbrian and Scottish hills. In part it may result from the droppings of icebergs detached from the termination of such a glacier.

River Deposits.—The valley of the river Ouse is narrow so long as it is confined by the oolitic limestones in its course through the north-western part of Bedfordshire as far as Oakley; here it enters on the softer tract of Oxford Clay, and we can trace its deposits of gravel over a width of from 2 to 4 miles. At the junction with the Ivel the gravel deposit again narrows, and continues about 1 mile wide, to the northward; the Ivel and Hiz have formed similar gravel beds, but of less extent, along their course. The Ouse seems to have cut itself down through the Drift, and near Bedford it has also cut through the Oxford Clay to the underlying Cornbrash. The material of the gravel appears to have been in the main washed out of the Boulder Clay, and it occurs in terraces at considerable heights above the present river bed. The Ouse has also deposited much alluvium or river mud, which forms rich pasturage ground.

Traces of Pre-historic Man.—It was in the old river-gravels of the Ouse at Biddenham, near Bedford, that Mr. James Wyatt, F.G.S., first found those rudely chipped flint implements which are the earliest indications yet known of the presence of man on the earth. At Biddenham the beds of drift-gravel form a capping to a low hill, which is nearly encircled by one of the windings of the river; the floor of the pits is 40 feet above the present stream. Numerous freshwater shells also occur, one of which, *Hydrobia marginata*, has never

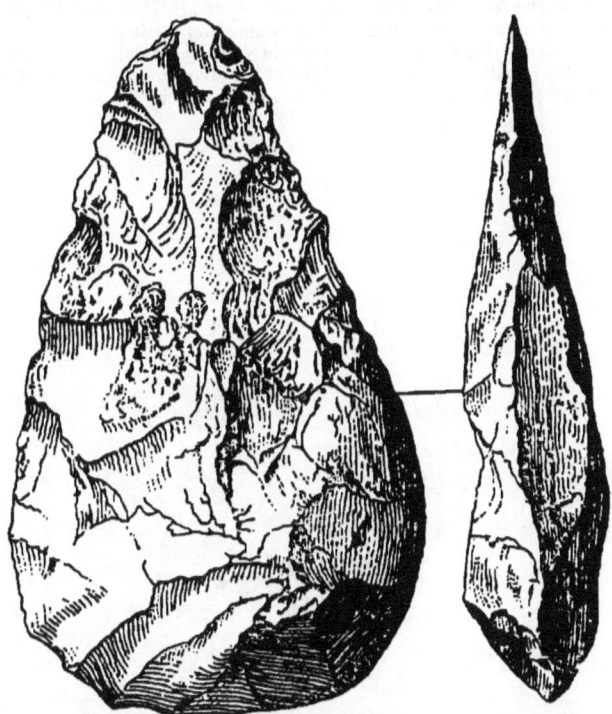

Fig. 7 —Palæolithic flint implement (one-half the natural size); front and side views From the old river-gravels at Biddenham near Bedford.

been found alive in this country. Numerous bones of large mammals, as the mammoth, rhinoceros, hippopotamus, reindeer, &c., are also found. Flint implements of this old kind have also been met with at Harrowden, Cardington, Kempston, Summerhouse Hill, and Honey Hill, all within a radius of 4 miles; they are usually pointed, and from 3 to 9 inches long. Possibly they were used as weapons, or for digging, or breaking holes in the ice for fishing, as the Esquimaux do at the present day; a similar tool was found near Henlow.

Of other flint and stone implements which belong to a later time, and which often exhibit great skill in their manner of formation, numerous examples have been found in the county. At Maiden Bower, near Dunstable, there is an old British Camp, and here Mr. John Evans met with a small, nearly triangular, flint hatchet, several scrapers, cores, flakes or knives, spear-heads, arrow-heads, and chisels. A celt, or axe-head, was found by Mr. Whitaker at Wanlud's Bank, near Luton, and others at the foot of Dunstable Downs, Leighton Buzzard, and in Miller's Bog near Pavenham; one formed of greenstone, and 4½ inches long, is from Kempston.

A thin perforated stone, 6 inches by 3 inches, found near Luton, may be a hammer-stone, used in the fabrication of implements from flint nodules or cores. A beautifully shaped flint skinning-knife, oval in form and neatly chipped, but not polished, was found at Kempston, with a blade or spear-head 6 inches long and 2¼ wide.

Thus from the important discoveries already made, it is evident that good work may be done in Bedfordshire in connection with the question of the history of man. Should these flint tools ever be found in or beneath the Boulder Clay itself, it would conclusively prove the existence of man in Bedfordshire prior to or during the last Glacial Epoch.

No. 2.

GEOLOGY OF BERKSHIRE.

NATURAL HISTORY AND SCIENTIFIC SOCIETIES.

Wellington College Natural Science Society; near Wokingham. Annual Report.
Newbury District Field Club.
Reading Microscopical Society.

MUSEUMS.

Newbury Museum.

PUBLICATIONS OF THE GEOLOGICAL SURVEY.

Maps.—Sheets: 7, Western part of London, St. Albans, Windsor, Uxbridge; 8, Wokingham, Croydon, Guildford, Reigate; 12, Newbury, Andover, Odiham; 13, Oxford, Reading, Wantage; 34, Chippenham, Swindon. Quarter Sheet: 45 S.W. Woodstock, &c.

Books.—The Geology of Parts of Berkshire and Hampshire, by Bristow and Whitaker, 3s. The Geology of Parts of Oxfordshire and Berkshire, by Hull and Whitaker, 3s. Geology of the London Basin, by W. Whitaker, 13s.

IMPORTANT WORKS OR PAPERS ON LOCAL GEOLOGY.

1856. Prestwich, Prof. J.—Gravel near Maidenhead in which Skull of Musk Buffalo was found. Journ. Geol. Soc., vol. xii. p. 131.
1875. Rupert-Jones, T., and King, C. C.—Sections of the Woolwich and Reading Beds at Reading. Journ. Geol. Soc., vol. xxxi. p. 451.
1881. Herries, W. H.—Bagshot Beds. Geol. Mag., p. 171.

See also General Lists, p. xxv.

THE rocks of Berkshire have been very carefully studied by the officers of the Geological Survey. Complete geological maps of the surface were published in 1860-61, and in the splendid memoir on the Geology of the London Basin, by Mr. W. Whitaker, a very full account will be found of all the strata except those which form the extreme north-west corner. The Tertiary beds of the south and east were first described by Professor Prestwich in a masterly series of papers published between 1840 and 1860 in the Journal of the Geological Society: Professor Rupert-Jones has also written on the subject.

As the longest axis of Berkshire extends nearly east and west, while the different beds of rock run across it in a slanting direction from north-east to south-west we naturally expect to meet with a good variety of formations in the county, and this we shall find to be the case. From the absence of disturbances however, and from the dip being gentle and coinciding with the general slope of the surface, the variety is not so great as might have been expected.

We shall commence with the oldest rocks, which occupy the extreme north-west of the county.

THE OOLITE. (1.) *The Oxford Clay.*—Entering at Coleshill and Lechlade, and running east by Thrupp Common and Newbridge and north-east by Hinksey, Wytham and Oxford, we have the Berkshire portion of a tract of stiff clay, which extends across the Thames into Oxfordshire and Gloucestershire. It contains frequent bands of limestone nodules or septaria, is of a blue colour when dug at any depth, but weathers yellow where exposed to the air; it forms a low tract of land bordering the Thames for about a mile or two on its southern side, and is of little economic value. It is mostly in pasture: a boring at Wytham passed through 596 feet of Oxford clay.

(2.) *The Coral Rag.*—This term is applied to a series of beds, clayey and sandy at the base, of which the middle portion is a rubbly oolitic limestone full of corals, capped in a few places by irony sands. The muddy sea in which the Oxford Clay was deposited must have cleared and shallowed, and in the warm waters coral reefs grew irregularly, resembling those now forming in tropical seas. Where present the thickness of the Coral Rag varies from 10 to 30 feet, and it constitutes a ridge overlooking the valley of the Thames. Entering the county near Shrivenham, we can trace it round Faringdon, and thence it occupies a tract three miles wide north of the river Ock as far as Abingdon and Cumnor: at Wytham it rises as an outlier to a height of 583 feet; it is largely quarried for road metal. The lower portion contains characteristic *Ammonites*; corals, and spines and plates of sea-urchins, occur in the middle portion in large numbers.

(3.) *Kimmeridge Clay.*—Another thick mass of blue clay with bituminous shales succeeds the coral rag; it forms flat wet land from Shrivenham station to Longcott. Passing under the well-known sponge gravels of Faringdon, it re-appears south of Shillingford and Stanford, and forms the "Fields of East and West Hanney and Drayton." Crossing the Ock at Abingdon, we can follow the Kimmeridge clay to Radley and Bagley Wood. Of the fossils *Ammonites biplex* is rather common and beautifully preserved; *Ostrea deltoidea* is very abundant. The clay is dug at several places for brickmaking.

(4.) *Portland Sand.*—There is a small outlier forming the hill on which the village of Bourton is built; it is well exposed in a large quarry there.

THE CRETACEOUS SYSTEM.—This term is derived from the Latin '*creta*' chalk, which is the best known and most conspicuous member.

(1.) *The Lower Greensand.*—This is a bed of loose reddish sand, often full of pebbles and very variable in thickness. At Fernham and Great and Little Coxwell it assumes considerable local importance, constituting the "sponge-gravels of Faringdon;" here it is largely quarried for gravel, which from its bright yellow hue is much sought after for walks and avenues. From Baulking eastwards to the Thames it is not seen, being overlapped by the Gault, but there is an outlier north of Sunningwell which forms a hilly tract; it makes a light dry arable soil.

(2.) *The Gault.*—This is a blue micaceous clay containing occasional nodules of limestone; it runs as a band about one to two miles wide between Ashbury and Stainswick to Uffington, where it turns due east and reaches through West Challow and Steventon to Wittenham; its upper boundary is well marked by a line of springs thrown out by the impermeable clay; forming a low plain at the foot of the chalk escarpment it is seldom exposed in sections, except in an occasional brick-pit; its thickness is about 100 feet.

(3.) *The Upper Greensand* or Chloritic Series.—From Wittenham Wood past Wallingford to Aston Tirrell the outcrop of this rock is not less than five miles broad. Following it westwards through Hagborne, Didcot, East Hendred and Wantage it rapidly narrows, until at Childrey, Sparshot, Compton Beauchamp and Ashbury it only forms the slope of the escarpment of chalk, which consequently becomes steeper as we follow it in this direction. Frequently the exact boundary is obscured by landslips. As the name implies, the rock is usually full of greenish grains. At Woolstone it is 60 feet thick, but above 100 at Didcot.

(4.) *The Chalk.*—This is perhaps the best known rock in England—lithologically speaking; it constitutes the central and most elevated portion of Berkshire. On the west the main mass spreads across from Hungerford to Compton Beauchamp, a distance of 12 miles. The strike, or direction, is here nearly east and west and continues so to the Thames. The dip is to the south-east at a very small angle, from one to three degrees only. In the south the chalk dips under the Tertiary beds of the valley of the Kennet, and rises up further south at a sharp angle along a line from Inkpen to Kingsclere: at Inkpen Beacon it attains an elevation of 1,011 feet, the highest point reached by the chalk in the south of England. The total thickness in Berkshire of this great mass of white soft limestone is probably about 900 feet; the chalk was eminently a deep-sea deposit, for when we examine it microscopically we find it to consist in large part of the tiny chambered shells of foraminifera, being very similar in composition, in fact, to the greyish-white ooze which numerous soundings have proved to form the floor of the North Atlantic Ocean.

The Lower Chalk has marly beds at the base about 80 feet thick, whose top is marked by a hard band called the *Totternhoe Stone*. This lower division may be seen in the Great Western Railway cutting at Wallingford Road station. Above it we get about 400 feet of chalk without flints, but containing marly partings which indicate the line of bedding: this division forms the comparatively low and flat table-land which extends from Moulsford and Streatley by Blewbury and Chilton, narrowing greatly as it goes westwards. Fossils are not very numerous, but *Ammonites varians* and *Turrilites* are characteristic. The top is formed by a hard cream-coloured band—the *Chalk Rock*—some 8 or 10 feet thick, which from its superior hardness usually forms the top of the chief chalk escarpment. Thus, we can trace it all along the northern brow of Ilsley Common and Childrey Warren; the Ridge Way runs along the edge, and sections are exposed at Cuckhamsley Knob, &c.

The Upper Chalk is characterised by the presence of flints, which occur most frequently in small irregular lumps, but also in flat sheets. The origin of flints is still a vexed question: many appear to have been formed by the deposit of siliceous matter on and around organic bodies, as sponges; the flat tabular masses of flint would seem to have been deposited in the bedding-planes probably after the consolidation of the rock. Such lines of flint may be seen in the railway cutting near Pangbourn, and in the chalk-pits at Courage north of Newbury, and at Cookham Dean, near Great Marlow. Fossils are of frequent occurrence both in the flint and the chalk; sea-urchins or *Echinoderns*, as *Ananchytes* and *Micraster*, abound, with sponges and such shells as *Terebratula*, &c. The main mass of the chalk passes in Berkshire as far east as Remenham, Wargrave and Maidenhead, but Windsor Castle is built on an inlying boss, probably elevated by some disturbance, of which there are also traces near Great Marlow.

The scenery of the Chalk Downs is very marked. The beautifully smooth swelling curves are covered with a short dense herbage which affords good pasturage for sheep and capital galloping grounds for horses. The higher part is, however, often covered with clayey deposits, as south of Ilsley, and then forms a soil on which the beech grows well; but as a rule the wide open nature of the country is always discernible. The valleys are often waterless, the rain being soon absorbed by the porous soil. White Horse Hill rises 893 feet above the level of the sea. Economically regarded, chalk is valuable as a dressing for clay lands; much is also dug to burn into lime, and it forms an ingredient of Portland cement. Whiting is solely made from chalk, Kintbury being the seat of manufacture. The soft upper chalk is here ground into a pulp with water and allowed to settle in tanks; about 2,000 tons are made per annum; it is mostly sent to Bristol by canal, and fetches about 8s. per ton.

Chalk was formerly more used for building than at present, and when carefully selected is well suited for inside work, being very easy to carve; examples may be seen in Sonning and Tilehurst churches. The chalk-rock affords a poor road-metal. Flints furnish an almost everlasting material for building; they have been used with good effect, dressed into a cuboidal form, in the beautiful little church at Shottesbrook; they are also ground up to use in the manufacture of glass and porcelain.

THE EOCENE SYSTEM.—A great break in the succession of the rocks intervenes here. In those now to be described, we find the remains of animals altogether different to those in the chalk, and we believe there was a great interval of time during which either (1) no deposit was formed, the country being a land surface, or (2) deposits were formed which were afterwards washed away—

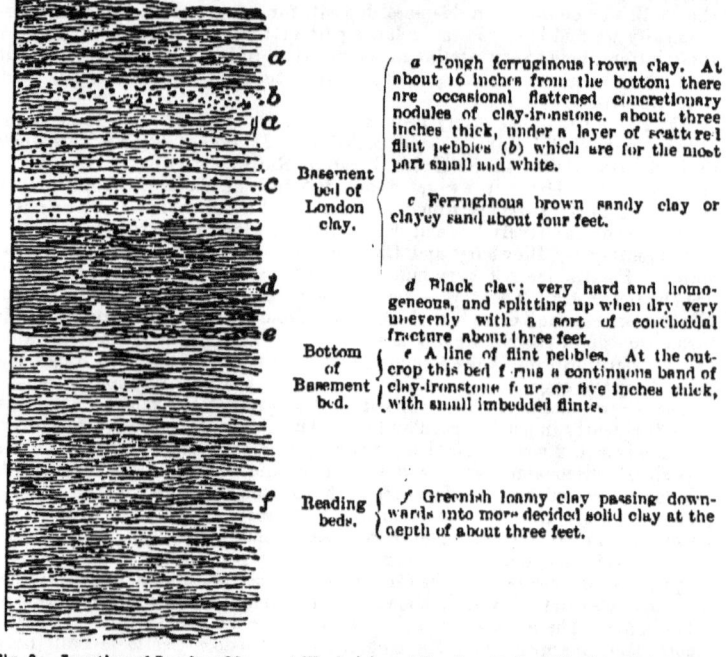

Basement bed of London clay.
a Tough ferruginous brown clay. At about 16 inches from the bottom there are occasional flattened concretionary nodules of clay-ironstone, about three inches thick, under a layer of scattered flint pebbles (*b*) which are for the most part small and white.
c Ferruginous brown sandy clay or clayey sand about four feet.

Bottom of Basement bed.
d Black clay; very hard and homogeneous, and splitting up when dry very unevenly with a sort of conchoidal fracture about three feet.
e A line of flint pebbles. At the outcrop this bed forms a continuous band of clay-ironstone four or five inches thick, with small imbedded flints.

Reading beds.
f Greenish loamy clay passing downwards into more decided solid clay at the depth of about three feet.

FIG. 8.—Junction of London Clay and Woolwich and Reading Beds at Kintbury Brickyard, north of Pebble Hill.

denuded off—before the rocks of which we now have to speak were laid down as sediment: perhaps both things happened.

(1.) *The Reading Beds* in Berkshire rest immediately upon the chalk, and were formerly known as the "Plastic Clay:" they consist of alternations of clays and sands of many colours, with rolled flint-pebbles. Fossils are few, *Ostrea bellovacina*, which much resembles the oyster of the present day, being the only one found in any number. Commencing at Prosperous Wood, about two miles south of Hungerford, we can trace these beds along the south of the Kennet to two miles east of Newbury: on the opposite side they continue nearly to Theale, then, after a gap formed by the connection of the Pang with the Kennet, we find the same sands, &c., spreading out east and west of Reading. Here they are largely worked at Katesgrove, Coley Hill, &c. The plastic clays are made into tiles, drain pipes, &c.; the sands are mixed with the clays in brick-making, or when white and clean are used in glass works. Thence the outcrop continues about half a mile wide through Sonning and the two Walthams towards Windsor. There are numerous outliers on the chalk hills north of the Lambourne and the Pang, as at Beedon Hill, Basildon, and Farnborough Copse. The junction with the chalk is often marked by swallow-holes, which are caverns or funnel-shaped pipes, into which the water, running down the slope of the Tertiary beds, enters and disappears.

(2.) *The London Clay.*—Where the junction is exposed with the Reading beds we see a "basement bed" of blackish flint pebbles which are traversed by cracks, so that they fall to pieces on receiving a gentle blow: they are embedded in a brownish loam; above this are sandy clays passing up into stiff blue and brown clay: this formation has a broad outcrop in Berkshire. There is a good section in Shaw brick-kiln, north of Newbury. Eastwards of Reading these clays occupy a district gradually widening from 3 or 4 to 10 or 15 miles

Fig. 9.—Gravel-pit on Inkpen Common, showing the "high-level gravel," composed of sub angular flints.
    a White angular flint gravel.
    b Brown flint gravel, containing large blocks of grey-wether sandstone.
    c Lower Bagshot sand.

in breadth, over which are scattered large outlying masses of rocks presently to be described. Brickyards and railway cuttings afford the only sections, as at Sonning, Windsor, Englefield, Frilsham, &c. *Pectunculus,* and *Ditrupa plana* are characteristic shells.

(3.) *The Bagshot Beds.*—These are sandy, forming ranges of barren heath-covered hills in the south of the county; sections and fossils are equally rare: they form the high grounds of Coldash and Bucklebury Commons, north of the Kennet: south of that river they extend from Inkpen Common by Greenham Heath to the commons of Tadley, Silchester, and Burghfield; then we come to the valley of the Loddon, where the beds have been denuded. Crossing this gap we find ourselves on the western end of the main mass near Finchampstead and Wokingham, and it continues by Ascot race-course to Egham: south of this line we pass over some more clayey beds, known as Middle Bagshot or Bracklesham Beds, and ascending Easthampstead Plain find ourselves on the northern extremity of the well-known Chobham Ridges, on the loose Upper Bagshot Sands.

Standing here on the latest formed of the stratified rocks, which constitute the county of Berks, we should, in imagination, sink a deep borehole and picture it passing in succession through all the beds we have named until, at a depth of perhaps, 2,000 feet, it entered the formation—the Oxford Clay which we found at the surface in the north-west corner near Lechlade.

A well lately sunk at Wokingham passed through London Clay, 263 feet; Reading Beds, 54 feet; Sand, 16 feet; and pierced the Chalk for 64 feet, obtaining a good supply of water.

GREY-WETHERS OR SARSEN STONES.—Blocks of a hard sandstone are frequently found on the surface of the chalk, &c.: they are used for building and mending roads. They would appear to be consolidated masses of either the Reading Beds or the Bagshot Sands, which have been let down, as it were, by the washing away of the looser matter in which they were once embedded (Fig. 9.)

SURFACE DEPOSITS.—Our examination of the strata at any point is often interfered with by the presence of beds of gravel, brick earth, &c., many feet thick, which cover over and hide the underlying rocks. The chalk is often covered with a stiff brown and red clay full of unworn flints. This would seem to be the residue left by the removal of the carbonate of lime, by the chemical action of water charged with carbonic acid; this "clay with flints" is well seen in the road cutting south of Remenham. Of brick-earth there is not much; some may be seen near Cookham.

The Flint-gravels occur at high levels as at Pebble Hill, south of Kintbury, and often capping the hills of Bagshot Sand; or at low levels, as from the foot of Enborne Hill to Newbury, and on both sides of the Thames.

Alluvium.—In former times the river Kennet deposited much mud for about a quarter of a mile on either side of its present course: this now forms valuable water-meadows, always green. Beds of peat occur in it, from 5 to 15 feet thick; this peat is largely dug at Newbury, and when burnt the ashes form a valuable fertilizer, probably from the quantity of gypsum they contain. In the Museum of the Newbury Institution, there is a fine specimen of the skull and horns of *Bos primigenius*, a large extinct species of ox, which was dug out of the peat in Ham Marsh. In a large excavation close to Reading a fine section of old river-gravels and loams is exposed, in which Mr. E. B. Poulton has found trunks of trees and remains of such extinct animals as the mammoth, rhinoceros, &c.

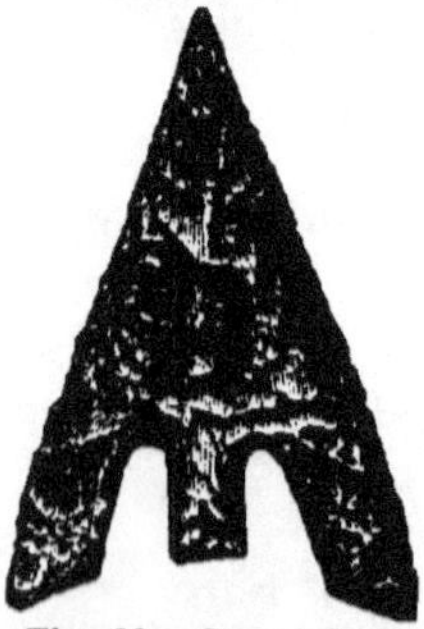

PREHISTORIC MAN.—Of those early dwellers in our islands to whom the use of metal was unknown, and who made their tools of flint, several relics have been found in Berkshire. From a "barrow" or interment on Lambourne Downs a beautiful dagger of white flint was obtained, which is now, with an arrow-head and other objects from the same locality, in the British Museum. Some beautifully formed arrow-heads of the same material, a scraper probably used for preparing skins, celts or flint axe-heads, and a nodule of iron-pyrites are also recorded by Mr. Evans, in his "Ancient Stone Implements of Great Britain," as having been found on the Berkshire Downs. A flint arrow-head was also found at Sutton Courtney. A large chipped but unground celt, 8 inches long by 2¾ broad, was found in the peat at Thatcham; and in the Geological Museum, Jermyn street, London, there is the fragment of a slender pointed flint "pick," which was picked up near Maidenhead. A perforated stone hammer-head was found at Sunninghill, and an oval flint blade near Long Wittenham. A triangular scraper of ochreous flint occurred

Fig. 10.—Stemmed and barbed Flint Arrow-head, found in a burial-mound on Lambourne Down, Berkshire, and now in the British Museum.

in the Thames near Windsor. Probably many more such objects would turn up if they were intelligently searched for, and there is no better preparation for the task than to endeavour with two flint nodules or a flint and a hammer to produce similar specimens; the task will give us some idea of the dexterity, and acquaintance with the properties of the material, which must have been possessed by our predecessors who lived in Berkshire, perhaps 10,000 years ago.

No. 3.
GEOLOGY OF BUCKINGHAMSHIRE.

NATURAL HISTORY AND SCIENTIFIC SOCIETIES.

High Wycombe Natural History Society.

MUSEUMS.

Eton College Museum.

PUBLICATIONS OF THE GEOLOGICAL SURVEY.

Maps. Sheets: 7, Western London, St. Albans, Wendover, Windsor, Staines, Uxbridge, High Wycombe; 13, Oxford, Reading, Wantage. Quarter Sheets: 45 N.E. Buckingham, Brackley; 45 S.E. Bicester, Brill; 46 N.W. Newport Pagnel, Woburn; 46 S.W. Leighton Buzzard, Aylesbury, Tring; 52 S.W. Northampton, Olney.

Books.—Geology of the Country round Banbury, Bicester, Woodstock and Buckingham, by A. H. Green, 2s. Geology of the London Basin, by W. Whitaker, 13s.

IMPORTANT WORKS OR PAPERS ON LOCAL GEOLOGY.

1861. Whitaker, W. On the "Chalk Rock" in Bucks. Journ. Geol. Soc., vol. xvii. p. 166.
1865. Whitaker, W.—The Chalk of Bucks and the Totternhoe Stone., Journ. Geol. Soc., vol. xxi. p. 398.
1867. Brodie, Rev. P. B. Purbeck Beds &c., at Brill. Journ. Geol. Soc., vol. xxiii. p. 197.
1881. Hudleston, W. H. Gasteropoda of the Portland Rocks. Geol. Mag., p. 385.

See also General Lists, p. xxv.

NOTWITHSTANDING the extensive and varied character of the rocks of this county, they do not seem to have received much attention from local geologists. We have admirable geological maps of the surface published by the Government Survey between 1861 and 1865, and mainly the work of Messrs. Whitaker, Hull, Bauerman, and Green; Mr. Whitaker has also well and minutely described the Chalk district, Dr. Fitton the Portland Beds and Lower Greensand, and the Purbeck Beds at Brill have been noticed by the Rev. P. B. Brodie; Mr. Macalister also published some notes on the northern portion in the *Geologist* for 1861, but there remains an extensive field to be filled by those who live upon the spot. Every quarry, brick pit, railway cutting, or opening of whatsoever kind into the earth, should be diligently examined, carefully measured and drawn, and continually examined for fossils; in this way the broad outlines we now possess must be filled up, and if only one person in each parish could be induced to undertake such work, the gain to geological inquiry would be of real and great value.

Commencing with the oldest rocks, we find in the north that the river Tove has cut down into and exposed the UPPER LIAS CLAYS, which just enter the county from between Grafton Regis and Ashton to Castle Thorpe, two or three miles north of Stony Stratford; the Ouse has also exposed small patches near Stoke Goldington and Weston Underwood.

THE OOLITE.—Of the sandy and slaty beds known as *Northampton Sand* there are no good sections; indeed, these beds have either thinned out or changed in character in this direction.

The Great Oolite.—The limestones of this well-known bed are exposed in many quarries; from Brackley and Buckingham it passes round by Stowe, Lillingston and Potterspury; Stony Stratford, Newport Pagnel and Cold Brayfield mark its eastern limit, but it stretches north-eastward from these points into Bedfordshire.

The limestones are of a compact and *not* oolitic character; they are often soft and marly, and in such beds fossils are most common; frequently we find only the moulds or casts of shells, but they are generally entire, and appear to have been deposited in still water. Clayey beds occur at the base, above which come thin-bedded, cream-coloured limestones and marls, whilst the white, hard, fine-grained limestones are more usually found near the top. Between Brackley and Buckingham there are many quarries, as near Westbury. At Shalstone Hill Farm there is a bed of very hard, white limestone, composed almost entirely of broken shells, and yielding a beautiful building stone. In the valley of the Thornborough Brook there are some large quarries, exposing a section of more than 30 feet; a fault, with a downthrow to the east of 10 feet, runs across the quarry; very interesting exposures of "faults," or dislocations of the beds, may also be seen in a quarry near South End. On Whittlewood Forest there are several quarries; here, on blocks of stone which have been long exposed to the weather, one finds standing out in relief portions of shells, plates and spines of sea-urchins, star-like joints of the stems of crinoids, with worm tracks and *serpulæ*.

From quarries at Gayhurst, Eckley, Salcey Wood and Olney many fossils have been obtained; the shell called *Terebratula maxillata* abounds here. The scenery of the tract occupied by the Great Oolite is undulating, hilly, and often well-wooded, with an eastward slope to the Ouse; the dip of the beds is also in this direction, and their total thickness is about 200 feet.

Forest Marble.—We seemingly have the end of this formation in Buckinghamshire; traces of it are seen in quarries near the brook east of Tingewick church, at Thornton and Lillingston Lovell; the beds consist of from 3 to 10 feet of blue or brown clay, with about 2 feet of hard grey slaty limestone underneath, almost made up of oysters.

Cornbrash.—We find the rubbly limestones which go by this name between Fringford, Tingewick and Buckingham, north of which town they extend past Maids Morton and Akeley; eastwards by Thornborough and Beachampton, the outcrop is 2 miles wide; near Stony Stratford it contracts to half a mile, again widening between Bradwell, Great Linford and Newport Pagnel; it is then obscured by the Ouse gravels, but re-appears as a narrow band, not a quarter of a mile wide east of Sherrington, and curves round thence to Newton Blossomville: it is largely quarried for road-making, being very hard. Near Buckingham, on the Bourton road, we see hard blue limestones associated with beds of black and blue clay; near Thornborough, on the Leckhampstead road, we have in the quarries good illustrations of the manner of weathering of these limestones: the top limestone bands are yellow and rubbly; then, under a bed of sandy clay, we have solid limestone, yellow outside, blue within; lastly, at the base we find solid blue limestone only. The change in colour is due to the iron, which exists in the lower bands as a carbonate, whilst upwards it gradually passes into a peroxide, under the influence of air and water; the total thickness here is about 40 feet. Near Wolverton there are several quarries; here the Cornbrash is of blue-grey colour, very hard and finely laminated; fossils are numerous but difficult of extraction. There are two curious inliers at West Stan Hill and Marsh Gibbon; the first is a very low hummock; a quarry here shows loose rubbly Cornbrash, with a bed of pale blue clay, full of oysters. Marsh Gibbon stands on a little hill of Cornbrash; there are several quarries here, which show loose rubbly stone on top, then soft marly clay, and hard blue limestone at the bottom.

The Oxford Clay. This is a light-blue clay, weathering yellow at the surface, and perhaps 500 feet thick; its western limit is defined by the beds of Cornbrash just described, but eastwards it stretches over a wide expanse to Muswell and Brill Hills, Quainton and Fenny Stratford; sections are not common, but may sometimes be seen in brick-pits, railway cuttings, or drain cuttings; towards the Ouse it occupies the top of the low escarpment, forming a heavy undulating country which rises with a short but steep slope above a plateau formed of Cornbrash. Near Goddington we have the base of the formation, here composed of hard, cream-coloured calcareous clay, containing *Ammonites Jason, &c.*; similar beds are exposed round Newport Pagnel, where *Ammonites Calloviensis* is found, with saurian bones, &c.; these, probably, are near the horizon of the Kelloway Rock of Wiltshire. The brick-pits near Padbury show shaly dark-blue clay, with large nodules and septaria of limestone. *Gryphæa dilatata* is common everywhere, with semi-transparent glassy crystals of selenite, called "fossil water" by the workmen, and yellowish lumps of iron pyrites.

The Coralline Oolite.—The sands and limestone at Headington and Stanton St. John, near Oxford, appear to pass eastwards into a clayey band containing *Ostrea sandalina*, which has been traced by Worminghall, Oakley, Boarstall, round Muswell Hill, and through Dorton to the base of Quainton Hill. At Studley there is a bed of grit, "a sort of argillaceous chert, rich in *Pinnæ, Ammonites*, and other organic remains;" this may mark the transition.

Kimmeridge Clay. - The Coralline Oolite having become itself a clay band, or being absent altogether, there is nothing over the greater part of Bucks to separate the two great clay masses named after Oxford and Kimmeridge from one another; it is however considered that the Kimmeridge Clay forms the plains and low grounds between Thame, Brill, Waddesdon and Aylesbury; the lower boundary line is drawn by the Geological Survey to end off near Stewkley, but this is marked "wholly conjectural;" it is a bluish-grey or yellow shaly clay, often bituminous, sandy towards the base and with irregular bands of limestone nodules: like the Oxford clay, it is best fitted for pasture land; oaks grow well upon it; from the impervious nature of this wide clay tract there are no springs. Near Hartwell and Aylesbury this Kimmeridge clay is largely worked for brick-making; here it is of a dull leaden colour, lignite (fossil wood) is common and fossils abound; *Ammonites biplex* occurs, with the original beautiful shelly lustre still preserved, and many of the fossils sparkle with iron pyrites and are found entire and in great perfection.

Portland Beds.—The lower part is sandy, above which come beds of lime-

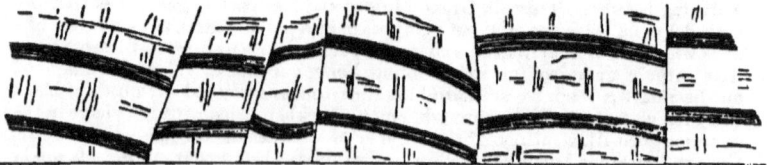

Fig. 11.—Diagram Section of small faults in the Portland Stone, Brill. This dislocation of the beds was seen in a small pit about two-thirds of a mile south-west of the church. The dark bands are two shaly layers; the greatest displacement, or "throw," only amounts to two feet. (After A. H. Green, Geol. Survey.)

stone; the main outcrop passes north-east from Thame by Cuddington and Dinton to Aylesbury and Bierton; outlying masses occur at Muswell Hill,

Brill, Ashendon, Whitchurch &c. it is quarried for building; here, green grains, and then limestones with bands of clay and grit, altogether about 25 feet thick; in one pit, south-west of the church, a number of small faults were seen, dislocating the beds about 2 feet at each step. Near Aylesbury the Portland limestone is called "pendle;" it is soft and sandy. *Trigonia gibbosa* with *Cardium dissimile* are the most common fossils.

Purbeck Beds.—These are thin-bedded limestones and clays or freshwater origin; they come above the Portland beds at Brill, also capping the ridge between Whitchurch and Oving, and they crop out east of Cuddington and south-east of Stone; the limestones are drab-coloured, close-grained, and in beds 6 or 8 inches thick; the total thickness is about 10 feet; fish teeth and scales occur, with the little crustacean known as *Cypris*.

THE CRETACEOUS FORMATION.—The sandy beds at the base, known as *Lower Greensand* are very irregular in occurrence; in the north-east they are called the Woburn sands, which come southwards from Great and Little Brickhill to Fenny Stratford, where they are overlapped by the Gault Clay; here they form dry, hilly ground; on a hill near Great Brickhill there is a section 30 feet deep, showing coarse reddish sands resting on Oxford clay; scattered through the sands are "red coprolites," or phosphatic nodules, for which the sands are sifted, the coprolites are then ground up for manure; the same bed occurs at Rushmoor brick-yard and at Potton in Bedfordshire, and Upware near Cambridge. More than twenty species of fossils native to the bed have been found here by Mr. W. Keeping, besides numbers washed out of lower formations,

There are several sections at Brill, where at the base, we see the Portland Sands with

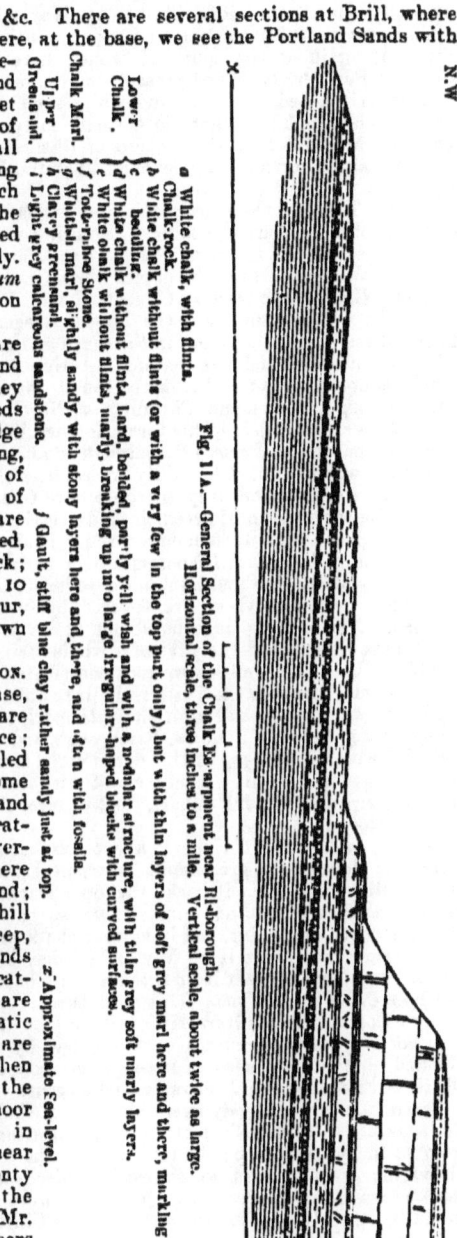

Fig. 11A.—General Section of the Chalk Escarpment near Bledlorough. Horizontal scale, three inches to a mile. Vertical scale, about twice as large.

especially the Kimmeridge Clay. At Wavendon there is an outlier containing seams of fullers' earth; passing southwards the Lower Greensand crops out again at the south of Stone; here it rests partly on Purbeck, partly on Portland beds, and presents indications of Wealden beds at its base. At Hartwell a bed of brown and white sand, about 8 feet thick, has been used for glass-making; there is a small outlying patch south of Oving, and others on Muswell Hill, Brill, north of Chearsley, &c.; the main mass also crops out again north of Haddenham. Quainton Hill is also capped by Lower Greensand.

Gault.—This is a thick mass of pale-blue clay, sometimes shaly, often with whitey-brown phosphatic nodules. Entering the county in a broad band 3 miles wide, extending from Towersey to Henton, we trace it by Weston Turville to Hulcott; here the outcrop broadens to 7 miles between Cubbington and Wing on the west and Buckland and Eaton Bray on the east. The manner in which it overlaps the Lower Greensand we have already seen; its thickness is about 130 feet in the south and 205 feet in the north-east, and it is dug for brick and tile making. A rich seam of "black coprolites" is found about 40 or 50 feet below the top of the Gault; it has been worked at Eddlesborough, Puttenham, Cheddington, Northall, Billington, &c.

The Upper Greensand.—This formation overlies the Gault, and extends from between Henton and Princes Risborough to about a mile north-east of Buckland, from which point it is hardly traceable, although it is once more seen in the brick-yard at Eaton Bray; here we have 6 feet of pale-grey marl, underlaid by the same depth of greenish sandy clay. At Buckland it is seen in the sides of a pond, and a stream cuts through it about half a mile south-east of Aston-Clinton church, at Risborough it is, perhaps, 60 feet thick; in the road cuttings about Bledlow and Henton we see beds of crumbling whitish sandstone, sometimes hard, and these are covered by clayey dark-green sand, seldom exposed. Numerous springs burst out at its junction with the Gault below, which have determined the sites of a whole string of villages.

The Chalk.—This well-known rock constitutes a large portion of the south of the county, rising in the Chiltern Hills to about 900 feet above the sea level; the lowest beds are of soft whitish *chalk marl*, about 80 feet thick, and form rising ground between Bledlow, Risborough, and the reservoirs between Tring and Marsworth. The top of this division is formed by the hard *Totternhoe Stone*, a rather brownish sandy chalk with dark grains; it is well seen in a small quarry near "Tring Wharf," and in several road cuttings along the line just mentioned.

Lower Chalk.—This division is about 400 feet thick and is *without* flints; it forms the slope of the great chalk escarpment, and is exposed along the sides of the valleys which cut it back; the top is marked by a thin but very constant bed of *Chalk rock*, hard and pinkish; we see this layer in a large chalk-pit one mile north of Henley, and in various chalk-pits between the latter town and Marlow, also near High Wycombe, Chesham, &c. Well-sinkers call it "rock," and have to blast it on account of its hardness.

Upper Chalk with Flints. This mass is some 300 feet thick and forms the eastern slope of the Chiltern Hills; it may be distinguished from the lower beds not only by the presence of flints, but by the character of the fossils. We find in it many species of sea-urchins, sponges, &c., whilst in the Lower Chalk whorled and spiral univalve shells occur.

Beech trees grow largely over this tract of chalk, and their wood is used for chair-making. The chalk itself is much burnt for lime; it is got in large pits from 50 to 150 feet deep; the men work at the top of the pits with crowbars, dislodging great masses, which break to pieces in their fall; fortunately the chalk is traversed by fissures or joints, for were it one homogeneous mass it would be scarcely possible to work it. The Chiltern Hills command an extensive prospect and rise sharply from the Oolitic plain which lies to the north-west.

TERTIARY BEDS—EOCENE FORMATION.—Occupying a large triangle between Colnbrook, Beaconsfield and Rickmansworth, we find sands and clays of an age altogether distinct from, and later than, those rocks of which we have hitherto been speaking.

Reading Beds.—These extend eastwardly from Woburn by Burnham, Beaconsfield, Hedgerley, Chalfont St. Peter's and Denham; they consist of variegated plastic clays with light-coloured and green sands; they are exposed in brickyards at Hedsor, Burnham, &c. There are several outliers on the chalk, as at Penn, Pollard Wood, &c. No fossils have been found.

The London Clay.—This rests on the Reading Beds round Fulmer, at Stoke Common, Iver, Red Hill, &c.; it is a stiff brownish clay, often containing large nodules called *septaria*, which are rounded lumps of impure limestone, traversed by fissures filled with carbonate of lime.

THE DRIFT.—In attempting to trace the beds of stratified rocks of which we have been speaking, the geologist finds that they are largely covered over and obscured by irregular, often thick, masses of clay full of pebbles, or by more or less extensive patches of sand or gravel. These deposits are relics of the *glacial period;* the clays were probably dropped by melting icebergs, or possibly pushed out from the end of a great glacier which travelled down from the north into Bedfordshire. The northern side of the Ouse valley between Brackley and Olney is tolerably free from Drift, but south of this line and spreading over the plain formed by the Oolitic Clays we find thick and varied deposits; clean gravel beds occur round Hardwick, between Finmere and Buckingham, and round Stowe and Akeley. In the gravel-pit at Tingewick we see false-bedded coarse gravels with beds of soft clean sand. Near Buckingham there are several examples in gravel-pits of the manner in which these beds are sometimes contorted and bent into curves; this is supposed to be due to the grounding of icebergs: similar cases occur at Foxcott and Maids Morton. The stiff clay drift is well shown round Leckhampstead, where it contains striated boulders of carboniferous limestone, &c. There is also a connecting link in the shape of clayey dirty gravels which are widely distributed.

Over the chalk hills there is much "clay with flints," probably a result of the decomposition of the chalk during long ages.

The Ouse river-gravels cover much ground; they abound in oolitic fossils, especially *belemnites*, which are sometimes collected by the villagers, who consider them, when pounded, an excellent cure for rheumatism!

Flint Implements, the tools of the early inhabitants of this country before they became acquainted with the use of metals, have only been found in one or two places. Pulpit Wood, near Princes Risborough, appears to have been the site of an ancient encampment, and flakes of flint occur in numbers; they were probably used as knives; rounded tools, called scrapers, which would be useful in cleansing skins, have been found here too; a rude flint arrow-head was found by Mr. John Evans in 1866, on the surface of a field between Eddlesborough and Tring, at the foot of the chalk hills. Many more would doubtless turn up if they were intelligently looked for.

No. 4.
GEOLOGY OF CAMBRIDGESHIRE.

NATURAL HISTORY AND SCIENTIFIC SOCIETIES.

Cambridge Philosophical Society. Proceedings and Transactions.
Cambridge Field Naturalists' Club and Entomological Society.
Cambridge Natural Science Club.

MUSEUMS.

The Fitzwilliam Museum, Cambridge.
Geological Museum, Cambridge.
New Museum, Cambridge.
Wisbeach Museum.

PUBLICATIONS OF THE GEOLOGICAL SURVEY.

Maps.—51 S.W. Cambridge. (Survey not completed.)
Books.—The Geology of Rutland and Part of Cambridge, by J. W. Judd, 12s. 6d. Geology of the Fenland, by Skertchly, 40s. Geology of the Country round Cambridge, by Penning and Jukes-Browne.

IMPORTANT WORKS OR PAPERS ON LOCAL GEOLOGY.

List of 121 works (to end of 1873) on the Geology of Cambridgeshire has been compiled by W. Whitaker, Esq., for the Woodwardian Museum, Cambridge.

1861. Notes on Cambridge Palæontology, by Prof. H. G. Seeley. Ann. and Mag. Nat. Hist., Ser. 3, vols. vii. and viii.
1862. The Elsworth Rock, by H. G. Seeley. Ann. and Mag. Nat. Hist., Ser. 3, vol. x. p. 97.
1863. Fossil Wood in the Oxford Clay near Peterborough, by Dr. H. Porter. Journ. Geol. Soc., vol. xix. p. 317.
1865. Ammonites from the Cambridge Greensand, by Prof. H. G. Seeley. Ann. and Mag. Nat. Hist., Ser. 3, vol. xvi. p. 225.
1866. On the Warp, by Rev. O. Fisher. Journ. Geol. Soc., vol. xxii. p. 553.
1867. Terebratulidæ from Upware, by J. F. Walker, Esq. Geol. Mag., vol. iv. p. 454.
1872. Upper Greensand of Cambridge, by W. J. Sollas. Journ. Geol. Soc., vol. xxviii. p. 397.
1873. Ventriculites of Cambridge Upper Greensand, by W. J. Sollas. Journ. Geol. Soc., vol. xxix. p. 63.
1875. Relations of the Cambridge Gault and Greensand, by A. J. Jukes-Browne. Journ. Geol. Soc., vol. xxxi. p. 256.
1875. Geology of Cambridgeshire, by Prof. Bonney. Deighton, Bell & Co.
1876. Drift Deposits in the Cambridge Valley, by W. H. Penning. Journ Geol. Soc., vol. xxxii. p. 196.

1877. Coral Reef at Upware, by Blake and Hudleston. Journ. Geol. Soc., vol. xxxiii. p. 313.
1877. Fauna of Cambridge Greensand, by A. J. Jukes-Browne. Journ. Geol. Soc., vol. xxxiii. p. 485.
1878. Post-Tertiary Deposits of Cambridgeshire, by A. J. Jukes-Browne. Deighton, Bell & Co., 2s. 6d.
1879. Dinosauria of Cambridge Greensand, by Prof. H. Seeley. Journ. Geol. Soc., vol. xxxv. p. 591.
1879. Mammaliferous Deposit at Barrington, by Rev. O. Fisher. Journ. Geol. Soc., vol. xxxv. p. 670.
1881. Seeley, Prof. H. G. Two Ornithosaurs from the Cambridge Upper Greensand. Geol. Mag., p. 13.

See also General Lists, p. xxv.

IN a small pamphlet published by the authorities of the Woodwardian Museum, at Cambridge, Mr. W. Whitaker gives the titles of 121 works or papers on the geology of the county, which had been made public up to the end of 1873. Among those who have done much to elucidate the nature of the strata and their fossil contents in this district, we may mention the names of Professors Sedgwick, H. G. Seeley, Bonney, and McCoy; the Revs. P. B. Brodie and O. Fisher, and Messrs. W. J. Sollas, A. J. Jukes-Browne, J. F. Walker, H. Keeping, J. J. H. Teall, &c. The surface of the county has been examined, and the position of the rock-beds laid down on maps by the officers of the Government Geological Survey. These maps, however, are not yet published, with the exception of a small portion round Whittlesea and Thorney, which is included in sheet 64; the southern division (51 south-west) is, however, nearly ready for publication. A geological map of the district round Cambridge was prepared by the late Mr. Lucas Barrett in 1859, of which a revised edition by Mr. Jukes-Browne was published in 1873. The Geological Survey Memoir by Mr. Skertchly, on "The Fenland," published in 1877, gives an excellent and complete account of the northern portion of the county. In the Woodwardian Museum, under the public library at Cambridge, the visitor will find a grand series of local rocks and fossils, the result of many years' work on the part of the energetic geologists who have been or are connected with this institution.

Cambridgeshire may be divided—geologically and topographically—into four areas or regions. The first of these is the small tract occupied by the Oolite Clays in the south-west of the county, extending from Fenny Drayton and Willingham by Papworth St. Agnes and Elsworth to Croxton, Eltisley and Caxton, being in fact the eastern edge of the great sheet of clay which forms nearly the whole of Huntingdonshire.

Secondly, we have the valley of the Cam and Rhee, excavated mainly in Neocomian and Lower Cretaceous beds. This valley runs from south-west to north-east, from Guilden Morden by Cambridge to Ely.

The third region is constituted by the chalk hills, which form the south and south-east of the county, running by Royston and Linton to Newmarket. These three divisions include all the old stratified rocks of Cambridgeshire, which are thus seen to belong to two of the great geological divisions only, viz. the Oolitic and Cretaceous. They have a *strike*, or line of direction or extension, along the surface from north-east to south-west, while they each incline or *dip* to the south-east, passing under the bed next above, so that the lowest and oldest rocks are the Oolite Clays just mentioned as occurring on the western side of the county.

The fourth division is a portion of the great fen district, and is composed of beds of clay, peat, and gravel, which rest upon and conceal the northward extension of the older strata which forms the three districts before described. The river Ouse, in its course from east to west across the county, defines pretty clearly the southern termination of the fens.

We shall now consider each district with as much details as our space permits, pointing out the nature and position of the rocks, and the character of their fossils, concluding with a brief statement of the evidence which is here afforded as to the antiquity of man.

THE OOLITE.—In the south-west of England, from Dorset to Oxford, and again in the north-east, in Yorkshire, we have two great beds of clay belonging to the middle and upper part of the Oolitic system, separated from each other by an intervening bed of limestone—more or less continuous—which is called the *Coral Rag*. This limestone has associated with it two sandy beds—one above and the other below—which are called the *Upper and Lower Calcareous Grit*. But from the quarries at Headington and Wheatley, near Oxford, to Acklam Wold, in Yorkshire—a distance of 240 miles—there is no good line of demarcation between the *Oxford* and the *Kimmeridge Clays*, with the exception of certain limestone bands in Cambridgeshire which we shall describe. Two great clay masses pass into each other gradually and insensibly, and it becomes impossible to lay them down separately on a map.

Fine sections in the *Oxford Clay* are to be seen in the brickyards at St. Ives and Peterborough, but these are both outside the county boundary. The same stratum passes under the fen beds and forms no doubt the foundation of all the low flat country round the Wash. At two or three points the Oxford Clay rises above the later deposits, as round Whittlesea and Thorney, forming what were once islands when the fens were undrained. There is a large pit to the north-west of Whittlesea, where the Oxford Clay is seen to be of a deep-blue colour, and to contain much pyrites and wood. There are also clay.pits round the town of Whittlesea, which have yielded many beautiful *ammonites*, together with *Gryphæa dilatata*, a large and very characteristic shell. At Eastrea brickyard clays, perhaps rather higher in the Oxfordian series, are worked. Passing farther south, we find the Oxford Clay extending from Fenney Drayton, by Papworth St. Agnes, to Eltisley and Croxton. Here it forms rather high flat ground; but being thickly covered by boulder clay, there are few sections visible. The region, however, is interesting from the occurrence of several bands of hard rock in the clay, which were discovered by Professor H. G. Seeley, and the exact position of which in the Oolitic series is very doubtful. The most important is at Elsworth, 8 miles north-west of Cambridge. Here are layers exposed in the brook-course of a homogeneous limestone, full of oolitic grains and (when unweathered) of a dark-blue colour. Whether this rock represents the *Lower Calcareous Grit*, or is simply a calcareous band high up in the Oxford Clay, is a point of much interest. Mr. Seeley is inclined, from the fossils he found in it, to assign to it the latter position. There are also thin bands of limestone at Tetworth and near Gamlingay and Boxworth. From the neighbourhood of these villages, as we pass eastwards, the "clay seems to graduate imperceptibly up to the Kimmeridge Clay of Cottenham." For the transition clays, which contain both Oxfordian and Kimmeridgian fossils, Mr. Seeley has proposed the name of Tetworth (or Ampthill) Clay; it would of course occupy the place of the Coral Rag of other districts.

The Coral Rag.—Near the little village of Upware, about 10 miles north-east of Cambridge, and 3 miles from Reach, there is a low plateau formed of limestone (capped by Neocomian Sands), which is probably a remnant of a coral reef, the only one on this line between Oxford and Yorkshire. The limestone is worked in two quarries, situated at the north and south ends of the plateau. The southern pit shows a creamy-white limestone, with layers of corals and many other fossils, as *Cidaris Florigemma*, &c. The northern pit is in rather lower beds, and shows a coralline Oolite with few fossils, mostly echinoderms.

The Kimmeridge Clay.—This deposit is very thin in Cambridgeshire: it can be traced by Cottenham and Haddenham to Ely, where, at Roslyn Hill, it is largely excavated for the purpose of tamping, or making watertight, the banks of the dykes in the fens: its thickness here is 60 feet, and it is a bituminous, purplish, or bluish-black laminated clay, containing large nodules of impure limestone. Many fossils have been obtained here, including nine or ten species of saurians, several of which were new species, and have been described by Professor Seeley. In a pit, sunk by Mr. Keeping near the Coral Rag Pit, at Upware, he found the Kimmeridge Clay curving up unconformably against the limestone, and at the junction the clay had mixed with it a quantity of rounded or broken fragments from the Coral Rag, "so that it actually presented the appearance of boulder-drift."

Of the highest Oolitic strata—the Portland Stone and Sand, and the Purbeck Beds—there is no trace in Cambridgeshire, neither are there any representatives of the Estuarine Series which assumes so much importance in the south-east of England, where it is known as the Wealden. It is probable that land existed in this area during the time of deposition of these strata elsewhere.

NEOCOMIAN FORMATION.—This important division is scantily represented by the beds commonly called *Lower Greensand*. They are of a sandy nature, and we can trace them from the well-known "diggings" at Potton in Bedfordshire, by Gamlingay, Bourn, Dry Drayton, and Oakington to Upware. Their thickness is variable; in the south of the county about 100 feet, but thinning northwards to only 8 feet at Wicken. Some of the sandy beds are sufficiently consolidated to furnish an inferior building stone, known as "Carstone." These beds lie under Cambridge at a depth of 100 to 150 feet, and are reached by artesian wells, yielding a good supply of excellent water. At Ely, the Lower Greensand is seen resting on Kimmeridge Clay and about 8 feet thick. At Haddenham it is a light-brown ferruginous sand 15 feet thick. At Upware there are extensive workings for *coprolites*. Properly speaking coprolites are the fossilized dung of saurians, fishes, &c.; but the term has been applied to all the more or less rounded dark-looking phosphatic nodules which occur in great numbers in these deposits. In the shallow diggings at Upware we see about 8 feet of Neocomian Sands, which contain two phosphate beds, an upper one, 6 or 7 inches thick, and a lower one forming the base of the deposit, and from 1 to 2 feet in thickness. This lower bed contains many pebbles, which have to be picked out. Fossils are numerous. They may be divided into two classes, placing in one division those which from their rolled and water-worn appearance, the marks of boring shells which they bear, and their usual dark colour, appear to be *derived, i.e.* washed out of some older deposits, as the Kimmeridge and Oxford Clays, &c.; and another set of a lighter hue and more perfect preservation, which are of the age of the deposit. Most of the so-called coprolites are pieces of wood, casts of shells, bones, and even lumps of mud, all of which have been penetrated by phosphate of lime, so as now to form when properly treated a most valuable manure.

CRETACEOUS FORMATION.—The term "galt," or "golt," is applied by workmen to any stiff clay, being often used in the Midlands to describe the Oxford Clay, or even the chalky Boulder Clay. By geologists, however, it has been adopted as a name for the lowest member of the Cretaceous Series proper, which is accordingly termed the *Gault*. Professor Bonney describes it as a "tenacious pale bluish-grey clay, in which are occasional concretions of iron pyrites, often more or less converted into limonite, together with small crystals of selenite and not a few rather light-brown phosphatic nodules." This stratum runs through Cambridgeshire from Guilden Morden and Kingston by Cambridge, Girton, Horningsea and Soham, forming flat ground for one or two miles on each side of the Cam. Its thickness is variable, and it appears to lie unconformably on the Lower Greensand, while its upper surface is also uneven and much eroded. At Guilden Morden its thickness is about 200 feet, and thence it thins northwards to 170 feet at Bassingbourne, 160 feet at Barrington, 150 feet at Haslingfield, 120 feet at Grantchester, and under Cambridge it is only a little over 100 feet in thickness. In fact in this county the Gault has suffered great erosion; all the upper portion has been worn away, and we shall find the residue of the denuded portion at the base of the next higher deposit—the Chalk-Marl. We can study the Gault in many large excavations where it is worked for brick and tile-making, as in the deep pit on the right of the Newmarket road near Cambridge, and at Barnwell. At the latter place lines of "coprolites" are seen. Many of these Mr. Sollas has found to be phosphatized sponges. Other fossils are not common near Cambridge, but further north, at Reach, &c., the usual common gault fossils, as *Belemnites minimus* and *Inoceramus concentricus*, are found. From an examination of the shells in this and in other areas, and a comparison of their habits with those of existing molluscs of the same genera, Mr. F. G. H. Price has come to the conclusion "that the Lower Gault was a near-shore or muddy estuarine deposit, and that in course of time, as the sea-bed gradually sank, so did that sea naturally become deeper and deeper," as is shown by the disappearance of shells which inhabit shallow water and the incoming of genera of deeper seas.

The Chalk-Marl.—To see the base of this division it will only be necessary to visit some of the numerous coprolite workings which occur along its outcrop for a distance of 50 miles, from Soham to Hitchin. The chief pits lie near Grantchester, Meldreth, Whaddon, Shepreth, Horningsea, Coton, and Royston. The sections here exposed show on the top perhaps 2 or 3 feet of gravel and sand, then 10 to 20 feet of chalk-marl, which is of a greyish tint, and is locally called "clunch." The lower part of this marl contains many green grains, and its base is "a stratum barely a foot in thickness, which is full of green grains, and black granules and nodules, looking like a sediment from the purer marl above. The green grains are the mineral *glauconite*, and the black nodules phosphate of lime, often but erroneously called "coprolites." This stratum is the famous *Cambridge Greensand*, and was long thought to represent the true Upper Greensand of the south of England. Messrs. Whitaker, Jukes-Browne, and others have now shown that this bed is really the base of the Chalk-Marl. The Upper Greensand proper comes to an end a short distance north of Tring, in Hertfordshire, and the Upper Gault similarly, near Barton, in Bedfordshire. These beds must have been washed away in Cambridgeshire, probably by strong marine currents, and we have their "riddlings," as it were, in the conglomeratic stratum which constitutes the base of the Chalk-Marl. In 1877 there were raised from this seam in Cambridgeshire and Bedfordshire 55,000 tons of phosphatic nodules, valued at £150,000. The trade, however, is falling off greatly, owing to extensive imports from abroad, especially from the port of Charlestown in America, whence 170,000 tons were exported in that year to our manure manufacturers. In 1879 only 25,000 tons of coprolites were got in Cambridgeshire, and the average price was £2 7s. per ton. A rent of from £100 to £140 per acre has been usually paid for the privilege of digging the nodules, the average yield being 300 tons per acre. The land has afterwards to be properly levelled, re-soiled, and returned to the owner at the end of two years.

After being dug out, the nodules are washed in a mill erected on the workings, and then conveyed to the chemical works, where they are ground to powder and treated with sulphuric acid, which converts the insoluble phosphate of lime into soluble super-phosphate.

When the bed is examined minutely, the green grains of glauconite are found to be the internal casts of the shells of foraminifera. Many of the nodules are phosphatized sponges; a fact which Mr. Sollas was enabled to demonstrate by examining thin semi-transparent slices of them under the microscope. Fossils are very numerous. There are a few bones of birds, and many of reptiles and fishes, while shells and remains of crustaceans are numerous. As in the Lower Greensand two sets of fossils can be distinguished: the first occur as dark phosphate casts, and have been washed out of the Gault; the second comprises those which lived at the time of the formation of the stratum in which they are found. Of the former, *Ammonites splendens*, *A.*

rostratus, and *A. auritus*, and of the latter *Exogyra laciniata*, and *Plicatula sigillina* may be mentioned.

Although pebbles are not so numerous as in the Lower Greensand coprolite bed, yet masses of stone of large size are sometimes found, of which there are some good examples in the Woodwardian Museum. They appear to be rocks which have been brought from the east of Scotland and the south of Norway, possibly by floating ice.

Large outliers of the *Chalk-Marl* occur west of the Cam; but the edge of the main mass of the bed lies about 2 miles east of the river. It is about 60 feet in thickness, and is of a light greyish tint.

The White Chalk.—The Lower Chalk without flints is worked for building-stone at Madingley, Burwell, Reach, &c.; and also furnishes excellent lime. It is rather too perishable for the outside of buildings, but works very easily, and is well adapted for interior carvings. It is impossible in many places to draw a precise line between the Chalk-Marl and the *Lower Chalk without flints*, and the latter again passes up gradually into the *Upper Chalk with flints*. At some points, however, we see a hard band, the *Totternhoe Stone*, on the top of the Chalk-Marl, and another hard bed called *Chalk Rock* between the Upper and Lower Chalk. In a chalk pit about a mile south of Litlington church, there is a hard cream-coloured nodular layer, about 6 inches thick, seen at the bottom of the section, and the chalk above is hard and massive. This pit is probably in the upper part of the "Lower Chalk." Mr. Jukes-Browne has lately published the following important classification of the Chalk of Cambridgeshire, based on the different fossils which the beds contain:—

Cambridge Greensand or Chloritic Marl ...	
Zone of *Rhynchonella Martini* (Chalk-Marl) ...	Lower
Totternhoe Stone (at Burwell)	Chalk.
Zone of *Holaster sub-globosus*	
Melbourne Rock (10 feet), ...	
Zone of *Rhynchonella Cuvieri*	Middle
Zone of *Terebratulina gracilis*	Chalk.
Zone of *Holaster planus*	
Chalk Rock	Upper
Zone of *Micraster cor-bovis*	Chalk.

The Upper Chalk forms the line of hills which runs by Royston and Linton towards Newmarket. In the rounded outlines of these hills, with their waterless valleys between, we recognise at once the characteristics of chalk scenery. In the various pits the usual chalk fossils are to be found—sea-urchins *(echinoderms)*, known by their rounded form, with such shells as *Terebratula* and *Inoceramus;* the lines of dark-coloured flints are visible enough, but their exact mode of formation is as difficult to explain as that of the phosphatic nodules called coprolites. Indeed, in the chalk, true coprolites are sometimes found: they may generally be told by their twisted exterior. The thickness of the chalk is here about 800 feet. The western slope of the chalk hills towards the Cam valley is somewhat abrupt; but the dip-slope eastwards is long and gentle. The Gog-Magog hills rise to a height of 302 feet above the sea; and the heights of a few other points on the chalk are Babraham (church), 83 feet; Little Abington (church), 104 feet; Linton (church), 133 feet; The Rivey (a hill near Linton), 350 feet; and Balsham 395 feet. When chalk is examined microscopically, the mass of the rock is found to be made up of the minute shells of foraminifera, which probably accumulated at the bottom of a deep sea comparable to the North Atlantic of the present day.

THE DRIFT—GLACIAL DEPOSITS.—Of the Tertiary Formations—the London Clay, the Crag, &c., which are found resting upon the chalk in Suffolk and Essex—we get no trace in Cambridgeshire. We pass at once from the Chalk —an oceanic deposit—to beds of gravel and boulder clay, which were formed during the last glacial period, a time of excessive cold, when great glaciers existed in England, and during which the land underwent great changes of level. The lowest glacial deposits in this county appear to be certain beds of gravel and sand termed *Middle Glacial* by Mr. S. V. Wood, jun. They are exposed near Quendon, Wicken Bonant, &c. The latest observations, by Mr. Jukes-Browne and others, tend to show that these gravels belong rather to the *Boulder Clay* which rests upon them. This clay is a stiff brown or blue mass, full of rocks of all sorts and sizes, but especially of chalk, so that it is often termed the *Great Chalky Boulder Clay.* It laps over the Chalk hills and stretches over the high ground which lies between the valleys of the Cam and Ouse. Its extent and position show (1) that it was once a continuous sheet covering the whole county, and (2) that it was deposited after the main physical features of the county had been formed, since it occurs on the valley slopes. The smaller lateral valleys have, however, been cut and the principal valleys deepened since its deposit. Northwards the Boulder Clay passes under the Fen Beds; but it stretches in all directions far beyond Cambridgeshire, terminating southwards on the brow of the Thames Valley. Its thickness in this county is very variable, but often above 100 feet. The varied rocks and fossils which it contains appear mostly to have come from the north, from Lincolnshire and Yorkshire, and some even from Scandinavia. These may have been broken off and dragged underneath, or pushed before a great glacier advancing southwards. Some of the masses of rock which have been carried in this way are of enormous size. At Roslyn Hole, near Ely, there is a mass of Chalk, Greensand, and Gault, 480 yards in length by 60 in width, which is surrounded by and rests upon Boulder Clay, the latter occupying a hollow scooped out of Kimmeridge Clay. This enormous mass must have been transported by ice from the eastwards.

Of later date than the Boulder Clay, from the washing of which they are indeed probably derived, are the *Hill* or *Plateaux Gravels* found capping Barrington Hill, the Gog-Magog Hills, Copley Hill, &c. Then, at lower heights, we have a series of *Old River Gravels* containing shells and bones, the latter often belonging to extinct species, as the Mammoth, Rhinoceros, &c. Lastly gravel and mud, the deposit of the existing streams, fringe their courses and form the flat "river meadows" along their sides.

THE FEN BEDS. The southern boundary of the Fen district follows an irregular line from Earith on the west, by Over and Willingham, to Swaffham Priory, and thence by Soham to Mildenhall. All to the north of this line is "a low plain, varying from 5 to 20 feet above mean sea-level. Much of this land is below the highwater level of ordinary tides, and would be overflowed but for the erection of great banks along the sea-board, which are known as the "sea walls." Slight elevations of from 20 to 80 feet stand out from the dead level of the fens like islands from the midst of the sea. Such are Ely, Whittle-

Fig. 12.—Section across Butcher's Hill, between Littleport and Welney. It is a small example of one of the " Fen islands " of former days. (After Skertchly, Geol. Survey.)

a Peat.
b Silt.
c Sand and Gravel.
d Boulder clay.
e Kimeridge Clay.
xx Water line.

sea, Thorney, Eastrea, &c.; with these exceptions there is nothing to break the monotony of the plain; and when the isles and bounding hills lie beyond the field of vision, the landscape is as even, and the horizon as circular, as is the sea-view when no land appears. Through this dreary land the Nene and the Ouse once found their sluggish way, interlacing at numerous points, almost losing themselves in meres and morasses, flooding the land in winter, drying into strings of stagnant pools in summer, and on meeting the salt-water of the sea, pushing it back till the tide, gathering strength, overcame the river waters and rushed up the estuaries as bores. Now the rivers are banked and travel direct to the sea, the winding estuaries are straightened, and bores are of very rare occurrence, every mere and morass is drained, the whole country is intersected with artificial dykes, and powerful engines pump the superfluous water into the arterial cuts. The land, which up to the time of Elizabeth furnished only coarse fodder to cattle and geese, and afforded a home to countless wild fowl and fish, upon which a half-savage scanty population subsisted, is entirely under cultivation, possesses but few acres of waste, and forms one of the richest agricultural districts of the kingdom." (Skertchly.)

The lowest Fen Beds are *gravels* of marine origin, which crop out all round the edge, and of which large patches also occur at March, Whittlesea, and Chatteris. As it furnishes a solid foundation and a fair water supply, we find many villages on it.

Peat occupies a much larger area. It rests upon the gravel, and as we follow it towards the sea, we find intercalated beds of *silt* or warp or mud, one kind of which has been called "buttery clay." The silt is a marine deposit, being the mud left behind by the tides, and in this way a width of land of three miles has at some points been formed since the Roman Period—a fact

Fig. 13.—Trees of the second and third "buried forests;" Fir clasping Oak, Wood Fen, near the old "Blue Boar," Isle of Ely.
    a Peat, 5 feet.   *c* Oak.
    b Kimeridge Clay.   *d* Fir.

which, together with the extensive enclosures which have been made, has transformed Cambridgeshire from a sea-coast to an inland county. The average thickness of the peat is about 6 feet. It is largely dug for fuel in the neighbourhood of Ely; it contains at least five "buried forests," or horizons on which stools and trunks of large trees are found. The land formed by it can be readily distinguished by its blackness and even surface and by the absence of hedge-rows.

PREHISTORIC MAN.—The earliest traces of the existence of man in this country consist of stone implements mostly made of flint. We can distinguish two distinct classes of these stone tools, both in their appearance and mode of occurrence.

1. *Palæolithic or Old Stone Age.*—The specimens which belong to this period are usually found in beds of gravel, sand, or loam. They are from 4 to 8 inches in length, and are either pointed or oval in form. They were evidently manufactured by striking off flakes from a suitable lump of flint with a stone pebble until the desired form was produced.

2. *Neolithic, or Newer Stone Age.*—The implements of this period are found on or close to the surface, or in the barrows or tumuli, which are the burying

places of ancient chiefs. Their forms are varied and elegant, and exhibit an immense advance in skill on those of the Palæolithic age. Moreover many of them have been polished by rubbing on stone slabs.

There are no connecting links in England between these two divisions of the great "Stone Age," and many theories have been advanced to account for the abruptness of the transition from one to the other. The discoveries of Mr. Skertchly have now made it almost certain that the break is due to the coming on of the Glacial Period. The Palæolithic implements belonged to men who lived in this country before the time of great cold, which came on perhaps a quarter of a million years ago. Driven southwards by the ice, they probably inhabited the south of France and similar latitudes for a long period of years. Then the cold having passed away, their descendants many thousands of years afterwards again returned to this area; but in the interval they had made

Fig. 14.—Trees of the third and fourth "buried forests;" Fir astride Fir, Wood Fen, between Ely and Littleport.

a Peat.   *b* Fir.   *c* Fir.

great progress, had attained great skill especially in the manufacture of stone tools, and to their relics we apply the term *neolithic*.

Then came the discovery of metals—first, copper and tin, of which the *Bronze Age* gives evidence, and then comes on the *Iron Age* in which we live.

Palæolithic implements have been found in the gravel pits at Barnwell, Chesterton, and the Observatory Hill, Cambridge.

Of neolithic tools many specimens have been obtained. The fen district, especially Burwell Fen, has yielded many celts, arrow-heads, flakes, scrapers, hammers, &c. At Ely a fine perforated axe-hammer made of greenstone has been found; some well-polished celts have been found at Coton, and altogether there are from thirty to forty localities in Cambridgeshire in which relics of the Stone Age have been met with. Of these full particulars may be found in Mr. John Evans' excellent book on the "Ancient Stone Implements of Great Britain."

No. 5.

GEOLOGY OF CHESHIRE.

NATURAL HISTORY AND SCIENTIFIC SOCIETIES.

Chester Society of Natural Science. Annual Report.
Historical Society of Lancashire and Cheshire; Liverpool. Transactions.

MUSEUMS.

Museum of the Mechanics' Institute, Chester.
Macclesfield Museum.
Stockport Museum.

PUBLICATIONS OF THE GEOLOGICAL SURVEY.

Maps.—Quarter Sheets: 73 N.E. Nantwich; 73 N.W. Whitchurch, Malpas; 79 N.E. Liverpool, Holywell; 79 S.E. Flint, Mold; 80 N.E. Altrincham, Knutsford; 80 N.W. Prescott, St. Helens, Warrington; 80 S.W. Chester, Tarporley; 80 S.E. Northwich, Middlewich; 81 N.W. Stockport; 81 S.W. Macclesfield; 88 S.W. Manchester, Oldham.

Books.—Geology of the Country round Altrincham, by E. Hull, 1s. Geology of the Country round Stockport, Macclesfield, Congleton, and Leek, by Hull and Green, 4s. Geology of the neighbourhood of Chester, by A. Strahan.

IMPORTANT WORKS OR PAPERS ON LOCAL GEOLOGY.

List by W. Whitaker, Esq., in the Proceedings of the Liverpool Geological Society, of 190 books, papers, &c., published up to 1873.

1835. Egerton, Sir P. G.—Gravel Bed containing Recent Marine Shells at "The Willington," and at Narley Bank. Proc. Geol. Soc., vol. ii. pp. 189 and 415.
1848. Ormerod, G. W.—The Salt Field of Cheshire. Journ. Geol. Soc., vol. iv. p. 262.
1865. Brodie, Rev. P. B.—Remarks on the Lias Outlier in South Cheshire. Proc. Warwick Field Club, p. 6.
1869. Hull, Prof. E.—On a Ridge of Lower Carboniferous Rocks, probably occurring beneath the Trias of Cheshire. Journ. Geol. Soc., vol. xxv. p. 171.
1870. Maw, G.—Rhætic Beds in Cheshire. Geol. Mag., vol. vii. p. 203.
1872. Mackintosh, D.—Sea-Coast Section of Boulder Clay. Journ. Geol. Soc., vol. xxviii. p. 388.
1873. Mackintosh, D.—On Remarkable Boulders. Journ. Geol. Soc., vol. xxix. p. 351.
1873. Ward, T.—The Cheshire Salt District. Proc. Lit. and Phil. Soc. Liverpool, No. xxvii. p. 39.
1874. Reade, T. M.—Shells of the Cheshire Low Level Drift. Journ. Geol. Soc., vol. xxx. p. 27.

1874. Shone, W.—Foraminifera in the Boulder-clays. Journ. Geol. Soc., vol. xxx. p. 181.
1877. Mackintosh, D.—New Sections of Boulder Clay. Journ. Geol. Soc., vol. xxxiii. p. 730.
1878. Shone, W.—Glacial Deposits of West Cheshire. Journ. Geol. Soc., vol. xxxiv. p. 383.
1879. Dawkins, W. B.—Mammoth Pre-glacial in Cheshire. Journ. Geol. Soc., vol. xxxv. p. 140.
1881. Strahan, A.—Lower Keuper Sandstone of Cheshire. Geol. Mag., p. 396.
See also General Lists, p. xxv.

GEOLOGICALLY speaking, the structure of the county of Chester is characterised by simplicity and uniformity. Nine-tenths of its area is composed of rocks belonging to one geological formation only—the Trias or New Red Sandstone; the remainder is composed of beds of Carboniferous age, which form the hilly region of the east and north-east.

The district cannot be said to have been neglected by geological writers. In a "List of Works on the Geology, &c., of Cheshire," compiled by Mr. W. Whitaker, and published in the Proceedings of the Liverpool Geological Society for 1875-6, we find the titles of 190 articles written by 85 authors. Of these, however, Professor E. Hull claims no fewer than 28, Mr. G. H. Morton, of Liverpool, 19, and Mr. E. W. Binney, of Manchester, 18.

The whole of the county has been examined by the officers of the Geological Survey, and their published maps, sections, and memoirs form a full and satisfactory guide to a complete knowledge of its structure.

Mr. Binney's papers were chiefly published in the Proceedings of the Literary and Philosophical Society of Manchester, and those of Mr. Morton in the publications of the Liverpool Geological Society, and of the Literary and Philosophical Society.

Cheshire forms an exception to the fact that in England the oldest rocks occur as a rule in the north-west part of any district whose geological structure we wish to describe. On the contrary, the first-formed and consequently oldest rocks occur in the east and north-east of Cheshire, where the boundary line extends up to the Yorkshire Moors. The cause of this is to be found in the upheaval of the Pennine Chain, so that rocks of Lower Carboniferous age have there been brought to the surface and exposed by subsequent denudation.

These Lower Carboniferous beds we also find, so to speak in their correct place on the west side of the river Dee, where a fine ridge of rocks of the age of the Carboniferous Limestone and Millstone Grit runs north and south for a distance of 21 miles, the town of Mold being the central point. Looking from these hills eastwards we have before us an extensive plain composed of "Red Rocks" of Triassic age, thickly overlaid with boulder clay, and composing, as has been said, nine-tenths of the county. This plain is over 40 miles in width, and its eastern boundary is formed by a repetition of the identical strata—the Carboniferous Limestone, Millstone Grit, and Lower Coal-measures—upon which we have taken our imaginary stand in Flintshire.

As the oldest rocks occur in the north-east, we must commence our geological description there.

CARBONIFEROUS FORMATION.—The lowest bed or member of this well-known series of rocks is the *Carboniferous or Mountain Limestone*, which in Derbyshire occupies an extensive tract of country between Buxton and Matlock. In Cheshire it only peeps up at one point, viz., the Astbury lime works, east of Congleton. It hardly occupies any surface area, and is said to have been discovered by a Derbyshire servant girl about 200 years ago, who noticed that the rock in the brook was the same as she had seen burnt for lime in her native county; it has been worked ever since, and is much sought after, as it yields a good hydraulic lime. The rock occurs just on the east side of a great dislocation, called the "Red Rock Fault," which we shall describe further on.

The Yoredale Rocks.—These beds rest upon the limestone just described and attain a total thickness in this region of 4,000 feet. They extend from Congleton Edge to St. James, south of Macclesfield, and thence eastward for three miles, and then run up the valley of the river Goyt to a point about half-way between Taxall and Pott Shrigley. They consist in the lower part of black shales, alternating with thin black earthy limestones (perhaps the equivalent of the *Upper Limestone Shales* of other districts); then comes a considerable thickness of hard fine-grained sandstone, surmounted by alternations of dark shales and sandstones. Fossils are scarce, but *Goniatites, Chonetes, Productus*, &c., have been found at Congleton Edge.

The Millstone Grit.— The beds which bear this familiar name have been subdivided by Professor Green as follows:—

First, Grit, or *Rough Rock*, with feather-edge coal.
    Shales, with a thin coal at the bottom near Buxton.
Second, Grit, or *Haslingden Flags*.
    Shales, with two or three thin coals.
Third, or *Roaches Grit*.
    Shales.
Fourth, or *Kinderscout Grit*, generally in two beds.

The total thickness of these beds is as much as 3,000 feet, in the north-east of Cheshire, near Marple; but they thin out southwards considerably. The sandstones are very massive, and of a coarse grain—often conglomeratic in fact.

The feather-edge coal is a very variable seam, about 3 feet in thickness:

it is finely exposed in the railway cuttings and river cliffs, near New Mills.

The Millstone Grit can be traced from near Macclesfield, running north-east by Rainow, to New Mills, and thence along the Goyt to Marple: it also forms the moorland region of the extreme north-east corner of Cheshire, between Longden Dale, Featherbed Moss and Holme Moss, and is pierced by the Woodhead tunnel of the Manchester and Sheffield Railway. Fine sections too are to be seen along the course of the river Tame. Fossil remains of such plants as *Lepidodendron*, *Sigillaria*, and *Calamites* are not uncommon, with the shells *Aviculopecten* and *Goniatites* in the roofs of the little coals.

The Coal-measures.—The Cheshire Coal-field is a prolongation of the extensive coal-field of South Lancashire, running due south from Oldham by Stalybridge and Hyde to Macclesfield. *The Lower Coal-measures* rest upon the Millstone Grit, and include some thin seams of coal, known as the *Gannister Coal*, the *Bakestone Dale Coal*, &c. The Middle Coal-measures contain the chief workable seams, which crop out about two miles to the east of Stockport and Poynton. There are from six to ten seams, of good thickness and quality; the lowest is the *Redacre mine*, which is identical with the famous *Arley*, or *Royley mine* of South Lancashire.

Fossils.—The ordinary Coal-measure plants, already enumerated for the Millstone Grit, occur in fair abundance.

Anthracosia, the common Coal-measure shell, also occurs, and the iron-stone nodules yield remains of fishes; everything shows that here, as elsewhere, the coal forests grew in mighty swamps near the sea, on an area which underwent frequent upheaval and depression.

In 1875 there were 37 collieries at work in Cheshire, and they raised 658,945 tons of coal, besides 8,200 tons of fire-clay, 1,000 tons of oil shales, and 1,500 tons of iron ore. The amount of coal remaining to be extracted was estimated in 1871 by Mr. Dickenson at 200 millions of tons; in 1879 there were 26 collieries at work, producing 720,350 tons of coal and 6,062 tons of fire-clay.

The Neston Colliery.—We must not dismiss the coal question without mentioning that coal is worked at Neston, on the north side of the river Dee. The beds here are evidently a continuation of the seams which form the Flintshire Coal-field. The measures dip to the north-east, and probably lie under all the peninsula of Wirrall, and pass under Liverpool itself, but at a considerable, and as yet unknown depth. Coal has been proved also on the east side of the Dee, north of Chester.

Scenery of East Cheshire.—The tract of land formed by the old Carboniferous rocks, which we have been describing, constitutes a bold and hilly country, forming the west flank of the Pennine Chain. Several hills rise to a height of above 1,000 feet, as Cloud Hill, near Congleton (capped by 3rd Grit), 1,190 feet; Tegg's Nose, near Macclesfield (3rd Grit), 1,300 feet; source of the Goyt, 1,600 feet; Eddisbury Hill, 1,000 feet; Northern Nancy, near Kerridge, 930 feet. The height of Marple Church is 624 feet; the Cage in Lyme Park, 882 feet; Park Moor, south side of Lyme Park, 1,350 feet, &c.

The *strike* or direction of extension of the strata is north and south, and in this direction we see lines of hills and escarpments to run with wonderful regularity, the "scarps" being formed of the hard sandstones, while the valleys between have been excavated by rivers out of the softer shales. The whole country rises at once from the Plain of Cheshire on the west: it is cut off from it in fact by a great line of dislocation known as the *Red Rock Fault*, which has been traced along the west side of Congleton Edge by Astbury lime works, the North Rode viaduct crossing the river Dane, Broad Oak Reservoir, Macclesfield, and thence nearly due north out of the county. On the east side of this great fault the beds have been heaved up several thousands of feet, and from the top of the elevated mass all the red sandstones and marls which once stretched continuously eastwards have been worn off.

The hard beds of the Carboniferous formation, however, have been better able to withstand denudation and now stand up in bold relief. Another dislocation, known as the *Anticlinal Fault*, runs parallel to the Red Rock Fault, about three miles to the east of it.

THE PERMIAN FORMATION.—The *Lower Red Sandstone* is found on the east of Stockport, forming a tract one and a half miles in width, and narrowing southward until it ends east of Poynton. Many sections are shown in the banks of the Mersey, near Stockport, where the rock is seen to be a bright red, sometimes striped and mottled sandstone, very soft and crumbly and without pebbles. The beds dip to the westwards. A little outlying patch of Permians also occurs at Torkington, near Hazel Grove: the beds here are only to be seen in the brook-courses.

THE TRIAS, OR NEW RED SANDSTONE.—It is in Cheshire that the Triassic formation attains its chief British development, both in thickness and in the extent of surface occupied by it.

At first sight it seems almost hopeless to fix any definite lines marking subdivisions, in the great mass of red sandstones and marls, which attain a total thickness in this county of over 3,000 feet. By long-continued and patient work in the field, Professor Hull has been able to construct the following table, showing the various beds from below upwards, and to trace and map them at every point to which the formation extends. The thickness of each bed in Cheshire is also given.

TABLE OF TRIASSIC ROCKS.

Keuper	Keuper Marls, with Upper Keuper Sandstone......	1,500 feet
	Lower Keuper Sandstone (Waterstones)	400 feet
	Muschelkalk..................... Wanting in England	
Bunter	Upper Mottled Sandstone.................................	600 feet
	Pebble Beds	800 feet
	Lower Mottled Sandstone.................................	400 feet

The Bunter Sandstone.—Bunter is a German word, and means variegated, in allusion to the bright and varied tints of the stone. As these are the lowest Triassic beds, we find them exposed in the west of the county, and they, as well as the overlying strata, have a gentle dip or inclination to the eastwards. The upper and lower beds are soft reddish sandstones, whilst the pebble beds between contain vast numbers of well-rounded quartz pebbles, and usually form the top of the first of the two series of escarpments which can be traced in Triassic districts; the second escarpment—further to the east—being due to the hard beds at the base of the Keuper division. Very fine sections of the Pebble beds are exposed in the banks of the Dee, at Holt; thence we can trace the Bunter beds by Chester, and the various subdivisions are well shown at several points in the peninsula of Wirral, both on the coast, as at Burton Point (junction of Pebble beds with Lower Mottled Sandstone), West Grange Hill, West Kirby, Eastham (ditto), Hilbre Point, the "Red Noses" at New Brighton (Upper Mottled Sandstone), and in various quarries inland at Shotwick, Birkenhead, &c. The Bunter is a splendid water-bearing formation, and wells sunk in it near Birkenhead are now yielding a total of more than seven millions of gallons daily, whilst Liverpool is deriving an equal quantity from the same rock, on the northern side of the Mersey. The town and works of Crewe are supplied with water of singular purity from a deep well in the New Red Sandstone, sunk at Whitmore station, on the recommendation of Professor Hull, affording another proof of the importance of geological knowledge to our every-day life and interests.

The Keuper Beds.—On the Continent we find thick beds of a shelly limestone—the Muschelkalk, resting upon the Bunter; but these are absent in England, and we infer that this region was probably elevated, and formed dry land for a time, to sink again when the Keuper strata were being deposited. Confirmation of this theory is obtained from the fact that where the junction of the Bunter and Keuper beds is exposed, the former are seen to present a worn, uneven, or eroded surface, telling of a long interval of time during which they were exposed to denuding influences.

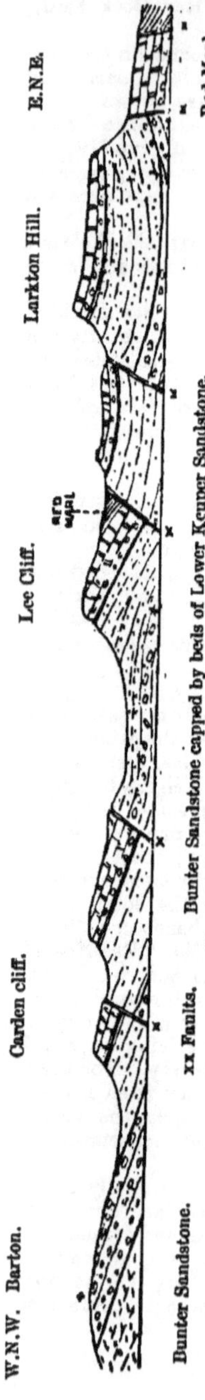

The lowest Keuper bed is a hard conglomerate, and above it come light red, yellow, or white sandstones, yielding building stone of good quality. Then come the "Waterstones," so named by Mr. Ormerod from an appearance on the surfaces like watered silk, although the term is now generally understood to apply to the copious flow of water which these beds yield when tapped by a boring. Next we get a great thickness—over 1,000 feet—of Red Marls, with beds of rock-salt near the base, and an irregular band, called the Upper Keuper sandstone, about 200 feet from the top.

All the plain of Cheshire east of the Peckforton Hills is composed of these Keuper beds, but the monotony of the surface is relieved by one or two striking escarpments, which we will first describe. Between Malpas, Tarporley, and Frodsham there exist several faults, running north and south, which with the unequal hardness of the strata have produced escarpments of considerable abruptness and altitude. The Peckforton Hills rise to a height of nearly 1,000 feet, as at Larkton Hill, Lee Cliff, Carden Cliff, Raw Head, and Beeston. Their western slopes are formed of Bunter sandstone, and the summits consist of the hard conglomerate basement bed of the Keuper. To the north and south they are cut off by cross faults, and on the east their boundary is a north and south fault, with a downthrow to the east which brings in the Keuper marls. Springs burst out all along the line of this east fault, from Beeston Castle to Malpas, and a lode of copper ore has been worked in it at Gallantry Bank.

At Alderley Edge in the north-east of the county we get another elevation formed of Keuper sandstone and conglomerate. The Edge runs east and west for about two miles from Chorley to Haresfield, and rises to a height of 650 feet. Carbonate of Copper occurs here also, and the ore is now worked: it is from the occurrence of copper ore in them that the *Keuper* beds derive their name, Keuper being derived from Kupfer, the German word for copper; some, however, would derive the term *Keuper* from another German word meaning *clay* or *marl*. Prof. Lebour, however, who is an authority on geological nomenclature, has kindly informed us that "Keuper is a local term (miner's *patois*, in fact), applied by the workmen, not only to the metal, copper (Kupfer), but also to the rock impregnated with coppery ore.

The *Lower Keuper Sandstones* are extensively quarried at Storeton Hill and New Brighton, near Birkenhead, and at Delamere, Manley, and in the Peckforton Hills. The beds are often traversed by suncracks and footprints of a remarkable reptile—the *Labyrinthodon*—are not unfrequent on the surfaces of the slabs.

Keuper Marls with Rock-Salt.—Beds of salt have long been known to exist in the Keuper Red Marls of Cheshire, beneath the towns of Northwich, Winsford, Middlewich, Sandbach, Lawton, Nantwich, &c. There

Fig. 16.—Section across the Peckforton Hills, Cheshire.

are two beds of rock-salt, the first is found below Northwich at a depth of 120 feet, and is 75 feet thick; it was discovered in 1670 during a boring for coal. About a century ago, an agent of the Duke of Bridgwater's bored 30 feet below the upper salt bed and struck another—lower—bed of salt, which turned out to be from 90 to 120 feet thick: of this fine bed about 15 feet in thickness of the purest part is now mined at Northwich. A great deal of brine is also pumped up, and being evaporated, about 25 per cent. of pure salt is obtained. This comes chiefly from rain-water which has percolated through the upper salt-bed. The effects of the removal of thick masses of rock from beneath the surface have made themselves very seriously felt at Northwich. The surface there is gradually sinking, so that houses have to be removed or rebuilt, and quite a lake is being formed at one point, along the course of a small brook that ran into the river Weaver; further serious subsidences occurred in December, 1880.

To explain the original formation of the salt we must imagine the Triassic rocks to have been deposited in great inland salt lakes. The excessive salt-

Fig. 17.—Cliffs of Lower Keuper Conglomerate, on Alderley Edge, Cheshire.

ness of the water, and the quantity of salts of iron, &c., conveyed into it by rivers, were inimical to animal life, of which accordingly we find scarcely a trace. Occasionally the waters were so concentrated as to deposit their excess of salt in beds more or less thick. Great quantities of carbonate of iron were brought down by the river waters, &c., and on exposure to the air this was rapidly converted into peroxide of iron, by the escape of the carbonic acid. Such was the amount of oxide of iron present that the minute particles of sand or mud, as they were deposited on the bottom, were each encrusted with a pellicle of it, and the marls and sandstones consequently have a more or less deep red colour.

The amount of salt produced in 1875 in Cheshire was:—

White Salt...1,255,500 tons
Rock Salt.. 105,000 tons

Total.. 1,360,500 tons

The various small lakes or meres which are dotted over the surface of

Cheshire are very probably owing to subsidences of the surface caused by the dissolving of beds of salt which once lay beneath. The Red Marl used formerly to be largely dug and spread over the land as a top-dressing, and the hollows of old marl-pits may still be seen in many fields. It forms a rich and good soil where it occupies the surface, but it is usually overlaid by a greater or less thickness of the surface deposits shortly to be described.

THE RHÆTIC BEDS AND LOWER LIAS. There is a well-known (by name) Liassic outlier between Whitchurch and Market Drayton, the northern portion of which is in Cheshire. A line drawn from Burley Dam to Audlem marks its northern edge. Here we see red and variegated rubbly marls forming the top of the Keuper. Then come the Rhætic beds, composed of very hard grey marls and sandstones, capped by sands, shales, and limestones of the Lower Lias. A fault runs along the eastern edge of this outlier, which extends southwards past Prees to Edstaston.

THE DRIFT. Good work has been done by Messrs. De Rance, Shone, Mackintosh, Reade, and others in investigating the deposits of clay, sand, and gravel, which are spread so thickly and continuously over the greater part of Cheshire and Lancashire. A triple division appears to have been made out, viz. an Upper and Lower Boulder Clay, with sands and gravels between. The Boulder Clays are full of fragments of foreign rocks, most of which have come from the Lake District, or the south of Scotland, often striated and polished, and appear to have been brought by glaciers or dropped from melted icebergs. The Middle Sands may mark an interval in the intense cold of this glacial period. From all three divisions of these drift deposits many shells and fragments of shells have been obtained by the gentlemen named above, and they are mostly of species which now inhabit the Northern or Arctic seas. Fine sections of these glacial deposits are exposed at Dawpool, and at various other cliff sections in Wirral, and they extend thence more or less all over the Red Marl of the Plain of Cheshire, usually completely masking the strata below.

PREHISTORIC MAN. The remains of early man appear to be very scarce in this county, but one or two important "finds" have occurred. A celt or stone axe was found near Tranmere which had part of its wooden handle still remaining; the greater part of the wood had perished, but enough remained to show that the handle had passed in a slightly diagonal direction towards the upper end of the stone; it is now in the Mayer Museum, at Liverpool. Again, the historian Stukeley states that in cleansing the moat at Tabley, near Knutsford, "they found an old British axe, or some such thing, made of large flint, neatly ground into an edge, with a hole in the middle to fasten into a handle; it would serve for a battle-axe." Another perforated axe, made of grit, and $7\frac{1}{2}$ inches long, was found at Siddington, near Macclesfield. Mr. De Rance records a quoit-shaped stone implement, 6 inches in diameter, from drift 20 feet below the surface, at Stalybridge Railway Station, and a smaller ring of stone was found in gravel near Macclesfield in 1860.

No. 6.

GEOLOGY OF CORNWALL.

NATURAL HISTORY AND SCIENTIFIC SOCIETIES.

Cornwall Royal Polytechnic Society; Falmouth. Annual Report.
Royal Institution of Cornwall; Truro. Journal.
Miners' Association of Cornwall and Devonshire; Truro. Papers and Proceedings.
Plymouth Institution and Devon and Cornwall Natural History Society; Plymouth. Annual Report and Transactions.
Royal Geological Society of Cornwall; Penzance. Transactions and Reports.
Penzance Natural History and Antiquarian Society.

MUSEUMS.

Falmouth Museum.
Natural History and Antiquarian Society's Museum, Penzance.
Museum of the Royal Geological Society of Cornwall, Penzance.
Museum of the Royal Institution of Cornwall, Truro.

PUBLICATIONS OF THE GEOLOGICAL SURVEY.

Maps. Sheets: 24, Coast from Bolt Head to East Looe, Plymouth; 25, Launceston, Tavistock, Callington, Dartmoor; 26, Bideford, Holsworthy, Hatherleigh; 29, Coast from Hartland Point to Cambeak; 30, Camelford, St. Columb, Bodmin; 31, St. Austell, Camborne, Redruth; 32, Lizard Head; 33, Land's End, Penzance.

Books. Report on the Geology of Cornwall, Devon, and West Somerset, by De la Beche, 14s. Palæozoic Fossils of Cornwall, Devon, and West Somerset, by Prof. Phillips.

IMPORTANT WORKS OR PAPERS ON LOCAL GEOLOGY.

List by Mr. Whitaker of 654 works by 237 authors in No. xvi. of the Journal of the Royal Institution of Cornwall. 1875.

1868. Holl, Dr. H. B. Older Rocks of East Cornwall. Journ. Geol. Soc., vol. xxiv. p. 400.
1870. Peach, C. W.—Cornish Fossils. Report Brit. Assoc. 1869, p. 99.
1871. Collins, J. H.—Handbook to the Mineralogy of Cornwall and Devon. Truro and Lond.
1871. Allport, S. Phonolite from the Wolf Rock. Geol. Mag., vol. viii. pp. 247, 336.
1872. Pengelly, W. Insulation of St. Michael's Mount. Journ. Roy. Inst. Cornwall, No. xiii. p. 81.
1872. Pattison, S. R.—Upper Limits of the Devonian System. Proc. Geol. Assoc., vol. ii. p. 277.
1872. Henwood, W. J.—Metalliferous Deposits of Cornwall. Journ. Roy. Inst. Cornwall, No. xii. p. 9.

1873. Collins, J. H. Mining District of Cornwall and West Devon. Proc. Inst. Mechan. Eng. for 1873, p. 89.
1875. Phillips, J. A. The Mining Districts of Cornwall. Journ. Geol. Soc., vol. xxxi. p. 319.
1876. Belt, T. The Drift of Cornwall. Journ. Geol. Soc., vol. xxxii. p. 80.
1876. Phillips, J. A. The so-called "Greenstones" of Western Cornwall, Journ. Geol. Soc., vol. xxxii. p. 155.
1878. Phillips, J. A. "Greenstones" of Central and Eastern Cornwall. Journ. Geol. Soc., vol. xxxiv. p. 471.
1878. Foster, Dr. C. Le Neve. Great Flat Lode south of Redruth and Camborne. Journ. Geol. Soc., vol. xxxiv. p. 640.
1878. Collins, J. H. The Hensbarrow Granite District. Lake, Truro, 2s. 6d.
1879. Ussher, W. A. E. Historical Geology of Cornwall. Geol. Mag., vol. xvi. p. 27, &c.
1880. Phillips, J. A. Concretionary Patches, &c., in Granite. Journ. Geol. Soc., vol. xxxvi. p. 1.
See also General Lists, p. xxv.

WHEN the rocks of any country are known to be rich in substances which are valuable to man, it is plain that they are likely to attract considerable attention and study, while the mines or works necessary to obtain the ores, &c., will themselves teach us much concerning the relations, structure, and modifications of the strata, which we could never otherwise have learnt.

In Mr. Whitaker's list, published in No. xvi. of the Journal of the Royal Institution of Cornwall, we find the titles of 654 books, papers, and maps, by 237 authors, which relate to the geology of Cornwall, and were written between 1602 and 1873; the great mass of these, however (633 out of 654), have been written during the present century, for the scientific study of rocks is of very recent origin.

Among the early workers we may name Borlase, W. Pryce, J. Carne, W. Phillips, J. J. Conybeare, and Professor Sedgwick; Messrs. Henwood, R. W. Fox, C. W. Peach, and R. Hunt connect the old school of geologists with the men of to-day; for they lived on from one epoch to another; the great work of (Sir Henry) De la Beche must never be forgotten, while of later workers we may enumerate Warington W. Smyth, Pengelly, A. H. Church, J. A. Phillips, J. H. Collins, Professor Bonney, S. Allport, and Dr. C. le Neve Foster.

The ordinary description of the rocks of Cornwall—"four or five islands of granite rising from a sea of clay-slate"—is sufficiently true to be taken as a starting point. A bare and wind-swept peninsula, the wealth of Cornwall lies under and not on the soil, and in the seas that lave her shores, as witness the old Cornish toast —"Fish, tin, and copper!"

In describing the rocks of Cornwall, we shall begin with the oldest or first-formed; and at first sight the task might seem to be an easy one. The fact, however, is that the stratified rocks of this district have been so interfered with, upheaved, dislocated, disturbed, and altered by the presence and intrusion of igneous rocks; their fossils are so few, indistinct or wanting, and they have been so little studied apart from the metallic ores which they contain, that their exact relations and order of succession are far from being thoroughly determined or worked out. The geological map of the county was executed by De la Beche forty or fifty years ago, before the establishment of the Geological Survey; and although it is a surprising work of skill for the time and as the result of one man's labour, yet it does not exhibit the detail which we now consider necessary. When maps of the region on the scale of six inches to a mile shall have been prepared by the Ordnance Survey, then our Government Geological Survey will take the matter in hand, and we shall have minute and accurate coloured maps showing the extent and position of every bed of rock.

SILURIAN FORMATION. On the south coast, from Chapel Point past the Dodman and round Veryan Bay to Gerran Bay, we find the cliffs composed of grits and grey quartzites. In these Rocks at Gorran Haven (Carn Goran) Mr. Peach was the first to find fossils, such as the brachiopod shells, *Orthis calligramma, O. elegantula, Strophomena grandis,* the trilobites *Homalonotus bisulcatus, Calymene parvifrons,* &c.; these show the beds to be of about the same age as the *Upper Bala Beds* (Lower Silurian of Murchison) of North Wales. The strata dip south-east, or towards the sea, and may be separated from the adjoining Devonian slates by a fault, or (as Sedgwick believed), the strata may be *inverted* by the great disturbances to which the country has been subjected and thus made to rest upon the Devonians instead of lying beneath them as they naturally would do.

DEVONIAN FORMATION.—The beds of this age consist largely of clay-slate, locally known as killas or shillet; near the granite this is usually of a green or violet hue, changing as the distance from the igneous rocks increases, to grey, blue, or buff; beds of red or grey grit and sandstone also occur, with irregular impure limestones and interbedded volcanic lavas and ashes. Although these slaty rocks must be some thousands of feet in thickness, yet we are not sure of the exact order of their superposition, or as to which are the older and which the newer beds. The difficulties, in fact, in the way of making out their order of succession are very great; to begin with, they are much broken up, dislocated, faulted, and altered by the disturbances to which they have been subjected since their formation: the thrusting-up of enormous masses of igneous rock (granite, &c.), through the slates, together with the heated waters and vapours which have since traversed the strata, has greatly interfered with and changed them. In the south-west part of the county, fossils—upon which we rely so greatly in classification—seem entirely wanting, while those which occur in the eastern and northern divisions are generally in a wretched state of preservation.

Lower Devonian.—The red grits and slates at St. Veep, Polruan, Fowey, and Polperro are well seen in cliffs along the coast; they dip northwards or inland, and seem to be the lowest rocks of the district. At Lantirit and Polperro scales of such fishes as *Scaphaspis* and *Pteraspis* occur; these were long thought to be the remains of fossil sponges, and were named *Steganodictyum Cornubicum* by Professor M'Coy, but of their real nature there is now no doubt; spines of fishes (ichthyodurolites) also occur; these fish-bearing beds may perhaps be of the same age as the *Lynton slates* of North Devon.

Middle Devonian.—The mass of the strata on the west coast, from St. Agnes Head to Padstow and thence inland, by Bodmin and Liskeard to St. Germans and Plymouth, is probably of Middle Devonian age. Fossils are fairly numerous in bands of limestone, near New Quay and Padstow; they include the brachiopod shells *Stringocephalus, Pentamerus brevirostris, Atrypa desquamata,* &c.

Upper Devonian.—The line of junction between the carboniferous and Devonian strata runs from the north-west coast near Forrabury eastwards by Lesnewth, Laneast Down, and South Petherwin. Whether the Carboniferous rocks rest conformably on the older beds is a disputed point; Dr. Holl thought they did not. At Petherwin there are about half-a-dozen beds of limestone which have yielded seventy-three species of fossils, including eight species of the characteristic Devonian cephalopod—*Clymenia.*

On the coast the Upper Devonians curve southwards to Tintagel and Delabole.

At the latter place are extensive quarries in grey and blue slates used for roofing purposes, paving, tombstones, &c.; about 120 tons per day are raised; these quarries have been worked since the sixteenth century.

Up to 1836 the older Palæozoic rocks were lumped together under the name of *Grauwacke* a miners' term imported from Germany. But Murchison and Sedgwick, after examining and defining the Silurian rocks of Wales, turned their attention to the south-west counties, and distinguished the rocks which there underlie the Carboniferous beds, as being the marine equivalents of the lacustrine *Old Red Sandstone* of Hereford and Scotland, a conclusion to which they were mainly led by the researches of Lonsdale, who stated that the fossils from these Devonian strata, constituted a group intermediate between those of the Silurian and Carboniferous Formations.

CARBONIFEROUS FORMATION. These beds —*Culm-measures* as they are often called—form the northern corner of the county; a line drawn from Boscastle to South Petherwin marking, as has been already indicated, their southern boundary line, where they rest upon the older Devonian strata.

The culm-measures consist of black shales or schists, with seams of grit and chert, and include beds of volcanic ash and dark limestone, generally dipping northwards, but much undulated or even contorted; this is well seen on the coast north of Boscastle, where the rocks are also traversed by many veins of white quartz. The fossils include ferns and other plant remains, with such shells as *Posidonomya, Goniatites*, &c.

IGNEOUS ROCKS. The crystalline rocks of Cornwall, *i.e.* those which have once been in a more or less fluid state from which they have cooled down, occupy a larger surface and are more important than those of any other English county. They may be divided into two classes, (1) those which are *contemporaneous* with the stratified rocks in which they occur, and (2) those which are *intrusive*, having forced their way through or between the slates.

Contemporaneous Igneous Rocks.—Volcanic action appears to have been rife in this district during the period of deposition both of the Devonian and Lower Carboniferous strata. We find beds of old lavas—the so-called greenstones—and sheets of volcanic ash interbedded with the sedimentary rocks. The crystalline rocks, formerly lumped together as "greenstones," are now known to need several different names for their correct designation: the majority are altered dolerites, others metamorphosed ash-beds, &c.; such rocks occur in long narrow bands west of Penzance, in the cliffs on the east of Mount's Bay, between St. Erth and Camborne, round Padstow, and between that place and St. Tudy and St. Teath, from the coast at Tintagel eastwards to Callington, and between Liskeard and Saltash. These sheets of trap rock are indeed very numerous; it is difficult to distinguish precisely between such as are contemporaneous and others which may be intrusive, but they are all older than the granite, which cuts abruptly through them at several places.

Intrusive Igneous Rocks. Granite.—The granite of Cornwall occurs in four large masses, connected by smaller protrusions and continued westwards by the Scilly Isles, which are entirely composed of this rock. These granitic masses form the back-bone of the peninsula, rising as small plateaux above the surrounding slates; and if we could strip off the stratified rocks we should doubtless find granite everywhere underlying them. The mineralogical composition of the rock is quartz, felspar, and mica; towards the exterior of the bosses a black mineral called schorl often replaces the mica.

The Cornish Granite is undoubtedly an igneous rock which,when in a melted state, rose up from below, invading, pushing upwards, and perhaps, partly consuming the sedimentary strata which originally lay above it. It never reached the surface in its melted state, or it would have cooled down much more rapidly, and would now present a very different appearance, but it has been exposed to view by the denudation of some thousands of feet of sedimentary deposits which once covered it over.

The beds of clay-slate or Killas, surrounding the granite bosses, are much altered by the great heat from the latter rock, and the effect can be traced to distances varying from a quarter to half a mile; close to the junction the slates have often been changed to micaceous schists.

Brown Willy District. This large granite mass has an area of about 65 square miles; the chief heights are Brown Willy, 1,368 feet (highest in Cornwall); Rough Tor, 1,296 feet; Caradon Hill, 1,208 feet. It is 16 miles west of Dartmoor, but the two are connected by the granite of Hingston Down halfway between.

Hensbarrow or St. Austell Granite.—Area about 35 square miles: heights, Hensbarrow Hill, 1,034 feet; St. Stephen's Beacon, 700 feet; Carclaze Tin Mine, 665 feet. This boss lies five miles south-west of the Brown Willy area. A remarkable rock named Luxullianite occurs in the parish of that name; it consists of large crystals of black tourmaline (schorl) and pink felspar embedded in a base of grey quartz; it is only known from blocks found on the surface, but these probably come from an east and west dyke, since they run in that direction. A fine mass furnished the material for the Duke of Wellington's sarcophagus in St. Paul's Cathedral.

Fifteen miles further in the same direction (south-west) bring us to the round patch of granite (area, 55 square miles), of which *Carn Menelez*, 822 feet, is the culminating point.

The Land's End Granite lies seven miles due west of that last named; its area is about 65 square miles: Heights Hill, north-west of Towednack, 805 feet; Castle-an-Dinas, 735 feet. Smaller masses of granite occur at St. Michael's Mount, Tregonning and Godolphin Hills, Carn Brea and Carn Marth (near Redruth), Cligga Head (near St. Agnes), and at Castle-an-Dinas and Belovely Beacon (near St. Columb Major and Roche).

Elvans.—These are dykes of quartz-porphyry which run out from the granite into the surrounding slates; they are of later formation than the granite, since they cut across it.

The granite is divided into blocks by the numerous joints which run through it; these joints are mostly shrinkage cracks produced in the mass when cooling; they greatly facilitate the quarrying of the rock. The largest works are at the Cheesewring near Liskeard, at Penryn, Lamorna, Mill Hill in Maldron, Falmouth, Carn Brea, St. Austell, Par. and Calstock; these have furnished the materials for many important public works, as the London Docks, Westminster Bridge, Thames Embankment (in part), &c.: the shaft of the Wellington Monument at Strathfieldsaye is of one solid block of granite, 30 feet in height, from Constantine.

Many of the elvans when polished form beautiful decorative stones for columns, &c., from their porphyritic character.

The exposed summits or tors of the granitic masses weather away most rapidly along the lines of the joints, and from this circumstance it not unfrequently happens that one block is left poised upon another at some point generally near its centre; it then forms a *logan* or rocking-stone; the best known is at Castle Treryn, St. Leven; it weighs 65 tons.

Serpentine.—This is a compact soft but tough rock of a green, grey, or red colour, sometimes traversed by veins of whitish steatite or soap-stone, and often veined or mottled, furnishing a valuable stone for decorative purposes; its chemical composition is a hydrous silicate of magnesia. It forms a large part of the Lizard peninsula, a plateau partly cultivated, partly moorland, with precipitous sea-cliffs 100 to 200 feet in height. Professor Bonney has shown this mass of serpentine to be an *altered igneous rock;* it is clearly intrusive in the slates. There are several quarries, and the rock is easily turned in the lathe into small columns, vases, &c.; it takes a high polish. Small quantities are exported to Bristol for the manufacture of carbonate of magnesia.

Another igneous rock called *Gabbro* (diallage rock of De la Beche) is found at St. Keverne and Crousa Down; this is of later date than the serpentine, through which it has burst.

Phonolite or Clinkstone has been named from its ringing sound under the hammer. Its only British locality is the Wolf Rock, a rugged isolated mass about nine miles south of the Land's End; it is composed of nepheline, felspar, and hornblende.

MINERALS.—The disturbances to which hard and somewhat brittle rocks like the slates have been subjected, by the upheaval through them of enormous masses of igneous rock, and the general strains resulting from those slow movements of the earth's crust which have caused continents to become seas and seas to become continents, changes which have not taken place once only but many times in the long history of the past, could not but have the effect of shattering and breaking the rocks to a considerable extent. The cracks or fissures so produced have since been filled with mineral substances brought up from below dissolved in hot water or by super-heated steam. These matters were gradually deposited on the sides of the veins until ultimately they became filled up. In the vein-stuff quartz is, perhaps, the commonest ingredient, but altogether about 150 distinct minerals are known to occur in the rocks of Cornwall, and most of these are found in the veins. When a vein contains metallic ores it is called a *lode*; the usual direction of the lodes is from a little north of east to a little south of west, and they dip at an average angle of 70 degrees; the richest lodes seem to occur at or near the junction of the slate with granite or elvan. All the mineral wealth of Cornwall is either contained in such lodes or has come from them, for the extensive surface deposits of stream-tin merely represent the wreck—the wearing away—of formerly existing lodes.

The following figures, showing the amount of minerals raised in the county, are taken from Mr. R. Hunt's invaluable "Mineral Statistics" for 1877:

Tin Ore.- Cornwall and Devon.

Year.	No. of Mines.	Tin Ore.	Metallic Tin.	Value.	Per Ton.
		Tons.	Tons.	£	£ s.
1872	162	14,266	9,560	1,459,990	152 15
1873	215	14,885	9,972	1,329,766	133 7
1874	230	14,039	9,942	1,077,712	108 8
1875	183	13,995	9,614	866,265	90 2
1876	135	13,668	8,500	675,750	79 10
1877	98	14,142	9,500	695,162	73 3

The ore is called "black tin;" it is the peroxide SnO_2. The greater part is obtained from lodes in the rocks, but a certain amount (about 750 tons of black tin) was got in 1877 from "streams, rivers, and foreshores."

The tin-ore is not equally distributed throughout the vein or lode, but generally occurs in masses called "bunches," which are connected with one another by thin strings of ore. Sometimes, however, we meet with the tin in a number of small veins which cross and interlace with one another, forming what is called a "stock-work;" the tin at the famous Carclaze mine, two miles north of St. Austell, occurs in this manner: this mine is now chiefly wrought for china-clay. The chief tin districts are (1) St. Austell, (2) St. Agnes, (3) St. Just and St. Ives.

Copper. Sixty-eight mines, 39,225 tons of ore (copper pyrites or yellow copper ores chiefly, a sulphide of copper and iron, Cu_2S, Fe_2S_3), valued at £169,553, and yielding 2,937 tons of pure copper.

The copper-bearing lodes occur (1) in the Gwennap, Redruth, and Camborne district, and (2) round Breage, Marazion, and Gwinear.

The working of copper is comparatively of recent date, not going back to before the year 1700; it is calculated that copper ore to the value of 13 millions sterling has been raised during the present century, yielding a clear profit of 2¾ millions.

Lead. The lead ore (chiefly galena, the sulphide of lead PbS), raised from eleven mines, amounted to 2,167 tons, yielding 1,674 tons of lead and also 23,035 ozs. of silver. Neither galena nor any other lead ore occurs in any quantity in or near the granite, but is confined to the slates.

Iron Pyrites (the bisulphide of iron FeS_2).—Fourteen mines yielding 14,289 tons, value £8,294.

Iron Ore (brown hæmatite).—Nine mines, 4,963 tons, value £2,920.

Bismuth. –Two mines; 8 cwt., value £15.

Wolfram (tungstate of iron and manganese).—One mine (East Pool), 15 tons, value £150.

Uranium occurs as the oxide U_3O_8, called *pitch-blende*. Two mines, 2 cwt., value £11 15s. 3d.

Ochre and Umber, 500 tons, value £250.

Manganese, 2,496 tons, value £6,084.

Arsenic, 1,718 tons, value £6,189.

Porcelain Clay or Kaolin, 200,345 tons. This is obtained from highly decomposed granite, and consists of the disintegrated and metamorphosed felspar of that rock. It is sent to the Potteries in North Staffordshire, and is largely used in the manufacture of calico and paper. The price is now about 15s. per ton. Often on the outside of the granitic masses of Cornwall the rock is so decomposed by the percolation of rainwater holding carbonic acid in solution, that the granite may be dug with a spade to the depth of 20 feet or more. Mr. Collins believes that the alteration of the felspar into kaolin is due to the action of water or steam containing hydrofluoric acid.

China Stone, 39,500 tons. This is chiefly raised in the neighbourhood of St. Austell; it also is derived from the decomposition of granitic rocks, but contains quartz as well as kaolin. It is sufficiently compact to be used for building purposes, and the churches of St. Stephens and Probus are partly built of it. It is chiefly used in the manufacture of glaze for earthenware.

Fire Clay, 505 tons. *Candle Clay or Miner's Clay*, 400 tons. This is a tough clay obtained at St. Agnes and used for the purpose of wrapping round and holding the candles used by miners in underground work.

For the last six or seven years the mining industries of Cornwall have been in a state of great depression, and large numbers of miners have emigrated. Yet in the past great riches have been extracted from the rocks. The history of tin mining in Cornwall dates back at least to the time of the Phœnicians—several centuries before the Christian era, and the total value of the metals raised since then has been estimated at £60,000,000.

Of the most successful mines we may mention those of silver and lead in the Menhenniot district, Trelawney and Mary Ann; the copper mines of the Caradon district, of the Fowey district, such as Fowey Great Consols and Crinnis, the united mines of St. Day, the Carn Brea series including Dolcoath, Cook's Kitchen, East Pool, and others, producing copper, tin, silver, wolfram, cobalt, arsenic, zinc, &c.; then further to the south-west the Helston and St. Just districts producing copper and tin at Levant, Botallack, Wheal Vor, &c. The Botallack mine is famous from the fact that its workings extend under the sea for more than half a mile.

SURFACE DEPOSITS. These would come under the Post-Tertiary or Pleistocene division of geologists. They include all the various beds of sand, gravel, and clay, with other deposits which occur irregularly and without any very definite order over the surface of the county. Flints are of not uncommon occurrence in these surface deposits, and large unbroken flint nodules occur in the Scilly Isles; these are probably derived from the chalk which once stretched far in this direction.

Quartz Gravels. - The quartz gravels on Crousa Down in the Lizard District and at Crowan occur at heights of about 400 feet above sea-level; these may be of Tertiary age, for they evidently belong to a system of drainage quite different from the river basins which now exist.

The Sands and Clays of St. Agnes occur at about the same height and may be of the same age as the quartz gravels mentioned above.

Of glacial deposits no clear evidence is to be obtained in Cornwall; there are extensive spreads of *boulder gravels*, which appear to have been formed by torrential waters, resulting perhaps from the melting of great masses of snow, but no striated surfaces and no erratic blocks are known.

The Raised Beaches, which are common on the coast, mark the period of extreme subsidence of the land in glacial times; in Cornwall this amounted to 15 feet only, while it was as much as 1,500 feet in Wales. Among many raised beaches we may name those at Pornanvon Cove, south of Cape Cornwall, at New Quay, Fistral Bay, St. Ives, Gerran's Bay, Falmouth Harbour, Coverack Cove, &c.

The Head is a stony clay seen above the last-named deposits; the *Stream-tin Gravels* are associated with it, and mark a period of great surface waste, the rainfall being greater than at present; they occur in the valleys on the southern side of the Cornish watershed; later still are the *Submerged Forests* containing stumps of the Oak, Alder, Hazel, &c., and occurring on the coast at or about low water mark at Looe, Fowey, Falmouth (Gyllyngvaes), Mount's Bay, Padstow, &c.

St Michael's Mount in Mount's Bay is an isolated mass of granite 195 feet in height; its distance from the nearest point of the mainland, Marazion Cliff, is 1,680 feet. It is, as is well known, an island at high tide but a peninsula at low water. The isthmus is composed of highly-inclined Devonian slates, into which the granite is seen to send veins and dykes, proving its intrusive nature. The Mount owes its present singular position to a subsidence of the coast-line, combined with denudation; the softer slates having been worn away by the sea more rapidly than the granite mass. The subsidence is proved by the submerged forest in Mount's Bay. Very little change has taken place, however, during the last 1,900 years, for the Roman author Diodorus Siculus, who wrote about 9 B.C., describes the Mount, under the name of Ictis, as a spot where tin was bought and sold, and his language precisely applies to its present general condition.

Blown Sands, forming hillocks or dunes, are chiefly seen along the north-western coast, as along the east side of St. Ives Bay, in Perran Bay, between Fistral and New Quay Bay, Constantine Bay, the east side of Padstow Harbour, &c.; along the coast much land has been spoilt and many houses and even churches devastated by the encroachments of the sand. At Bude Haven the sand has been compacted into stone by the deposition of a cementing substance (carbonate of lime) between its grains by land-springs.

The sea-beaches, river gravels, mud-deposits, and modern peat growth bring the geological record down to the present day.

PREHISTORIC MAN. Of the early or *Palæolithic* tribes no traces have yet been found in Cornwall; the absence of caverns and the probability that most of the gravel beds of glacial or pre-glacial age have been swept out to sea, combine to account for this want of evidence.

The *Neolithic* workers in flint and stone are scantily represented by the following "finds:"—rough celts or axe-heads made of greenstone were found in barrows (burial mounds) at St. Just, and are now in the Truro Museum, which also contains a highly polished celt of jadeite found near Falmouth; a large celt, 11¾ inches long, by 4 wide and 1⅜ in thickness, "from Cornwall," is in the Antiquarian Museum of Edinburgh; another of serpentine, found near Truro, is in Canon Greenwell's collection. An axe-head, formed of greenstone and perforated for a handle, was found in the middle of a cairn at Pelynt; a bronze dagger was found in a barrow in the same field.

Stone querns for grinding corn have also been found, with mortars (at Kerris Vaen); and at St. Agnes, Truro, a flint arrow-head. Spindle-whorls of stone, which were used in spinning, are not uncommon; they are called by the natives "Pixy's grindstones;" they are small round stones an inch or two in diameter with a hole in the centre.

Besides these stone tools, some monuments of early races of men exist in the cromlechs, monoliths, stone circles, avenues, hut dwellings, and rude stone camps called cliff or hill castles which still stud the wilder portions of the county.

No. 7.

GEOLOGY OF CUMBERLAND.

NATURAL HISTORY AND SCIENTIFIC SOCIETIES.

Cumberland Association for the Advancement of Literature and Science. Carlisle. Transactions.
Keswick Literary Society.

MUSEUMS.

Keswick Museum.

PUBLICATIONS OF THE GEOLOGICAL SURVEY.

Maps.—98 N.W. Wastwater; 99 N.E. Ravenglass; 101 S.E. Keswick; 102 S.W. (Survey not completed).
Books.—Geology of the Northern Part of the English Lake District, by J. C. Ward, 9s.

IMPORTANT WORKS OR PAPERS ON LOCAL GEOLOGY.

1859. Binney, E. W.—Lias Deposits near Carlisle. Q. Journ. Geol. Soc., vol. xv. p. 549.
1861. Wallace, W.—Lead Ore of Alston Moor. Stanford.
1863. Harkness, Prof. Skiddaw Slate Series. Q. Journ. Geol. Soc., vol. xix. p. 113.
1864. Murchison and Harkness.—Permian Rocks of the North West of England. Q. Journ. Geol. Soc., vol. xx. p. 144.
1865. Harkness, Prof.—Lower Silurian Rocks of the South East of Cumberland. Q. Journ. Geol. Soc., vol. xxi. p. 235.
1868. Nicholson, Prof. H. A.—Geology of Cumberland and Westmoreland. Hardwicke, 3s. 6d.
1869. Bolton, Jno.—Geological Fragments. Atkinson, Ulverstone.
1872. Aveline, W. T.—Continuity and Breaks of the Silurians of the Lake District. Geol. Magazine, vol. ix. p. 441.
1874. Nicholson, Prof. H. A.—Silurian Rocks of the English Lake District. Proc. Geol. Assoc., vol. iii. p. 105.
1874. Wurzburger, P.—Geology of the Cumberland Iron Ore District. Journ. Iron and Steel Institute, No. 2, p. 287.
1875. Ward, J. C.—Granitic, Granitoid, and Metamorphic Rocks of the Lake District. Q. Journ. Geol. Soc., vol. xxxi. p. 568; vol. xxxii. p. 1.
1875. Ward, J. C.—Glaciation of Southern Part of the Lake District. Q. Journ. Geol. Soc., vol. xxxi. p. 152.
1875. Kendall, J. D.—Hematite Deposits. Manch. Geo. Soc., vol. xiii., p. 231.
1876. Bird, —.—The Red Beds at the Base of the Carboniferous Limestone in the North West of England. Proc. Geol. Soc., W. Riding Yorks, N.S., pt. ii. p. 57.

1876. Russell and Holmes.—Raised Beach between Whitehaven and Bowness. Rept. British Assoc., p. 95.
1876. Green, Prof. A. H.- Basement or Old Red Conglomerates. Physical Geol. p. 398.
1876. Kendall, J. D.—Hematite in the Silurians. Q. Journ Geol. Soc., vol. xxxii. p. 180.
1877. Ward, Rev. J. C. Lavas of Eycott Hill. Microscopical Journ. vol. xvii. p. 239.
1877. Martin, R. F.- On some Peculiarities of the West Cumberland Coal Field. Transac. Cumb. Assoc., Pt. ii. p. 63.
1877. Topley, W. and Lebour, G. A. Intrusive Character of the Whin Sill. Q. Journ. Geol. Soc., vol. xxxiii. p. 406.
1877. Postlewaite, Jno.- Mines and Mining in the Lake District. Leeds.
1878. Hodgson, Miss E. - Visits to Trap Dykes. Trans. Barrow Field Club, vol. ii. p. 89.
1879. Jenkinson and Ward, J. C. —Geology of Cumberland. Encyc. Britannica, vol. vi. p. 697.
1879. Ward, Rev. J. C.— Physical History of the Lake District. Geol. Mag., pp. 49, 110
1879. Mackintosh, D. Erratic blocks of Cumberland, &c. Q Journ. Geol. Soc., vol. xxxv. p. 425
1881. Holmes, T. V. Permian, Trias and Lias of Carlisle Basin. Q. Journ. Geol. Soc., vol. xxxvii. p. 286.
1881. Kendall, J. D.- Interglacial Deposits of West Cumberland and N. Lancashire. Q. Journ. Geol. Soc., vol. xxxvii. p. 29.

See also *General Lists, p. xxv.*

The rocks of a county like Cumberland can scarcely fail to attract a full share of notice; the beauty and variety of the surface, with the mineral treasures which lie beneath, lead direct to inquiries concerning the strata which form the one and contain the other.

It is to Professor Sedgwick that we owe the first scientific investigation of the older rock beds of this district, aided by Mr. (now Professor) McCoy in the description of the fossils. Among other early writers we may name Messrs. J. Otley, D. Sharpe, William Hopkins, and Professor Phillips; in later times Professors Harkness and Nicholson stand prominently forward. All amateur work, however, must give place to the minute accuracy which we expect from the officers of the Geological Survey; only a small part of the results of their examination of the north of England have yet been published, but the map and memoir of the Lake District by the late Mr. J. Clifton Ward is an admirable piece of work, and Messrs. Aveline, Russell, Holmes, Goodchild, Dakyns, &c., may be trusted to complete the task in the same satisfactory manner.

With reference to its rocky structure it seems best to divide Cumberland into two portions; the first of these would include all the old rocks forming what we usually term the "Lake District"; the second would comprise all the newer rocks which surround this region on the east, north, and west, forming what Mr. Clifton Ward used to term the "framework," the setting or border, as it were, of the older central rocks; it will be more natural to take the former division first.

LOWER SILURIAN FORMATION. —The rocks which in Cumberland can be seen to underlie all others, and which, therefore, must be the oldest or first-formed, have been named the *Skiddaw Slates* because they form that lofty mountain. They are black or dark grey rocks, sometimes soft and shaly, with interstratified sandy or gritty beds, which yield good flagstones. These rocks occupy the surface of the central part of the county from Egremont, Lamplough, Cockermouth, Ulldale, and Hesket Newmarket on the north, to a line drawn from Dent Hill by Crummock Water, and south of Newlands and

Keswick on the south; their total thickness is not less than 10,000 feet. In the extreme south of the county the Skiddaw Slates again rise to the surface at Black Combe, west of the estuary of the Duddon. Fossils are few and far between, and the search for them is only likely to be successful where the cleavage is undeveloped, or where it coincides with the bedding. The masses of fallen stones or screes found on the mountain sides should be diligently examined.

About forty species of graptolites have been found, with two brachiopods, twelve trilobites, and several plant remains; worm-tracks and burrows are not very rare. Altogether these fossils would seem to indicate that the Skiddaw Slate is of about the same age as the Arenig Beds (Lower Llandeilo) of Wales; they must have originally been deposited as beds of mud, sand, or grit, on the bottom of a rather shallow sea. Their strike is from north-east to south-west, and their dip, though they undulate greatly, is very generally to the south-east.

In the neighbourhood of the igneous rocks the Skiddaw Slates have been metamorphosed, being converted into chiastolite slate or mica-schist; at some points, as in Skiddaw Forest, where the beds are much altered, although no igneous rocks are seen near on the surface, yet doubtless granitic masses exist beneath at no great depth.

The following subdivisions have been indicated by Mr. Ward:-
6. Passage beds to volcanic series.
5. Black slates of Skiddaw.
4. Gritty beds of Gatesgarth (Buttermere), Lutterbarrow, Tongue Beck (Skiddaw), Watch Hill, and Great Cockup.
3. Dark slates.
2. Sandstone series of Grasmoor and Whiteside.
1. Dark slates of Kirk Stile, between Loweswater and Crummock.

A bed of grit in No. 4 Mr. Ward considers like one which occurs at the base of the Arenig Slates of Wales; if this be so then the lower strata may be of Cambrian age, Nos. 1 and 2 representing the Lingula flags, and No. 3 the Tremadoc Slates; the finding of more fossils in these lower beds is necessary to decide this question.

Some beds of the Skiddaw Slates have been worked for slate pencils at Keswick, but the chief economic value of the rocks lies in the numerous mineral veins by which they are traversed, and which are generally also lines of fault, where fracture of the rocks has taken place, the fissures produced having been filled by minerals deposited by water.

In 1877 of zinc ore 75 tons, value £225, were obtained near Keswick; from the Force Crag mine, in Coledale, 622 tons of barytes were raised, value £1,033.

Iron ore (a red hematite) has been worked in the hills between Buttermere and Ennerdale Lake, and in Eskdale. In former times copper was largely worked at Newlands, west of Derwentwater.

As a general rule the Copper Lodes run east and west, and the Lead and Iron Lodes north and south; the former appear to be the older, being frequently displaced or shifted by the latter.

The mountains formed of Skiddaw Slates have soft outlines and smooth grassy slopes, except where intrusive igneous masses occur; thus the doleritic boss of Castle Head stands out as a bold crag above the surrounding and softer slates, and the igneous dyke forming Friar's Crag serves as a strong lake wall protecting the softer rock behind it.

VOLCANIC SERIES OF BORROWDALE (Green Slates and Porphyries of Sedgwick). —All along their southern boundary the Skiddaw Slates pass under a great series of beds of volcanic ash and breccia, alternating with sheets of lava, and the whole traversed by dykes and masses of igneous rock. It is most remarkable that along all this line of junction there runs a zigzag fault or dislocation, so that the passage from the one formation to the other is nowhere distinctly seen. Elsewhere, however, at Black Combe, and on Eycott Hill, beds of slate and volcanic ash are seen to alternate at the junction, and there is no doubt but that the higher series rest conformably on the lower one. The Borrowdale beds are finely seen in the valley from which they are named; they occupy all the south and south-east of Cumberland, extending over the border into Westmoreland; their principal heights are:—Harter Fell, 2,141 feet; Scaw Fell Pikes, 3,229 feet; Great Gable, 2,949 feet; High Stile, 2,643 feet; Causey Pike, 2,082 feet; Glaramara, 2,560 feet; Ullscarf, 2,370 feet; Great Dodd, 2,807 feet.

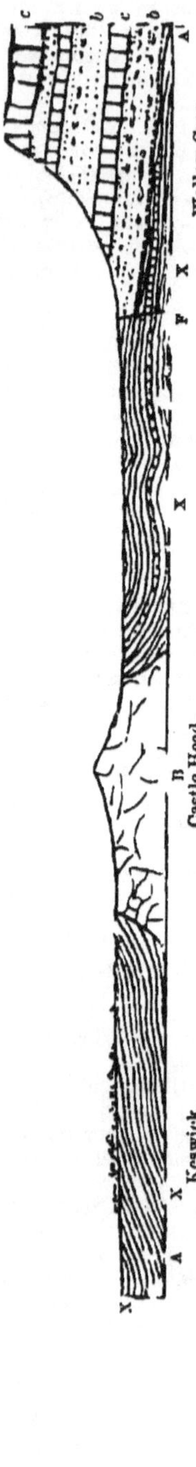

The total thickness of the Volcanic Series is from 12,000 to 15,000 feet; formerly it extended right over the surface area now occupied by the Skiddaw Slates, for we find it resting on the latter, not only on the south as already described, but also on the north at Caldbeck Fells, Eycott Hill, Ulldale, and Sunderland.

It is plain that towards the end of the Skiddaw Slate period submarine volcanoes burst into action, and for a time beds of lava and ash were deposited on the sea-bottom alternately with layers of mud and sand; but elevation of the sea floor taking place the whole area became land, and all the superincumbent beds consist of similar volcanic matter which fell upon and around the volcanic cones. One of the main centres of eruption is probably represented by the round boss of intrusive igneous rock (dolerite), which forms Castle Head, near Keswick, and from which probably flowed the lavas of Wallow Crag. All trace of the actual cones or craters is of course gone long since; they have all been denuded, swept away by the sea, by rain, and rivers, &c.; what we now see for instance at Castle Head is the stump or plug of the volcano, rock matter which solidified, perhaps, thousands of feet below its actual summit.

A strong proof of the subaerial origin of the Borrowdale Series is the entire absence of fossils; such a thickness of strata could scarcely accumulate in the sea without enclosing some evidences of the life of the period.

A fine section of these beds is exposed on the sides of Falcon Crag, Brown Knotts, and Bleaberry Fells, two miles south of Keswick. The strike is the same as that of the underlying Skiddaw Slates, viz., from north-east to south-west.

The following table by Mr. J. C. Ward shows the succession of the beds, the oldest being at the base:--

9. Bedded, mostly fine flinty ash, of Great End, Esk Pike, and Allen Crags.

8. Unbedded, coarse ash and breccia, of Broad Crag and Long Pike.

7. Bedded, and rough ash of Scafell Pikes, Glaramara, and Ullscarf.

FIG. 18.—A to A¹ Section from Greta Hall, Keswick, to Wallow Crag. Wallow Crag. c Lava beds. B Dolerite, probable source of the lava-flows, produced most likely long after volcanic action had ceased. Scale, 3 inches to 1 mile. x Skiddaw Slate. b Ash and breccia of Wallow Crag. Fault, at the junction of the Skiddaw Slates with the Volcanic Series.

6. Partially bedded, fine flinty ash of Base Brown and Rosthwaite Fell.
5. Well-bedded ash of Seathwaite.
4. Contemporaneous traps (or lava beds) of Watendlath Fell, High Seat, and Bleaberry Fell.
3. Breccia and bedded ash, of Brand Fell, and Watendlath.
2. Contemporaneous traps (lavas) of Honister, Dale Head, Gate Crag, and Falcon Crag.
1. Purple breccia, ash, and some contemporaneous trap.
Transition series to Skiddaw Slates.

Comparing these rocks with those of Wales, they appear to represent both the Llandeilo Flags and the Lower Bala Beds; the chief difference being that the volcanic masses of Cumberland, unlike the lavas and ashes of North Wales, were accumulated on the land and not in the sea.

Southwards in Westmorland the Volcanic Series is seen to be overlaid by the Coniston (=Bala) Limestone, and that again by a great thickness of beds of Upper Silurian age. These beds just enter Cumberland in its south-west corner, on the west side of the Duddon estuary.

It may seem strange at first that such compact well-bedded rocks as many of the slates in the Borrowdale Series can be composed of volcanic ashes which have fallen on land surrounding volcanic vents, but examination of existing craters, as in the neighbourhood of Vesuvius, shows precisely similar phenomena.

Most of the rocks of this series, whether ash or trap, yield good material for wall-making or road-mending; the slates have been worked at several points. Many rare minerals occur, especially in the rocks of Carrock and Caldbeck Fells, as Wulfenite, Scheelite, Brochantite, Apatite, Wolfram, Caledonite, Malachite, &c. There are also many metallic lodes, though these have not been much worked of late years; of lead ore the Caldbeck mines in 1877 yielded 96 tons, containing 72 tons of lead and 720 ozs. of silver, total value £1,197; also 5½ tons of copper ore, value £30.

The famous plumbago or black-lead mine of Borrowdale is well known. It occurs in pipes, strings, or detached masses in and close to a dyke of igneous rock (diorite and diabase), which comes up through highly-altered ashy beds belonging to the volcanic series. This Borrowdale Graphite (wrongly called black-lead) is nearly pure carbon (it always contains a little iron); it is extremely well fitted for the manufacture of the best lead pencils; its present value is about 40s. per pound. Early in the present century an immense mass of plumbago, weighing about 70,000 lbs., was discovered at the Borrowdale Mine, but for the last thirty years it has been but little worked.

The hard, massive, and well-jointed rocks of the Borrowdale Series form mountains with rough and hummocky outlines and steep and craggy sides, in strong contrast to the flowing outlines of the Skiddaw Slate hills. Sometimes the hill sides show horizontal lines or steps, due to the alternations of beds of ash and lava of varying hardness, as on the east side of Derwentwater, and in St. John's and Shoulthwaite Vales. When dykes of igneous rocks cross the beds, if harder they form craggy and conspicuous lines across the country, but if softer or more jointed than the surrounding rocks they give rise to what are called "doors," *i.e.*, squarish openings in mountain ridges, as Mickle Door, between Scafell Pikes and Scafell, Comb Door, &c.

AN INTERREGNUM.—At the close of the Silurian Period there is a great break in the history of the stratified rocks of Cumberland; no deposits of the age of the old Red Sandstone occur. What happened during this interval? It must be remembered that at this time the Skiddaw Slates were buried beneath 26,000 feet, and the volcanic series below 14,000 feet of higher Silurian beds. At this depth they would be under great pressure, and subjected to intense heat. Then a lateral pressure was brought to bear on the beds from the north-west and from the south-east, gradually elevating them into an anticlinal curve or fold, whose axis ran from north-east to south-west of course at right angles to the upheaving force. As the top of this curve rose above the sea-level it would suffer extreme denudation from the action of the waves; all the top of the curve was worn away, so that lower and lower beds were exposed to view, until even the Skiddaw Slates were uncovered, the 26,000 feet of rocks above them having been in this neighbourhood entirely removed.

The following illustration from a paper by Mr. Ward may help to make this clear:— "If we pile up black cloth in layers to a thickness of 10 inches, red cloth upon this to a thickness of 12 inches, and blue cloth upon the red 14 inches thick, then by bringing some powerful force to bear upon the two ends of the pile, the whole may be thrown into curves and contortions by the lateral pressure, and the centre portions consequently raised above the level of the ends. Imagine, then, a large pair of shears brought forward which shall cut off or pare down all the upraised central portion; in this way the uppermost blue cloth may be removed altogether from the centre of the low arch, a less amount of the red cloth layers would be removed, while perhaps only the few topmost inches of black cloth would be touched; but the consequence would be that, at the centre of the cloth dome, an arch of black cloth would appear, on either side of this would lie inclined layers of red cloth, to be flanked in their turn by similarly inclined layers of blue cloth." Now these three layers of cloth represent the three great divisions of the Silurian strata of the north-west of England, and the cutting of the shears represents the planing action of the sea.

INTRUSIVE IGNEOUS ROCKS.— It was at the commencement of the epoch of disturbance and denudation just referred to, when the Skiddaw Slates lay most deeply buried beneath the surface, that the effects of the internal heat of the earth came most strikingly into action. Large masses of melted rock forced or rather ate their way upward, metamorphosing the neighbouring rocks, and sending dykes and quartz veins amongst those still higher, as all were contorted and cleaved by the intense lateral pressure called into play during the slow upheaval of the district. Of these masses of igneous rock the granite of Eskdale is by far the largest, forming the rugged hills on both sides of the Esk and Mite, for a distance of about fourteen miles from north to south. It is a reddish rock, and sends off dykes of quartz-felsite (corresponding to the Cornish elvans) into the surrounding beds of Green Slates and Porphyries.

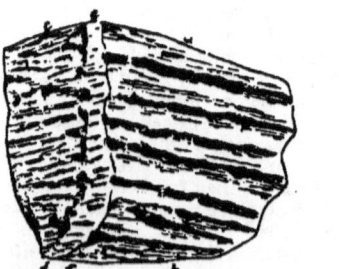

Fig. 19.— Banded and altered Skiddaw Slate, near the Syenite of Scale Force, Buttermere. Near the syenite the slate has been converted into a compact, flinty-looking rock, with occasionally a banded structure, weathering on the outside like lines of bedding; *a b* is part of the outer surface, showing weathered ridges and hollows; and *c d*, a section of the slate cut at right angles to *a b*, showing the extension of the weathering inwards at *e f*.

The "Skiddaw Forest Granite" is exposed at two or three points in the bottom of the Calder Valley; lastly a small area of granite occurs in connection with the Skiddaw Slates of Black Comb, and has been worked at Holebeck.

The Syenite (or syenitic granite) of Ennerdale and Buttermere differs from a true granite in containing hornblende instead of mica. It forms a widespread mass, ranging north and south for nine miles, and adjoining the Eskdale Granite at the foot of Wastwater. Through a large part of its course it forms the boundary between the Volcanic Series and the Skiddaw Slates, but in the south occurs wholly among the former; both these divisions of bedded rocks are much altered in its neighbourhood, the alteration sometimes extending to a distance of more than a mile from the mass.

The Quartz Felsite of St. John's Vale may be considered as a compact granite; it gives off a dyke of a beautiful porphyritic character, which can be traced for many miles towards Armboth and Helvellyn. Many sheets or dykes of other igneous rocks occur; we can only mention the minettes or mica-traps of Sale Fell and Dodd, the diorites of Seathwaite How, Little Knott, Hindscarth, and Burtness Comb; and the dolerites of Wythop Fells, Castle Head, Keswick, and Longstrath in Borrowdale.

CARBONIFEROUS FORMATION.—The basement beds of the Carboniferous Formation are thick irregular deposits of reddish conglomerates and coarse sandstones, which were formerly assigned to the Old Red Sandstone, but which further research has shown to belong to a later age; they rest quite unconformably on the Silurian rocks, and are not continuous, appearing only in detached masses or patches, which thin out with great rapidity. For these reasons it has been suggested that they accumulated in valleys, or at all events along the coast lines of the Carboniferous sea. Professor Ramsay thinks that glacial action had much to do with their origin, and certainly many of the stones are smoothed and striated; the included pebbles are of all sizes up to two or three feet in diameter, and are chiefly grits or sandstones from the Upper Silurian rocks of Westmorland.

These red conglomerates form the mass of Mell Fell (1,760 feet high), and are well exposed along Dacre Beck; they also crop out at one or two other points from beneath the Mountain Limestone

The Carboniferous (or Mountain) Limestone.—East of Mell Fell and extending to Penrith, we find thick beds of grey limestone, alternating with thin layers of shale and sandstone; this is the Mountain Limestone. We trace it northwards to Hesket Newmarket and High Hesket, the river Petterill marking its eastern boundary; all along this line it dips eastwards, passing under the Permians of the Vale of Eden. There is a fault, however, running north and south along the line of junction.

On the east side of the Vale the Mountain limestone is brought up abruptly by the Great Pennine Fault, a mighty dislocation of the strata, which commences in Dumfriesshire and runs southwards to Brough; the rocks on the east side of this fault have been heaved up from 3,000 to 6,000 feet above their proper position.

Here the Carboniferous Limestone forms a grand escarpment at Crossfell, and thence northwards by Alston Moor and Brampton to Bewcastle, Haythwaite, and Shield.

From the west flank of Crossfell to east of Brampton the aqueous rocks are traversed by a great intrusive sheet of igneous rock (basalt) called the Great Whin Sill.

Under Alston Moor we find twenty different beds of limestone, altogether 470 feet thick, between which are sandstones and shales 1,686 feet thick.

Returning to the west side of the Eden, the Carboniferous Limestone extends westwards from Hesket to Bolton and Ireby, and then southwards by Pardshaw and Lamplugh to near Egremont. Here it is covered over by newer rocks, but again crops out further south at Millom.

The Mountain Limestone is very rich in fossils; it contains immense numbers of marine shells, and is sometimes almost made up of the joints of encrinites, or "sea lilies," marine animals which lived attached to the ocean floor.

Of late years immense quantities of iron ore (red hæmatite) have been obtained from this rock; in 1877 there were raised 1,351,441 tons, valued at £965,302. The ore lies in pockets, or fills cavities in the limestone; there can be no doubt but that it has been deposited in these places from water as it trickled through the beds of the limestone, probably during the Permian period. The chief mines lie about six miles east of Whitehaven, but the Millom area is also very productive.

In the district of Alston Moor veins of lead ore (galena) and zinc ore are largely worked; 3,000 tons of lead ore were got in 1877, containing 14,091 ozs. of silver, the total value being £37,000; of zinc ore 1,700 tons were obtained, value £5,400; of barytes 25 tons, value £31.

In the Pennine Chain (at Alston Moor, &c.) the name *Yoredale Series* is applied to limestones, sandstones, and shales about 600 feet thick, which rest upon the Mountain Limestone, and still higher beds of sandstone are called the *Millstone Grit*, but these divisions have not been recognised west of the Eden Valley.

Coal-measures. These consist of shales, grits, and sandstones, containing seven or eight workable seams of coal. The coal-field extends along the coast from Whitehaven by Workington to Maryport, and reaches for about five miles inland; a long strip of Lower Coal-measures runs from Maryport to Welton. The dip of the beds is on the whole to the west and north; in the first direction they pass under the sea and have been followed at Whitehaven for 3,200 yards beneath the water; northwards they dip under Carlisle and Silloth, but must lie here at a very great depth, probably above 3,000 feet. At Whitehaven massive red sandstones 150 feet thick form the top division of the coal-measures.

In 1877 the coal raised at thirty-five collieries was 1,515,783 tons; at this rate of consumption the present district will be exhausted in about three centuries. Of fire clay 30,762 tons were obtained, and 9,471 tons of oil shales.

PERMIAN FORMATION. In Cumberland red sandstones of great thickness rest unconformably on the Carboniferous strata, and form the Plain of Cumberland, extending all along the Eden from Penrith to Hayton and Longtown, and westwards to the sea at Bowness, Silloth, and Allonby.

At Penrith the *Lower Permians* are thick-bedded bright red sandstones, 5,000 feet thick, in which fossil footprints have been found. The *Middle Permians* near Longwathby, consist of red clays, containing gypsum, and these are overlaid eastwards by dark red fine-grained sandstones the *Upper Permians*; the two last divisions may be 3,000 feet in thickness.

East of Kirkoswald and Carlisle, and again between Aspatria and Wigton, beds of magnesian breccia occur in the Lower Permians, which are locally called "brockram;" these have much the appearance of an old glacial deposit.

Another area of Permian beds forms a narrow strip along the coast from St. Bee's Head, past Ravenglass to the Duddon estuary. In the fine section at St. Bee's Head we see at the base three feet of breccia (resting on coal-measure sandstone), then 11 feet of magnesian limestone (containing the shells *Gervillia* and *Schizodus*), 30 feet of red and green marls with gypsum, and on top 50 feet or more of red sandstones.

THE TRIAS. On the north-east of Carlisle red marls and sandstones of *Keuper* age can be traced round Scaleby and Brunstock.

To the south-west of Carlisle the Triassic beds above noted are overlaid first by representatives of the RHAETIC BEDS (black shales), and then by blue shales and then limestones belonging to the LOWER LIAS; the latter occur at Aikton, Orton, Oughterby, &c., but the country is so overlaid by drift that there are no good sections.

Of later stratified rocks such as elsewhere form the Oolitic, Cretaceous, or Tertiary Formations we find no trace in this county; probably during all these later geological epochs the Cumbrian area was above the sea level, and was undergoing the wear and tear of atmospheric denudations, its mountains being slowly carved out, and its present surface configuration gradually brought into shape.

SURFACE DEPOSITS.—Over a great part of the surface of Cumberland it is easy to trace the work of ice during the last *Glacial Period*. The mountains of the Lake District were then buried deeply under ice, which passed outwards in all directions, smoothing and grooving the rocks over which it flowed, and carrying stones of all sizes from the high to the low grounds, where they now form the deposit called Till, or boulder clay. The softer

Fig. 20.—Perched blocks at a height of 2,000 feet on Glaramara, between Longstrath and Seathwaite Vale, Borrowdale.

rocks, especially along lines of fault, were hollowed out, and form the present lake-basins.

Then depression of the land took place, and the whole slowly sank 2,000 feet below its present level; to this period belong the sands and gravels which lie above the Till. Finally, elevation succeeding, glaciers again occupied the mountain ranges, but they were of diminished size, and the climate continuing to ameliorate, the ice soon disappeared altogether.

The raised beach, 25 feet above the present sea-level, which can be traced from Workington to Bowness, marks an elevation of the land in *post-glacial*

Fig. 21.—Moraine mounds in the upper part of the Wyth Burn Valley, west of Thirlmere.

times; many of the smaller lake-basins too have since been filled up by matter transported from the higher lands, and now form stretches of rich alluvial land in the valleys or peat mosses on the fells.

PRE-HISTORIC MAN.—Many specimens of the polished stone tools used by the Neolithic tribes have been found on or near the surface. Large celts or axe heads made of the highly altered felstone-like or flinty ashes of the Borrowdale Series are indeed known to be characteristic of this district; they have been found at the Druid's Circle Keswick, Loughrigg Tarn, Penrith Beacon, and Great Salkeld; only three instances are known in England where the wooden handles of these celts have been found still attached to them, and two of these are from Cumberland. One was found by Mr. R. D. Darbishire, F.S.A., in peat, which had once formed the bed of a small lake or tarn; with it were found another haft of the same character, and several stone celts, one of them nearly fifteen inches long; some wooden paddles and clubs, pottery, and other objects were also found; the other specimen was found in the Solway Moss near Longtown, it is now in the British Museum.

Perforated stone axes have been found at Plumpton near Penrith, at Red Dial near Wigton, and on Ousby Moor: a perforated hammer head made of serpentine was found on Hallguard Farm near Birdoswald, and a hammer stone at Melmerby.

Fig. 22 —Stone celt in its original wooden handle. Found by Mr. Darbishire in peat which had once formed the bed of a tarn or small lake in Cumberland.

Flint flakes or knives are known from Irthington, and Canon Greenwell found a beautifully wrought blade of flint with a burnt body in a barrow or tumulus at Castle Carrock.

Rude stone circles occur on the mountains, and many remains of what appear to be stone pit-dwellings. On a rock close to an old camp on Lazenby Fell there is a rock scored with about 70 deep grooves from four to seven inches long and about one inch deep and wide, pointed at each end as if from sharp-ended tools or weapons of stone having been ground in them.

No. 8.

GEOLOGY OF DERBYSHIRE.

NATURAL HISTORY AND SCIENTIFIC SOCIETIES.

Chesterfield and Derbyshire Institute of Engineers. Transactions.
Derbyshire Naturalists' Society; Derby.

MUSEUMS.

The Free Public Library and Museum, Derby.
Geological Museum, Castleton.

PUBLICATIONS OF THE GEOLOGICAL SURVEY.

Maps.—Quarter Sheets: 62 N.E. Lichfield, Tamworth; 63 N.W. Ashby-de-la-Zouch; 71 N.W. Belper, Wirksworth; 71 S.W. Castle Donington, Derby; 71 S.E. Loughborough; 72 N.E. Ashbourne; 72 S.E. Burton-on-Trent, Tutbury, Uttoxeter; 81 N.E. High Peak; 81 S.E. Buxton, Bakewell, Winster; 81 N.W. Stockport; 81 S.W. Macclesfield; 82 N.E. Tickhill, Worksop; 82 S.E. Mansfield, Ollerton; 82 S.W. Chesterfield, Matlock; 82 N.W. Rotherham, Sheffield, Dronfield; 88 S.W. Manchester, Oldham; 88 S.E. Holmfirth, Penistone.

Books.—The Lower Carboniferous Rocks of Derbyshire, by A. H. Green, 4s.; Geology of Parts of Notts and Derbyshire, by Aveline, 1s.; Ditto of Parts of Notts, Derby, and Yorkshire, by Aveline, 1s.

IMPORTANT WORKS OR PAPERS ON LOCAL GEOLOGY.

List of works on the Geology of Derbyshire in the Survey Memoir "On the Lower Carboniferous Rocks of Derbyshire," by A. H. Green, 1869.

1860. Jones, Professor Rupert, and Parker, W. K.—Fossil Foraminifera from Chellaston, near Derby. Journ. Geol. Soc., vol. xvi. p. 452.
1870. Mello, Rev. J. M.—An Altered Clay-Bed and Section in Tideswell Dale. Journ. Geol. Soc., vol. xxvi. p. 701.
1875-77. Mello, Rev. J. M.—The Bone-Caves of Creswell Crags. Journ. Geol. Soc., vols. xxxi.-xxxii.-xxxiii. (three parts).
1877. Dawkins, W. B.—Ossiferous Deposit at Windy Knoll. Journ. Geol. Soc., vol. xxxiii. p. 724.
1877. Pennington, Rooke.—Barrows and Bone-Caves of Derbyshire. Macmillan, 6s.
1878. Painter, Rev. W. H.—Castleton: Its Extinct Fauna and Physical Surroundings. Midland Naturalist, vol. i. p. 63.
1879. Dawkins, W. B., and Mello, J. M.—Further Discoveries in Creswell Caves. Journ. Geol. Soc., vol. xxxv. p. 724.
1880. Wilson, E.—Forams in the Mountain Limestone. Midland Naturalist, vol. iii. p. 220.
1880. Wilson, E.—Fossil Fish from Ticknall. Midland Naturalist, vol. iii. p. 172.

1881. Woodward, C. J. Minerals of the Midlands. Midland Naturalist, vol. iv. pp. 87, &c.
See also *General Lists*, p. xxv.

PHYSICAL FEATURES. -In examining the geological structure of any district, it is well to first view it as a whole, noting its hills, valleys, and plains, the direction of its rivers and brooks, its occupations and industries, for these will all be mainly attributable to the rocks which compose it.

Considered in this way, Derbyshire falls into three natural divisions.

The first of these is situated in the north and west, extending a little east of the river Derwent, and westwards across the Dove into Cheshire and Staffordshire. Its hills of deeply-fissured limestone and the deep longitudinal valleys, whose eastern sides rise in bold abrupt "scarps," forming the watershed of central England, give a wild beauty and picturesqueness to this region of old rocks which render it the delight of tourists and travellers.

The Derbyshire coal-field occupies the east and north-east of the county, and forms another well-marked division: its streams flow north-east into the Idle or Don, and so ultimately into the Trent. Lastly, we have the country round the Trent, in the south, marked by a rich red soil and ranges of low sandstone hills. "Drift" clays and gravels, river deposits, &c., lie indiscriminately upon the older rocks, and will be considered separately.

Taking the most ancient rock first, we will examine more minutely the CARBONIFEROUS or MOUNTAIN LIMESTONE, which underlies all others in Derbyshire, and must, consequently, have been formed before them.

This bed occupies the surface from Matlock and Porwich, by Winster, Elton, Hartington, Bakewell and Eyam, to Castleton and Buxton: its thickness is estimated at from 4,000 to 5,000 feet: it is a pale grey, pure limestone, containing a few thin partings of shale and clay, and also two or three beds of what is locally called "toad-stone," which is a true volcanic rock. The remains of shells, corals, &c., prove, without doubt, that this limestone was formed at the bottom of a deep and clear sea. The well-known black marble of Ashford belongs to the upper beds of this formation; other lighter-coloured marbles which occur show, when polished for mantel-pieces, very beautiful sections of such fossils as *encrinites*, *orthoceras*, &c. The beds of toadstone mark great submarine eruptions, when masses of molten rock and ashes were spread over the sea-bottom. There is a very interesting section at Tideswell Dale, and in several of the lead-mines two, if not three, separate beds of this igneous rock have been cut through. In many places water, aided by the carbonic acid dissolved in it, has cut deep fissures through the massive limestones. Near Matlock, and in Dove Dale, Miller's Dale, Monsall Dale, at Cromford, and in numerous other spots, these lovely ravines, with steep, almost precipitous sides, and a rapid trout stream dashing along the bottom, delight the eye, which has become wearied with the monotony of the moors above. In several places remarkable caverns have been hollowed out by water in the rocks, especially near Matlock and Castleton. A good example is the Blue John Mine near the latter town; "Blue John" is a variety of fluor spar: the mine produces some twelve or thirteen tons per annum, valued at £40 per ton; the spar is turned into vases, &c. - a purpose for which it is said to have been held in esteem as far back as the occupation of this country by the Romans. Veins of *galena* (sulphide of lead) are frequent. In 1874 there were 118 lead mines in Derbyshire, but 35 of these raised less than one ton of ore. The total quantity of ore raised was 4,301 tons, which yielded 3,572 tons of lead, and from which 800 ounces of silver were also extracted. *Of zinc ores* 120 tons were raised, value £250. That rare mineral *elaterite* (elastic bitumen) occurs at Windy Knoll, near Castleton. The mountain limestone again crops out about Crich, where it is very rich in minerals.

In 1879 there were 105 mines at work, producing 5,439 tons of lead ore, value £42,018; 79 tons of zinc ore, value £276, were also obtained.

DERBYSHIRE.

YOREDALE ROCKS AND MILLSTONE-GRIT.—On all sides (except where faults break the order) we can trace the Mountain Limestone passing under certain thick beds of sandstone and shale; these in turn are overlaid by the Coal-measures, and these again by the red marl of the plains.

These newer strata were all once continuous, right over and above the limestone. Deep-seated forces, which we usually describe as volcanic, upheaved these beds from below, forming the ridge called the Pennine Chain. The highest central portion would be the part most fissured and shattered in this process, and would subsequently be most exposed to the denuding influences of the weather. From that part all the upper beds have been, as it were, planed off, by ice and frost, and running water, so that the large dome-shaped mass of Mountain Limestone now constitutes there the surface rock.

Surrounding it on all sides come the *Yoredale Rocks*, an alternation of shales and sandstones about 500 feet in thickness. Owing to the unequal hardness of the beds, landslips frequently occur, as at Alport Tower, and on the flanks of Mam Tor, the "Shivering Mountain."

The Millstone-Grit is the name applied to certain massive coarse sandstones, separated by very thick beds of shale, which rest upon the Yoredale Rocks, and are in turn surmounted by the lower Coal-measures. Their line of outcrop runs, on the whole, from north to south. There are four or five beds of hard sandstone, each from 90 to 150 feet thick, whilst the shales that intervene are about equally thick. The latter have, of course, worn away much faster than the sandstones, and we consequently find deep valleys running from north to south, whose bottoms and sides are in the shales, whilst continuous and lofty escarpments formed by the "grits" stretch parallel to them in the same direction.

The fourth, or lowest bed of gritstone is called the Kinderscout Grit; it forms a range of moorlands stretching from Glossop and Hayfield eastwards across the watershed. In the centre, separated from the main mass as an outlier, it constitutes the high tableland of the Peak. This high flat patch is about 2,000 feet above the sea, and is triangular

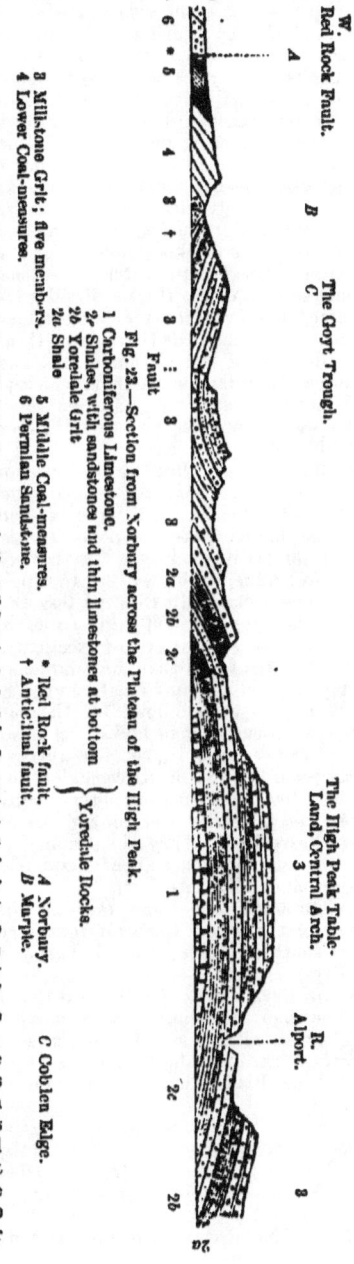

Fig. 23.—Section from Norbury across the Plateau of the High Peak.
1 Carboniferous Limestone. 2r Shale, with sandstones and thin limestones at bottom. 2b Yoredale Grit. 2a Shale. 3 Millstone Grit; five members. 4 Lower Coal-measures. 5 Middle Coal-measures. 6 Permian Sandstone.
* Red Rock fault. † Anticlinal fault.
} Yoredale Rocks.
A Norbury. B Marple. C Cobden Edge.

in shape, the vertex pointing eastwards. Much of its surface is covered by peat-mosses, but every here and there bosses of the Millstone-Grit are seen of the most curious and fantastic shapes, like the caverned and hollowed rocks on a wild sea-coast. The sides of the Peak are scored by ravines or cloughs, as they are called, worn out by streams that course down the hill-side to join the little rivers Ashop and Noe.

Walking eastwards from the Peak we cross the various escarpments formed by the sandstones, until we reach the "Rough Rock," the highest member of the Millstone-Grit series. Turning northwards or southwards we can trace each escarpment running in an almost unbroken line into the far distance. If, however, we attempt to follow any single bed, we shall find it alter rapidly, thinning out or thickening as the case may be. Southwards the beds diminish rapidly in thickness, until near Duffield they disappear under newer rocks. South of the Trent a patch occurs near Stanton and Melbourne. It is here only some 200 feet thick. Besides the millstones, from which it obtained its name, this formation yields valuable building stones. On top of each of the great sandstone beds there occur thin beds of coal, and the whole points to a shallowing of the Carboniferous Limestone sea, with sometimes even the formation of land surfaces on which plants grew.

THE COAL-MEASURES. During the formation of the Mountain Limestone, deep-sea conditions prevailed in Derbyshire. In the Yoredale Rocks and Millstone-Grit, we have evidences of frequent upheavals and depressions of a shallow sea-bed. But the Coal-measures present us with the clearest evidence of ancient land surfaces, on which vegetation flourished with the rankest luxuriance. The transition, however, is not a sudden one.

The Lower Coal-measures or Ganister Beds consist of "thick sandstones, with intervening shales, a few thin coals, and beds of fire-clay of great economical value, largely worked to supply the Sheffield steel works with crucibles and fire-bricks." In 1874 no less than 23,164 tons of fire-clay were raised. The coals are poor and thin; but each bed usually rests on a layer of ganister, or fine hard-grained grit of excellent quality. Formerly used for road mending, "it was found that the scrapings of roads mended with it served admirably to puddle up any cracks in the Sheffield furnaces; and the rock itself is now largely ground down for the same purpose, and for coating the inside of the furnaces; it is laid on in a soft state, and baked by the heat into a good fire-resisting lining: it is also used, mixed with pounded fire-bricks, in the Bessemer process of steel-making, for lining the inside of the 'converters.'" These lower Coal-measures run northwards from Morley and Stanley, by Belper, Heag, Trinity Chapel, and Brampton, and again rise to the surface cast of Chesterfield. They dip steadily eastwards, passing under the true Coal-measures. They are about 1,000 feet thick; and in the north of the county form much hilly moorland.

The Middle and Upper Coal-measures constitute the most valuable part to man of the whole Carboniferous series. The Derbyshire Coal-field is only the southern continuation of that of Yorkshire; extending also into Notts, it comprises an area of above 700 square miles, of which about 230 square miles are in Derbyshire. On the east it is bounded by an escarpment of Magnesian Limestone, extending from the north by Bolsover, to Kirkby-in-Ashfield: the thickness of the Middle Coal-measures is here 2,500 feet. The principal coals worked are the Top Hard (5 to 8 feet); the Soft Coal (4 feet); the Clod or Black Shale (5 to 7 feet); and the Kilburn Coal (3 to 5 feet thick).

In 1874 there were 243 collieries at work in Derbyshire; and during the year 7,150,570 tons of coal were raised. The beds of iron ore (argillaceous carbonate) associated with the coal-seams are also of great value. Of this there were raised in the same year 239,292 tons, valued at £128,172 4s. 0d.

In 1879 the coal got amounted to 7,450,370 tons from 234 collieries; the iron ore had decreased to 146,341 tons, valued at £73,170.

DERBYSHIRE.

As elsewhere in this Carboniferous formation, which was evidently one of great volcanic activity, numerous *faults* or breaks in the beds occur; these have generally a north-west and south-east direction. The vegetable remains, which constitute the coal seams, have been so compressed into one mass as to have commonly lost all trace of their origin; but in the intervening shales and sandstones impressions of ferns, of long slender reeds or *calamites*, and of trunks of strange trees, are often met with in profusion. If a very thin section of coal itself be examined under the microscope, traces of vegetable tissues, or of the spores and spore-cases of ferns, are almost sure to reveal themselves. A plant named *Sigillaria* is very common: it attained large dimensions, for an individual 72 feet in length was found at Newcastle: it may be recognised by its longitudinal flutings, and the vertical arrangement of the markings, or leaf-scars, on its trunk: its roots are called *Stigmaria*. *Lepidodendron* was an allied species, but may readily be distinguished by the spiral arrangement of its markings. The plants that now form coal probably grew in low flat marshes near the mouths of mighty rivers, in a cool, equable, and moist climate. The growth and decay of vegetation in such a place for a thousand years would yield, perhaps, sufficient coal to form a seam a foot thick when the land there had sunk, and thick deposits of sand, mud, &c., had been laid upon the vegetable matter. Altogether, the British Carboniferous Rocks have yielded 2,409 species of fossils, of which 339 are plants.

A small portion of the Leicestershire Coal-field extends into the south-east of Derbyshire, but it belongs, geologically, to the former county.

THE PERMIAN FORMATION is principally represented by the *Magnesian Limestone*, which occupies a narrow strip on the north-east of the county from Ault Hucknall and Pleasley, by Bolsover and Langwith northwards. Here it is above 100 feet thick, and forms an escarpment overlooking the Coal-Measures: it produces some excellent building stones, which are largely quarried at Pleasley, Bolsover, &c.

THE TRIAS, or *New Red Marls and Sandstones*.— Passing now to the south of Derbyshire, we find a very different region from those already mentioned. The rocks are much more recent than those of the Carboniferous or Permian Formations, and they contain fossils altogether different, showing a great break or interval of time between, during which this country may have been above the sea. The lowest bed of the Trias is the *Bunter (variegated) Sandstone:* it is a light-red and yellow sandstone, containing numerous rounded quartz pebbles. There are some good sections exposed between Sandiacre and Cloud Hill, at Ashbourne, Morley, &c. About Derby it occupies the high ground to the north near Beardsall, Muggington, &c. The Bunter is chiefly important as a water-bearing formation, the

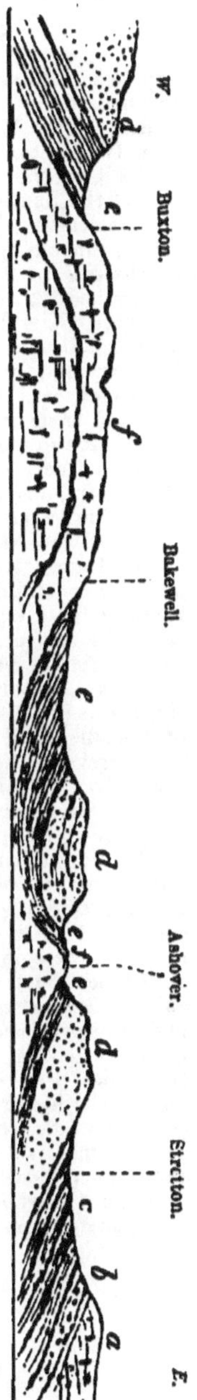

Fig. 24.—Section from west to east across Derbyshire.

a Permian.
b Lower Red Sandstone.
c Coal-measures.
d Millstone Grit.
e Yoredale Rocks.
f Mountain Limestone.

supplies obtained from it in several parts of England appearing almost inexhaustible.

Keuper Red Marl and Sandstone.—In England this formation rests directly upon the Bunter, but in Germany a bed of shelly limestone, the Muschelkalk, comes between. This interval is marked here by the worn, eroded upper surface of the Bunter, as exposed in the railway cuttings at Morley, near Derby, and elsewhere. The Red Marls stretch for some miles north and south of the Trent, running up the valleys of Charnwood Forest on the south, and reaching northwards to Sandiacre, Kedleston, and Shirley. In the sandstones (called waterstones) at the base there are important quarries at Stanton, near Ashbourne, at Kirk Langley, and at Donnington Park. The Upper Keuper Sandstone has been quarried near Chilwell. The Red Marl itself forms an excellent corn land, being rich in phosphates. Casts of salt-crystals often stud the freshly-exposed surfaces, whilst important beds of gypsum occur: at Chellaston this last-named mineral is largely quarried; when burnt it forms plaster of Paris, but the pure white variety, called alabaster, is made into chimney ornaments, statuettes, &c. Much of the gypsum got in England is obtained in Derbyshire, amounting to 56,668 tons in 1879; the value being about 6s. per ton. The scarcity of fossils in the Red Marl made the discovery of many species of *Foraminifera* in the Chellaston clays of special interest.

THE DRIFT. Vestiges of the glacial period, in the shape of beds of gravel and clay containing numerous kinds of rocks, occur in the south and east; but there is little on the west and north of the county. The limestone rocks, too, have not preserved the striations and ice-markings with which they were doubtless once scored.

RECENT DEPOSITS.—At the foot of many of the limestone hills there are thick beds of calc-tuff, or tufa, or travertin, a light porous deposit of carbonate of lime, dissolved out from the limestone fissures and caverns by the streams which traverse them, and again re-deposited in the open air. In the caverns themselves the stalactites depending from the roof and sides, and the stalagmite covering the floor, have had a similar origin. Some of these caverns too, have yielded interesting relics of pre-historic times. Thus Mr. Pennington has found bones of the bison, the mammoth, &c., with flint and bronze implements, at certain caves and fissures near Castleton; and the Rev. J. M. Mello announces similar discoveries at Creswell Crags, in the east of the county. The Creswell Caves are hollowed out of the Magnesian Limestone; in their lowest deposits we find the rude stone tools of the Palæolithic tribes, the earliest human inhabitants of these islands, associated with the bones of animals now extinct, including the mammoth and woolly rhinoceros; in the higher beds of sand, red clay, &c., which form the floor of the caves, we find better-made flint implements belonging to the later savage tribes called Neolithic. Professor Boyd Dawkins has rendered excellent service in working out the history of the deposits in these caves.

The *Peat Bogs* covering the moors are formed by the continued growth and decay of mosses; whilst the alluvial deposits, gravels, &c., of the rivers Trent and Derwent conclude a series of geological events which may be described in words, but of whose long-continued duration the mind is scarcely able to form an adequate conception.

No. 9.

GEOLOGY OF DEVONSHIRE.

NATURAL HISTORY AND SCIENTIFIC SOCIETIES.

Miners' Association of Cornwall and Devonshire; Truro. Papers and Proceedings.
Teign Naturalists' Field Club; Teignmouth. Report.
Devonshire Association for the Advancement of Science, Literature, and Art; Plymouth. Transactions.
Plymouth Institution and Devon and Cornwall Natural History Society; Plymouth. Annual Report and Transactions.
Torquay Natural History Society.
Exeter Naturalists' Club and Archæological Association.

MUSEUMS.

Devonport Museum.
The Albert Memorial Museum, Exeter.
Athenæum Museum, Plymouth.
Torquay Museum.

PUBLICATIONS OF THE GEOLOGICAL SURVEY.

Maps. Sheets: 20, Weston-Super-Mare, Bridgwater, and the coast to Culborne, Penarth, Lavernock; 21, Ilminster, Taunton, Honiton, Tiverton, Dulverton; 22, Lyme Regis, Torbay, Exeter; 23, Berry head to Start Point, Dartmouth; 24, Bolt Head to East Looe, Plymouth; 25, Launceston, Tavistock, Dartmoor; 26, Bideford, Holsworthy, Hatherleigh; 27, Foreland to Barnstaple Bay.
Books. Geology of Cornwall, Devon, and West Somerset, by De la Beche, 14s. Palæozoic Fossils of Cornwall, Devon, and West Somerset, by Prof. Phillips. Eruptive Rocks of Brent Tor, by F. Rutley, 15s. 6d.

IMPORTANT WORKS OR PAPERS ON LOCAL GEOLOGY.

List by W. Whitaker, Esq., of 300 works by 129 authors, in Transactions of Devonshire Association for Advancement of Science, Literature and Art, 1870; Supplementary List of 115 more in ditto, 1872.

1866. Jukes, Prof. Beete.—Devonian Rocks of North Devon. Journ. Geol. Soc., vol. xxii. p. 320.
1867. Etheridge, R. E.—Physical Structure and Fossils of North Devon. Journ. Geol. Soc., vol. xxiii. p. 568.
1867. Hall, T. M.—Relative Distribution of Fossils in North Devon Rocks. Journ. Geol. Soc., vol. xxiii. p. 371.
1868. Holl, Dr. H. B.—Older Rocks of South Devon. Journ. Geol. Soc., vol. xxiv. p. 400.

1869. Whitaker, W. New Red Sandstone of South Devon. Journ. Geol. Soc., vol. xxv. p. 152.
1862. Pengelly and Heer. The Miocene Deposits at Bovey Tracey. Phil. Trans. 1019.
1870. Ormerod, W. G. Granite of Dartmoor. Report Brit. Assoc. 1869, p. 98.
1870. Godwin-Austen, R. A. C. The Devonian Group. Report Brit. Assoc. 1869, p. 88.
1871. Whitaker, W.—Chalk of Devon. Journ. Geol. Soc., vol. xxvii. p. 93.
1875. Foster, C. Le Neve. -Haytor Iron Mine. Journ. Geol. Soc., vol. xxxi. p. 628.
1876. Lavis, H. J. J.—Triassic Strata near Sidmouth. Journ. Geol. Soc., vol. xxxii. p. 274.
1876. Woodward, H. B. Superficial Deposits of Newton Abbot. Journ. Geol. Soc., vol. xxxii. p. 230.
1876. Ussher, W. A. E. Trias of Devon. Journ. Geol. Soc., vol. xxxii. p. 367.
1877. Lee, J. E. —Upper Devonian Fossils in Shales of Torbay. Geol. Mag., p. 100.
1877. Woodward, H. B. Devonian Rocks near Newton Abbot and Torquay. Geol. Mag., p. 447.
1877. Woodward, H. B. Geology of Plymouth. Science Gossip, p. 169.
1878. Champernowne, A. Devonians of North and South Devon. Geol. Mag., vol. xv. p. 193.
1878. D'Urban, W. S. M. – Palæolithic Implements from the Valley of the Axe. Geol. Mag., p. 37.
1878. Hall, T. M.—Geology of Devonshire in White's Directory.
1878. Ussher, W. A. E. -Terminal Curvature in the South-Western Counties. Journ. Geol. Soc., vol. xxxiv. p. 49.
1879. Champernowne, A. - Devonian Stromatoporidæ from Totnes. Journ. Geol. Soc., xxxv. p. 67.
1879. Duncan, Prof. P. M. Upper Greensand Corals of Haldon. Journ. Geol. Soc., vol. xxxv. p. 89.
1880. Champernowne, A. Upper Devonians in Devonshire. Geol. Mag., p. 359.
1881. Champernowne, A. The Ashburton Limestone. Geol. Mag., p. 410.
1881. Davidson, Thos. -- Brachiopods from Budleigh Salterton Pebbles. Palæontographical Soc.

See also General Lists, p. xxv.

In Mr. Whitaker's list we find the title of 415 books, papers, &c., which had been written concerning the rocks of Devon between 1671 and 1870; of these, however, only nine date back to before 1800, a fact which shows the modern origin of geological inquiries. Of the workers in the first quarter of the present century we may name Dr. Berger, the Revs. R. Hennah and J. J. Conybeare, with Messrs. Allan, Jones, Taylor, Mawe, Prideaux, and Whidbey; in the next twenty-five years the grand work of De la Beche, who mapped almost single-handed the broad divisions of the rocks, stands out boldly, but Sedgwick and Murchison claim high honour, while Professor Phillips, the Rev. D. Williams, Messrs. Lonsdale, Weaver, and others, did good work; in the last thirty years we note the names of Jukes, Etheridge, Pengelly, Ormerod, Godwin-Austen, Meyer, Vicary, Townshend M. Hall, Salter, Worth, Champernowne, Valpy, H. B. Woodward, Ussher, &c.

It is not very difficult to understand the arrangement of the strata of Devonshire if we look at them in a broad way only. The oldest rocks, those called *Devonian*, form two broad bands along the north and south of the county, extending inland from the coast for some 10 or 12 miles in each case; these two bands are connected underneath, forming a great trough or fold, which is filled up by rocks of *Carboniferous* age, which consequently occupy all the centre of Devon, resting upon and in the hollow formed by the Devonian strata; through the southern Devonian and Carboniferous rocks the great *igneous mass* of the granite of Dartmoor has burst; lastly, fringing the east of Devon we find beds of Secondary or Mesozoic age ranging from the *trias* to the *chalk* inclusively.

When we come to examine the various strata in detail, we shall find immense difficulties, chiefly among the older rocks, arising mainly from their complexity of structure and arrangement, and from the manner in which they have been altered and dislocated by the volcanic outbursts and great earth-movements which formerly were most actively at work in this region.

Before the geology of Devonshire can be decisively made clear, it will be necessary that a new survey of the county shall be made by the Ordnance Survey on the scale of six inches to a mile; the existing maps on the one-inch scale date back to 1809, and are too small to permit of that minute mapping of the strata which their complex nature requires. When the six-inch map is complete, the Geological Survey will take the county again in hand, and doubtless many questions about which there is now much dispute in the geological world will be finally and thoroughly settled. We shall now describe the rocks in succession, beginning with the lowest or oldest, and taking the views regarding their succession which are at present most generally entertained.

THE DEVONIAN (OR OLD RED SANDSTONE?) FORMATION.—The beds of this age occupy two well-defined areas in the north and south of the county respectively; the arrangement of the strata of the northern portion is usually considered to be the simpler and more complete of the two, and we shall accordingly describe that region first.

North Devon.—Here red sandstones, slates, and limestones occupy a tract of land from 9 to 14 miles wide, extending from the coast as far south as a line drawn from Barnstaple to Anstey near Dulverton; the beds run or strike across this region from west to east, and, as they incline or dip towards the south, it follows that the bottom or oldest beds must be those which come to the surface or crop out furthest north. These are the *Foreland Sandstones*, red and brown in colour, and destitute of organic remains, except a few plant-markings found in yellow sandstones at Countisbury. In the cliffs the strata are seen to have a high northerly dip, but they roll over to the south and pass under the *Lynton Slates*, a fault, however, running along the valley of the Lyn at or near the line of junction. From Watersmeet by the Valley of Rocks to Woodabay the Lynton Slates are seen to be grey and reddish grits and slates, with some thin bands of impure limestone and ironstone, total thickness 1,500 feet; a few fossils occur, as *Chonetes Hardrensis, Orthis arcuata*, &c.

Next come the *Hangman Grits*, red, grey, and yellow grits and sandstones, 1,500 feet thick, forming the Little Hangman and Hangman Hill, and extending thence eastwards past Trentishoe and Martinhoe to Oare Oak Hill and Ex Head; they are almost unfossiliferous: upon these grits rest the *Ilfracombe Slates* (4,000 feet), silvery grey in colour, and containing thick beds of limestone, quarried at many points for lime-burning, and traceable from the west of Combe Martin Bay by Kentisbury to north-east of Challacombe; the fossils include many corals *(Favosites cervicornis, Stromatopora concentrica)* and brachiopod shells *(Stringocephalus Burtini)*: the *Morte Slates* may be considered as forming the upper portion of the Ilfracombe Group; they are 4,000 feet thick and unfossiliferous; they form Morte Point and Bull Point on the coast, and stretch inland along the *northern* slope of the high ground which extends from Pickwell Down to Span Head, and which is formed of red, brown, and grey, unfossiliferous sandstones—containing beds of hematite (iron ore) and manganese called the *Pickwell Down Beds*; their thickness is estimated at 3,000 feet. On the southern side of these hills we find the *Baggy* and *Marwood Beds*, also called the *Cucullea Zone* from the abundance of a fossil shell of that name; a quarry in these beds at Sloly, 2 or 3 miles north of Barnstaple, has yielded plant remains *(Calamites* and *Lepidodendron)*, with the shells *Lingula Mola, Avicula Damnoniensis, &c.*

The highest Devonian strata are termed the *Pilton Beds*, and are seen at Croyde Bay, Saunton, Braunton, Pilton, and thence eastwards to Anstey near Dulverton; they are purple and grey slates and shales, with thin irregular beds of limestone, containing numerous fossils, such as corals, shells, a star-fish, crinoids, &c.

Whether the arrangement of the strata of North Devon is really so simple as above described is a question which has excited much controversy. Professor Jukes believed that the Pickwell Down sandstones were the same as those of the Foreland, repeated by a fault with a downthrow to the north, or brought in by an inverted anticlinal; the great objection to this view is furnished by the fossils, which, according to Mr. Etheridge, differ widely in the different beds.

It is usually considered that these *Devonian* sandstones, slates, and limestones were deposited in the open seas (they contain *marine* fossils), at the same time that the *Old Red Sandstone* rocks of Herefordshire and Scotland were being formed in fresh-water lakes. Professor Jukes combated this view, stating that, from his work among similar rocks in the south of Ireland, he believed the lower red sandstones (of the Foreland, &c.) were of *Old Red Sandstone* age, but that the slates and limestones resting on them really belonged to the *Carboniferous* period; the generality of geologists have not yet, however, been converted to this opinion.

South Devon. —Very little is *certainly* known with regard to the order of succession of the old rocks of South Devon. The coast at Start Point, Prawle, Bolt Head, and Bolt Tail, is composed of metamorphic rocks, such as mica slate, chlorite slate, and chlorite rock, which extend inland to Marlborough and South Pool. The age of these rocks is doubtful; they may be *Lower Silurian*, in which case the gneiss rock on which the Eddystone Lighthouse is built may furnish a connecting link between them and the strata round Veryan Bay in Cornwall, which are certainly of Lower Silurian age. More probably, however, these South Devon rocks are metamorphosed Devonian slates, the change in character having been effected by a mass of some igneous rock which is hidden beneath them or is below sea-level; the great number of granitic pebbles which are thrown up on the coast here might well come from such a mass, while several blocks of granite have been brought up in the trawling-nets from the Salcombe fishing grounds.

All the rest of South Devon, as far north as Newton Abbot and Tavistock, may be said to be formed of slates and limestones, with some red sandstones; which, speaking generally, are of the same age as those of North Devon. The lowest beds may be the red sandstones seen at Cockington west of Torquay, at Staddon Point south of Plymouth, &c. The well-known limestones of Plymouth, Yealmpton, Berry Head, Torquay, and Chudleigh, may be correlated with the Ilfracombe Beds. They are largely worked for marble, building and paving-stone; a walk in the streets of Plymouth on a wet day shows the visitor innumerable beautiful corals (Fig. 25) in the stones upon which he is treading. Immense quantities of limestone were quarried at Oreston for the construction of the Plymouth breakwater. Of the marble quarries perhaps those at Ipplepen are best known; they furnished the fine polished columns used in the National Provincial Bank of England, Bishopsgate Street, London.

The slates which constitute the great mass of South Devon are so rolled, contorted, faulted, and interfered with by igneous rocks that their study is rendered very difficult. A patch of slate at Werrington, north of Launceston, is believed to be of Upper Devonian age, and fossils of this period have been found by Mr. J. E. Lee at Saltern Cove, Torbay. In 1873, Mr. Champernowne found a star-fish (*Helianthaster filiciformis*) in slate near Harberton which resembles those from the Pilton Beds. The flat coiled-up shell of a cephalopod called *Clymenia* is very characteristic of the Devonian rocks.

Roofing slates are got at Ashburton, Ivy Bridge, Kingsbridge, Bickley, and Tavistock; the striped slates of Yeolm Bridge, north of Launceston, are worked for chimney-pieces.

Mr. Etheridge enumerates 544 species of fossils from the British Devonian strata; 125 of these are fishes, and 116 brachiopod shells.

CARBONIFEROUS FORMATION.— The Culm-measures, as the beds of Carboniferous age are locally termed, occupy a trough in the centre of Devonshire; on the west they form the fine cliffs at Hartland Point and Clovelly; eastwards they pass under Triassic strata between Tiverton and Exeter, while the Devonian slates form their boundary to the north and south, as already described, except when they abut on Dartmoor. The lowest culm-measures contain black limestones seen at Fremington, South Molton, Swimbridge, Landkey, Bampton, and Holcombe Rogus; these beds dip southwards and rise to the surface again on the south side of the trough at Lew Trenchard, Bridestow, South Tawton, and Drewsteignton: they may represent part of the *Carboniferous Limestone*; *Posidonomya Becheri* is a common fossil.

Next come beds of hard white shale and black grits, with veins of chert, an impure variety of flint; these beds are 1,500 feet thick, and may represent the *Millstone Grit*; they form Codden Hill (628 feet) south of Barnstaple, and St. Stephen's Hill near Launceston.

The remaining country from Bideford, Chittlehampton, and Tiverton to Torrington, Chulmleigh, Broadbury, and Exeter, is formed of strata which may represent the *Lower Coal-measures*, resembling especially those of Pembrokeshire. They are hard grits and shales containing such plant-remains as ferns, *Calamites, Lepidodendron,* &c.; two species of fish belonging to the genera *Cœlacanthus* and *Elonichthys* have been found in ironstone nodules at Instow; but the Carboniferous beds of Devon are very barren of life, having yielded altogether not more than 50 species of fossils, as compared with nearly 2,000 from the rest of England.

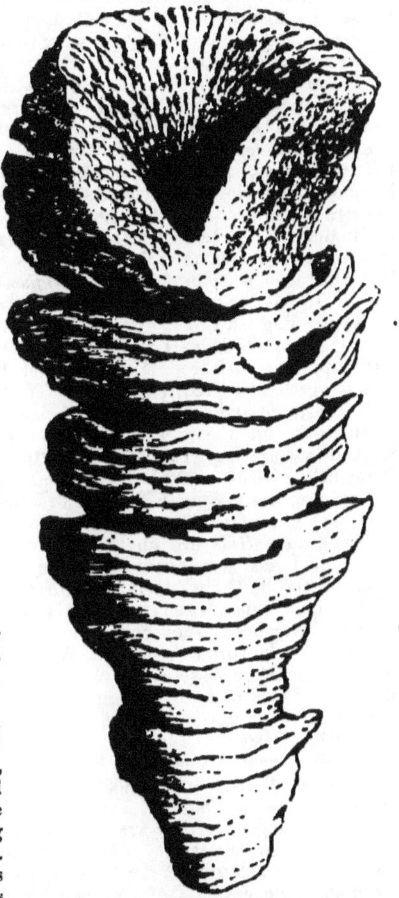

Fig. 25.—*Cystiphyllum vesiculosum*, showing a succession of cups produced by budding from the original coral. Devonian formation.

Irregular beds of *anthracite*, or culm, or stone-coal, can be traced from Luke's Nose, in Bideford Bay, eastwards through Bideford to Hawkridge

Wood, near Umberleigh station; near Bideford some soft veins of anthracite are now worked up into a paint called "Bideford Black." The sandstones or "grits" are quarried for building and road-mending; they are frequently ripple-marked.

From their lack of workable coal-seams or other minerals, the Carboniferous strata of Devon compare very unfavourably with those of other counties; they form a poor, rather sterile, and scantily populated country.

THE TRIAS OR NEW RED SANDSTONE. After the formation of the culm-measures a long period of time elapsed, during which the *Permian strata* of the centre and north of England were deposited; these are wanting in Devon, which may then have been above the waters. Of the succeeding rocks known as *Triassic*, we have, however, a fine development; these beds consist of breccias, sandstones, and pebble-beds surmounted by marls; all are more or less of a red tinge, and may attain a total thickness of 3,000 feet.

Bunter Sandstone.—To this division belong the unfossiliferous breccias and sandstones, well seen along the coast from Torbay by Dawlish to Exmouth; the (lower) red marls of Dawlish also belong to this age, whose top stratum is probably the remarkable bed (100 feet thick) of pebbles seen at Budleigh Salterton, and traceable inland to Burlescombe: most of the pebbles are of quartzite, and they contain fossils both of Silurian and Devonian ages, such as *Orthis Budleighensis, Trachyderma serrata, Spirifera Verneuilii*, &c.; they may have been derived from rocks which once stretched across the English Channel from Cornwall to Normandy.

The Bunter beds extend inland by Topsham, Exeter, Ottery St. Mary, Cullumpton, Kentisbere, Tiverton, and Burlescombe; a long narrow strip goes westward by Crediton to North Tawton and Jacobstow, an outlying mass at Hatherleigh and a patch at Portledge on Bideford Bay, marking a former great extension of the Trias in this direction; small outliers also occur at Slapton in Start Bay, Thurleston in Bigbury Bay, and at Cawsand south of Plymouth. Large crystals of *Murchisonite* (felspar tinged a yellowish-red by oxide of iron) occur in Triassic breccias between Teignmouth and Exeter; they have come from the granite of Dartmoor.

The Keuper Beds are finely exposed on the coast from Sidmouth to Axmouth, and are seen along the valleys at Colyton, Axminster, Stockland, Honiton, and on the western flanks of the Blackdown Hills.

The lower beds are red sandstones ("waterstones"), which rest conformably on the Bunter beds; they form the cliffs between Otterton Point and Sidmouth. At the former place Mr. Whitaker found the jaw of a remarkable reptile - the *Hyperodapedon*; while at Picket Rock Cove, east of Sidmouth, Mr. Lavis found numerous fragments of bones, including those of the *Labyrinthodon*. Red Marls of Keuper age form the rest of the coast from Sidmouth to beyond Axmouth; in these, Mr. Hutchinson found a reed-like fossil plant near Sidmouth. Altogether the coast section, for above 30 miles, from Torbay to Axmouth, is decidedly the finest exposure of the Triassic rocks of England; the beds all dip gently to the east at angles of from 2 to 5 degrees; they are intersected and repeated by several faults.

From the great scarcity of fossils, the presence of oxide of iron, which more or less colours all the beds, and the presence of beds of gypsum and pseudomorphic crystals of rock-salt, it is probable that both Bunter and Keuper beds were deposited in inland salt lakes of great extent.

The "red ground," as the Triassic strata generally are called, forms the best land in Devon: the soil is rich and fertile, and all the best apple orchards stand on it. At Watcombe, near Torquay, a fine clay is very largely worked for terra-cotta and art pottery.

THE RHÆTIC FORMATION consists of green marls, black shales, and white limestones (the "White Lias"), which form passage beds between the Trias and Lias; their thickness is about 100 feet. They are well seen in the cliffs east of Axmouth; one layer near the base of the black shales is so full of the teeth and bones of fishes and saurians as to be called the "bone bed."

THE LIAS.—The blue limestones with interbedded shales, which form the *Lower Lias*, are seen in the cliffs beyond Culverhole Point, above the Rhætic beds; they also occur inland east of Axmouth.

CRETACEOUS FORMATION.—The whole of the Oolitic and Neocomian strata being absent, the Cretaceous series reposes directly, first upon the Lias, and then upon the New Red Sandstone. In the fine section at Beer Head, west of Seaton, we see (1) Sands and Clays, 50 feet, representing the *Gault*; (2) Light-coloured sands with chert beds, 100 feet = *Upper Greensand*; (3) *Chalk Marl*, 10 feet, with layer containing phosphatic nodules (*Chloritic Marl*) at base; (4) *Lower White Chalk* (without flints), 25 feet; (5) *Upper White Chalk* (with flints), 140 feet.

The bottom layer of the Lower White Chalk is largely quarried for building-stone; it is of a yellowish-white tint, soft when extracted, but rapidly hardening on exposure to the air.

The sandy beds of the *Gault* and *Upper Greensand* can only be separated with great difficulty; they form the capping of the Blackdown Hills, and of all the hills which run thence to the coast between Sidmouth and Lyme Regis; near Broadhembury and Kentisbere the hard concretionary lumps which occur in layers in the sands have been largely worked to make scythe-stones (Devonshire Batts), and the hill-sides are disfigured by heaps of rubbish. Outliers of the Greensand occur on the hills called Great and Little Haldon, and on Milber Down; there is a remnant of angular chert as far west as Orleigh Court, only 3 miles from Bideford Bay. Outliers of *White Chalk* cap the hills east of Axmouth, also south of Chard and at Membury, Brice Moor, and along the coast as far west as Salcombe Mouth.

TERTIARY PERIOD.—After the deposition of the chalk, a great interval occurred unrepresented by any strata in Devonshire. The next beds belong to the

MIOCENE FORMATION.—Between Bovey Tracey and Newton Abbot, on the east of Dartmoor, we have a tract of land about 6 miles in length by 2 in breadth, which really represents a filled-up lake. Lignite, or "Bovey Coal," has been worked here for more than a century by means of deep cuttings and tunnels; it is used chiefly in the neighbouring potteries, its sulphurous smell rendering it unfit for domestic use. In 1861, Mr. Pengelly made a careful examination of the Bovey deposit, the expense being borne by Miss Burdett-Coutts. He found alternating beds of clay and lignite, with some sand, to the depth of 125 feet, below which the workings did not extend. The thin sand-beds disappeared as they were followed down the valley, showing that the probable source of both clays and sands was the decomposed granite of Dartmoor, while the lignite was formed of trunks of trees and matted vegetable matter derived from the plants which grew on the shores of the old lake. The fossil plants obtained were examined by Professor Heer of Zurich, who determined 50 species, 26 of which were new, while 19 were characteristic of the *Lower Miocene* age of the continent; they included coniferous trees *(Sequoia Couttsiæ)* allied to the giant Wellingtonias of California. ferns, water-lilies, custard-apples, palms, oak, laurel, &c., indicating a warm climate.

The lignites and clays of Bovey Tracey are unique in England, as a deposit of true Miocene age, if such they be; within the last year or two, however, Mr. J. S. Gardner, who has paid great attention to the *Upper Eocene* flora of the Bournemouth beds, has stated his opinion that the Bovey Beds may be of Eocene age. With the exception of a beetle's wing no animal remains whatever have been found at Bovey, and this absence of shells, bones, &c., which we should naturally expect to find in an old lake deposit, is remarkable and difficult to account for.

IGNEOUS ROCKS.—Both the Devonian and Carboniferous strata include beds of igneous rocks which appear to be contemporaneous with the rocks in which they occur; they alter the stratum lying *beneath* them only, showing that they had cooled down and solidified before the unaltered *overlying* stratum was deposited. To this class belong the bed of porphyritic felstone which can be traced at the base of the Upper Devonians between Morte Bay and Exmoor, the similar rock at Kentisbury in Ilfracombe the slates, and the masses of greenstone and volcanic ash which frequently occur in South Devon, running in parallel bands from east to west as at Saltash. Other rocks of very similar character appear to be *intrusive*, having been forced in a melted state between and through previously existing sedimentary rocks, which they have baked and hardened; of this nature are the bosses between Newton and South Brent, the rock which intrudes in the limestone at Yealmpton, &c.

At or about the commencement of the Carboniferous epoch considerable outbursts of a volcanic nature seem to have occurred, of which we find traces north of Tavistock. The well-known hill of Brent Tor, 1,114 feet in height, is composed of devitrified basalts and pitchstones, with beds of volcanic ash, and tells undoubtedly of the neighbourhood of an old volcano (*see frontispiece*).

Dartmoor.—*After* the close of the Carboniferous period and *before* the deposition of the Triassic strata (which contain rolled pebbles of the granite), the climax of volcanic action was reached in the upward movement of melted rock from below, which, bearing before it and, doubtless, to some extent, incorporating with itself the overlying rocks, has resulted in the formation of the great granitic tract having an area of 200 square miles, and known as Dartmoor. We must always bear in mind that the present surface of Dartmoor was never exposed to the air in a liquid state: the crystals of quartz, mica, and felspar which constitute the granite were formed under a depth of several miles of superincumbent rocks, which have since been removed by the agents of denudation— the sea, rain, rivers, and ice. The Carboniferous and Devonian rocks in the immediate neighbourhood of the granite have been altered by it, though not to the extent that one would have expected; near Chagford veins of granite from a few inches to 18 feet in width are seen like fingers penetrating the carboniferous slates. Large crystals of orthoclase felspar, sometimes 2 or 3 inches in length, occur in the granite, and black tourmaline (or schorl) is a common ingredient.

The rock is much traversed by joints, along which the weathering proceeds; this gives it a wall-like or even stratified appearance, and is the origin of the curious piles of rock known as "Tors" or "Cheese-wrings." Dartmoor forms a waste heather-covered region, with many bogs and peat mosses. The principal heights are—Yestor, 2,050 feet; Cawsand Beacon, 1,802 feet; Amicombe Hill, 2,000 feet; Newlake Hill, 1,925 feet. The granite from Hey Tor (where there are extensive works) was used in the construction of London and Waterloo Bridges.

Elvans.—These are veins and dykes of a whitey-brown rock termed quartz porphyry, which traverse the granite

Fig. 26.—Ideal Section from the west edge of Dartmoor to Ridge.—The amphibolite (or hornblende rock) and the gabbro are igneous rocks of a type distinct from the granite, although occurring in its immediate neighbourhood.

nd extend for long distances into the surrounding sedimentary rocks; those south of Modbury and at Roborough Down and Morwell, may be mentioned. The Roborough stone has been quarried for centuries for building purposes.

Lundy Island.—This is mainly formed of granite resembling that of Dartmoor, but with few large felspar crystals; Devonian slates occupy the south-east corner of the island. The large granite quarries here furnished much stone for the Thames Embankment.

Again in the *Triassic Period* we find felstones, basalts, and volcanic ashes, which occur at Washfield, near Tiverton, under Exeter Castle, south of Silverton, at Posbury Hill near Crediton, between Exeter and Great Haldon, and at Cawsand Bay. All these igneous rocks, from their superior hardness, have been much used for road-metal and as building stones.

Minerals.—The rocks of Devon yield many useful minerals: these occur (1) as regularly stratified beds (as the anthracite); (2) in veins or lodes which traverse both the igneous and sedimentary rocks, and which were originally cracks or fissures in the rocks, but have become filled up by mineral matter deposited from water trickling or rising through the fissures; (3) in surface deposits, as the stream tin and china clay, which result from the denudation and decomposition of the older rocks. The chief or master lodes run east and west, and are intersected by others, called cross-courses, running north and south.

The killas, shillet, or clay-slate of the Devonian beds near Tavistock, constitute the most important mineral-bearing rocks of the county. Copper is the chief product, but iron, manganese, and arsenic are also obtained: from the mine here called Devon Great Consols there was raised between 1840 and 1871 copper ore to the value of $3\frac{1}{4}$ millions sterling, and this from an area of 140 acres. Lead ore, rich in silver, has been worked in Devonian slates at Combe Martin in North Devon, and at Beer Alston in South Devon. Gold has been found in a lode of iron ore at North Molton, and in small quantities in the stream tin works on Dartmoor.

The mineral products of Devon for 1878 were—Tin Ore, 21 tons, value £744; Copper Ore, 16,980 tons, value £50,484; Lead Ore, 337 tons, containing 4,948 oz. of silver; Iron Pyrites, 1,380 tons, value £2,662; other ores of Iron, 6,434 tons, value £3,219; Umber (from Ashburton), 1,151 tons, value £950; Ochre, 374 tons, value £212; Arsenic, 3,091 tons, value £24,231; Clay, 55,844 tons, from the Miocene beds near Kingsteignton; Barytes, 1,843 tons, value £4,323.

SURFACE DEPOSITS.—These consist of gravels, sands, and clays, resting on any or all of the older rocks, and are of comparatively recent origin. Thus at Bovey the top 10 feet is formed of a sandy clay containing angular stones: this is called the "head;" it includes white clays, which contain leaves of the willow and dwarf birch *(Betula nana)*, indicating a cold climate. Great heaps of stones and débris spread over parts of Dartmoor and Exmoor, which recall the moraines of the glaciers of Switzerland, but the rocks beneath do not show the grooving and polishing which are the usual effects of the presence of great masses of ice; these marks may, however, have since been destroyed by weathering. Of erratic blocks or boulders we may note one of red granite (10 tons) at Saunton cliffs, near Braunton; one of felstone (13 tons) at Langtree, near Torrington; masses of trap-rock at Harberton, Druid near Ashburton, and Fremington near Barnstaple; red sandstones at Waddeton Court, near Dartmouth; all these are masses of rock which appear to have been transported from some distance, probably during the last glacial period, by the action of floating ice.

Great spreads of flint and chert gravel occur in East Devon, the result of the weathering of the Cretaceous strata. On Dartmoor there is much sand and gravel, the result of the decomposition of the granite; these frequently contain "stream tin."

Caves.—Several famous caves occur in the limestones of South Devon.

From fissures at Oreston, near Plymouth, many bones of extinct animals (mammoth, rhinoceros, bear, cave-lion, &c.), were obtained between 1816 and 1858; Kent's Hole, near Torquay, has been most systematically explored by Mr. Pengelly, aided by the British Association; the same extinct species of animals occur here, associated with flint and bone implements fashioned by man. In the lowest deposits in the cave (breccia) we find very rough flint tools, with bones of the bear; in the upper stalagmitic and red earth layers, teeth of hyænas are numerous; lastly, the bones of domestic animals, as the sheep, occur in the surface layer of black mould. It would seem certain that man has at intervals inhabited this cave for many thousands of years. The cavern at Windmill Hill, Brixham, has yielded similar evidence.

Raised Beaches.—A bed of sand and pebbles occurs at Northam, Saunton, and Croyde on Barnstaple Bay, it is 15 feet above sea-level and contains recent shells. On the south coast a line of raised beaches can be traced about 30 feet above sea-level at Hope's Nose and the Thatcher Rock near Torquay, and at Berry Head, Sharkham Point, Dartmouth, Start Point, and the Hoe at Plymouth. All these tell of a recent (geologically speaking) and general elevation of the land.

Buried Forests.—Trunks of trees, with peat, &c., are sometimes exposed at or about low-water mark, by violent storms which remove for a brief period the overlying deposits of sand and mud. At Northam Burrows, near Appledore, the stems of seventy or eighty large trees were exposed in 1864; great quantities of hazel-nuts were then found in the peat, with bones of the ox, wild boar, wolf, goat, &c., and flint flakes. On the south coast similar beds have been noticed at Thurlestone, in Bigbury Bay (1866), Blackpool near Dartmouth (1869), Torbay, Sidmouth (1873), &c.

Landslips. The coast near Axmouth has suffered much from landslips. On Christmas Day, 1839, 22 acres of land at Dowlands fell seawards, creating a chasm 1,000 yards in length and 300 in breadth; it was caused by the numerous springs which issue at the base of the *Greensand*.

PREHISTORIC MAN.— The evidence offered by caverns as to the great antiquity of man in this district we have already alluded to. The lowest deposits in Kent's Hole may belong to the early or *Palæolithic Stone Age*; a rude flint implement of this period, $7\frac{1}{2}$ inches long by $3\frac{1}{2}$ broad, was found at Colyton; four tools made of chert were found between Chard and Axminster during the erection of the telegraph posts; from a low hill of chert gravel at Broom, near Axminster, a large number of implements of Palæolithic forms, and made of dark brown chert, have lately been obtained; they have been placed in the Museums at Salisbury and Exeter; in 1879 the Rev. W. Downes found one of chert at Kentisbere Moor, in the valley of the Culm.

Of the later or *Neolithic Stone Age* we have several objects which have been found, either in the barrows or burial-places of the early savage tribes who inhabited Devon, or in the surface soil; we may note a celt or axe-head of flint from a barrow at Hartland, another from Bridge Farm, near North Tawton, a perforated adze of greenstone from North Bovey; a naturally perforated flint pebble which had evidently been used as a hammer was found by the Rev. R. Kirwan in a barrow at Thorverton, near Exeter; in barrows on Broad Down, near Honiton, two beautifully turned cups of Kimmeridge shale were found, with a bronze dagger, flint flakes, and nodules of pyrites and reddle; a lozenge-shaped flint arrow-head was found at Princetown, on Dartmoor; flint flakes occur in great numbers at Croyde.

No. 10.

GEOLOGY OF DORSETSHIRE.

NATURAL HISTORY AND SCIENTIFIC SOCIETIES.
Dorset Natural History and Antiquarian Field Club; Sherborne. Proceedings.
Purbeck Society.

MUSEUMS.
The County Museum, Dorchester.
Poole Museum.

PUBLICATIONS OF THE GEOLOGICAL SURVEY.
Maps.—Sheets: 15, Salisbury, Blandford, Ringwood, Cranborne; 16, Poole, Isle of Purbeck; 17, Bridport, Isle of Portland, Weymouth, Dorchester; 18, Sherborne, Wincanton, Beaminster, Yeovil; 21, Honiton, Tiverton, Dulverton; 22, Lyme Regis to Torbay.

IMPORTANT WORKS OR PAPERS ON LOCAL GEOLOGY.

1856. Falconer, Dr. H.—Description of two Fossil Mammals of the genus *Plagiaulax* from Purbeck. Journ. Geol. Soc., vol. xiii. p. 261.
1858. Phillips, Prof. J.—On a Fossil Fruit from Swanage Bay. Journ. Geol. Soc., vol. xv. p. 46.
1858. Fisher, Rev. O.—Natural Pits on the Heaths of Dorset. Journ. Geol. Soc., vol. xv. p. 187.
1860. Damon, R.—Geology of Weymouth and Portland. Stanford, 3s. 6d. or (with map) 5s.
1863. Day, E. C. H.—Middle and Upper Lias of Dorset Coast. Journ. Geol. Soc., vol. xix. p. 278.
1869. Egerton, Sir P. de M.—On a *Gyrodus* from Kimmeridge. Journ. Geol. Soc., vol. xxv. p. 379.
1869. Hulke, J. W.—On a large Saurian Humerus from Kimmeridge. Journ. Geol. Soc., vol. xxv. p. 386.
1870. Whitaker, W.—On the Chalk of Dorset. Journ. Geol. Soc., vol. xxvii. p. 93.
1873. Mansel-Pleydell, J. C.—Geology of Dorset. Geol. Mag. vol. x. pp. 402,438.
1875. Prestwich, Prof. J.—Quaternary Period in Portland and Weymouth. Journ. Geol. Soc., vol. xxxi. p. 29.
1877. Buckman, Prof. J.—Cephalopoda Beds of Gloucester, Dorset and Somerset. Journ. Geol. Soc., vol. xxxiii. p. 1.
1877. Kinahan, G. H.—On the Chesil and Cahore Beaches. Journ. Geol. Soc., vol. xxxiii. p. 29.
1880. Blake, Rev. J. F.—Portlandian Rocks of England. Q. Journ. Geol. Soc., vol. xxxvi. p. 189.
1881. Buckman, J.—Terminations of some Ammonites from the Inferior Oolite of Dorset and Somerset. Q. Journ. Geol. Soc., vol. xxxvii. p. 57.

See also General Lists, p. xxv.

SOME portions of Dorsetshire—more especially the coast—offer such unrivalled opportunities for geological study, that they early attracted the attention of writers on that science. The papers by Berger, Cumberland, De la Beche, Webster, Lyell, Buckland, Fitton, Clarke, and Thomson, published some forty years ago in the "Transactions &c. of the Geological Society," may still be usefully referred to. Later on, in the "Journal" of the same society, we have the results of the work of the Revs. P. B. Brodie, Dennis, Fisher, and Blake, Dr. Falconer, Professors Phillips and Buckman, and Messrs. Day, Whitaker, Hulke and Hudleston.

The maps and sections of the Government Geological Survey, executed about 1850 by Mr. H. W. Bristow, are simply indispensable to every student of the rocks.

The Dorset Natural History and Antiquarian Field Club has lately published the first two volumes of its "Transactions," edited by Professor Buckman, and containing valuable geological papers by Messrs. Mansel-Pleydell, Thomas Davidson, Rev. H. H. Wood, S. Buckman, and the Editor. Hutchin's "History of Dorset" may be consulted with advantage, and for the Weymouth district Mr. Damon's little work is useful; in the "Proceedings of the Geologists' Association of London" we find papers by Mr. H. G. Fordham on the "Wealden in Swanage Bay," and by Mr. J. S. Gardner on the "Eocene Strata"; for numerous other facts we are indebted to scattered notices in the "Geological Magazine."

GENERAL STRUCTURE OF THE COUNTY. The surface of Dorsetshire may be divided into four regions, which owe their characteristics to the rocks of which they are composed.

(1). In the north, north-west and west we have the Vale country, consisting of much clay land formed of liassic and oolitic strata and extending to the foot of the chalk hills; this is a fertile region and is one of the many parts of our Isle for which the term "Garden of England" has been claimed.

(2). The chalk forms a tract of country in the centre of the county, and east of it we have (3) sands (with some clay beds) of Tertiary age, forming a rather barren region. Lastly (4), a band of oolitic and cretaceous rocks forms a fringe along the coast, producing beautiful scenery and of high geological interest.

In the main the beds of rock have an inclination, or dip, towards the south-east; it results from this, that the oldest or lowest rocks are those which rise to the surface, or crop out, on the west and north-west of the county. In our description of the beds we shall commence with these older strata and gradually advance eastwards, rising in the geological series to newer and newer formations.

The heights of a few points above sea-level are Pillesdon Pen, 934 feet; Lewesdon Pen, 927 feet; Lyme Regis Guildhall, 28 feet; Charmouth Church, 92 feet; Stanton St. Gabriel Church, 411 feet; Chideock Church, 102 feet; Bridport Market House, 51 feet; Winterborne Church, 323 feet; Dorchester Church, 224 feet; Wareham Old Church, 38 feet; Poole Guildhall, 13 feet; these are taken from the "Abstract of the Levelling by the Ordnance Survey."

THE RHÆTIC BEDS.—Until 1860 these beds were classed with the Lias by English geologists, but their fossils on being closely examined proved to be of different species to those of the Lias, a fact which led to the separation of these Rhætic deposits from the great mass of the Lias above. They consist of greenish marl, black shales and compact whitish limestones, the latter known as the "White Lias"; the total thickness is only about 100 feet. These beds just enter the county near Lyme Regis; their fossils include the shells *Avicula contorta* and *Pecten Valoniensis*, while in the black shales there are one or two sandy layers containing such numbers of scales and bones of fishes and saurians as to be entitled "bone-beds."

THE LIAS.—This well-known formation is magnificently exposed at Lyme Regis, and extends eastwards along the coast past Charmouth and Seatown to Burton Bradstock. The middle and upper beds of this grand coast section have been well described by Mr. Day, but of the Lower Lias here no detailed account has yet been made public. The *Lower Lias* contains in its lower part, where it surmounts the Rhætic beds, many layers of limestone and these yield the fine specimens of saurians and fishes for which Lyme Regis has always been noted; the grand skeletons of *Ichthyosaurus* and *Plesiosaurus* extracted by Miss Mary Anning, Mr. J. W. Marder and Mr. Hawkins, are now deposited in our national museums; these remarkable reptiles must have been when full grown from 9 to 25 feet in length, and are of types unlike any now living; they probably inhabited shallow seas and estuaries, preying chiefly on fishes, as we may judge from the numerous remains, scales, &c., of the latter found in and near the skeletons of the great saurians. The upper portion of the Lower Lias is composed of marls and clays, and the thickness of this division is about 700 feet; it extends eastwards along the valley of the Char, past Marshwood to Pillesdon and Bettiscombe. The most characteristic fossils are the *Ammonites*, different species of which characterise different beds, forming "zones" in the following order from below upwards:—(1), *Ammonites planorbis;* (2), *angulatus;* (3), *Bucklandi;* (4), *Turneri;* (5), *obtusus;* (6), *oxynotus;* (7), *raricostatus.* The fossil fishes, for which Lyme Regis is famous, are believed to occur chiefly in the dark marls of the Bucklandi and Turneri beds.

A small oyster, *Ostrea liassica*, occurs in great numbers just at the junction of the Lower Lias with the Rhætic Beds.

Middle Lias.—This division is, according to Mr. Day, thicker on the coast of Dorset than anywhere else in England, as he estimates it at 500 feet. The lower part is marly, the upper sandy, and of the well-known irony limestone, which, under the name of marlstone, usually forms the top of the Middle Lias, there is only a very imperfect representative. The characteristic ammonites are *A. Jamesoni, Henleyi, margaritatus, spinatus.* Fine sections are exposed at Black Ven, Stonebarrow Hill, Westhay Cliff, Golden Cap, Down Cliffs, &c. The fossils lie very much in particular beds, and unless these are found the chance of collecting good specimens is not great; in the ledges of the lower part of the beach at Golden Cap immense numbers of *Belemnites* occur; a beautiful star-fish, *Ophioderma Egertoni*, is common at the base of a thick bed of sandstone about half way up Golden Cap and Down Cliffs. Inland the Middle Lias extends past Bridport to Netherbury, and thence westwards by Hawkchurch into Devon.

Upper Lias.—This division includes 90 feet of clays surmounted by sands whose exact position in the geological series is rather doubtful; they are above 100 feet in thickness, and were mapped by the Geological Survey as belonging to the Upper Lias, but it now seems probable that they belong in part or altogether to the inferior oolite; it will be best perhaps to adopt the views of Professor Ramsay and consider them, like the Rhætic beds, to be beds of passage linking together the great liassic and oolitic formations; we can trace these sands east of Bridport, past Milton to Beaminster and Burstock, and thence curving round by Crewkerne and Yeovil to Nether Compton. The fossils include *Rhynchonella cynocephala, Ammonites Jurensis, A. aalensis, &c.*

THE OOLITE.—Dorsetshire offers a fine opportunity for the study of the beds of this epoch; they occur in two distinct sets, viz., in the north-west of the county, and has a narrow tract along the south coast.

The Inferior Oolite can be traced from Oborne through Bradford Abbas to Berwick, and thence by Wayford and Beaminster to Burton Bradstock; at the latter point it consists of ferruginous limestones, but further north, at Powerstock and at Ham Hill, near Sherborne, we get thick beds of shelly oolite, which yield an excellent building stone: one stratum at Bradford Abbas, about three feet thick, is so charged with ammonites as to be called the Cephalopoda Bed, but Professor Buckman has lately shown that it does not correspond with the bed of the same name in the Cotteswold Hills, with which it had been confounded, but that its true position is at least 100 feet higher in the strata.

The fossils of the Inferior Oolite of this district include *Ammonites Humphriesianus, A. Parkinsoni, Terebratula Phillipsii,* several species of *Astarte, Trigonia,* &c.; the soil is of a reddish tinge and brashy.

The Fuller's Earth.—This division consists of blue and yellow marls and clays about 400 feet thick, divided about the middle by a bed of rubbly limestone, called the Fuller's Earth Rock, which is largely quarried for lime-burning; this bed only occurs in the northern or vale portion of the county. We can trace the Fuller's Earth running east of Sherborne in a long narrow strip south-west to Clifton Wood; after being interrupted by faults it spreads widely between Halstock and South Perrott, and then curves round by Misterton, and runs north of Beaminster to Toller Porcorum, whence it passes southwards, although interrupted by several faults, until it forms the coast for four or five miles east of Burton Bradstock. The Fuller's Earth is chiefly under pasturage; it does not contain many fossils, but a small oyster, *Ostrea acuminata,* is pretty common. The *Great Oolite,* which, further north, near Bath, furnishes such excellent stone, does not occur in Dorsetshire, being found to thin out and disappear as it is followed southwards.

The Forest Marble.—The beds of this name consist of clays containing layers of thin fissile limestones, altogether 450 feet thick; it is possible, however, that the lower portion may represent the Great Oolite of Gloucestershire. The Forest Marble forms a poor, wet soil, mostly in pasture; we can trace it between Stalbridge Park and Holt Hill, again between Sherborne Park and Lillington, and at Melbury Wood and West Chelborough; in the southern region it extends from Radipole to Langton Herring, and also along the valley of the Bredy. The limestone layers are quarried for flagstones, and at Brotherhampton near Bridport, are burnt into lime; at Long Burton near Sherborne, the limestone has been polished for chimney pieces, and is called "Yeovil Marble." Of the fossils which occur such as *Terebratula maxillata, Rhynchonella concinna,* &c. hardly any are peculiar to this formation.

Cornbrash.—This name implies a rock breaking up, under the influence of the weather, into small irregular fragments, producing a "brashy" soil on which corn grows well: it is a bed of cream-coloured limestone, varying in Dorset from 20 to 40 feet in thickness, and everywhere present in its proper position beneath the Oxford Clay, not only in this county but across England to Yorkshire. Entering on the north, at Stalbridge, it forms a band half-a-mile wide, through Stourton Caundle to Bishop's Caundle: here it is displaced by a fault, and we find it again between Haydon and Long Burton, and again round Yetminster, Ryme, Closeworth and Melbury: in the coast tract it occurs at Puncknowl, and between Radipole and West Chickerel. The fossils are not in fine condition, but the Rev. H. H. Wood has obtained about 180 species from the neighbourhood of Closeworth, including *Avicula echinata, Ammonites Herveyi,* &c.; and Mr. Darell Stephens here found a great rarity — the skull of a crocodile—*Steneosaurus Stephani.* A remarkable point about the fossils is, that many of them are Inferior Oolite species, which are wanting in the intermediate beds, thus giving an example of a recurrent fauna, which is of great interest. According to Professor Buckman, the Cornbrash contains more phosphate of lime than the other Oolitic strata, a fact which would account for its greater fertility.

The Oxford Clay.—This is a very thick bed of bluish clay, containing rounded masses (septaria) of limestone, called pudding-stones, which have been cut and polished at Weymouth, Melbury, &c. A common fossil is a large broad oyster, called *Gryphæa dilatata.* The Oxford Clay forms the low ground on each side of the river Lidden, between Stalbridge and Marnhull, and thence extends in a strip two or three miles wide by Lidlinch, Pulham and Holnest to Chetnole; this is its westward termination, but on the south we find it again in the upper part of the Bredy Valley, and also between Melcombe Regis and East Fleet: its thickness is perhaps 700 feet: being heavy, and difficult to cultivate, the land is mostly in permanent pasture. The hard bed found elsewhere at the base of the Oxford Clay, and known as the *Kellaway's Rock*, has not been noted in Dorsetshire.

Coral Rag.—In North Dorset, we see, above the Oxford Clay, some beds of oolitic and rubbly limestone, associated at the top and bottom with sands and marls, altogether about 100 feet thick. They stretch from Silton, by Stower, Marnhull, Sturminster Newton, Fifehead Neville, Haselbury Bryant, to Mappowder, Glanville Wooton and Hermitage, whence they are overlapped by the cretaceous rocks. There is a fine section of the strata shown by the Sturminster railway cutting, and the lower beds yield a good oolitic building-stone at Marnhull and Todbere.

At Weymouth the Coral Rag is finely exposed on the coast, and stretches across to Wyke Regis: it also runs as a long narrow band from Ringstead Bay and Radcliff Point, by Jordon Hill and Broadway, to Abbotsbury. At the latter place is a thick local deposit of iron-ore (hydrated peroxide).

The characteristic fossils are spines of a sea-urchin—*Cidaris florigemma*—with such shells as *Trigonia clavellata, Ammonites cordatus*, &c.; corals are frequent in Ringstead Bay only.

Kimmeridge Clay.—This formation takes its name from a little Dorsetshire village on the south-west side of the Isle of Purbeck; beds of the same age cap the Coral Rag in Ringstead Bay and extend thence westwards by Upway Street to Portisham; they also occur in the northern part of the Isle of Portland. In North Dorset the Kimmeridge Clay has an outcrop four miles wide at Gillingham and Motcombe, and thence passes southwards by West Orchard, Hammoon and Shilling Okeford to Melcombe Park. South of Weymouth, about 20 feet of sandy clays and grits are seen, containing *Ostrea deltoidea, Trigonia Meriani*, &c.; these are the Passage beds of Mr. Blake; in Ringstead Bay about 400 feet of blue sandy clays, with nodular limestone concretions, constitute the *Lower* Kimmeridge; the common shells are *Exogyra virgula, Thracia depressa, Ammonites biplex*, &c. The Upper Kimmeridge beds are 650 feet thick in Kimmeridge Bay, and consist of bituminous shales and cement-stones; these contain many fossils, as *Discina latissima, Cardium striatulum*, &c., and all the large bones of saurians, which have been described by Professor Owen, Mr. Hulke, and others, have come from this division; one paddle of a Pliosaurus, now in the Dorchester Museum, measures 6 feet 9 inches in length. Some of the shales are so bituminous as to be used in the neighbourhood instead of coal, when the latter substance is dear; attempts, too, have been made to distil mineral oils, and to make alum and gas, from these shales, but not, hitherto, with any great success. There is a good deal, too, of the yellow mineral called iron pyrites, and the spontaneous decomposition of this substance at Ringstead Bay, in 1826, produced so much heat as to inflame the shales, which continued to smoulder for several years. From the same cause similar ignitions have occurred in the lias beds at Charmouth, usually after heavy rains, or when the sea has dashed over the cliffs.

The Portland Beds are called after the isle, or rather peninsula, of that name: they consist of 70 feet of grey and yellow sands, surmounted by an equal thickness of shelly and oolitic limestone, containing nodules and bands of chert, an impure variety of flint: besides the locality from which they take their name, we find these beds in the Isle of Purbeck, circling round from St. Alban's Head by West Hill to Gad Cliff: they fringe the coast from Mewp's Bay to Durdle Door, and are also traceable between Osmington and Portisham: they are not visible in North Dorset, being overlapped and concealed by cretaceous strata, but they peep out again at Tisbury, in Wiltshire. The upper limestone bands yield some of the best building stone in England: the *roach bed* lies near the top; it is shelly and not suitable for fine work: the best stone is that of the *whit bed*, lower down, though it is not so thick. St. Paul's Cathedral and many of the churches and other public buildings of London, especially those erected in the reign of Queen Anne, are built of Portland oolite.

The fossils mostly occur as casts; *Cerithium Portlandicum* (the "Portland screw"), *Ammonites giganteus*, and *Cardium dissimile* may be mentioned; yellow crystals of sulphate of barytes are not uncommon.

The Purbeck Beds.—These consist of limestones, clays, and marls, altogether 300 feet thick: in the Isle of Purbeck they extend from Swanage to Worth Matravers, and thence in a narrow band through Kingston to Worbarrow Bay: as a narrow contorted band they are visible between Mewp's Bay and Man-of-War Cove: in the Weymouth district they cap the Isle of Portland on the south, and are found on the north between Osmington and Poxwell, and west of Upway. In 1850 Professor E. Forbes divided the Purbecks into Lower, Middle, and Upper: the Lower Purbeck contains "Dirt Beds," which are the remains of old land surfaces composed of vegetable soil, and containing stumps of trees, named *Mantellia*, which are allied to the living Cycads: in the Middle Purbecks we have "Cinder Beds," composed of vast masses of oysters: the upper beds are freshwater limestones full of *Paludina* and other freshwater shells: these limestones were formerly polished and used, under the name of "Purbeck Marble," to make the slender shafts in Gothic churches. Remains of insects are frequent in the Purbecks, and *Cyprides* (small freshwater crustaceans) are also very abundant. In 1854 Mr. W. R. Brodie found remains of small mammals in the Middle Purbecks, at Durlston Bay, and in 1856 the energetic researches of Mr. S. H. Beckles exhumed fragments (chiefly lower jaws) of 25 species from the same bed in the suburbs of Swanage. These have been described by Professor Owen as marsupials of small size, resembling the Kangaroo Rat of Australia.

The Purbeck strata close the great Jurassic series, as the Lias and Oolites are termed when considered as a whole: the nature of the sediment and the fossils they contain show them all to be of marine origin, until we come to the very top, where in the Purbeck Beds we have evidence of freshwater, probably estuarine, conditions.

NEOCOMIAN, OR LOWER CRETACEOUS PERIOD. The grits and sandy clays which lie conformably on the Purbeck series, are also of freshwater origin, and are known, the lower part as the *Hastings Beds*, and the upper division as the *Weald Clay*: they occupy the whole recess of Swanage Bay and extend due west to Worbarrow Bay, which is also excavated in them: their dip is due north at a high angle: they thin westwards in a remarkable manner from 1,800 feet at Swanage Bay to 725 feet at Worbarrow Bay, 660 feet at Mewp's Bay, 462 at Lulworth Cove, and 172 at Man-of-War Cove: inland they can be traced from the north side of East Chaldon to Osmington. In Swanage Bay bones of two gigantic reptiles, the *Iguanodon* and *Megalosaurus*, have been found, and here, too, Mr. Beckles has found gigantic footprints, 15 inches long, of these or some other huge monsters.

The Punfield Series was named by Professor Judd from the place of that name, on the north side of Swanage Bay. Here we have about 200 feet of laminated shales and sands containing bands of limestone and ironstone: these contain brackish water and marine fossils, showing a depression of the Wealden area which admitted the waters of the open sea.

The Lower Greensand consists of grey clays and sands about 60 feet thick, resting on the Punfield bed, and containing *Exogyra sinuata*, &c.

CRETACEOUS FORMATION.—Of the lowest division of this series the *Gault* —there is a poor representative seen in Punfield Cove: it is a dark sandy clay with few fossils, and is about 40 feet thick: it has been traced westward to Lyme Regis, where, at Black Ven, it rests upon the Lias. On the west side of Shaftesbury the Gault again crops out: here it is thicker and assumes its normal character of a blue micaceous clay.

The Upper Greensand varies from 100 to 200 feet in thickness: it is a yellowish-brown sand, containing beds of sandstone and layers of chert, and near the base is speckled green by silicate of iron: it runs everywhere at the base of the chalk hills, except where cut off by faults; from Punfield it passes westwards through East Chaldon, Abbotsbury Castle and Askerswell to Cheddington, then curves north-east by Maiden Newton, Evershot, Melcombe, Bingham, and Sutton Waldron to Shaftesbury, east of which town it has a broad outcrop reaching to Berwick St. John; generally its outcrop is less than half-a-mile wide. Many outlying patches of the Upper Greensand form the capping of the hills in West Dorset; among these are Eype Down, Golden Cap, Stonebarrow Hill, Haddon Hill, Lewsdon Hill, Pillesdon Pen, Conic Castle, Thorncomb, &c. The great unconformity of the cretaceous series to the underlying strata is well shown by this overlap of the Upper Greensand upon bed after bed of the lower rocks. The Upper Greensand is an excellent water-bearing stratum, the rain which has percolated through the chalk being arrested by the clayey base of the sands: hence springs issue all along its outcrop, and the sites of villages and towns have been determined by it. Near Shaftesbury a bed of sandstone is quarried for building purposes. Fossils are not numerous, but a branching sponge and the shells *Pecten asper*, *P. quadricostatus*, &c., show the strata to be of marine origin.

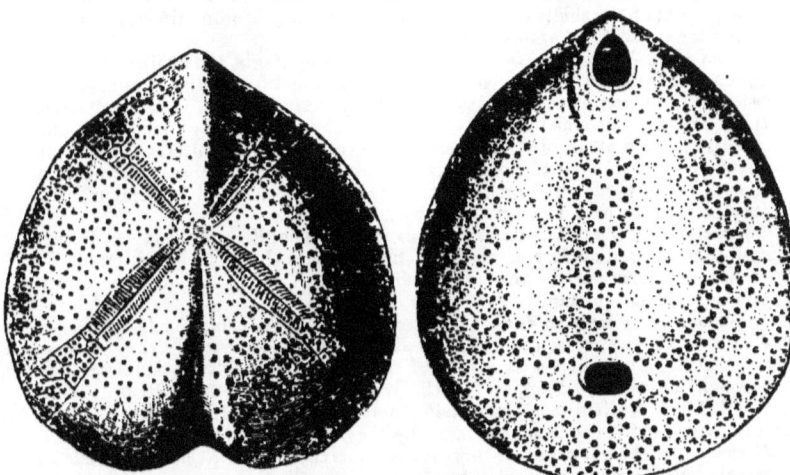

Fig. 27.—Upper Surface of *Micraster coranguinum*, a sea-urchin from the chalk.

Fig. 28.—Lower surface (showing the mouth and vent) of *Ananchytes ovatus*, a sea-urchin from the chalk.

The Chalk.—All the ordinary subdivisions are recognisable in Dorsetshire. First we have the *Chalk Marl*, about 60 feet thick, greyer than ordinary chalk, and containing a slight admixture of clay; at its base is a bed about 18 inches thick, containing phosphatic nodules and many fossils; this layer is known as the *Chloritic Marl*; next comes the *Lower Chalk*, or chalk without flints, from 50 to 150 feet thick; its top is the chalk-rock, a hard thin band of cream-coloured chalk; the *Upper Chalk*, or chalk-with-flints, is 600 feet thick on the east of Dorset, but it thins westward.

As a whole the chalk forms a broad band 10 miles in width, extending from Cranborne and Cranborne Chase to the south-west, past Blandford, Sturminster Marshall, Bere Regis, Piddletown, Hilton, Cerne Abbas, and Dorchester. The Downs above Beaminster are separated from the main mass, and other outliers occur further west near Mosterton, Cricket St. Thomas, &c.; along all the line now mentioned the dip is gentle, and to the south-east.

On the south coast the chalk has been upheaved so as to stand nearly vertically from Durdle Cove by Bindon Hill, Purbeck Down, and above Corfe

Castle to Ballard Down, where the Foreland marks its present termination, although originally it stretched rightacross to the Needles in the Isle of Wight.

In Dorset the chalk presents its usual characteristics—high, bare, and rounded hills covered with short herbage, but possessing little or no soil, dry valleys and few trees. On the southern slope of the Down at Owermoigne, facing Weymouth, a gigantic figure of George III. on horseback, 174 feet in height, has been made by cutting away the turf; near Cerne there is a similar figure of a giant holding a club More than half a million sheep still feed on the Dorset Downs, but a good deal of the land has of late years been brought under plough. The chalk may be examined in many pits where it is quarried for lime burning: the flints are used for road mending. The fossils of the lower chalk include the brachiopod shell, *Rhynchonella Cuvieri*, and several species of *Inoceramus*; in the Upper Chalk we find the rounded bodies of many sea-urchins, the commonest species being *Micraster cor-anguinum* (fig. 27) and *Ananchytes ovatus* (fig. 28), with the shell *Spondylus spinosus*, *Rhynchonella octoplicata*, fish-teeth, &c. The chemical composition of chalk is nearly pure carbonate of lime, and the microscope shows it to be composed in large part of the shells of microscopic animals, called foraminifera, belonging to the lowest scale of animal life.

Land-slips.—Along the Dorset coast, near Lyme Regis, the chalk and Upper Greensand rest on Rhætic and Liassic clays, and dip slightly towards the sea, while numerous springs issue at the junction of the sands and clays. Here are all the conditions favourable to the production of land-slips, and they frequently occur, and are sometimes of great magnitude. In 1839, about 22 acres of land at Bindon slipped down towards the sea, producing a chasm 1,000 yards in length by 300 in breadth: a model illustrating this may be seen in the Jermyn Street Museum, London.

TERTIARY EPOCH: THE EOCENE FORMATION.—The Eocene beds of Dorset rest on an eroded surface of chalk. They occur in the eastern portion of the county, and form part of the "Hampshire Basin," which includes the strata of the same age in Hants and the Isle of Wight.

Woolwich and Reading Beds.—These consist of sands, pebble-beds, and clays, with sometimes a layer of green-coated flints at the base where they rest on the Chalk. They form a narrow band from the south side of Studland Bay through Lulworth, Wool, Winfrith, Warmwell, Piddletown Heath, Bloxworth, Corfe Mullen, and Edmonsham, their outcrop being from half-a-mile to a mile wide, and it can be often traced by the oak trees which grow well on this formation. The old name for this division was the "Plastic Clay," and red and mottled clays are of frequent occurrence; at Crendle Common, near Cranborne, coarse pottery is made from a bed of this kind; fossils are wanting, except that leaves are found in a pale grey sandy pipe-clay near East Bloxworth.

Outliers of the Reading Beds occur on the chalk at Bincombe Down, and on Black Down above Portisham, &c. On Affpuddle Heath and Piddletown Heath, east of Dorchester, large natural pits occur, like inverted cones, measuring from 30 to 100 feet in diameter; these have been formed by the removal of the underlying chalk by water charged with carbonic acid gas, the upper sands, &c., have then fallen in.

The London Clay (or Bognor Beds) forms a similar surface feature to its Reading Beds, which it closely accompanies along the line just noted; the outcrop between Wimborne Minster and Fordingbridge is rather wider, occupying from one to two miles at the surface; it consists of brown sandy clays containing irony concretions and flint pebbles.

The Lower Bagshot Beds form all the land round Poole Harbour, reaching westwards through Wareham and Moreton to Woodsford, and stretching east and north-east across the boundary of the county by Kingston and Woolbridge; they are finely exposed in cliffs along the coast from Studland to Bournemouth and eastwards. They consist of sands of varied colours, containing beds of pipe-clay and brick-clay. Mr. Gardner estimates the total thickness at 1,000 feet. The clays are largely dug at Wareham, Poole, and elsewhere. In 1877 there were exported from Poole Harbour 56,345 tons of clay, most of it for British potteries in Staffordshire, &c., but 4,971 tons went to foreign ports; 16,570 tons of clay were also used at works in the district, where ornamental tiles, tesselated pavements, &c., are largely made; in 1879 the quantity of clay exported from Poole had risen to 67,019 tons. The clays contain many remains of plants, leaves, &c., indicating a rather warm climate.

It is not known when the Poole clay was first worked, but there is an "Order in Council," as far back as 1666, directing that no dues were to be paid on "tobacco pipe-clay" at Poole.

SURFACE DEPOSITS.—The chalk is in many places thickly covered by flints, the debris of higher beds of chalk which have been denuded by weathering agencies. There are thick gravels, too, on the table land between Poole and Bournemouth, containing many pebbles of quartz, granite, &c., which have probably been brought from the old Cornish rocks of the west; these are the high-level gravels of Mr. Codrington, and he considers that they were formed in the valley of a river running from west to east, and previous to the separation of the Isle of Wight from the mainland.

Low-level gravels—*i.e.*, such as are not more than 40 or 50 feet above the existing rivers, occur in most of the deep combes and along the sides of the valleys. In these remains of the extinct mammoth *(Elephas primigenius)* and of *Rhinoceros tichorhinus* have been found, especially in the gravels of the Stour, the Piddle and the Avon.

Evidence of an elevation of the coast line is afforded by the well-known raised beach at the southern end of the Isle of Portland; it is composed of sea-worn pebbles, at its highest point is 53 feet above sea level, and contains shells such as inhabit the shores of the English Channel; in Portland, too, many bones have been found in caverns and fissures in the limestone. At many places great masses of a coarse sandstone, called Druid or Sarsen stones, or grey wethers, are found lying on the surface; one called the Agglestone, near Studland, is computed to weigh 400 tons; these are now believed to be the remnants of some hard bed of the tertiary series, probably in the Bagshot Beds, which formerly extended far to the west of their present limits. Of glacial action, no undoubted signs have yet been observed in Dorsetshire.

Masses of blown sand, forming low mounds, occur on the coast in Portland Roads, and at the entrance of Poole Harbour; a famous mass of pebbles— the Chesil Beach—stretches for 16 miles between Portland and Burton Bradstock, the stone diminishing regularly in size from four inches in diameter at Portland to little more than sand at Bradstock.

PREHISTORIC MAN.—Of the earliest kind of stone implements which are assigned to the *Palæolithic Stone Age* only two specimens have yet been found in Dorsetshire, one from the gravels at Wimborne Minster, and another at Dewlish, three miles north of Piddletown; both are of flint. Of the later or *Neolithic Period* many specimens have been found; these are mostly chipped out of flint and include celts or axe-heads, arrow-heads, flakes or knives, hammers, scrapers, &c. They occur in the surface soil, or in the mounds which are the burial places of the early tribes who inhabited this country. Many are found in and near the famous entrenched camps on the chalk hills at Badbury Rings, Hod Hill, Maiden Castle, &c., and at Afflington, Bradford Abbas, Poundbury, Ridgeway Hill, &c. In a barrow at Afflington a ring of Kimmeridge shale was found with a bronze ring, flint-flakes and arrow-heads, a perforated whetstone, and glass and bone beads; at Porrington a shallow one-handled saucer or stand of Kimmeridge shale was found; at Winterbourne Steepleton a perforated axe-head, made of greenstone, was found in a barrow with some burnt bones. All these relics belong to a period which closed some thousands of years before the Christian era, for when Cæsar landed in Britain the inhabitants had long been acquainted with the use of iron.

No. 11.

GEOLOGY OF DURHAM.

NATURAL HISTORY AND SCIENTIFIC SOCIETIES.

Bishop Auckland Naturalists' Club.
Natural History Society of Northumberland and Durham; Newcastle-on-Tyne. Transactions.
North of England Institute of Mining and Mechanical Engineers; Newcastle-on-Tyne. Transactions.
Seaham Natural History Club.

MUSEUMS.

The Museum, Durham.
Athenæum Museum, Sunderland.

PUBLICATIONS OF THE GEOLOGICAL SURVEY.

Maps.— 103 N.W. Wolsingham; 105 S.W. Gateshead; 105 S.E. South Shields, Sunderland. (Survey not completed.)

IMPORTANT WORKS OR PAPERS ON LOCAL GEOLOGY.

1857. Kirkby, J. W.—Permian Fossils from Durham. Journ. Geol. Soc., vol. xiii. p. 213.
1858. Bate, C. Spence.—Permian Crustacean and Amphipod. Journ. Geol. Soc., vol. xv. p. 137.
1859. Kirkby, J. W. Permian Chitonidæ. Journ. Geol. Soc., vol. xv. p. 607.
1860. Kirkby, J. W.—On *Lingula Credneri* in Coal-measures of Durham, and on Permian System. Journ. Geol. Soc., vol. xvi. p. 412.
1864. Kirkby, J. W.—Permian Fish and Plants. Journ. Geol. Soc., vol. xx. p. 345.
1868. Tate, Geo.—Geology of Durham. Nat. Hist. Trans. of Northumberland and Durham, vol. ii. p. 1.
1876. Clough, C. T.—Section at High Force, Teesdale. Journ. Geol. Soc., vol. xxxii. p. 466.
1878. Clough, C. T., and Gunn, W.—Silurian Beds in Teesdale. Journ. Geol. Soc., vol. xxxiv. p. 27.
1880. Clough C. T.—Whin Sill of Teesdale. Geol. Mag., p. 433.

See also *General Lists, p. xxv.*

THE important economic products of the rocks of this county have led to their close examination and to repeated discussions as to their relations and origin, in consequence of which our knowledge of the structure of the district is tolerably full and complete. The Government Survey has published geologically-coloured maps of the northern part of the county, and the survey of the remaining portion is now being executed by Messrs. Cameron, Burns, &c., under the direction of Mr. Howell. An excellent horizontal section running from Allenheads on the west to near Sunderland on the east, and on a true scale of six inches to a mile (the work of Messrs. Howell and Burns), has also been published by the Geological Survey. Many valuable papers have been issued by the Tyneside Naturalists' Field Club and by the North of England Institute of Mining Engineers. For our knowledge of the Carboniferous rocks, we are largely indebted to Messrs. Hutton, Winch, Buddle, Wood, and Professor Lebour, whilst Sedgwick, King, Kirkby, and Howse have especially studied the Permian Formation. When the Government Survey is completed, its officers will no doubt issue a most full and minute account of the rocks of the district, but this cannot be expected for some three or four years. There is a good general account of the coal-field, with a map, in Professor Hull's " Coal-fields of Great Britain ;" the Report of the British Association Meeting at Newcastle (1863) contains some information; the other papers are either in the Geological Society's Journal or the Geological Magazine, with the exception of Professor King's account of the Permian fossils (with plates), which was published in 1850 by the Palæontographical Society.

The geological structure of the county of Durham is, considered broadly, simple enough. The beds of rock run from north to south, and they have a general dip, or slant, or inclination to the east. The oldest rocks occur in the west, forming part of the Pennine Chain, the highest point reached being 2,196 feet at Killhope Law. Byerhope Head is 1,720 feet, and Knucton West Fell 1,770 feet above the sea. The exact altitude of a few other points fixed by the Ordnance Survey may also be given: Barnard Castle (Church), 520 feet ; Staindrop (Church), 360 feet ; Ferry Hill (Windmill), 587 feet ; Darlington (Market Cross), 154 feet ; Stockton-on-Tees (New Church), 47 feet ; Durham Cathedral, 215 feet ; Chester-le-Street (Church Tower), 73 feet ; and Gateshead (Trinity Church), 154 feet.

In describing the various beds of rock, whose upturned edges form the county of Durham, we will, then, begin with the oldest rocks, which, as stated above, are those which form the high and barren moorlands constituting the eastern slope of the Pennine Chain.

SILURIAN FORMATION.—In the upper part of Teesdale, near Cronkley Scar, shales of Silurian age have lately been detected near the Old Pencil Mill. They are crossed by quartz veins and by several igneous dykes, and probably extend for some distance under the boulder clay. They lie just on the west or upthrow side of the Burtree Ford Fault.

THE CARBONIFEROUS FORMATION.—This is the name applied to a great series of beds, which further south, in Yorkshire, Derbyshire, &c., admit readily of a triple division into (1) Carboniferous Limestone at the base, (2) Millstone Grit in the centre, and (3) Coal-measures on top. But as we follow these beds northwards very considerable changes take place, and it becomes a matter of great difficulty to adapt the strata of Durham and Northumberland to this classification. When we get into Scotland we find the base of the Carboniferous Formation to be, not a great mass of limestone some thousands of feet thick, such as exists in Derbyshire, but a set of beds of sandstone and shale with workable seams of coal, to which the name *Calciferous Sandstone Series* is applied. As Durham lies midway between these two districts, we might expect to find in it an intermediate state of things, and this would in the main be true. In this county, however, we have not a complete series of the lower Carboniferous strata. The lowest beds occur outside the county, on the top and western slope of the Pennine Chain.

The Bernician Beds.—This is a term applied by Professor Lebour to beds in Durham and Northumberland which represent both the *Scar Limestone* of Professor Phillips and his *Yoredale* series : it was independently applied to the same beds as long ago as 1856 by the late Dr. S. P. Woodward. The term is derived from Bernicia, the ancient name for Northumberland.

If, however, we adhere to the older classification of Phillips, then the sheet of basalt, known as the *Great Whin Sill*, does doubtless in Teesdale and the neighbourhood form a useful line of division. All the beds below it may here be termed *Scar Limestone*, and those above it up to a bed called the "Fell-top limestone" would constitute the Yoredale series. The total thickness of these beds is about 2,500 feet, of which 1,500 feet belong to the Scar Limestone, and 1,000 feet to the Yoredale series. The various beds of limestone may be traced for considerable distances, and are distinguished by local names. Thus, commencing with the lowest in the Yoredale series, we have the "Tyne-bottom Limestone," resting on the Great Whin Sill, next comes the "Scar Limestone," then the "Five-yard Limestone," then the "Three-yard Limestone," and next the "Four-fathom Limestone;" this last is a very well-known bed, and is distinguished by the presence of the fossil *Saccamina Carteri*, a foraminifer of which the rounded segments are sometimes as much as one-eighth of an inch in diameter, and which stand out in relief on the weathered surfaces of the stone. It has been fully described and figured by Mr. H. B. Brady, F.R.S. (Carboniferous Foraminifera, Palæontographical Society, 1876). Higher up, we have the "Great limestone," which is a very continuous bed, from 30 to 40 feet thick, and is largely quarried. Fifty feet higher comes the "Little limestone," which is remarkable for its underlying seam of Coal. Then comes a considerable thickness—perhaps 350 feet—of shales and sandstones, above which we find the highest limestone band, called the "Fell-top limestone," which also has a thin Coal-seam beneath it. The sandstone-beds, which occur between the limestone, are locally termed *hazle*, and the bituminous shales *plate*. Fossils are not uncommon. In the shales plant-remains (such as form the Coal-seams) are well seen, and in the limestones marine shells like *Productus giganteus* (the "large cockle" of the miners) abound. The limestones are largely quarried in Weardale, being sent to the iron furnaces, where they are used as a flux to aid in the melting and removing of impurities from the iron. Near Frosterley, a bed called the Bishopley limestone yields an excellent building stone.

The lead veins which traverse the limestone beds, are also of great commercial importance; they usually run east and west, and appear to have originally been fissures, or cracks in the stone, which have been filled up with *galena* (sulphide of lead) and other minerals deposited from heated waters traversing these fissures. In 1875, the total amount of lead ore raised in Durham and Northumberland was 22,304 tons; this yielded 16,525 tons of lead, and 70,191 ounces of silver; this was principally raised in the Allendale, Weardale, and Teesdale mines. In 1879 there were raised 14,187 tons of lead ore, valued at £150,000. Valuable deposits of iron-ore (spathose carbonate and siliceous hematite) are also worked. In 1875, Weardale produced 34,829 tons of iron-ore, worth £21,626; but in 1879 only 16,679 tons were raised, value £10,007.

The Millstone Grit.—This division is about 500 feet thick in Durham; it includes all the beds of grit, sandstone, &c., between the Fell-top limestone and the Brockwell coal-seam; its main outcrop runs from Barnard Castle on the south, by Wolsingham and Stanhope to Shotley and Sicaley, but detached patches of it occur further west, on the most conspicuous hill-tops or fells. Fossils are few, chiefly plant remains, with a few marine shells, such as *Orthis, Spirifera,* &c. The sandstones are quarried to make into millstones or grindstones. "At Carr's Craggs millstones are lying, cut by Simpson, of Langdale, of very large size, and there are stones at present which may be cut to any size as far as 13,000 solid feet, or 39 feet in length." The uppermost beds of the division consist of fine-grained siliceous clays, to which the term *gannister* is applied.

The Coal Measures.—Durham contains the most important coal-field in the British Islands. On the west it is defined by the outcrop of the millstone grit, as already explained, while on the north it passes into Northumberland; to the east the coal-seams extend under the sea to a distance of perhaps 10 or 12 miles, rising at last in that direction to form part of the sea-floor, just as they rise and crop out in the contrary direction in the west; southwards, too, the various beds rise up in the same manner. Staindrop marks the south-west corner, and a line from this place to Seaton Carew, north of the mouth of the Tees, would mark the southern limit of the Coal-measures underneath the newer, later-formed Permian and Triassic strata. The fact that the coal-seams passed under and could be got by sinking through the magnesian limestone, was first pointed out at Haswell, by William Smith, the "Father of English Geology," in 1821.

At the time of its formation, the Durham coal-field was probably continuous with those of Yorkshire on the south, and of Whitehaven and Lancashire on the west; but at the close of the Carboniferous period, Professor Hull has shown that the crust of the earth in England was thrown into long east and west folds, one such fold running across from the mouth of the Tees to Morecambe Bay; from the top of this fold the coal-measures were in time denuded off by weathering influences, the sea, &c.; then subsidence followed, and the Permian beds were deposited. At the close of the Permian epoch terrestrial movements again occurred, but now along north and south lines, resulting in the elevation of the Pennine Chain, and dividing the coal-fields of Yorkshire, Durham, &c., from those of Lancashire and Whitehaven. Thus the present basin-shaped form of most of our coal-fields is the result of the action of subterranean forces along lines at right-angles to each other. It follows that it would be hopeless to bore for coal anywhere south of the line we have mentioned (from Staindrop to Hartlepool) to as far at least as York.

The total thickness of the Coal-measures of Durham is about 2,000 feet, and this is composed of fine or coarse-grained sandstone (locally termed *post*), interstratified with beds of shale *(metal)*, and about one hundred seams of coal, varying in thickness from a few inches to several feet. The Brockwell seam (3 feet) is the lowest important one; then come the Three-quarter (2½ feet), the Five-quarter (3 feet 4 inches), the Stone coal, Jelly coal, Townley or Harvey coal, Ruler coal, Five-quarter, Crow coal, and the Hutton or Low-Main seam. This last seam is 6 feet thick, and in Durham is moderately soft, and excellent for household use and coking; higher still we have the Five-quarter, the Bensham coal (6 feet thick, and extensively worked under the magnesian limestone near Sunderland), the Yard coal, Stone coal, Metal coal, and the High Main coal. This last was the original Wallsend seam, but is almost worked out; towards the valley of the Wear it is split into two seams by the intercalation of sandstone and shale.

The yearly output of coal is very large; for South Durham, in 1875, it amounted to 19,456,534 tons from 177 collieries; and for North Durham and Northumberland, 12,640,789 tons from 170 collieries. Of fire-clay got from the beds just beneath every coal-seam, a large quantity is also raised—603,226 tons in 1875; it is used for the manufacture of seggars, or for any purpose where capacity to resist great heat is required. There are also beds of ironstone, generally called mussel-bands, because they are full of the mussel-like shell, called *Anthracosia*. Of iron pyrites the total quantity obtained in Northumberland and Durham in 1875 was 1,520 tons, valued at £760. In 1879 the output of coal had decreased in South Durham to 17,306,482 tons; while of fire-clay, only 286,407 tons were raised.

In the shales, remains of plants are of frequent occurrence; ferns named *Sphenopteris*, *Neuropteris*, &c., portions of gigantic reeds called *Calamites*, the stems of giant club-mosses, called *Sigillaria* and *Lepidodendron*, which attained the dimensions of large trees, although no living club-moss exceeds 3 feet in height; these composed the chief part of the vegetation of the coal-period. The coal-seams themselves are formed of the matted remains of the same plants, squeezed into a small part of their original bulk. Fish remains are common in a thin shale bed which forms the roof of the Low-Main seam, but only one marine shell is known, *Lingula mytiloides*, and this appears dwarfed or stunted, as if it had lived under unfavourable conditions. All things combine to prove that each coal-seam grew as a forest on low marshy flats near the sea, the "under clay," full of rootlets (called *Stigmaria*, they belonged to the plant whose stem is named *Sigillaria*), being the soil. In these swamps a great thickness of vegetable matter accumulated, when, the land slowly sinking below the sea-level, mud and sand were deposited upon it, on which new forests grew; so seam after seam was formed.

In 1871, before the Royal Coal Commission, Mr. Forster estimated the coal which may still be got in Durham, at 3,738,750,000 tons, which at the present rate of consumption will be exhausted in less than 200 years. But if coal can be worked under the sea to a distance of $3\frac{1}{2}$ miles from the coast, a further quantity of 734,500,000 tons may be obtained.

Faults.—The coal-field is intersected by several dislocations or faults, which interrupt the continuity of the seams; they nearly all run from east to west, the best-known being the "Butter Knowle Dyke," which runs across the southern part of the basin. The strata to the south of it are 700 feet lower than those on the north, and thus all the coal seams from the "Five-quarter" downwards, are brought in again under the Permian rocks, as proved at Leasingthorne, Black Boy, and Eldon collieries. The Burtree-ford dyke runs north and south along the moorland country between the upper courses of the Tyne and Tees: it affects the Yoredale rocks only.

IGNEOUS ROCKS.—In Durham those rocks which have been melted or are "fire-formed," are intimately connected with the Carboniferous Formation; indeed most of them seem to have been injected into the positions which they now occupy at about the close of that period. *The Great Whin Sill* is a well-known bed of basalt which forms the bed and bank of the Tees for several miles above Middleton: it was plainly thrust into its present position, for it has baked or altered the beds which rest upon it as well as those upon which it rests. *The Cockfield Dyke* runs past that place to Bolam and crosses the Tees at Preston, whence it can be traced to near Scarborough. A similar dyke runs by Hamsterley to Sunderland Bridge. In the north of the county the *Hebburn Dyke* runs across the Tyne near that place. The *Little Whin Sill* is an interstratified basaltic mass which occurs in the Three-yard limestone of Weardale, near Stanhope. These basaltic masses are quarried at several points and yield good road-metal.

THE PERMIAN FORMATION.—These rocks derive their name from the kingdom of Perm, in Russia, where they were examined by Sir R. I. Murchison. They enter Durham near Pierce Bridge, and their outer edge runs past Midderidge and Boldon to South Shields. Along this line they form a well-marked escarpment with a steep slope to the west. On the east they pass under the Triassic rocks, but the junction is so overspread by gravel, boulder-clay, &c., that it is not easy to trace: its general direction, however, is along a line drawn from near Pierce Bridge to Hartlepool.

Five subdivisions have been made out in the Permian rocks of Durham. At the base, resting unconformably on the coal-measures, we have the *Lower Red Sandstone:* it is about 100 feet thick, incoherent and false-bedded. It is seen in quarries at South Shields, Claxheugh, Newbottle, &c. This sandstone yields an almost inexhaustible supply of pure water, with which the towns of Sunderland, South Shields, Jarrow, and Seaham are supplied. It is a source of much trouble in sinking colliery shafts. The plant-remains which occur, such as ferns, calamites, &c., are all of species also found in the Coal-measures beneath. It is indeed more than probable that these red sandstones will have to be included with the carboniferous system, but further researches are needed to finally settle this point. On these sands rests the *Marl Slate*, a thin yellowish laminated deposit only 3 or 4 feet thick. It is, of course, not a true slate; but the name is commonly applied by workmen to any tolerably hard stone which will split into thin layers. At Ferry Hill, Midderidge, Hartley's quarry, Claxheugh, &c., some very finely preserved fossil fishes have been obtained from this bed of the genera *Palæoniscus, Platysomus,* &c., also traces of plants, and in the south of the county a few fossil shells.

The Magnesian Limestone.—The three remaining subdivisions constitute the Lower, Middle, and Upper beds of this well-known rock: the total thickness is about 400 feet: at many points the rock is a true dolomite, *i.e.* composed of equal proportions of carbonate of lime and carbonate of magnesia, and the stone is then of a yellowish hue. At Marsden it is largely quarried and sent to Sunderland: there it is treated with sulphuric acid, the magnesia is dissolved out, and from the liquor obtained Epsom salts (sulphate of magnesia) readily crystallizes out: a considerable portion of the Epsom salts now sold is obtained in this way. The Upper Limestone is well seen in Roker Cliffs, on the coast north of Monkwearmouth, and in the quarries on Fulwell Hill: it is remarkable for its concretionary structure. The Middle beds can be best studied at Humbleton and Tunstall Hills, and the Lower Limestone at East Thickley, Whitley, &c. Fossils occur most frequently in the Middle Limestone: they are chiefly such marine shells as *Productus horridus*, *Spirifera alata*, &c. Bones of labyrinthodonts have been found in the Midderidge quarries by Joseph Duff, Esq.

The Trias, or New Red Sandstone.—This forms the low ground of the south-east corner of the county, on which Darlington, Stockton, Sedgefield, and Seaton Carew are situated. Red marls with gypsum occur, and beds of reddish sandstone are seen on the coast near Seaton Carew and in the banks of the Tees, but there are few good sections. In a boring on the opposite side of the Tees at Middlesborough, Messrs. Bolckow and Vaughan went through 1,300 feet of red marls and sandstones, and left off in a thick bed of rock-salt; so there can be no doubt of the great thickness of the triassic beds in this district. At Oxenhall, about three miles from Darlington, are some large natural pits known as "Hell Kettles": they vary from 114 to 75 feet in diameter, and appear to be due to the sinking in of the surface, in consequence of the dissolution of beds of gypsum (which is soluble in water) which lie underneath. No fossils occur in these triassic strata, which appear to have been formed in great salt lakes or inland seas.

Glacial Deposits.—There are clear traces in Durham of the former existence of glaciers. A great ice-sheet seems to have come down from the Cheviots and Pennine Hills, smoothing and scoring the rocks it passed over, and finally, as it melted, leaving behind a thick deposit of blue or grey boulder clay, full of fragments of old rocks, such as granite, porphyrite, basalt, limestone, &c. Grooved and polished surfaces may be seen at the Trow rocks, near South Shields, and at Ryhope Snook, south of Sunderland. Boulder clay is well exposed on the coast at Hendon and the Blue House, south of Sunderland, but it extends more or less over the entire coal-field. Beds of sand and gravel, usually false-bedded, rest on the boulder clay, and are consequently of later date. They probably indicate a depression of the land when the boulder clay was in part washed up and the materials resorted by marine currents.

Recent Deposits.—Along the coast peat-beds or "submarine forests" occur at Whitburn, West Hartlepool, &c.: they have yielded antlers of the red deer and Irish elk, a tusk of the mammoth (an extinct species of elephant), hazel nuts, &c. Beds of blown sand or *dunes* are also common along the coast. The river-gravels of the Wear and its deposits of fine mud (brick earths), also belong to this period. Bones of the lynx have been found in a fissure of the Mountain Limestone in Teesdale.

Pre-historic Man.—Stone implements of the later or neolithic type used by the early inhabitants of these islands, before they became acquainted with the use of metals, have been found at several places in Durham. A stone axe-head, made of basalt and $5\frac{3}{4}$ inches long, was found at Sherburn; another of the same material was dug out of a peat-moss at Cowshill-in-Weardale; it is now in the fine collection of Canon Greenwell, of Durham, as is also another, perforated to admit a handle, which comes from Millfield, near Sunderland. At Coves Houses, near Walsingham, a stone hammer-head has been found, and flint arrow-heads at Newton Ketton and Lanchester Common. The interesting remains found in the Heathery Burn Cave, near Stanhope, belong to a rather later period: they included ornaments of jet, with many bone pins or awls, and several bronze objects.

No. 12.

GEOLOGY OF ESSEX.

NATURAL HISTORY AND SCIENTIFIC SOCIETIES.
Epping Forest and County of Essex Naturalists' Field Club: Transactions.

MUSEUMS.

Chelmsford Museum.
Museum of the Essex Archæological Society, Colchester.
Maldon Museum and Library.
Saffron Walden Museum.

PUBLICATIONS OF THE GEOLOGICAL SURVEY.

Maps.—Quarter Sheets: 1 N.E. Maldon; 1 S.E. Tilbury; 1 N.W. Epping; 1 S.W. London; 48 S.E. Walton-on-the-Naze; 48 S.W. Colchester. Sheets: 2, Sheerness; 47, Saffron Walden, Thaxted.

Books. Geology of the Eastern End of Essex (Walton Naze and Harwich), by W. Whitaker, 9d. Geology of Colchester, by W. H. Dalton, 1s. Geology of the London Basin, by W. Whitaker, 13s. Geology of the N.W. of Essex and N.E. of Herts, by Whitaker, 3s. 6d.

IMPORTANT WORKS OR PAPERS ON LOCAL GEOLOGY.

1871. Prestwich, Prof. J.—Red Crag of Essex. Journ. Geol. Soc., vol. xxvii. p. 324.
1868. Wood, S. V., Jun.—Pebble Beds of Essex. Journ. Geol. Soc., vol. xxiv. p. 464.
1848. Brown J.—Pleistocene Deposits near Copford. Journ. Geol. Soc., vol. iv. p. 164; vol. viii. p. 184.
1876. Penning, W. H.—Physical Geology of East Anglia. Journ. Geol. Soc., vol. xxxii. p. 191.
1880. Walker H.—Elephant Hunting in Essex. Trans. Epping Forest Field Club, vol. i., p. 27.
1881. Corder, H.—Stone Implements from Chelmsford. Trans. Epping Forest Field Club, vol. ii., pp. 29, 31.
1881. Dalton, W. H.—The Blackwater Valley. Trans. Epping Forest Field Club, vol. ii., p. 15.

See also General Lists, p. xxv.

THE rocks of Essex have attracted a fair amount of attention from geological observers. The Government Geological Survey has published coloured maps of the southern half of the county, prepared by Messrs. Whitaker, Dalton, Penning, and others, and the examination of the northern portion is also, we believe, completed, although the maps are not yet quite ready for issue: on these maps the various layers or strata of rock are laid down with the utmost accuracy, and indicated by means of various colours.

The geology of the southern portion bordering on the Thames and Lea has been described by Mr. W. Whitaker in his "Geology of the London Basin," and the same author has also published a description of the strata in the north-eastern corner of the county, round Harwich and the Naze. Numerous papers by Professor Owen, Messrs. Brown, Mitchell, Clarke, Charlesworth, Searles V. Wood, Harmer, Penning, Prestwich, &c., have appeared in the Journal of the Geological Society and the "Geological Magazine." The deposits at Grays and Ilford are favourite hunting grounds of the members of the Geologists' Association, and have frequently been described in their Proceedings.

Geological Formations in Essex.—The county affords a fair diversity of rocks. The strata rest one upon the other with a general inclination or dip to the south-east: a consequence of this is that the lowest (and therefore oldest) beds crop up and occupy the surface in the north-west of the county, and here accordingly we find the white chalk or upper cretaceous strata. Resting upon the chalk, but formed at a much later period, is the lowest Tertiary stratum, known as the Woolwich and Reading beds, above which comes the London clay, and lastly the sandy Bagshot beds. Subsequent comparatively recent deposits of boulder-clay, sand, brick-earth, gravel, &c., repose irregularly upon the eroded edges of the true rock-masses enumerated above.

THE CRETACEOUS FORMATION.—Commencing then with the beds of chalk which we have already stated to be the lowest and oldest of the strata which form the county of Essex, we find in the north-west of the county hilly and undulating ground, which presents, though in a less marked manner, the characteristic features of the downs of the south of England. This chalk district extends from the north and west boundaries of the county as far south as a line drawn through Bishop's Stortford, Castle Hedingham, and Sudbury: it is only along the boundary line with Cambridgeshire, however, that the chalk hills make any feature in the landscape, or that we find the white earthy limestone known as chalk actually forming the surface. The Royston downs rise to a height of 400 feet, and from this point to Linton and Haverhill the average height is between 300 and 400 feet. The top of the Rivey, a hill just outside the county near Linton, is 325 feet above the river Cam and 365 above the sea-level. The *Lower Chalk* is exposed along the western slope of this low escarpment and along the valleys of the little streams that run northwards into the Cam.

The *Upper Chalk* is distinguished by the presence of numerous layers of flints, which are absent or rare in the beds below. A microscopical examination of chalk reveals the fact that it is almost entirely composed of minute shells of *foraminifera*, a low class of animals which are, even now, forming a deposit very similar to chalk on the bed of the North Atlantic Ocean. The chalk dips to the south-east and undoubtedly underlies the whole of Essex, rising again to the surface on the south of the Thames to form the hills known as the North Downs. One little patch, however, crops out in the south of the county between Purfleet and Grays Thurrock; here it is largely worked for the manufacture of "whiting," and the flint nodules are employed in making porcelain. Numerous fossils, as sea-urchins of the genera *Ananchytes*, *Micraster*, and *Galerites*, shells, as *Terebratula carnea*, *Inoceramus*, &c., and in fact all the ordinary fossils of the Upper Chalk, are fairly plentiful. At Grays the chalk has been proved by boring to be about 660 feet thick, and at Harwich 888 feet.

THE TERTIARY PERIOD.—The chalk was probably deposited at the bottom of a rather deep ocean, and was hardened, elevated to form dry land, eroded, and again depressed, before the next bed of rock was deposited upon its surface. Evidently then there must have been a great interval of time between the two, and proofs of this interval are further furnished by the fact that the fossils found in the beds we are now going to describe are of an entirely different nature to those of the chalk, showing that a complete change in the life existing in this region of the earth had taken place. Here then geologists draw one of their great boundary lines, placing the chalk, with certain beds beneath it, in a group denominated the Secondary; beds lower still (coal, slate, &c.) are known as Primary, but the newer, higher strata are denominated Tertiary, or *third* in order of the stratified rocks.

The lowest Tertiary strata are known as the EOCENE FORMATION, and are subdivided into

(1.) *The Thanet Sands.*—Resting on the mass of chalk near Grays we find 20 or 30 feet of fine light-coloured sands, in which few or no fossils occur; they rest on a layer of green-coated flints known to the workmen as the "Bullhead Bed." Mr. Whitaker has lately shown that the Thanet sands also occur at Ballingdon, a suburb of Sudbury, where they are 15 feet thick, and from this point they may extend along the edge of the chalk to Bishop's Stortford.

(2.) *The Woolwich and Reading Beds.*—These are mottled clays and sands, also seen at Sudbury, above the Thanet sands: thence they run by Shalford and Easton park to the south of Stortford, where they are well exposed in several brickyards. In South Essex they run from Wennington by Orsett to Stanford. Their thickness is about 40 feet.

(3.) *The Blackheath (or Oldhaven)* beds also occur near Orsett: they are composed of well-rolled flint pebbles embedded in sand, and are only a few feet thick. Mr. Bristow remarks: "At Hassenbrooke, east of Orsett, the farmer has some very productive strawberry beds, and the fruit, well known for its good quality, fetches a high price. I was told that the produce of 12 acres fetches £100 a week in Covent Garden Market, a price much above what the neighbouring growers can get. The explanation of this difference in quality is, I believe, that the sandy pebble-bed upon which these strawberries are grown makes a soil especially favourable for their cultivation, and that good gardening has little to do with the matter."

(4.) *The London Clay.* This is the chief Essex formation. When we descend the gentle eastern slope of the chalk, we see before us a second escarpment, rising in low hills to heights of between 200 and 300 feet above the sea. The lowest portion of the northern slope is composed of the Reading beds, but on the top and stretching thence right away to the mouth of the Thames and the North Sea, we find beds of stiff brownish and bluish clays, in which layers of *septaria* occur: the latter are rounded masses of impure carbonate of lime, usually traversed by cracks, which are filled up with crystals of the same material. In the south-east of Essex the London clay is 480 feet thick, but it thins out considerably as we follow it westwards. Near Harwich the London clay was formerly worked for the septaria, which were used in the manufacture of cement. Here, too, it contains numerous small nodules of iron pyrites, which are still collected on the beach near Walton, for the manufacture of copperas, to the amount of about 150 tons per annum. At many places in Essex the London clay is worked for the manufacture of bricks, and sections may be seen near Brentwood Railway Station, Galleywood, Ingatestone, Maldon, &c.

There is a spring at Hockley, the water of which, like much of that from the London clay, contains sulphate of magnesia (Epsom salts). An endeavour was made some years ago to construct a "spa" with "pump rooms," and a woman was employed to dispense the waters whose strong healthy appearance visitors were led to believe was the result of their medicinal effects. In spite, however, of such a strong corroboration of the efficacy of the Hockley waters, the public refused to be cured, and the speculation proved a failure; all that now remains of the spa being the buildings which are still known as "Hockley Spa." Teazles grow well on the London clay, and also the elm, oak, and ash.

(5.) *The Bagshot Beds.*—These sandy unfossiliferous strata only occur as outlying patches in the southern half of the county: they contain pebble-beds, formed of well-rolled flints only, and are about 25 feet thick: their position is invariably on the high ground and hill-tops, and springs issue from their junction with the London Clay below. They occur at High Beech, on the hill of Havering-atte-Bower, and there is a rather large mass which stretches southwards from Shenfield by Brentwood to Warley Common: they form the upper layers of the Brentwood brickfields, are also exposed in the railway cutting there, and the pebble-beds have been extensively dug on Warley Common: other patches occur near Frierning, Stock, Galleywood Common and Rayleigh: they also form the upper part of Langdon Hill, which rises to a height of 388 feet, and can be seen from long distances.

Now the five divisions of the Tertiary Epoch which we have described, viz., the (1) Thanet, (2) Reading, (3) Blackheath, (4) London Clay, and (5) Bagshot beds, are considered by geologists to form the lowest division of the Eocene formation. After the deposition of the Bagshot beds, the area seems to have been elevated so as to become dry land; at all events, there are no traces in Essex of strata of Upper Eocene age, nor of the succeeding Miocene period, which we find well represented in the Isle of Wight and on the Continent.

PLIOCENE FORMATION.—In the Eocene beds only a very few of the fossil shells (about $3\frac{1}{2}$ per cent.) are identical with species now living: in the Pliocene strata, on the contrary, we find in the lowest beds that one-half of the fossil shells belong to living species, while in the upper portion nine-tenths are precisely similar to existing forms. These Pliocene beds only occur in the eastern counties, where they have received the name of "crag," from the Celtic word *creggan*, a shell, in allusion to the fact that the beds are almost made up of a mass of comminuted shells.

The Red Crag. —The Lower, White, or Coralline Crag does not occur in Essex, so that the Red Crag rests upon the London clay: it only occurs in the north-east corner, where it is well exposed to the fine cliff sections at Walton Naze. Here we see it to be a coarse, reddish brown and grey sand, full of shells, and showing much false bedding. Harwich Cliff had also once a capping of Red Crag, but this has been removed by the encroachment of the sea. Of shell-fish, 148 species have been obtained from Walton, of which 75 are Mediterranean species, showing probably a rather warmer climate than at present.

Coprolite Bed.—At the base of the Red Crag we find a layer of phosphatic nodules, whose valuable fertilizing properties were first pointed out by the late Professor Henslow: this bed is worked at Wrabness, Little Oakley, &c., and about 5,000 tons altogether of these coprolites have been raised in Essex.

Chillesford Beds. —At the Naze we see 6 or 7 feet of sandy clay resting on the Red Crag, which may possibly belong to these beds.

PLEISTOCENE FORMATION.— *The Drift or Glacial Beds.*—Beds of Upper Pliocene age found in Norfolk and Suffolk give plain indications that a colder climate was then beginning to prevail. The Mediterranean shells disappear, and are replaced by Arctic forms: still later we enter on a period of severe cold, when icebergs detached from glaciers, which then covered all the north of England and the whole of Scandinavia, came sailing southwards laden with rocks detached from the land over which they rubbed, or the glaciers themselves may have pushed southwards as far as Essex and Middlesex. The Lower Boulder clay then formed in Norfolk, &c., does not occur in Essex: here the first glacial deposits are pebbly gravels and sands—*the Mid-glacial Beds*. These may mark a temporary diminution in the cold, together with a subsidence of the land: they are overlaid and overlapped by the Upper or Chalky Boulder clay, which reaches southwards to the northern edge of the Thames Valley. Thus it is spread as a surface covering over all the stratified rocks of Essex, except along the bottoms of the river valleys, where it has been washed away, exposing the Mid-glacial gravels beneath. It is a stiff clayey mass of a brownish hue, full of chalk-pebbles, flints, and pieces of rock, containing, too, numerous fossils brought from afar: it often forms a stumbling-block to the young geologist, who fails to distinguish it from the stratified rocks which it conceals.

Post-Glacial Beds.—The valleys were probably filled up during the Glacial period, and have since been re-excavated by the rivers which now occupy them. Most of the material so removed has been swept out to the sea; but here and there along the river's course we find beds of fine loam or brick-earth, and again beds of gravel or sand, at varying heights, which have been deposited by the streams at different periods. The brick-earths at Stratford, Ilford, Grays, Copford, &c., have yielded numerous remains of fresh-water mollusca, mainly such as still inhabit the adjoining district. At Grays, however, a shell known as *Cyrena fluminalis* occurs, which is now only known in the Nile; at Copford sixty-nine species of shells have been found, three of which (of the genus *Helix*) no longer inhabit England; from peat-beds exposed in the old brick-yard at Lexden, remains of beetles, mostly of non-British genera, have been obtained. Bones and teeth of the mammoth, of a rhinoceros, hippopotamus, and other extinct species also occur in these brick-earths. A very fine collection of the teeth, bones, &c., of these extinct mammals, made at Ilford by Sir Antonio Brady, has lately been presented by him to the British Museum.

Extensive deposits of sand and mud occur all along the Essex coast, from Shoeburyness to Harwich. At Clacton stumps of trees forming a submerged forest are to be seen at low water, which indicates a subsidence of the coast-line. The slope of the land as it passes under the sea is a very gentle one, and indeed the deepest part of the North Sea between Essex and Holland is only 180 feet, so that a very slight elevation would suffice to render England again, as it has several times been since the deposit of the Eocene beds, a part of the continent of Europe. A well-sinking for water at Clacton-on-Sea in 1878 passed through Post-Glacial Sands, 18 feet; London Clay, 194 feet; Reading Beds, 56 feet; entering the chalk at a depth of 270 feet; the same rock was found to lie 210 feet below the Castle Brewery, Colchester.

Does Coal exist beneath Essex?—It has been pointed out by Messrs. Prestwich, Godwin-Austen, and others, that the coal-beds of Belgium pass westwards to near Calais, trending in a direction which, if continued, would bring them underneath the south of England to the Somerset and Bristol coal-fields, which are supposed to be a continuation of them, as the coal-seams are much alike in quality, number, and thickness. Now a well-boring at Harwich struck dark slaty rocks which we know to lie beneath the coal, at a depth of 1,026 feet, and these rocks had a rather high inclination towards the south. Borings at Kentish Town (1,114 feet), and Meux's brewery in London (1,000 feet), show on that side also old rocks of Devonian age. Now between these, and consequently underneath Essex, it is possible that a coal basin occurs, hidden deep down under about 1,500 feet of chalk and Eocene strata. This is a most important question for the landowners of

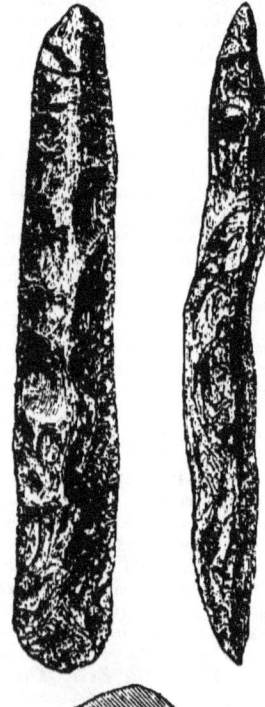

Fig. 29.—Flint "pick" or chisel, found at Great Easton, near Dunmow, Essex; length 6¾ inches. The lower figure shows the cross section.

Essex, and it is highly desirable that a boring should be made at some suitable point to test the question.

PREHISTORIC MAN.—Several examples have been found in Essex of the flint and stone implements used by the early inhabitants of this country, when as yet metals were unknown. A chipped flint celt or axe, 9 inches long and 3¼ broad, was found at Blunt's Hill, near Witham: another specimen, ground at the edge and 6½ inches long, occurred at Stifford, near Grays Thurrock: the same spot also yielded a perforated stone hammer, circular in outline and 3 inches in diameter. At Audley End, a stone pestle like a small club, 9¼ inches long, was found in a gravel pit, with a Roman cinerary urn. A remarkably large flint chip or flake was found in digging the foundations of a house on Windmill Hill, Saffron Walden. Mr. John Evans remarks concerning this specimen: "One face is somewhat flatter than the other, but both faces are dexterously and symmetrically chipped over their whole surface. The small flakes have been taken off so skilfully and at such regular intervals that, so far as workmanship is concerned, this instrument approaches in character the elegant Danish blades. The form seems well adapted for a lance head, but on examination the edges appear to be slightly chipped and worn away, as if by scraping some hard material. It would appear then more probably to have been used in the hand." At Great Easton, near Dunmow, a flint pick or chisel has been found, and a perforated axe-head made of greenstone near Colchester. Such were the tools and weapons used by our predecessors in this country up to perhaps 5,000 or 6,000 years ago, when the use of metals was discovered. Doubtless many more specimens would be found if gravel pits and other excavations were watched by intelligent observers.

No. 13.

GEOLOGY OF GLOUCESTERSHIRE.

Natural History and Scientific Societies.

Bristol Naturalists' Society. Proceedings.
Cotteswold Naturalists' Field Club; Gloucester. Proceedings.
Clifton College Scientific Society; Bristol. Transactions.
Cheltenham Natural Science Society.
Stroud Natural History Society.
Bristol Microscopical Society.
Cheltenham School Natural History Society. Report.
Cirencester Microscopical and Natural History Society.

Museums.

Bristol Baptists' College Museum.
Bristol Library and Museum Society.
Bristol Library, King Street.
Bristol Museum and School of Art.
The Normal Training College Museum, Cheltenham.
The Corinium Museum, Cirencester.
Royal Agricultural College Museum, Cirencester.
Clifton Museum near Bristol.
The Museum and Schools of Science and Art, Gloucester.

Publications of the Geological Survey.

Maps. Sheets: 19, Bath, Frome, Wells; 34, Stroud, Fairford, Swindon, Chippenham; 35, Bristol Coal-field to Forest of Dean. Quarter Sheets: 43 N.E. Woolhope to Malvern; 43 S.W. Forest of Dean to Monmouth; 43 S.E. Forest of Dean by May Hill to the Severn. Sheet 44, Evesham, Worcester, Cheltenham, Burford.

Books.—Geology of Parts of Wiltshire and Gloucestershire, by Prof. Ramsay, &c., 8d. Geology of the Country round Cheltenham, by E. Hull, 2s. 6d. Geology of East Somerset and the Bristol Coal-fields, by H. B. Woodward, 18s.

Important Works or Papers on Local Geology.

List of 750 works on Geology of Gloucestershire and Somersetshire in "Geology of Bristol Coal-field," by H. B. Woodward.

1865. Austin, Fort-Major Thos. The Millstone Grit. 8vo, London and Bristol.
1869. Thomas, A.—The Forest of Dean. Trans. S. Wales Inst. of Engineers, vol. vi. p. 200.
1870. Etheridge, R. Dolomitic Conglomerate of Bristol. Journ. Geol. Soc., vol. xxvi. p. 174.
1870. Tate, R.—Palæontology of Junction Beds of Lower and Middle Lias in Gloucestershire. Journ. Geol. Soc., vol. xxvi. p. 394.
1871. Tate, R.—Marine Invertebrate Fauna of the Lias. Geol. Mag., vol. viii. p. 4.

GLOUCESTERSHIRE. 93

1873. Anstie, John.—Coal-fields of Gloucestershire and Somersetshire, and their Resources. 8vo, London.
1881. Longe, F. D.—Oolitic Polyzoa. Geol. Mag., p. 23.
1881. Wethered, E.—Grits and Sandstones of Bristol Coal field. Midland Naturalist, vol. iv. pp. 25, 59.
1881. Wright, Dr. T.—Physiography and Geology of the Country Round Cheltenham. Midland Naturalist, vol. iv. p. 145.

See also General Lists, p. xxv.

THE strata which compose the county of Gloucester have so long and repeatedly been carefully examined by competent geologists, that we are well acquainted with, at all events, their general relations and mode of occurrence; it will, indeed, be impossible to name one-half of the workers who have contributed to the literature of the subject, for in the preparation of even this brief account, we have consulted some hundreds of books, papers, notes, &c.

Among the earlier workers we may name William Smith, G. Cumberland, T. Weaver, W. Lonsdale, J. Buddle, Murchison and Strickland, Brodie and Buckman. Professor Phillips has described the old rocks of May Hill and Tortworth; of the coal-fields Professor Hull gives a good general account in his "Coal-fields of Great Britain," and the same author, as a member of the Geological Survey, mapped and described much of the Cotteswold district. Dr. Wright and Dr. Lycett have added greatly to our knowledge of the fossils of the Liassic and Oolitic strata, and each has also published masterly descriptions of portions of the county—the former in his address to the Geological Section of the British Association (Bristol meeting, 1875), and the latter in his "Handbook of the Cotteswold Hills" (published in 1857).

Fort-Major Austin has diligently studied the Millstone Grit; but in connection with Carboniferous Rocks the palm must be awarded to Mr. W. Sanders, whose map of the coal-field, on a scale of four inches to a mile, was a marvellous achievement for a single private individual to accomplish. In the fine museum of the Bristol Philosophical Institution, there is a grand series of local rocks and fossils, whose collection and display are mainly due to the energies of Messrs. Etheridge, Tawney, and Sollas; and the Natural History Society of Bristol, together with the Cotteswold Field Club, have done much good in exciting a love for the study of geology.

The whole county has been geologically mapped on the scale of one inch to a mile by the Government Geological Survey, and these maps, with the accompanying memoirs by Professor Hull, Mr. H. B. Woodward, &c., form a thorough and complete guide to the subject. To the Rev. W. S. Symonds, Rev. F. Smithe, Professors Rupert-Jones, Judd, Jukes, and Tate, and Messrs. R. Etheridge, W. W. Stoddart, J. W. Salter, W. C. Lucy, and many others, we can only express our general obligations.

As in other regions, the present configuration of Gloucestershire depends mainly upon the nature of the rocks which form its various parts, and on the way in which these have been acted on through long periods of time by agents of denudation—the sea, rivers, rain, ice, chemical action, &c. The surface of the county plainly divides itself into four regions, but two of these are closely related. First, we have on the west of the Severn a forest region, including the Forest of Dean Coal-field and the old Silurian rocks of May Hill. The second region much resembles this, but lies on the other side of the Severn; it includes the Bristol coal-field, with the Silurian rocks of Tortworth. The flat lands round the Severn forming the well-known "Vale of the Severn," constitute the third region, and the fourth consists of the Oolitic rocks, which, in the east of the county form a well-marked escarpment with an old tableland on the top, gently sloping eastwards.

The elevations of a few points are Pewsdown Hill, 1,200 feet; Cleeve Cloud, 1,081 feet; Leckhampton Hill, 978 feet; Painswick Hill, 929 feet; Stinchcombe Hill, 725 feet; Tewkesbury Market House, 45 feet; Gloucester Cathedral, 58 feet; Stone Church, 100 feet.

In describing the strata it will be the simplest plan to begin with the oldest or first-formed, and then to take in turn the overlying beds.

UPPER SILURIAN FORMATION.—There are two uncovered areas of Silurian rocks in this county: they are both exposed at the surface through having been thrust up by volcanic upheavals from below, which produced faults or dislocations in the rocks; thus it has been calculated that the rocks of May Hill are 9,000 feet above their proper position. We can trace a line of upheaval running northwards by Woolhope, and southwards from May Hill to Purton Passage and then on the other side of the Severn, bringing up rocks of the same age round Berkeley and Tortworth.

May Hill.—Here the central dome or mass is formed of the Upper Llandovery, or May Hill Sandstone: then comes a slight encircling ridge composed of Woolhope Limestone; then a depression formed by the Wenlock Shales, with a ridge beyond of Wenlock Limestone, and on the outside again, the Ludlow Beds: the area here composed of Silurian rocks measures about 6 miles north to south, by 3 miles east to west, and includes the eminence known as Huntley Hill. The summit of May Hill (which is also known as Yartledon Hill) is clad with firs, and affords a very fine view in all directions: there are several old quarries from which the ordinary Silurian fossils may be obtained, such as the corals *Petraia, Favosites*,&c., the shells *Chonetes lata, Strophomena depressa*, with fragments of trilobites, &c. At Dymock, the Downton Sandstones are quarried: these are passage beds between the Silurian and Old Red Sandstone formations.

The Tortworth District.—Here again we have the same beds as at May Hill: they form a long strip, reaching 12 miles due south from Tites Point on the Severn: their extreme width near Stone is 2¾ miles: the district is the least picturesque of all the Silurian tracts which lie on the eastern borders of Wales, no portion of the surface attaining the height of 200 feet above sea-level. The Upper Llandovery Beds stretch from Tortworth to Stone, and are again seen at Charfield Green: they are chiefly Red Micaceous Sandstones, with a westerly dip of about 25 degrees, and are traversed by several sheets of igneous rock (basalt) which is quarried for road metal. Beds of Wenlock Shale with two bands of Wenlock Limestone extend from Cinderford Bridge southwards to Whitfield: they dip to the east and are overlaid conformably by Ludlow Shale, Sandstone, and Limestone: the latter strata also occur further north round Berkeley and run up to the banks of the Severn: this tract of Silurian rocks is traversed by several faults, and is an intricate and difficult region to understand.

THE OLD RED SANDSTONE.—Beds of this age cover a great surface area in Herefordshire, and stretching eastward, they pass completely round the Silurians of May Hill, encircling also, save at one point on the south, the Forest of Dean Coal-field: they form the banks of the Severn for about a mile north and 2 miles south of Purton Passage, and crossing the stream, pass southwards through Berkeley to Cinderford Bridge: again in this region they extend from Thornbury to Milbury Heath and thence come round in a narrow band south of Tortworth to Wickwar Common. In the extreme south of the county we again see them in the gorge of the Avon, at Clifton, rising up from beneath the Carboniferous Limestone: where exposed, they are usually seen to consist of red marls and red and brown sandstones, with some thin bands of impure nodular limestone, called cornstone: they form a poor stony soil in this district, and fossils are very scarce; a few fish scales and spines have, however, been found: near the top is a bed of conglomerate; the well-known Logan, or rocking-stone, called the "Buckstone," near Monmouth, is formed of this conglomerate: sections are exposed along the Wye, on the banks of the Severn, in road-cuttings near Drybrook, Forest of Dean, and at many points near Tortworth Hill and Milbury Heath.

THE CARBONIFEROUS FORMATION.—As Gloucestershire includes two distinct coal-fields it will perhaps be better to describe them separately.

1. *The Bristol Coal-field.*—Under this term we include all the Carboniferous rocks lying between Bristol and Kingswood on the south, and Tortworth on the north. The beds called *Lower Limestone Shale* form the base of the entire series, and are 500 feet thick in the gorge of the Avon, where they contain the Bone or fish-palate bed, only 4 to 6 inches thick, but containing an immense quantity of fossils. The *Carboniferous*, or *Mountain Limestone* is 2,600 feet in thickness. From Clifton and Durdham Down it runs north to Westbury and King's Weston: after a break we again see it at Almondsbury Hill, and thence follow it along the Ridgeway to Olveston, and north-east to beyond Cromhall, whence it curves southwards past Wickwar to Chipping Sodbury. It forms bare and rugged scenery, with a thin reddish soil, and makes an excellent white lime: it is also used for road metal and for building purposes: fossils are numerous, especially brachiopods, crinoids, and corals; indeed, much of the rock seems to be composed of the fragmentary joints of crinoids. *The Upper Limestone Shale* is the term applied to the alternations of shales and sandstones with impure limestones and thin seams of coal which form the passage beds between the subjacent beds of pure limestone and the overlying *Millstone Grit:* the latter is usually a hard close-grained grit or quartzite; miners call it the "Farewell Rock," inasmuch as it underlies all the workable coal-seams: its thickness is about 1,000 feet: at Clifton it has been largely quarried for building-stone; it is again exposed south of Cromhall, and can be traced southwards to Yate, but elsewhere it is covered over and concealed by Secondary Rocks.

The Coal-measures are divided into an upper and lower series, each about 2,000 feet thick, and composed of shales and sandstones with bands of ironstone and seams of coal: these two series are separated by a thick mass of sandstone, the Pennant Grit, which varies from 1,000 to 2,000 feet in thickness; it contains much iron-ore, and is largely quarried for building purposes. There are 20 seams of coal above 2 feet in thickness, and the outcrops of the principal seams can be traced over an area extending southwards from Cromhall through Iron Acton, Westerleigh, Stapleton, and Mangotsfield to near Kingswood, where an anticlinal line, accompanied by a fault, causes the beds to roll over and dip to the south, thus forming a partly separate basin between the Avon and the Mendip Hills. The quantity of coal raised in the Gloucestershire basin in 1879 was 471,290 tons.

2. *The Forest of Dean Coal-field.*—Here the strata form a very perfect "basin;" on the outside we have encircling rings of Mountain Limestone (600 feet thick) and Millstone Grit (500 feet) which form high ridges, and in the centre the Coal-measures form poor, sandy, and peat-covered soil, on which, nevertheless, some of the finest oaks and beech trees which England has ever produced have grown. The Coal-measures are 2,765 feet in thickness, and contain 15 seams of coal, of which only 8 are of a thickness of 2 feet and upwards: the area of the coal-field is about 34 square miles: the amount of coal raised in 1879 was 779,428 tons from 55 collieries; of ironstone (brown hematite) 58,000 tons were obtained. In one seam, the Coleford High Delf, there is a fine illustration of what colliers call a "Horse," where the coal has been swept away and replaced by sandstone; it can be traced for over two miles, and varies in breadth from 170 to 340 yards: these "Horses" evidently mark old river channels which were cut through the mass of vegetable matter, then in a comparatively soft state, which has since been compressed and hardened into coal.

Coal beneath the Severn.—The two coal-fields we have described are connected by a third, which lies between them. The railway cuttings near Almondsbury and the borings in the bed of the river at English Stones and other points, show that Coal-measures exist beneath the Severn and the adjoining low lands, from Denny Island up as far as Berkeley: from the disturbed state of the beds, however, and the difficulty of access to them, it is not likely that this area will admit of profitable working.

THE PERMIAN FORMATION.—Only a very small area is occupied by the Permians in this county: they are seen as breccias, resting on the southern flank of the Malverns, north of Bromesberrow; and the Rev. W. S. Symonds has detected them on the east side of May Hill.

THE TRIAS.—*The Bunter Sandstone* is seen at Bromesberrow to rest unconformably on the Permian breccias: further south this stratum has thinned out altogether, and the *Keuper Beds* are seen to rest directly upon Carboniferous and Devonian strata: in the Bristol district, a deposit called the *Dolomitic Conglomerate* forms the base of the Keuper series: it is usually under 30 feet in thickness, and is composed of rounded and angular pebbles of Carboniferous Limestone, cemented together by carbonates of lime and magnesia: it appears to have been formed as a pebble-beach fringing the land, whilst the Keuper marls and sandstones were being deposited further out at sea: at the Yate Rocks, near Chipping Sodbury, it is of a yellow tint; it forms irregular patches south of Durdham Down, and round Henbury, Thornbury, and on the opposite side of the Severn west of Tidenham: near Durdham Down, bones of two species of saurians were found in this bed in 1836, and were described by Dr. Riley and Mr. Stuchbury; they referred them to the genera *Palæosaurus* and *Thecodontosaurus;* Professor Huxley states that they both belong to the *Dinosauria*.

The New Red or Keuper Marls are of a red tint, with many streaks or patches of a bluish-green: they form low ground, and run from Bristol to Tytherington, and then further north form part of the vale of the Severn, between Gloucester and the Malvern Hills: here they dip steadily to the east at a low angle; their thickness at Bristol is 300 feet, but north of Gloucester it is perhaps double this amount: they form a rich loamy soil; veins of gypsum are common, and the mineral *celestine* is abundant in the Bristol district: fossils are scarce or absent; and, as elsewhere, these Triassic Beds appear to have been deposited in salt-lakes of great extent.

THE RHÆTIC OR PENARTH BEDS.—These are very thin, usually under 50 feet in thickness; but from the number and nature of their fossils, they are of high interest to the geologist: they form a considerable spread of land round Patchway station, and can be traced as a very narrow band running due north from the Avon to Tites Point, and thence north-east along the Severn to Tewkesbury: the sections at Garden Point, Frethern, Aust Cliff, and other points along the Severn are well known: the Rhætics usually consist of 15 to 20 feet of buff marls at the base, then 10 feet of black and light-coloured shales, with one or more "bonebeds," and lastly the white earthy limestones (10 to 15 feet) called White Lias: at Cotham we fine the well-known *Landscape stone*, a thin bed of hard limestone, whose beautiful markings are due to the infiltration of oxide of manganese: the shells *Avicula contorta* and *Cardium Rhæticum*, are everywhere common; but these Rhætic Beds will well repay the local collector for his most assiduous attention.

THE LIAS.—Stiff brown and blue clays of liassic age enter Gloucestershire near Stratford-on-Avon, where they attain a thickness of 500 feet and extend in a south-westerly direction by Cheltenham and Gloucester, here forming the eastern part of the vale of the Severn, and passing on by Dursley and Wickwar to Bitton; there are also two large liassic outliers north of Bristol, one round Horfield and Filton, and the other round Alveston, besides small ones further north: at the base of the lias several bands of limestone are interstratified with bluish shales, and these are worked near Tewkesbury and elsewhere: the Lower Lias forms rich pasture land, furnishing material for the famous "double Gloucester" cheese.

The Middle Lias or Marlstone runs along the slope of the Cotteswold Hills, thickening as we follow it northwards from a few feet near Bitton to 120 feet at Leckhampton. The *Upper Lias Shales* which come next behave in a similar manner, but are as much as 300 feet thick at Cleeve Cloud: they contain nodular layers of limestone which have yielded fine fossil fishes. The Lias generally is rich in fossils, but the various *Ammonites* which characterise different zones or levels, and the gigantic bones of extinct saurians, are especially noteworthy. Like the Red Marls beneath, the Liassic strata dip eastwards, usually at from 2 to 5 degrees only, so that in small sections they appear quite horizontal.

THE OOLITES.—*The Midford Sands* are passage beds well seen at Frocester Hill, Nailsworth, Wotton-under-Edge, &c.: their thickness is very variable, but never above 150 feet. The *Inferior Oolite* lies upon the above-named strata: it is 264 feet thick at Leckhampton, but thins both to the south and east: it yields an excellent building-stone and is largely quarried at Guiting,

Fig. 30.—A, *Ammonites bifrons*, Upper Lias; B, *Ammonites margaritatus*, Middle Lias; C, *Ammonites bisulcatus*, Lower Lias.

Stanway Hill, Cleeve Cloud, Leckhampton Hill, Painswick Hill, &c. The *Fuller's Earth* is a blue and yellow clay, 128 feet thick at Wotton-under-Edge, 70 feet at Stroud, and absent altogether in the extreme north-east of the county: it was formerly worked for cleansing woollen cloth, but its use for this purpose has now almost gone out. The *Great Oolite* is another excellent building-stone, much worked at Minchinhampton; above it, between Cirencester and North Leach, we find the *Forest Marble* 40 feet thick and composed of shelly limestones and thin beds of clay; then comes a thin rubbly limestone—the *Cornbrash*—and lastly, occupying a small area south of Fairford and Cirencester, the *Oxford Clay*. Almost every stratum of the Oolites is fossiliferous, and the corals, sea-urchins, shells, &c., show the beds to have been deposited in a moderately deep sea studded with islands, and enjoying a warm, almost tropical climate. At present these Oolitic beds form the Cotteswold Hills, having a steep escarpment facing to the west and a gentle slope coinciding with the dip of the strata to the eastward. Formerly, however, there can be no doubt that they stretched westward right across the vale of the Severn, and abutted on the Malverns, and Welsh Hills.

RECENT DEPOSITS.—The existence of glacial deposits in Gloucestershire is still rather a doubtful point: pebbles of quartzite, grit, hornstone, &c., are freely scattered over the surface and are termed "Northern Drift:" doubtless they came from the north, but their carriage by glaciers or icebergs cannot be said to have been definitely proved: beds of gravel and sand containing marine shells occur at elevations which show a submergence of the district to the depth of at least 500 feet since the glacial period, and then doubtless the

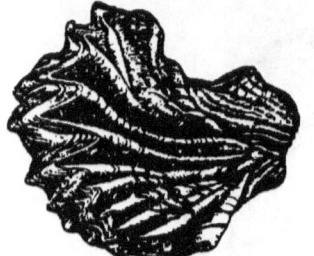 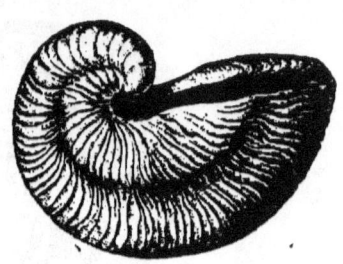

Fig. 31.—*Ostrea Marshii*, an oyster with both valves of the shell plaited. *Cornbrash*.

Fig. 32.—*Gryphæa incurva*, a thick shelled oyster from the *Lower Lias*.

sea swept between the Cotteswolds and the Malverns, forming the Straits of Malvern: of still later date are the alluvial deposits of the Severn, the muds and gravels which form flat meadows bordering the existing stream.

PREHISTORIC MAN. Of the stone tools used by the early inhabitants of this country before they discovered the art of smelting metallic ores, we have instances of celts, or axe-heads made of flint, found at Cherbury Camp, Pusey, Faringdon, and among the relics found at the Roman villa at Great Witcombe was a British hatchet of flint, another made of greenstone was found near Cirencester: a hammer stone of quartzite is recorded from Whittington Wood, and flint flakes or knives from Oakley Park, Rodmarton, and Micheldean: at the latter place and at Turk Dean arrow-heads of the same material have been found.

No. 14.

GEOLOGY OF HAMPSHIRE.

NATURAL HISTORY AND SCIENTIFIC SOCIETIES.

Winchester and Hampshire Scientific and Literary Society; Winchester Journal.
Winchester College Natural History Society. Report.
Bournemouth Natural History and Antiquarian Society.
South of England Literary and Philosophical Society; Southampton.

MUSEUMS.

Museum in Haslar Hospital, Gosport.
Museum of the Hartley Institute, Southampton.
Winchester Museum.
Alton Museum.

PUBLICATIONS OF THE GEOLOGICAL SURVEY.

Maps.—Sheets: 8, Farnham, Guildford, Dorking; 9, Chichester, Midhurst, Horsham; 11 Winchester, Southampton, Portsmouth, Petersfield; 12, Andover, Basingstoke, Odiham; 14, Marlborough, Amesbury, Westbury; 15, Salisbury, Ringwood; 16, Hurst Castle to Lulworth.
Books.—The Geology of Parts of Berkshire and Hampshire, by Bristow and Whitaker, 3s. Geology of the London Basin, by W. Whitaker, 13s. Geology of the Weald, by W. Topley, 28s.

IMPORTANT WORKS OR PAPERS ON LOCAL GEOLOGY.

List, by Mr. Whitaker, of 315 works on the Hampshire Basin, in Proceedings of Winchester and Hants Scientific and Literary Society, 1873.

1866. Nicolls, Lt.-Col. W. T.—Sarsen Stones near Southampton. Geol. Mag. vol. iii. p. 296.
1867. Wise, J. R.—The New Forest. 2nd Ed., Lond.
1870. Codrington, T.—Superficial Deposits of South Hants. Journ. Geol. Soc., vol. xxvi. p. 528.
1872. Prestwich, Prof. J.—Raised Beach on Portsdown Hill. Journ. Geol. Soc., vol. xxviii. p. 38.
1872. Evans C.—Geology of Portsmouth. Proc. Geol. Assoc., vol. ii. pp. 61 149.
1876. Seeley, Prof. H. G. -A Zeuglodon from Barton Cliff. Journ. Geol. Soc.. vol. xxxii. p. 428.
1879. Gardner, J. S.— Bournemouth Beds. Journ. Geol. Soc., vol. xxxv. p. 209.
1880. Judd, Prof. J. W.—Oligocene Strata of Hampshire Basin; Journ Geol. Soc.,vol. xxxvi. p. 137.

See also *General Lists, p.* xxv.

IN the Proceedings of the Winchester and Hampshire Scientific and Literary Society for 1873 we have a list (compiled by Mr. W. Whitaker,

F.G.S.) of 315 papers, books, &c., which relate to the geology of this county and its immediate neighbourhood. Foremost among these are the excellent coloured maps of the Government Geological Survey, showing the exact position of the various strata, on the scale of one inch to a mile. Mr. Whitaker's own "Memoir on the Geology of the London Basin" describes minutely the north and north-east of Hants, and the eastern district round Petersfield is described in the "Memoir on the Wealden District," by Mr. W. Topley.

In the Journal of the Geological Society there are valuable papers by Sir Charles Lyell, Professors Owen and Prestwich, Messrs. Clarke, Codrington, Meyer, Woodward, and others; Mr. Caleb Evans has well described the neighbourhood of Portsmouth in the Proceedings of the Geologists' Association, vol. ii.

The general arrangement of the beds of rock which form Hants is not difficult to understand. The chalk forms a great fold or curve, or roll, occupying a large tract in the centre of the county, and dipping or inclining both to the north and to the south. In both these directions it is overlaid by much newer strata—sands and clays of Tertiary age—those on the north forming part of what is called the London Basin, and those on the south the Hampshire Basin. In the east of the county, however, the chalk has been so worn away by the various agents of denudation—rain and rivers, frost, ice, and snow, the sea, &c.—as to expose certain rocks which lie beneath it, and which are therefore the oldest that occur in the county. These old rocks form the termination of a considerable tract called the Wealden, which stretches through Kent and Sussex to the sea.

We shall now describe the strata in succession, commencing with the lowest or first-formed.

NEOCOMIAN OR LOWER CRETACEOUS FORMATION. Four subdivisions have been traced in the beds of this age by the officers of the Geological Survey: the lowest of these, called *Atherfield Clay*, just touches the county boundary along the railway east of Bramshot and south of Hind Head. Next are the *Hythe Beds*, composed of sandstones and limestones about 300 feet thick; these form the high ridge extending by Hind Head, Greyshot Down, Bramshot, and Rake Common: they dip to the west and are overlaid by the *Sandgate Beds*, composed of sands and clays about 60 feet in thickness; these occupy low ground about half-a-mile in width, and extending from Petersfield to two miles north of Headley, beyond which point they cannot be traced; they are exposed in the railway cutting north of Rake. In the cutting at Petersfield we see the *Folkestone Beds* overlying the last-named strata; they are sandy, with a bed of coarse grit, coloured red by iron and known as carstone: the thickness of this last division is about 100 feet; it can be traced from Frensham Common through Woolmer Forest and Greatham to the west of Petersfield, whence it curves eastwards.

Fig. 33.—*Trigonia lata*, Neocomian Formation: *a*, external cast; *b*, internal cast.

The above strata are collectively known as the *Lower Greensand*; their fossils, such as ammonites, nautili, &c., indicate their deposition in a sea of moderate depth. Economically they are not of much importance; the carstone is used for pavements and building walls. In his "History of Selborne," Gilbert White writes: "From a notion of rendering their work the more elegant and giving it a finish, masons chip this (Woolmer Forest) carstone into small fragments, about the size of the head of a large nail; and then stick the pieces into the wet mortar along the joints of their freestone walls; this embellishment carries an odd appearance, and has occasioned strangers sometimes to ask us pleasantly 'whether we fastened our walls together with tenpenny nails?'"

THE UPPER CRETACEOUS FORMATION.—The first stratum of this division is called the *Gault;* this is a stiff blue clay about 100 feet thick, which is worked for brick-making in some pits half-a-mile south of Petersfield, and extends thence by Stroud Common, Greatham Mill and Worldham, through Alder Holt Wood towards Farnham. At the junction of the Gault with the Lower Greensand a bed of phosphatic nodules occurs.

The Upper Greensand. This rock is not true to its name in this neighbourhood; it is here calcareous and of a whitish hue, resembling indeed the lower beds of the Chalk: it extends from Binstead, through Selborne, to Buriton and East Meon. The beds are well exposed in quarries and deep road cuttings, and the Malm rock (a hard sandstone) forms a well-marked cliff 30 or 40 feet in height. Of fossils, *Pectens* and *Nautili* are most common; there is a good collection (made by Mr. W. Curtis) in the museum at Alton.

In the north of Hants there is an exposure of the Upper Greensand at Kingsclere and Burghclere, running east and west for about 5 miles; this outcrop is geologically termed an *inlier,* as it is surrounded by higher strata (chalk), by whose denudation it has been exposed to view. A similar inlier occurs further west at Shalbourne, which, however, barely enters Hampshire. These two inliers owe their exposure to a line of elevation running east and west; the beds curve over, dipping on one side to the north and on the other

Fig. 34.—View of the Malm rock (or Upper Greensand) Escarpment at Hartley, between Alton and Selbourne. A hard calcareous sandstone (Malm rock or firestone) forms a cliff 40 ft. high; softer beds of the Upper Greensand undulate lower down; while the gault forms the flat foreground. Woods called "hangers" grow on the sides of the cliffs.

to the south; they consist here of greenish sands and irregular beds of gritty sandstone overlaid by yellowish-white malm rock.

The Chalk. This is a soft, white, earthy limestone, perhaps the most widely known of all the British rocks. It forms the centre of Hants, stretching from Inkpen Beacon, Kingsclere, and Odiham on the north to Landford, Hursley, and Hambledon on the south, a distance of about 22 miles. The mass of chalk here exposed and extending westwards into Wilts, is the largest expanse of this rock in the British Islands; from it the North and South Downs run eastwards, and the Berkshire Downs and Chiltern Hills north-east.

Forming the base of the chalk we find a stratum 2 or 3 feet thick, called the *Chloritic Marl,* because it is speckled with green grains: it contains a band of (so-called) coprolites, which, when ground up and chemically prepared, form a valuable manure. This bed extends from Alton to Selborne and thence to East Froyle, where it has been largely worked. Above this comes the *Chalk Marl,* 40 or 50 feet thick, which is simply chalk with a slight admixture of clay; next we get the *chalk without flints* 400 or 500 feet thick, and lastly the *chalk with flints,* about 200 feet thick. All these chalky strata lie in the form of a great arch; on the north they dip sharply to the north, then about Andover they lie horizontally, but following them southwards we soon find them inclining in that direction, and they pass down underneath newer rocks, beneath the Solent, rising up again in the Isle of Wight. In the isolated chalk ridge called Portsdown, we have a repetition of this arch on a smaller scale.

K 2

Chalk is so homogeneous a rock that it would not be easy to discover its dip, were it not for the nodules of flint in the upper beds, which usually occur in lines along the planes of bedding. Flint is nearly pure silica: it existed dissolved in the water of the sea in which the chalk was formed, and in the water which has since percolated through it; it was deposited on and around the various sponges, echinoderms, &c., which lived in such numbers on the floor of the chalk ocean, and has also filled up fissures, and replaced chalk and organic matter at intervals since that period. The Lower Chalk is harder than the Upper, since the silicious matter, which in the latter is aggregated into nodules, is in the former equally dispersed throughout the rock, forming about 4 per cent. of its mass.

Chalk contains many fossils, visible enough to the naked eye. Large *Ammonites* are common near Selborne, *Belemnites* looking like pointed pencils abound, globular sea-urchins of the genera *Ananchytes, Micraster,* and *Galerites* are common, with bivalve shells like *Pecten nitidus, Inoceramus Cuvieri,* &c., but these are not all; when we disintegrate the rock and examine it microscopically we find it largely formed of the tiny shells of creatures called Foraminifera, whose descendants now inhabit the North Atlantic, their débris forming the soft greyish-white ooze on which the Anglo-American telegraph cable lies.

Chalk is quarried at many points in Hants to burn into lime; the rock in Harewood Forest near Andover, which is very soft and white, is especially suited for this purpose. It absorbs water rapidly, so that the hills, or downs as they are called, are always dry, and even the valleys between are usually without the running streams which are found on other formations. By the carbonic acid gas contained in the rain-water the chalk is dissolved and carried away in solution, so that no thickness of soil is formed, and the beautiful curves which characterize the scenery are formed of the solid rock covered only with a short, sweet herbage, which forms admirable pasturage for sheep. Swallow-holes are hollows connected with fissures in the chalk; in the pond at Alresford, said to have been formed by Bishop Lucy in King John's time, the water sometimes disappears, being carried off by such an opening. The pot-holes, filled up with gravel, which are so noticeable in the sides of railway-cuttings through the chalk, have also been formed by the dissolving of the rock by the percolation of carbonated water.

THE TERTIARY EPOCH.—EOCENE FORMATION.—The strata which rest upon the chalk are of very much newer age than it. The chalk, as we have indicated, accumulated on the bottom of a deep sea; it must have been consolidated and elevated so as to become land, and then again depressed, before the beds of clay and sand of Tertiary age, which we are now about to describe, were laid down on it. These beds once covered it over everywhere, but after the great, though slow, earth-movements took place, which resulted in throwing the rocks of the south of England into long folds or undulations running east and west, the newer beds were worn off the anticlinal curves, or tops of the Hampshire folds, and so the beds which occupy the northern hollow or synclinal fold were separated from the similar strata which lie in a similar hollow in the south, between Portsmouth, Romsey, &c. The occurrence of patches of Tertiary strata—outliers as they are termed—which are found at various points between the London basin on the north, and the Hampshire Basin on the south, confirm the fact of their former connection. We shall describe these two areas separately.

LONDON BASIN.—In the absence of the Thanet Sand, we find the *Woolwich and Reading Beds* resting directly upon the chalk; they have a high northerly dip, and so cover a narrow tract of country from Inkpen, by East Woodhay, Old Basing, and near Odiham. These beds are mottled sands and plastic clays about twenty feet thick; they are sometimes worked for brickmaking. Except a few oyster-shells (*Ostrea bellovacina*), fossils are scarce or absent.

The London Clay is bluish-grey or brown in colour, and about 50 feet thick;

it has a rather broad outcrop between the Woolwich and Reading Beds on the south, to the Thames on the north. Mr. Bristow records numerous fossils from the steep sides of a brook on the southern corner of Kingsclere Common, including such shells as *Pectunculus brevirostris, Turritella imbricataria,* &c.

The *Bagshot Beds* are of a sandy nature, but with some thin bands of clay. They form much heath and waste land at Silchester Common and West Heath, and again on the east side of the river Loddon, to the north of Aldershot and Winchfield, extending as far as the Blackwater on the north and east, and forming a region of sandy flats and ridges, on which the troops frequently exercise. Fossils are all but unknown in these beds.

THE HAMPSHIRE BASIN.—Here again the Tertiary strata occur in the same order, but with a southerly dip, and covering more ground; moreover, beds are found at the top of the series which are newer—later formed—than any which occur in the north or London Basin.

The *Woolwich and Reading Beds* extend from near West Bourn by Bishops Waltham, Michelmarsh and Tytherley, then curving round to the south-west, past Sherfield English; these strata are commonly termed the "Red," or "Plastic" clay; they are worked at many points for brick and tile-making, and preserve their character unchanged over great distances.

The *London Clay or Bognor Beds* occur between West Bourn and Havant, and extend thence to Timsbury. On the south side of the Portsdown ridge we find a repetition of these two series of Tertiary strata, forming North Portsea, Fareham and South Hayling: the beds were once continuous over the top of Portsdown, but all the elevated portion has been swept away. In the Dockyard Extension Works at Portsmouth, in 1870, the London clay was finely exposed in sections 500 feet long and 60 feet or more in depth, and the "Plastic Clay" lying beneath it was also reached. Large collections of fossils were made by Messrs. C. J. A. Meyer and C. Evans, some of which were new to science, and have been since described by Dr. H. Woodward (see Quart. Journ. Geol. Soc., vol. xxvii. p. 74 and p. 90).

The *Bagshot Beds.*—These are sands and clays extending over a considerable area, forming, indeed, all the south-west of Hants between Southampton and Romsey and the sea, including the New Forest. Several subdivisions can be traced in them: (1.) *The Lower Bagshot Beds.* From clays of this age, at Bournemouth, Messrs. J. S. Gardner and W. S. Mitchell have obtained large numbers of well-preserved plant-remains, chiefly leaves of the cinnamon, fig, the fan palm, beech, maple, elm, acacia, ferns, &c.: on the whole these indicate a warmer climate than now prevails in this latitude. These beds are of fresh-water origin, and were deposited by a large river running from the west.

Next come sands and clays called the *Bracklesham Beds,* and above them the *Barton Beds*; the latter form the north part of the New Forest, and extend to the sea, forming the cliffs at Barton, so noted for the fossils they have yielded. Remains of a few fishes, of a crocodile, and of about 200 species of marine shells have been found, with large *Nummulites,* foraminifera which are named from their resemblance to pieces of money (*nummus,* a coin).

The *Headon Beds* form the south part of the New Forest, reaching from Beaulieu Heath to Hordwell Cliff; at the latter place they include beds of clay which are highly fossiliferous; bones of reptiles and mammals (the *Palæotherium,* &c.) occur with both brackish-water and marine shells. The Headon strata are sometimes called the *Fluviomarine Series,* to indicate their formation in and near the estuary of a great river.

In the *Brockenhurst Series,* which occurs at the place of that name in the New Forest, and at Roydon, Whitley Ridge, and Lyndhurst, we find numerous corals and many marine shells.

Professor Judd would term the Barton Clay *Upper Eocene,* and all the other Tertiary beds here named he places in the OLIGOCENE FORMATION, a term first introduced by the German geologist, Professor Beyrich, for strata, which on the Continent overlie, and are not closely related to Eocene strata.

OUTLIERS OF EOCENE AGE.- At Horsdon Common, Conholt, Windmill Hill, East Stratton, and two or three other places, detached patches of the *Woolwich and Reading Beds* occur, resting on the exposed tract of chalk which now separates the Tertiary rocks of the London and Hampshire Basins. These outliers are but the remnants of beds, certainly some hundreds of feet in thickness, which have been denuded by the combined agencies of the weather and the sea.

PLEISTOCENE OR POST-TERTIARY DEPOSITS. Hampshire has none of the Boulder clay, or erratic blocks, which in counties north of the Thames tell of the passage of glaciers or icebergs over the surface. But the thick gravel beds, with which the hills are often capped and the plateaux covered, are perhaps of quite as early a date. These occur on ground east of Southampton, 420 feet high, and also cover the New Forest at a considerable altitude; on Portsdown Hill, at a height of 300 feet, we get an old sea-beach. These facts indicate a considerable elevation of the south of the county since the time of formation of these gravel beds at or a little below the sea-level. The chalk of the western part of Hants, bordering Wiltshire, is thickly strewn with blocks of a coarse sandstone, the well-known "grey-wethers," or sarsen-stones: these are probably the remains of a bed of sandstone forming part of the *Bagshot Series*, which formerly extended far in this direction. In the "Geological Magazine" for 1866 (vol. iii. p. 296), there is an account, by Col. Nicolls, of some large sarsen-stones found in the neighbourhood of Southampton, and preserved in the court-yard of the Hartley Institute. At some points, as at Upton Grey, there is much "clay with flints" lying on the chalk, representing indeed its insoluble residue, and formed from it: where this occurs the oak grows well, while on a chalky subsoil we seldom see this tree. Lastly, we get deposits of gravel on the sides and at the bottom of the existing valleys, formed and still forming by the rivers which have scooped out these valleys.

PREHISTORIC MAN.- In gravel beds on Southampton Common and near Bournemouth, occurring at heights of 120 to 160 feet above the sea, and far above the level of the neighbouring streams, flint implements have been discovered by Messrs. Read, Jas. Brown, Stevens, Dr. Blackmore, and others, most of which are now in the Blackmore Museum at Salisbury: they are roughly chipped oval or pointed lumps of flint, and belong to the *Palæolithic* or Older Stone Period: the men who used them were quite ignorant of the use of metals. Other flint tools of lighter and more elegant form, such as arrow-heads, celts or axe-heads, and flakes or knives, have been found at Horndean, St. Mary Bourne, Portsmouth, Andover, Bishopstow, &c.: these mostly occur on or near the surface, and are usually rubbed or polished: they are assigned to the *Neolithic* or Newer Stone Period.

No. 15.

GEOLOGY OF ISLE OF WIGHT.

NATURAL HISTORY AND SCIENTIFIC SOCIETIES.
Isle of Wight Philosophical and Scientific Society.
MUSEUM.
Ryde Museum.
PUBLICATIONS OF THE GEOLOGICAL SURVEY.
Map.— Sheet 10 (price 4s.).
Books.—The Tertiary Fluvio-Marine Formation of the Isle of Wight, by E. Forbes, 5s.; The Geology of the Isle of Wight, by H. W. Bristow, 6s.

IMPORTANT WORKS OR PAPERS ON LOCAL GEOLOGY.

1851. Mantell, Dr. G. A.—Geological Excursions round the Isle of Wight. 2nd Edit. London.
1859. Wilkins, Dr. E. P.—Geology of the Isle of Wight (with a relief map). 8vo.
1862. Fisher, Rev. O.- Bracklesham-Beds. Journ. Geol. Soc., vol. xviii. p. 65.
1865. Whitaker, W.–Chalk of the Isle of Wight. Journ. Geol. Soc., vol. xxi. p. 400.
1867. Mitchell, W. S. Alum Bay Leaf-Bed. Report Brit. Assoc., p. 146.
1868. Codrington, T.–Strata in Whitecliff Bay. Journ. Geol. Soc., vol. xxiv. p. 519.
1870. Codrington, T.—Superficial Deposits of the Isle of Wight. Journ. Geol. Soc., vol. xxvi. p. 528.
1871. Browne, A. J.- Valley of the Yar. Geol. Mag., vol. viii. p. 561.
1880. Judd, Prof. J. W. —Oligocene Strata of the Hampshire Basin. Journ. Geol. Soc., vol. xxxv. p. 137.
1881. Keeping, H., and Tawney, E. B.—Beds at Headon Hill and Colwell Bay. Journ. Geol. Soc. vol. xxxvii. p. 85.

See also *General Lists, p.* xxv.

THE rocks of the Isle of Wight offer to the student an epitome of the geology of the south-east of England. In the splendid sections exposed along the line of coast we are enabled to examine the strata, and trace their relations to each other in a manner which is impossible in inland districts, while the beauty of the scenery, originated by and dependent upon the arrangement of the rock-masses, lends an additional charm to the investigation.

Such a region would be likely to attract, and has attracted, much attention from geologists. In the map of the Island (Sheet 10), constructed by the officers of the Geological Survey, we have an exact representation of the position at the surface occupied by each formation, while the horizontal (Sheet 47) and vertical (Sheet 25) sections also published by the Survey, show the lie of the beds, their thickness, &c., with the greatest accuracy; descriptive of this map and the sections we have an excellent memoir by Mr. Bristow, published in 1862; Professor E. Forbes in 1856 well explained the structure of the northern half of the island in his memoir on "The Tertiary Fluvio-marine Formation of the Isle of Wight."

Of earlier workers we must note Mr. Thomas Webster, who wrote some good descriptions of the strata in Sir H. C. Englefield's "Description of the Isle of Wight," published in 1816.

Dr. Mantell's book, "Geological Excursions round the Isle of Wight" (2nd Edition, 1851), has numerous plates of fossils and is well known, whilst Mr. E. P. Wilkins has also written a capital general account of the geology of the Island.

In the volumes of the Palæontographical Society the shells which occur in the Tertiary beds have been admirably figured and described by Messrs. Edwards and Searles Wood, and the reptiles of the Wealden Formation by Professor Owen; while Mr. J. S. Gardner has undertaken the description of the Eocene plant remains. In the earlier publications of the Geological Society of London, many valuable papers on special points have appeared by Dr. Fitton, Sir Charles Lyell, Professor Owen, Messrs. F. W. Simms, J. S. Bowerbank, and S. P. Pratt, while Professor Prestwich aided greatly in clearing up the relations of the Tertiary strata to each other. Of later date, in the "Quarterly Journal" of the Geological Society, we have Professor Judd's paper on the Punfield Beds (vol. xxvii. 1871), Mr. W. Whitaker on the chalk (vol. xxi. 1865), Mr. Codrington on the gravels (vol. xxvi. 1870), and Mr. J. W. Hulke on the great reptilian bones so diligently collected by the Rev. W. Fox. Quite recently Professor Judd has put forth a fresh classification of some of the newer strata, but his views have been opposed by Messrs. H. Keeping, Tawney and others, and the question is still under discussion.

The dweller in London can see specimens of almost all the rocks and fossils which occur in the Isle of Wight, arranged in the Geological Museum in Jermyn street; here too he can examine the beautiful models of the island constructed by the late Captain Ibbetson, which afford not only a faithful representation of the topography, but express the geological structure in the most accurate manner. New and excellent maps of the Isle of Wight on the scales of 6 inches and 25 inches to the mile have lately been published by the Ordnance Survey.

In explaining the position and nature of the strata it seems the simpler plan to commence with those which are the oldest or first formed. These will be found underlying all the others, and from them we can ascend gradually, passing over rocks some thousands of feet in thickness, until finally we arrive at such deposits as the river-gravels or shingle-beaches which are now in course of formation.

NEOCOMIAN (or LOWER CRETACEOUS) FORMATION. THE WEALD CLAY.—The beds called by this name are exposed on the south-west side of the island along the coast between Compton Bay and Atherfield, and they appear at the surface as far inland as Mottestone and Brixton; on the east coast they are again seen over a small area round Sandown; altogether they perhaps form five square miles of the surface. These two areas of exposed Wealden strata are of course connected, and the beds run uninterruptedly across the Island, but over all the central part they are covered over and concealed by newer beds; a boring made at Godshill, Arreton, Kingston, &c., would pierce these later-formed rocks and enter the Wealden strata at a depth of a few hundred feet.

The lowest Wealden strata are seen on the shore at Brook Point; here at the base is a thick bed of pale sandstone, above which come red and green marls; these dip or incline on one side to the north, and on the other to the south and east, marking in fact that bending or arching of rock-beds which geologists call an anticlinal curve. In the variegated marls on the shore there occur numerous fragments of the trunks of pine-trees, forming what has been called the "*Pine raft*"; the stems are of a black colour and have been completely mineralized, yet the woody structure, the bark, and the rings of annual growth can still be traced; thin slices examined under the microscope show the dotted vessels characteristic of pine-trees. No roots are found, and it is not probable that these trees grew where we now find them; they were probably swept here by the current of a mighty river, just as thousands of similar trees are now carried down by the Mississippi and other large rivers until on reaching the estuary they become water-logged and sink to be buried underneath the sand, mud, &c., which go to form the delta. Iron pyrites is a common mineral at Brook Point, its yellow crystals filling up cracks in the tree-trunks, some of which are 2 or 3 feet in diameter and 20 feet in length.

Walking southwards along the shore past Chilton Chine and Grange Chine we note a continuance of the variegated marls, clays and sandstones, and also see the inconstant character of the different layers, the sandstone beds especially thinning out and disappearing altogether in very short distances. The thickness of the true Wealden beds exposed here is about 300 feet, which is of course not the total thickness inasmuch as the base is not seen. Mr. Bristow believes them all to represent the "Weald Clay" of Sussex. In the Sandown neighbourhood similar beds can be seen at low tide on the shore, and the trunks of pine-trees occur there also. The fossils of the Wealden Beds are of fresh-water species, and include the minute crustaceans called *Cyprides*; shells, as *Unio Valdensis* and *Paludina Sussexensis*; scales and teeth of a fish, *Lepidotus*; and, most remarkable of all, great bones of extinct reptiles. Of one of these, the *Iguanodon*, Mr. Beckles has found the footprints on a bed of sandstone on the shore west of Brook Point, and at Brook Chine, Sandown, and several other localities, the Rev. W. Fox has found bones both of the *Iguanodon*, the *Plesiosaurus*, and a reptile new to science—the *Hypsilophodon*. Thus in these deposits of Wealden age we have evidence of a mighty river which flowed from west to east, draining a continent which stretched far into the Atlantic; the strata formed in its estuary now extend from the Vale of Wardour in Wiltshire to the Boulonnais in France, a distance of 200 miles, and from the Isle of Wight to Quainton in Bucks, about 100 miles; the delta of the Ganges affords us at the present day an instance of a tract of land similar in origin and of about the same area.

The Punfield Series. These were so named by Professor Judd from their occurrence in Punfield Cove on the north side of Swanage Bay in Dorsetshire: in the Isle of Wight they consist of grey shales with some beds of limestone and sandstone, altogether about 200 feet thick. They are seen at Brixton Bay, Atherfield Point, and Sandown Bay to overlie the Wealden Beds proper. They contain both freshwater and marine fossils and show the coming on of an open sea, the old delta sinking and the coast line of that period slowly receding westwards. Fibrous carbonate of lime, commonly called "beef," is common in the limestone bands, which are used locally for rough paving and which are sometimes quite made up of small oysters; in Compton Bay the Punfield series have a high northerly dip and are traversed by several small faults; they are also well seen at Cowleaze Chine and Barnes High: the sandy beds at the latter place have yielded many reptilian bones to the local geologists.

THE LOWER GREENSAND. -The strata known by this name are about 900 feet in thickness; they are of marine origin, and show a continuance of that depression of the earth's surface in this district which we have just alluded to. At the junction with the Punfield Series is a stratum about 2 feet thick, full of a large bivalve shell called *Perna Mulleti*; next come 60 feet of clayey strata called the *Atherfield Clay*, because it is well exposed at the place so named; then come alternations of sandy clays and concretionary sands called the "Crackers" from the noise the sea makes in breaking over the ledges of this rock which, near Atherfield Point, stretch far out into the sea; they contain numerous remains of crustaceans (*Astacus*), for which reason two layers are known as the "lobster beds;" above these come 400 feet of sandy clays with the shells *Terebratula sella*, &c., and the uppermost beds are ferruginous sands about 250 feet thick, in which fossils are scarce: these upper sands were formerly worked for glass-house purposes at Rocken End. The Lower Greensand is well exposed on the coast from Atherfield Point by Whale Chine, Walpen Chine, and Blackgang Chine to Rocken End; it stretches across the Island from Arreton and Shorwell on the north by Newchurch, Godshill, and Kingston to Shanklin and Chale on the south, extending westwards in a narrow outcrop, past Brook Church to Compton Bay; it is the "iron sand" of Webster, and the "Shanklin sand" of the natives; it also forms the basis of the "Undercliff." At Blackgang Chine a chalybeate spring issues from a bed of soft sandstone, 130 feet above the sea. In his memoir Mr. Bristow gives a list of 137 species of fossils from the Lower Greensand of the Isle of Wight, 111 of which were shells.

THE CRETACEOUS FORMATION. The lowest division of the great series which takes its name from its most characteristic member, the chalk (Latin *creta*, chalk), is called the *Gault*: this is a bed of blue clay, sometimes sandy, and with minute spangles of mica; its thickness here is 100 feet; it is known in the Island as the "blue slipper," from the tendency of the overlying strata to slip or slide over its surface. "The beautiful and romantic scenery of the Undercliff or 'Back' of the Island has been mainly produced by the foundering of the superincumbent strata over the Gault clay, when the latter has been rendered unctuous by the water which (after percolating through the overlying beds) furnishes the land springs which break forth at its surface." It is these springs, too, which cut narrow ravines or gullies, locally termed "chines," in their short and rapid passage to the sea. The Gault can be traced from Compton Bay curving round to Gatcombe, and thence just north of Yaverland to the sea; along this line it dips to the north. Crossing the broad valley excavated in the Lower Greensand we again find the Gault as a narrow band with an outcrop about 300 yards wide encircling the southern range of Downs, but here it dips to the south: evidently these northern and southern outcrops were once connected, and of the mass which once stretched right across, we have a fragment left in the outlier at Gossard Hill. The Gault clay is recognisable by the damp and rush-grown nature of its soil, which is everywhere in pasture. Fossils are few, but in the gully west of the hotel at the top of Blackgang Chine, such shells as *Inoceramus sulcatus* and *I. concentricus* have been found.

The Upper Greensand. This division is composed of sandstone and sandy beds containing chert (an impure variety of flint), about 150 feet in thickness: it is also known as the Firestone and Malm Rock. We can trace it as a narrow band from Compton Bay running eastwards along the southern foot of the chalk hills, to the cliffs in Sandown Bay beneath Bembridge Down; then on the south side of the Island the Upper Greensand forms the long lower slopes of the Downs between Chale, Niton, and Shanklin, having too an outcrop half-a-mile broad for some distance north and south of Gatcombe; it is finely exposed in the cliffs on the right-hand side of the road in going from Ventnor to Blackgang. On Shanklin Down the higher beds are quarried for building-stone; when first raised the stone is soft and easily worked, but it rapidly hardens when exposed to the atmosphere. Most of the old churches on the south side of the Island are built of this material: the chert furnishes an excellent stone for road-mending, being much tougher than the chalk flints. Fossils are not numerous, but the branching arms and stems of a sponge (*Siphonia*) are frequent on the surfaces of the sandstone beds; the shells, such as *Nautilus Fittoni*, *Trigonia caudata*, &c., are all of marine species.

The Chalk.— The base of this formation is marked by a bed of yellowish marl 5 feet thick, called the *Chloritic Marl* from the presence of dark green grains of the mineral *glauconite;* it contains many fossils, as *Scaphites*

Ammonites varians, &c., with many rounded phosphatic nodules called coprolites.

The chalk of the Isle of Wight has been minutely studied by the well-known French geologist, Dr. Charles Barrois, and the following table, showing the thickness of the sub-divisions and their characteristic fossils, is taken from his valuable work.

		Feet	
Chalk-with-flints	Zone of *Belemnitella* 265 " *Micraster coranguinum* 525 " *Micraster cortestudinarium* 165 " *Holaster planus* 65		1,020
	Chalk Rock, 8 or 10 inches.		
Chalk-without-flints	Zone of *Terebratulina gracilis* 65 " *Inoceramus labiatus* 130		195
Chalk Marl	*Turrilites*, and *Scaphites æqualis*		115
Chloritic Marl	Zone of *Ammonites laticlavius*		6
	Total thickness		1,336

The Chalk-Marl is of a greyish tint; it is dug for lime-burning and sometimes for building-stone: there is a large quarry at the west end of Brixton Down, and the interior of Brixton church is partly built of this stone. Stems of a plant *Clathraria Lyellii* are found in it.

The common *White Chalk* which succeeds is a pure earthy limestone, which is divisible into a lower portion—without flints—and an upper, in which black nodules of that substance are seen at intervals of a few feet, forming lines parallel to the planes of bedding of the rock. These two divisions are separated by the bed which Mr. Whitaker has called the *chalk-rock*; this is a hard, cream-coloured band of limestone about a foot thick; it often has a layer of green-coated nodules on top, with lumps of iron pyrites: it is seen in a chalk-pit on the south-east corner of Shalcombe Down, in pits on the south side of Mottestone Down, in the quarry east of Carisbrooke Castle, in a disused pit on the western end of Arreton Down, and in the pits about a mile south-west of Brading Church.

The White Chalk forms the ridge of hills which stretches east and west right across the Island, from Culver Cliff to the Needles, isolated masses of rock standing in the sea and marking the former extension of the chalk, which once was continuous with that of Dorsetshire. At the extremities of the ridge the chalk is seen to dip very steeply, almost vertically where in contact with the Tertiary Beds, to the north, so that the outcrop is very narrow, from a quarter (Afton Down) to half a mile wide, but going inland the dip decreases and the outcrop widens to three miles between Mottestone and Carisbrooke. The hills on the south side of the Island, called St. Catherine's Down (830 feet above the sea) and Shanklin Down, are outliers of chalk having a slight southerly dip; these hills were once continuous with the main central ridge, but in the course of long ages the beds forming the top of the arch or curve which connected them have been worn away by the agencies of the weather and by the sea. The sea-cliffs formed by the chalk are striking objects from their whiteness and verticality: the latter is due to the joints or cracks by which the rock is traversed, and by which it is divided into a series of more or less regular blocks; the sea attacks and wears away the base of the cliffs, and then the upper portion falls down; the hard flint nodules are used by the waves as missiles, which they hurl with great force against the face of the rock, thus making it to aid in its own destruction.

The chalk hills, or downs, as they are called, are easily distinguished by the bold, rounded outline which they present: scantily furnished with soil, they are covered by dense short herbage on which sheep do well, but woods are wanting, though the beech in places grows fairly well: the surface is dry, for the chalk is a very porous rock and rapidly absorbs the rain. Good sections are exposed in the numerous pits where the chalk is extracted for lime-burning, while the flints furnish fair road-metal. Fossils are frequent, such as the sponges called *Ventriculites*, sea-urchins of the genera *Micraster*, *Ananchytes*, &c., shells like *Inoceramus*, *Pecten*, *Terebratula*, &c., but, in fact, the mass of the white chalk is made up of the minute shells of *foraminifera*, which are only visible under the microscope, so that geologists class chalk as an *organically-formed* rock. The general opinion is that it accumulated on the floor of an ocean, in a similar manner to the whitish ooze, which deep-sea soundings bring up from the bed of the North Atlantic. The so-called Agates, or Isle of Wight pebbles, which are common on the beach between Ventnor and Sandown, are really rolled chalk-flints, which owe their beauty to the silicified sponges they often contain: the brown markings on them are due to the presence of peroxide of iron: the "Isle of Wight diamonds" are fragments of rock-crystal or quartz.

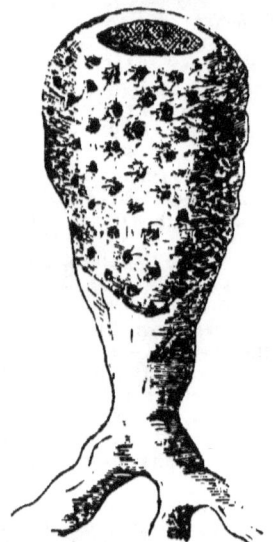

Fig. 35.—*Ventriculites simplex*, a sponge from the chalk.

The Tertiary Epoch. Of the great interval of time which elapsed after the deposition of the chalk and before the formation of the lowest Tertiary strata, we really know little or nothing. We are certain that the time-interval must have been immense, for in the newer beds, which we are now going to describe, we find the remains of an entirely new set of creatures, resembling more nearly those of the present day, with which, indeed, a few of them are identical. Hence Sir Charles Lyell called the lowest Tertiary beds the Eocene, a word which means the "dawn of new life."

Eocene Formation. The lowest strata of this division are called the *Woolwich and Reading Beds;* they were formerly known as the Plastic Clay. They are seen on either side of the Island resting on a rather uneven surface of chalk, in Whitecliff Bay on the east, and in Alum Bay on the west: at the latter place their thickness is 84 feet, but they thicken eastwards to 200 feet in Whitecliff Bay; they have been so upheaved as to be now standing on end, or nearly vertical, inclining to the north; they consist of red, white, and blue mottled clays, with some brownish sands: fossils are wanting: at Newport coarse pottery and tiles are made from the clays of this age.

London Clay or Bognor Beds. - These vary from 200 feet in thickness in Alum Bay to 300 feet in Whitecliff Bay; they are divided from the Reading Beds by a band of flint-pebbles, and consist of grey and brown sandy clays, from which bricks are made at Newport. The fossil shells, such as *Pinna affinis*, *Panopæa intermedia* and *Pholadomya margaritacea* show the London Clay to have been formed in a sea some 600 feet deep, and probably in the proximity of land.

The Bagshot Beds. This is the term applied by the Geological Survey to a great thickness of strata (altogether about 1,200 feet) composed of sands and clays, the sub-divisions of which were first made by Mr. Trimmer in 1850.

Lower Bagshot Beds.—These are 660 feet thick in Alum Bay, but only 142 feet in Whitecliff Bay. They are mainly sands, but contain beds of pipe-clay, in which, at Alum Bay, the remains of beautifully-preserved leaves of land plants occur. Four species of *Leguminosæ* (the pea and bean order) have been found, long thick leaves of fig trees, and the palmated leaves of an *Aralia*.

The *Middle Bagshot Beds* are 411 feet thick in Alum Bay and 710 feet in Whitecliff Bay: they have been sub-divided into a lower or *Bracklesham Series*, (111 feet thick), and an upper series—the *Barton Clay* (400 feet): these are separated by a bed of flint-pebbles. In Alum Bay there are beds of wood-coal, or lignite, 2 feet thick, in the Bracklesham Beds, and the rootlets can be traced in seams of clay which underlie these lignite beds. The Rev. O. Fisher has minutely studied the same strata in Whitecliff Bay (see Journal of Geological Society, vol. xviii.) and found *Cardita planicosta* to be the most common fossil shell.

The *Upper Bagshot Beds* are pure white and yellow sands, about 150 feet thick, which have long been worked in Alum Bay for glass-making purposes: between 1850 and 1855 there were shipped from Yarmouth 21,984 tons of sand mostly to Bristol and London.

All these Bagshot Beds can be traced with some difficulty right across the Island, but it is in Alum Bay that they are best seen: here the rich and variegated colours of the sands and marls, combined with their almost vertical position, form a scene of surpassing beauty.

UPPER EOCENE OR FLUVIO-MARINE SERIES.—The lower and middle Eocene strata are all of marine origin, but the upper beds indicate an elevation of the sea-bottom, and are of a fresh-water and estuarine character, having been deposited in and near the mouth of a great river: their divisions and relations were thoroughly worked out by Professor E. Forbes between 1848 and 1853.

The *Headon Series* form the central and lower part of Headon Hill, where they are 180 feet thick: they extend thence past Newport to Whitecliff Bay; east of the Medina river they occupy a large area, having been elevated by a fault or dislocation of the strata, which runs nearly north and south along the bed of the river: the "throw" of this fault is about 100 feet, so that the Headon Beds at East Cowes have been elevated 100 feet above their proper position. At Headon Hill the lowest beds contain such fresh-water shells as *Planorbis euomphalus* and *Cyrena cycladiformis*, with a land shell, *Helix labyrinthica*: higher beds contain brackish-water shells, as *Potamides cinctus*, and even marine forms as *Venus incrassata*: higher still are beds of fresh-water limestone, with *Paludina* and *Cyrena*.

The *Osborne, or St. Helens Series*, are red and green sands and clays, about 80 feet thick, containing fresh-water shells and the seed-vessels of *Chara*, a fresh-water plant: a bed of sandstone, called the Nettlestone Grit, has been much used for building purposes near Ryde.

The *Bembridge Series*.—This is the most important division of the Fluvio-marine Formation. As we go northwards from the chalk we find the high dip of the strata gradually to decrease until the Bembridge Beds are nearly horizontal; as a consequence of this they occupy a large surface area, forming the greater portion of the surface of the Island north of the chalk downs: they consist of 20 feet of soft cream-coloured limestone, overlaid by 40 feet of marls: these two divisions are separated by an oyster bed. The limestone occurs at Binstead, Bembridge, Cowes, Gurnet Bay, Sconce, the upper part of Headon Hill, and running out to sea forms Hempstead Ledge: the quarries at Binstead have been worked for centuries. The most remarkable fossils are the bones of large mammals, chiefly of the genus *Palæotherium*, a creature resembling the Tapir.

MIOCENE FORMATION.—In considering the *Hempstead Beds* as belonging to the Miocene Epoch, we follow the classification of Sir Charles Lyell: these strata are 170 feet thick at Hempstead Cliff near Yarmouth, and they also form Parkhurst Forest: at the base is a band of black laminated clay, containing a Miocene shell—*Rissoa Chastelii*, then come green and brown marls with estuarine shells, and the top 10 feet are of marine origin, containing the shells *Corbula pisum* and *Voluta Rathieri*. These beds form a poor soil, usually waste, pasture, or woodland.

We have given above the usual classification of the Upper Tertiary strata of the Isle of Wight. Very recently, however, Professor Judd has put forth a new arrangement of the beds, which is well worthy of careful consideration; the Barton Clay he classes as Upper Eocene; all the strata between the Barton Clay and the top of the Hempstead Beds he places in the *Oligocene Formation*, and divides them into (1) Headon Group (including the Upper Bagshot Sands); (2) Brockenhurst Series; (3) Bembridge Series; and, (4) Hempstead Series. The term Oligocene was introduced by Professor Beyrich in 1854, as a name for certain deposits on the Continent, which lay between the Eocene and Miocene Formations, and were not closely related to either. Professor Judd clearly shows that there are three deposits containing the remains of marine animals (*viz.* the Barton Clay, Brockenhurst Series, and Hempstead Series), while the intervening Headon Group and Bembridge Group are of estuarine origin.

THE DRIFT. Of the boulder clay of the Northern and Midland Counties we find no trace in the Isle of Wight, neither are there any signs of the passage of glaciers over its rocks. The gravels, however, which cap the hills on the northern side of the Island, are certainly of high antiquity: they occur on Headon Hill (390 feet above the sea), Hempstead Cliff (200 feet), &c. Mr. Codrington has shown that these *high-level gravels* were once connected, and that they were formed on an old table-land, whose surface sloped to the north: at that time the Isle of Wight was not separated from the mainland, and the Solent was a river flowing into the sea somewhere near Spithead, and of which the Test, the Medina, &c., were tributaries: the high-level gravels were deposited in the estuary of this river, and the land

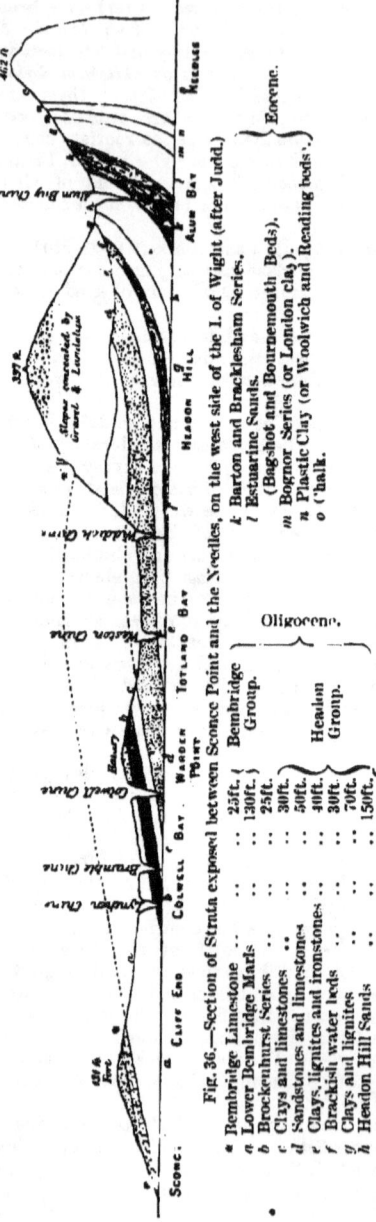

Fig. 36.—Section of Strata exposed between Sconce Point and the Needles, on the west side of the I. of Wight (after Judd).

a Bembridge Limestone ...... 25ft.
a' Lower Bembridge Marls ...... 130ft.
b Brockenhurst Series ...... 25ft.
c Clays and limestones ...... 30ft.
d Sandstones and limestones ...... 50ft.
e Clays, lignites and ironstones ...... 40ft.
g Brackish water beds ...... 70ft.
h Headon Hill Sands ...... 150ft.

k Barton and Bracklesham Series.
l Estuarine Sands.
 (Bagshot and Bournemouth Beds).
m Bognor Series (or London clay).
n Plastic Clay (or Woolwich and Reading beds).
o Chalk.

has since undergone a considerable elevation. Mr. S. V. Wood believes that these gravels date back to the early part of the Glacial Period, and mark a submergence of the land which then occurred.

Of later date are the brick-earths and marls at Tollands Bay, Wotton Creek, Brook Chine, and the Foreland. In beds of sand and peat at Brook and also near Bembridge Point trunks of trees and hazel-nuts have been found: the latter are called "Noah's nuts" by the natives: in the gravels at Freshwater Gate teeth of the Mammoth (an extinct species of elephant) have been found.

PREHISTORIC MAN.—But few traces of those early dwellers in our islands who were unacquainted with the use of metals have occurred in the Isle of Wight; in a bed of brick-earth at the Foreland, about 80 feet above the sea, Mr. Codrington found an ovate *palæolithic* flint implement, about 4 inches long by 3 in width: of the later, or *Neolithic* Stone Age, there is a polished flint celt or axe-head, five inches long, in the British Museum, labelled "from the Isle of Wight."

No. 16.

GEOLOGY OF HEREFORDSHIRE.

NATURAL HISTORY AND SCIENTIFIC SOCIETY.
Woolhope Field Club ; Hereford. Transactions.

PUBLICATIONS OF THE GEOLOGICAL SURVEY.

Maps.—Quarter Sheets : 42 N.E. Hay, Talgarth ; 42 S.E. Abergavenny, &c. 43 N.W. Woolhope ; 43 N.E. Woolhope to Malverns ; 43 S.W. Forest of Dean to Monmouth ; 43 S.E. Forest of Dean to the Severn.

IMPORTANT WORKS OR PAPERS ON LOCAL GEOLOGY.

1856. Banks, R. W. On the Downton Sandstones near Kington. Journ. Geol. Soc., vol. xii. p. 93.
1860. Symonds, Rev. W. S.—On the "Passage Beds," at Ledbury. Journ. Geol. Soc., vol. xvi. p. 93.
1871. Brodie, Rev. P. B. - On the "Passage Beds" near Woolhope. Journ. Geol. Soc., vol. xxvii. p. 256.
1872. Murchison, Sir R. I.— Siluria. 5th edition ; Murray, 18s.
1880. Callaway, Dr. C.—Pebidian Rocks in Malvern Hills. Q. Journ. Geol. Soc., vol. xxxvi., p. 536.

See also *General Lists, p.* xxv.

THE entire surface of the county of Hereford has been surveyed and mapped by the Government Geological Survey. These maps are on the scale of one inch to a mile, and are simply indispensable to every working geologist ; but amateur or at least private students have also made known to us much about the structure of the rocks which here form the crust of the earth." To Sir Roderick I. Murchison we owe the recognition of the Silurian rocks as a distinct system, and with his name must be associated that of the Rev. T. T. Lewis. Messrs. Salwey and Lightbody of Ludlow, and Henry Brookes of Ledbury, have been indefatigable collectors of fossils ; Professor Ray Lankester has described many of the Old Red Fishes in the volumes of the Palæontographical Society, while the Revs. W. S. Symonds and P. B. Brodie, with Messrs. H. E. Strickland, R. W. Banks, and Curley, have done good local work. Dr. Grindrod has collected in his museum, at Townsend House, Malvern, a most interesting and indeed splendid collection of local fossils and rocks, and the Malvern College museum also contains a good series.

Herefordshire may be styled the county of the Old Red Sandstone, for this rock constitutes nine-tenths of its surface, spreading in long undulations right down from west to east, and having an average height of from 300 to 400 feet above the sea ; from the Ordnance Survey we find the height of Hay (Parish Church) to be 304 feet, Weobley 325, Leominster (Town Hall) 250 feet, and Kimbolton near Leominster (Church) 415 feet. Through the plain of Old Red still older rocks break at three points, viz., the Malvern Hills on the east, near Woolhope in the south-east, and between Ludlow and Presteign in the north-west; lastly over the surface there are spread beds of clay and gravel called Drift, which geologists have shown within the last few years to be the relics of a glacial period, when ice covered the high grounds, or came sailing over the seas in the form of bergs.

We shall describe in order all these rock-strata, commencing with the oldest and first-formed, which, although they lie below all the others, yet now, like a geological paradox, constitute the highest elevations.

PRE-CAMBRIAN, OR LAURENTIAN FORMATION. — The Malvern Hills, which bound Herefordshire on the east, are a long ridge running north and south, and attaining at the Worcestershire Beacon a height of 1,396 feet; they are formed of gneissic, granitoid and schistose rocks, whose strike is from north-west to south-east. Dr. Holl has long stated his conviction that these Malvern rocks are truly stratified though highly altered deposits, and that they are to be classed with the oldest rocks known, perhaps with those which in Canada are called Laurentian; but at all events, as they clearly underlie Cambrian strata, they may with safety be termed "Pre-Cambrian."

On the east the Malvern ridge is bounded by a great fault or dislocation, which brings Triassic strata on a level with beds which should, in a normal state of things, lie many thousands of feet below them. It is owing in part to such great movements of the earth's crust, and in part too to their superior power of resisting wear and tear, that the old Malvern rocks now stand up

Fig. 37.—Section from Ledbury to the Malvern Hills, length about four miles.

j Old Red Sandstone.
i Shaly Sandstones.
h Aymestry Limestone.
g Lower Ludlow Shales.
f Wenlock Limestone.
e Wenlock Shale.
d Woolhope Limestone.
c Upper Llandovery Sandstone and Conglomerate.
} Silurian.
b Hollybush Sandstone and Black Schists with Olenus.
} Cambrian.
x Granitic Rocks, flanked by (a) Crystalline Felstones, Schistose, and Gneissic Rocks.
} Pre-Cambrian.

high above the much newer or later-formed beds by which they are surrounded. It is probable also that a line of fault runs along the west side of the ridge; but it has not as yet been so clearly traced as that on the eastern side.

THE CAMBRIAN FORMATION.—On the western slope of the Malverns there is a large quarry, in Raggedstone Hill, where grey and brown sandstones with a quartz conglomerate at the base, are seen resting unconformably on the pre-Cambrian gneiss. These beds are termed "*Hollybush Sandstone*," after the adjoining valley, and are from 200 to 600 feet in thickness; they have yielded a few fossils—an annelid or boring worm, *Trachyderma antiquissima*, and shells named *Lingula squamosa*, *Obolella Phillipsi*, &c.

Above the sandstones come the *Malvern black shales*; their thickness is from 500 to 1,000 feet, and they include some beds of trap-rock, doubtless ancient lava-flows, which spread from submarine vents over the ancient sea-bottom. These shales extend westward by Coal Hill to the edge of Howler's Heath; of fossils they have yielded several species of trilobites, *Agnostus*, *Conocoryphe*, *Olenus*, &c., but the closest examination is needed to detect these evidences of the presence of life; in fact, the best way is to take home quantities of the shale in a basket, and examine it at leisure with a lens. The upper shales are of a greenish tint and contain *Dictyonema sociale*, a fossil of a net-like appearance; they are exposed near Hayes Copse. The two divisions of the Cambrian beds which we have now described are considered to be of the same period as the Tremadoc beds of North Wales.

L

THE SILURIAN FORMATION.—Further west still, and resting on the Cambrian strata just named, we find rocks of Upper Silurian age, the entire Lower Silurian series being absent. First, we see the *Upper Llandovery* or *May Hill Sandstone*, composed of grey and purple sandstones, shales, and conglomerates, about 1,000 feet in thickness; from Cowleigh Park they run southwards through West Malvern, but just west of the Herefordshire Beacon they are cut out altogether by a fault; re-appearing west of Swinyards Hill, they stretch westwards past the Obelisk, and form Howler's Heath. Above these beds, we find the *Woolhope Limestone*, here rough and impure, which is exposed north of Crumpend Hill and near the Wych; it is 150 feet thick; then come the *Wenlock Shales* (640 feet), and the *Wenlock Limestone* (280 feet), surmounted by the *Ludlow Beds*, which contain a nodular representative of the *Aymestry Limestone*. In all these beds the usual Silurian fossils - trilobites, brachiopods, crinoids, &c., occur. Both Cambrian and Silurian strata run on the whole in a north and south direction, parallel to the Malvern ridge, but opposite each end of this elevation they bend round to the westward; notwithstanding minor undulations, they have an inclination or dip at varying angles to the westward, in which direction they are soon overlaid and concealed from view by the Old Red Sandstone. The passage beds between the two great formations were well exposed during the excava-

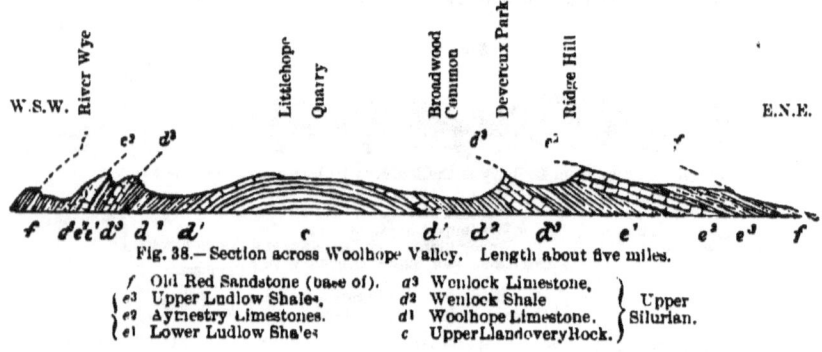

Fig. 38.—Section across Woolhope Valley. Length about five miles.

f Old Red Sandstone (base of).	a^3 Wenlock Limestone, ⎫
*e*³ Upper Ludlow Shales,	d^2 Wenlock Shale ⎬ Upper
*e*² Aymestry Limestones.	d^1 Woolhope Limestone. ⎭ Silurian.
*e*¹ Lower Ludlow Shales	*c* Upper Llandovery Rock.

tion of the Ledbury tunnel, on the Worcester and Hereford Railway, in 1860.

The Woolhope District.—Here exactly the same series of Silurian beds which we have just been noting at Malvern is brought up in what Sir R. Murchison called a valley of elevation. They have been bent into a dome-shaped mass, of which the top has since been worn off by those active agents of denudation—rain, rivers, frost, ice, &c., so that the oldest rock is now seen in the centre, while the other beds form ridges or valleys running round it, according to their respective hardness. The *May Hill Sandstone* forms Haugh Wood in the centre (570 feet above the sea). This is encircled by a broad band of *Woolhope Limestone*, then there is an inside slope of *Wenlock Shale*, with a ridge beyond of *Wenlock Limestone;* next the *Lower Ludlow Shales* form a narrow valley, beyond which is a lofty encircling series of *Aymestry Limestone*, and the outside slope of all is formed of the *Upper Ludlow Shales*. The Old Red wraps round the whole. The symmetry of the valley would be perfect, but for a fault which runs across the north-west corner below Backbury Hill. There are large quarries in the *Wenlock Limestone* at Dormington, where many fine corals, as *Halysites catenulatus* and *Favosites gothlandicus*, may be obtained. The junction beds with the Old Red are seen near Perton.

At Hugley, about three miles north of the Woolhope district, the *Ludlow*

Beds are again exposed to view over a small area. Altogether the neighbourhood of Woolhope is to the geologist most interesting and instructive: the very name seems to have an inspiring influence, for is not the "Woolhope Naturalists' Field Club" one of the most energetic and hardworking in the west of England!

Ludlow and Presteign District.—Here again the same Upper Silurian rocks are found as at Malvern and Woolhope. The *May Hill Sandstone* is here called "Corton Grit": its base is a conglomerate, blocks of which have been carried (probably by ice) far to the eastward. The *Woolhope Limestone* in this district has been greatly altered or metamorphosed by intrusive masses of igneous rock. From their superior hardness the latter now form conspicuous hills, as Yat Hill, Hunter Hill, Stanner Rocks, &c., the valleys between having been excavated out of the softer Silurian strata: in fact the district between Kington and Old and New Radnor is a most picturesque and remarkable one.

At Aymestrey the limestone bed, which takes its name from that village, is finely exposed in the banks of the river Lugg and in the Garden House quarries. *Pentamerus Knightii* and *Lingula Lewisii* are here the common fossil shells.

Very good Silurian sections are exposed near Ludlow in the banks of the Teme. Here the passage beds, or *Downton Sandstones*, by which the Ludlow Shales are connected with the Lower Old Red Sandstones, are well seen, and the famous Upper Ludlow "Bone Bed," may be found at Ludford lane, &c.; it is a thin stratum (one or two inches thick), literally made of the remains of fishes, crustaceans, &c.

THE OLD RED SANDSTONE.—For England Herefordshire is pre-eminently the county of the Old Red: the total thickness of the formation is here not less than 10,000 feet. The lower beds are red marls with impure limestone concretions called "cornstones," and red and grey sandstones; above these are chocolate-coloured sandstones, red marls and cornstones, and towards the top conglomeratic beds with more red marls. The change of colour from the prevailing grey tint of the Silurians to the deep red of the marls and sandstones is very striking; there is no unconformability, however, between the two series of strata. The fossils of the Old Red are chiefly fragmentary fish remains and traces of both land plants and fucoids, with crustaceans. The fishes belong to the order of *Ganoids*, of which the sturgeon is a good living specimen: species named *Pteraspis, Scaphaspis, Cephalaspis*, &c., characterise the lower beds, while *Pterichthys* and *Holoptychius* belong to the upper strata; the crustaceans *Pterygotus, Stylonurus*, &c., also characterise the lower strata.

From the evidence of the fossils, the colour and nature of the rocks, &c., it is probable that the "Old Red" of Herefordshire and the adjoining parts of Wales was deposited in a large inland sea or lake. On the south this lake was bounded by a ridge running east and west, beyond which was the open sea, in which were being deposited at or about the same time marine strata, containing shells, corals, &c., which now form the limestones, slates, &c., of the Devonian formation: thus we speak of *Devonian* or *Old Red Sandstone*, as the two sets of strata may have been formed contemporaneously or nearly so, but under different conditions. The lower beds of the Old Red Sandstone are well exposed near Leominster, where, in the quarries at Leyster's Pole and Puddlestone, many of the enamelled plates which formed part of the armour or external hard covering of such fishes as *Scaphaspis* and *Cephalaspis* may be found. From quarries at Putley, Perton, Tarrington, &c., all in the Woolhope district, the Rev. P. B. Brodie obtained many interesting crustaceans, relics of plants, &c. As we walk south and west past Hereford to the Black Mountains, we pass over an undulating country, the rounded ridges and hills being formed of the harder beds of sandstone, which have resisted denudation better than the softer marls. On gaining the summit of the Brecknockshire Beacon (2,862 feet), we find the highest beds of Old Red occur, the carboniferous strata being just denuded off. It is of this spot that Sir R. I. Murchison writes in "Siluria:"— "The grandest exhibitions, however, of the Old Red Sandstone in England and Wales appear in the escarpments of the Black Mountain of Herefordshire. . . . In no other tract of the world visited by me have I seen such a mass of red rocks (estimated at a thickness of not less than 10,000 feet) so clearly intercalated between the Silurian and the Carboniferous strata."

By its decomposition the Old Red Sandstone usually produces a strong, loamy, fertile soil: where the Cornstones occur, the lime which they contain results in the production of the best land in the county; many orchards, and a few hop yards, are situated on the red land.

The sandstones are largely quarried for building purposes; the Three Elms Stone near Hereford is well known: there are also quarries near Ledbury and at many other places. Chepstow Castle and Tintern Abbey are both built of Old Red Sandstone. The beds of conglomerate yield millstones for cider-making processes, and, with some of the cornstones and sandstones, are also broken up for road mending. The sandstones are often ripple-marked and false-bedded, indicating shallow waters.

CARBONIFEROUS FORMATION. *The Mountain Limestone* just enters the county in the south-east, forming the picturesque cliffs through which the Wye has cut its way: it is a bluish-grey rock, forming precipitous cliffs or "scars," often penetrated by caves and fissures: iron ore and a little lead ore occur here.

THE DRIFT. Between the time of accumulation of the strata we have now described as forming the "solid geology" of Herefordshire, and the deposit of the beds of gravel, clay, and sand which at many points cover over the subjacent rocks and hide them from our view, there elapsed a period of time so great that we can scarcely define it in years. During this interval all the strata which form the central and eastern counties of England -the coal measures, Triassic, Oolitic, Cretaceous and Tertiary beds were being in succession laid down as sediment on ocean-floors, and this area was being alternately raised far above or depressed below its present level. If any Secondary or Tertiary strata were ever present in Herefordshire they have long since been swept off by the forces of denudation rain, rivers, ice, &c. which have moulded and sculptured the surface of the county into the aspect it now wears.

The Glacial Period. But some of the surface deposits of clay, gravel, &c., bear witness of the last period of intense cold which prevailed in these latitudes, when glaciers covered all the high grounds, and entering the sea sent off detached masses, icebergs, laden with mighty cargoes of mud and rock, to be dropped at a greater or less distance from their native spot. In the *boulder-clay* we have a deposit, either the product of a glacier pushed before its advancing face or accumulated beneath its moving mass, or in some cases dropped from melting bergs: it is a stiff clayey mass, full of stones or boulders of all sizes and shapes: many of these are scratched or striated, and formed of rocks which occur *in situ* at a greater or less distance, often many miles from the spot where they are now found. Unfortunately as yet in Herefordshire these glacial deposits have not been much studied, and we know little of their thickness, contents, or distribution. Mr. Curley has described them in the neighbourhood of Hereford and Ludlow as containing large boulders and with overlying and intercalated beds of gravel and sand, but the subject is still almost a virgin one.

Alluvial Deposits. The rich water-meadows which border the Lugg and other streams have been formed by successive deposits of mud, gravel, &c., during times of flood: horns of the red deer and the old British ox, *Bos primigenius*, are sometimes found in these deposits.

PREHISTORIC MAN. -- In the hill of Carboniferous Limestone called the Great Doward, which rises above the River Wye on its right bank near the limestone escarpment called Symond's Yat, there is a well-known cavern called "King Arthur's Cave": this was explored by the Rev. W. S. Symonds in 1871. In the cave he found the present floor to consist of black soil, with fragments of Roman pottery. Below this came a layer of cave-earth, 3 feet in thickness; it contained implements made of flint, together with bones of the mammoth, long-haired rhinoceros, cave-lion, cave-bear, and other animals now extinct. Beneath this was a deposit of stratified red sand, with rolled Silurian and greenstone pebbles: this Mr. Symonds believes to be part of an old river-bed of the Wye, which then flowed on a level with the mouth of the cave; the river now runs 300 feet down below the cave, so that the period when this bed of sand was formed in the cave must indeed lie far back from the present time: but beneath the sand is a layer of stalagmite, and then, lower down still, more cave-earth containing more stone stools and bones of animals. The evidence which the deposits in this cave give as to the antiquity of man is indeed remarkable and convincing.

No. 17.

GEOLOGY OF HERTFORDSHIRE.

NATURAL HISTORY AND SCIENTIFIC SOCIETIES.
Watford Natural History Society and Hertfordshire Field Club; Watford. Transactions.

MUSEUMS.
Haileybury College Museum.
Watford Free Library and Museum.

PUBLICATIONS OF THE GEOLOGICAL SURVEY.
Maps.—Sheets, 1; S.E. Corner of Herts; 7, South Herts; 47, East Herts; Quarter-sheets, 46 S.E. Hatfield; 46 S.W. Tring, &c.; 46 N.E. Hitchin, &c.
Books.—Geology of the London Basin, by W. Whitaker, 13s.; ditto of north-west part of Essex, and north-east of Herts, by W. Whitaker, 3s. 6d.

IMPORTANT WORKS OR PAPERS ON LOCAL GEOLOGY.
List by W. Whitaker, Esq., of 58 works in Trans. Watford Natural History Society for 1876.
1850. Prestwich, Professor J. On the Basement Bed of the London Clay. Journ. Geol. Soc., vol. vi. p. 252.
1851. Prestwich, Professor J. Water-bearing Strata of the Country round London. 8vo. London.
1854. Prestwich, Professor J. The Woolwich and Reading Beds. Journ. Geol. Soc., vol. x. p. 75.
1858. Prestwich, Professor J. Boulder Clay at Bricket Wood, Watford. Geologist, vol. i. p. 241.
1865. Clutterbuck, Rev. J. C. Water Supply. Journ. Roy. Agric. Soc., Series 2, vol. i. p. 271.
1868. Hughes, Professor T. McK. Plains of Herts and their Gravels.
1868. Wood, S. V., jun. The Pebble-Beds of Herts. Journ. Geol. Soc., vol. xxiv. p. 464.
See also General List, p. xxviii.

IN the Transactions of the Watford Natural History Society (vol. 1 p. 78) there is a very useful list, compiled by Mr. W. Whitaker, of about sixty papers which have been written on the geology of this county. If, however, the student provides himself with the geological maps of the surface issued by the Government Survey, together with the "Memoir on the Geology of the London Basin," by Mr. Whitaker, he will be in possession of almost everything that is known, up to the present time, about the rocks of the district. The Rev. J. C. Clutterbuck has made many valuable observations bearing on the question of the water supply, and Professors Prestwich and McKenny Hughes, Messrs. John Evans, S. V. Wood, jun., Hopkinson, and others, have done good work.

In describing the rocks of Hertfordshire we will begin in the north-west of the county, as we shall there find the oldest beds.

THE CRETACEOUS FORMATION. Westwards of Ivinghoe, and two or three miles to the north of Hitchin and Baldock, the county boundary just touches a stiff clayey bed called *The Gault*, and includes a small portion of a soft white and green sandstone which rests upon it, and is known as the *Upper Greensand*. These beds form part of the plain which runs along the western foot of the chalk hills, and as they hardly enter the county may be dismissed with a mere mention.

The Chalk.—Mr. Whitaker has subdivided this well-known formation as follows:—

	Feet.
Chalk with flints	300
Chalk rock	4
Chalk without flints	400
Totternhoe stone	6
Chalk marl	80

The lowest bed or *Chalk-marl* rests upon the Upper Greensand and forms part of the same low ground; its top bed is a sandy limestone, called the Totternhoe stone; this can be traced from Tring by Miswell, Marsworth, Pirton, and Cadwell to Ashwell. At the *bottom* of the chalk marl we find a bed of phosphatic nodules, wrongly called coprolites; this bed is largely worked at several places between Hitchin and Cambridge.

The *Lower White Chalk*, or chalk without flints, rises in a rather steep slope, above the beds already described; it forms, in fact, all the western slope of the well-known Royston and Luton Downs, and their continuation, the Chiltern Hills. There are comparatively few flints in the Lower Chalk, but they are not altogether absent. Fossils are fairly numerous, but not of many species, the shells *Terebratula* and *Inoceramus* being most common.

Fig. 30.—Section in a Chalk-pit north of Barkway.

a Whitish Boulder-clay; occurs in a hollow at south-west corner
b Upper Chalk, with scattered flints.
c Chalk-rock. Hard cream-coloured crystalline chalk, irregular beds, 2 to 2½ ft. thick.
d Lower Chalk, with such fossils as *Inoceramus, Scalpellum*, &c.

This pit is interesting, as showing a considerable disturbance in the chalk, the beds dipping northwards, or down the escarpment (contrary to the general dip), at as much as 60 degrees.

The *Chalk Rock* is a hard cream-coloured bed which forms the top of the Lower Chalk; it contains layers of green-coated nodules; it is crossed by numerous joints, and rings when struck by the hammer. Owing to its superior hardness it is found near or at the top of the line of chalk hills. We can trace it by Berkhamsted Castle, Boxmoor, Apsley, south-west of Dunstable, Kensworth, at Jack's Hill Windmill, south of Baldock, and thence north-eastward to Lannock Farm.

The Upper Chalk.—This division forms the long and gentle south-east slope, reaching from the hill-tops down to Rickmansworth, Watford, Hatfield, and Hertford, and thus really forming the greater part of the county. It is characterised by numerous layers of black flints which indicate the lines of bedding and show the dip to be very small, not more than three or four degrees, and to the south-east. The *Upper* is also much softer and whiter than the Lower Chalk, its fossils are also different; sea-urchins, as *Micraster, Ananchytes,* &c., and sponges abound. A small "inlier" of chalk is found at Northaw, having been exposed by the denudation of the overlying strata.

To the naked eye chalk appears a white homogeneous rock; but when it is broken up in water and examined under the microscope, the whole mass is seen to be composed of the minute shells of a class of animals called *foraminifera*. Countless millions of these tiny creatures still inhabit our oceans, and as they die their shells sink to the ocean-floor, forming, by their soft incessant rain, great thicknesses of whitish mud which, if hardened and compressed, would form a rock indistinguishable from our chalk. We assign, therefore, to the chalk a similar origin.

Chalk scenery is too well-known to need description. The chalk district of Hertfordshire is, however, much more fertile and better wooded than the tracts of the same rock in the south of England, which form the North and South Downs, &c., and this is due to the covering of clay, brick-earth, sand, gravel, &c., which usually conceal the chalk from view in this county, and which we shall presently describe. On the summit and western slope of the Chiltern Hills the characteristic rounded downs and hollow combes are, however, well shown, as at Ivinghoe, Dunstable, between Sundon and Pirton, &c. Deep wells in the chalk yield large supplies of pure though hard water. The Colne Valley Water Works Company has sunk a large well in the Bushey Meadows, near Watford, to a depth of 235 feet. The water obtained is softened by Dr. Clark's process, and then pumped up to a reservoir on Bushey Heath, 518 feet above the sea, from which it can be conveyed to any point between London and Watford by gravitation alone.

Chalk is burnt into lime both for building and agricultural purposes. It was formerly more used for building than at present, and St. Albans Abbey contains much chalk, or Totternhoe stone, together with many rough flints. For indoor use it is well fitted from its lightness, but when exposed to the weather it rapidly decays, as witness the noble western front of Dunstable Priory church.

The height of a few places on the chalk tract above the level of the sea may here be given: Dunstable toll-house, 558 feet; Great Offley Church, 524 feet; Hitchin Church, 216 feet; Stevenage, 306 feet; St. Peter's Church, St. Albans, 402 feet; St. Andrew's Church, Hertford, 137 feet.

TERTIARY PERIOD.—EOCENE FORMATION.—The south and east of Hertfordshire are composed of beds of clay and sand, between the time of whose formation and that of the chalk there elapsed a period to be measured, perhaps, in millions of years. In these Eocene beds the fossils are entirely of different species to any which we find in the chalk, and some of the shells are identical with those now living.

The Woolwich and Reading Beds.—These are the lowest Eocene beds which occur in Hertfordshire, and they rest directly upon the chalk; they consist of alternations of bright-coloured plastic clays with sandy beds; their thickness here is about 25 feet. We can trace them from Harefield Park to Watford, where there are good sections at Bushey Heath and Watford Heath Kilns; thence they run by Hatfield Park to Hertford, where again there are good sections in the brickyards. At the junction with the chalk there is usually a layer of green-coated flints. Fossils, with the exception of one or two bands of oysters, are of rare occurrence.

The London Clay.—This is of a brownish hue at the surface, but blue when dug at any depth. It covers over the Reading Beds, and is indeed seen occupying the *upper part* of the brick pits already named near Watford, Hatfield, and Hertford. From these points it extends to the south and east until it passes out of the county; it rises gently from its northern boundary into hills of some elevation, forming a second escarpment parallel to that of the chalk and about 16 miles distant from it. The lowest stratum of the London clay is known as the "Basement Bed;" it contains a layer of flint pebbles.

Eocene Outliers.—Detached patches of the Eocene beds above described occur at several points on the broad chalk tract, thus indicating the former extension of these beds over a much larger area to the north and west than they now occupy. These outliers are found at Micklefield Hall, Micklefield Green, Sarratt, Abbots Langley, Bedmont, Bennet's End, Leverstock Green, and on the St. Peter's side of St. Albans.

Deep Borings in Herts.—In two borings lately made by the New River Company (for water), we have most interesting evidence of the existence of rocks of great geological age, lying far below the surface:

I. Boring near Ware:		II. Boring at Turnford, nr. Cheshunt:	
Gravel	14 feet	Gravel	20 feet
Chalk	544 ,,	London Clay	45 ,,
Upper Greensand	77 ,,	Reading Beds	45 ,,
Gault	160 ,,	Chalk	625 ,,
Wenlock Shales	50 ,,	Upper Greensand	25 ,,
(Upper Silurian)	...	Gault	160 ,,
		Upper Devonian...	...

THE DRIFT. With the exception of the upper part of the chalk ridge, the strata we have been describing above, are hardly ever found forming the actual surface of the county. They are generally covered over by irregular beds of gravel, sand, and clay of comparatively recent origin, and which mask and obscure the solid beds that lie below. These surface beds are termed *drift* because they are believed to have been brought from a distance, chiefly by the aid of ice, during the *Glacial Period,* a time when England was more than once submerged, and when icebergs and glaciers came downwards from the north, laden with stones and all kinds of débris from the rocks they had grated over.

Upon most of the chalk district we get a thick layer of clay with flints. This appears to be the result of the decomposition of the chalk during long ages, the soluble part carbonate of lime—having been carried away by rain-water. Gravelly beds which occur at Hertford, Barnet, &c., are of the *Middle Glacial* age, and these are overlaid by the great chalky boulder clay, which is *Upper Glacial;* this is well seen at Bricket Wood, about half way between Watford and St. Albans, where it is twenty feet thick. Lastly beds of gravel and brick-earth occur along the river courses, deposited by the streams in later times.

We must not omit mention of the masses of "Hertfordshire pudding-stone" which are scattered over the surface. These consist of flint pebbles in a siliceous matrix, and come from the Woolwich and Reading beds. The stratum which probably yielded them may still be found between Aldenham and Shenley.

PREHISTORIC MAN.— Several instances are recorded by Mr. John Evans in his great work, "The Ancient Stone Implements of Great Britain," of discoveries in this county of the chipped and polished flint and other stone tools and weapons, which were used by the inhabitants of this country before the discovery of the properties of metals.

In these we can distinguish two periods: an earlier or *Palæolithic* stone age, when the tools, &c., were merely roughly chipped nodules of flint, and a later or *Neolithic* period, when considerable skill in the manufacture of such objects had been attained, as is evidenced by the more delicate and varied forms, and by the fact that most of the objects have been rubbed or polished.

Palæolithic implements have been found by Mr. Evans at Bedmont, near Abbots Langley, in the valley of the Gade; by Mr. Penning, of the Geological Survey, in the neighbourhood of Bishop's Stortford; and by Mr. W. G. Smith in high-level gravels near Hertford and Ware.

Of neolithic tools, the flint axe-heads called celts have occurred at Ware, Albury ($6\frac{3}{4}$ in. by $1\frac{1}{8}$ in. and polished all over), Abbots Langley ($4\frac{1}{2}$ in. long, with the edge intentionally blunted by grinding, so that it was possibly a battle-axe), and Panshanger. Flint flakes or chips, evidently the work of man, have been found near St. Albans, Abbots Langley, and Ware. At Tring grove in 1763 a skeleton was found, with which flint arrow-heads, some polished thin stones (bracers), a ring of jet and two urns had been interred. But with the advent of Man the geological record merges into that of History.

No. 18.

GEOLOGY OF HUNTINGDONSHIRE.

PUBLICATIONS OF THE GEOLOGICAL SURVEY.

Maps. –Quarter Sheets : 52 N.E. Huntingdon ; 52 S.E. St. Neots ; Sheet 64 Peterborough. (Survey not completed.)
Books.--The Geology of Rutland and Part of Huntingdon, by J. W. Judd, 12s. 6d. Geology of the Fen-land, by Skertchley, 40s.

IMPORTANT WORKS OR PAPERS ON LOCAL GEOLOGY.

1847. Mitchell, Dr. J.—On the Drift of Huntingdon. Journ. Geol. Soc., vol. iii. p. 3.
1853. De la Condamine, Rev. H. M. -On a Freshwater Deposit in Huntingdonshire. Journ. Geol. Soc., vol. ix. p. 271.
1861. Porter, Dr. H.---Geology of Peterborough and its Neighbourhood. 8vo, Peterborough.

See also General List, p. xxviii.

THIS county does not offer an inviting field for ordinary geological work, which is probably the reason why little has been published concerning it.

The southern and central portion of the county was examined by Mr. Howell, and his map was published by the Government Geological Survey in 1864. The northern portion was surveyed by Professor Judd, whose map, with an admirable descriptive memoir, appeared in 1872. A geological map of the county, however, was published as long ago as 1821 by William Smith, the "Father of English Geology."

The alluvium of the Bedford Level was described in the *Geologist* by Mr. C. B. Rose, and papers on this subject by the Rev. H. M. De La Condamine, Dr. Mitchell, and Professor Seeley have appeared in the Geological Society's Journal.

The surface of Huntingdonshire is low and undulating, with a gentle slope towards the Fens. On the west Keyston Toll-house is 245 feet above the sea, Elton 107 feet, Spaldwick 91 feet, Chesterton 64 feet, and Peterborough only 28 feet ; Huntingdon is 45 feet, St. Ives 26 feet, and Fenny Stanton 42 feet respectively above sea-level.

In examining the geological features, if we begin in the north-west, we shall find there the oldest and also the most varied rocks.

THE OOLITE.—Between Peterborough, Wansford, and Elton, we have a very interesting expanse of several beds of the INFERIOR OOLITE on the Huntingdonshire side of the Valley of the Nene. Crossing the river at Wansford we find ourselves on the—

Northampton Sand.- Following this bed southwards past the Mill, we get a fair section in a small pit at New Close Cover. Here at the top is some brown and yellow sand full of plant-remains, with white sand underneath. Returning, we trace the same sands past Stibbington to Water Newton, where, with some associated beds of clay, they are dug in the brickyards.

The Lincolnshire Oolite Limestone.—In the Wansford stone-pits we have some good sections of this fine bed, which in Lincolnshire is 200 feet thick, and which is so largely worked for building purposes at Ketton, Ancaster, &c. From the "Wood-pit," at Stibbington, the late Dr. Porter collected about forty species of fossils: at Water Newton brickyard there is 4 feet of fine-grained oolitic limestone visible, with nodules of ironstone resting upon it. Here we are evidently near the "thin end of the wedge," for eastward of this point the limestone dies out altogether.

THE GREAT OOLITE. All four members of this series are present. The lowest, or *Upper Estuarine series*, is clayey; above it we have the *Great Oolite Limestone*, then the *Great Oolite Clays*, and lastly, another limestone band the *Cornbrash:* all of these come on one above the other, overlying the beds of inferior oolite just described. The Upper Estuarine series is dug for brickmaking at Water Newton, and is well exposed in the Sibson tunnel. At the western end of the tunnel, near Wansford station, the whole series of beds, from the Great Oolite Limestone down to the thin representative of the Lincolnshire Limestone and Northampton Sand, may be seen.

The Great Oolite Limestone forms the steep escarpment of Alwalton Lynch, where it contains numerous bands of fossil oysters. The "Alwalton marble" was formerly dug here, but the pits are now closed: the stone was a hard, blue, shelly limestone, which took an excellent polish, but was not very durable. There is a petrifying spring here, which has formed a deposit of travertin, often encrusting tufts of grass, snail-shells, &c., and covering them up; the same beds are traceable round to Elton in one direction, and on the east are finely exposed in the railway cuttings near Bottlebridge. The Great Oolite clays are often seen resting upon the limestone; they contain bands of ironstone balls.

The Cornbrash is a hard, blue limestone when dug at any depth, but it weathers light-brown at the surface; it is exposed in the valley of the Billing Brook, especially about Water Newton Lodge, and is here very fossiliferous; it is very hard, and is used for road metal; the top is usually marked by a band of large rugose oysters—*Ostrea Marshii:* it forms a light red soil. There is a small inlier between Stilton and Yaxley, which is brought up by a fault. No fewer than 68 species of fossils have been found here, and the most delicate shells, often retaining their pearly lustre, are found in a perfect state.

MIDDLE OOLITE. *The Oxford Clay.* This great clay deposit, probably more than 600 feet thick, occupies the county to the exclusion of almost every other rock: to the north-east it passes under the Fens, of which, however, it constitutes the substratum, rising above it in island-like masses here and there; but even in the rest of the district it is generally so covered by the superincumbent boulder-clay and gravels as to be seldom exposed, even in the river valleys. The land it forms is mostly devoted to grazing, although there is some woodland, while of late years an increasing area has been brought under the plough.

Of the hard sandy beds at the base, known as the *Kellaway's Rock*, there are no good exposures. At Luddington brickyard light-blue Oxford clay is dug under 5 or 6 feet of drift to a depth of 30 feet. Those characteristic fossils, *Gryphæa dilatata* and *Belemnites Puzosianus*, occur here: similar beds are seen at Great Gidding, Haddon, Standground, Woodstone, and Fletton. In the various brick-pits south of Peterborough the dark-blue clays contain large quantities of fossil wood, either converted into jet or mineralized by iron pyrites: it is sometimes found in masses of large size, having evidently floated about on the deep sea, at the bottom of which the Oxford Clay was being deposited, until, becoming water-logged, it sank to the bottom and was covered up. The teeth and spines of fishes also occur here. The same Clay extends right down by Huntingdon, past Kimbolton and St. Neots, and eastward by St. Ives into Cambridgeshire. In the Fen district deep trenches are dug through the surface peat and silt into the clay beds below, and the latter are then dug and spread out on the surface. At Ramsey, near the Railway station, it was formerly got and burnt for ballast, a use to which the clays of this formation are frequently applied. At Ramsey Heights there are several brickyards at work, the Oxford clay being locally termed "galt;" in one of these peat is used for fuel. At Forty-foot Bridge a band of hard rock 8 or 10 inches thick is found at a depth of 15 feet: *Ammonites Lamberti* is common here. Connington and Holme brickyards, and indeed the brickyards generally over the region we have described, afford other sections of the bluish clays.

In the extensive excavations at St. Ives we find a rather pale bluish-grey fine clay, with occasional calcareous concretions, small crystals of selenite and lumps of pyrites: about 64 species of fossils from these pits are shown in the Woodwardian Museum at Cambridge, including 19 species of *Ammonites; Gryphæa dilatata* is also most abundant. Here may be seen one of those bands of impure limestone which occur in this district at various horizons in the Oxford Clay, and some of which have been considered representatives of the *Coral Rag*, of which the nearest undoubted exposures are at Upware, near Cambridge, and Stanton St. John's, near Oxford.

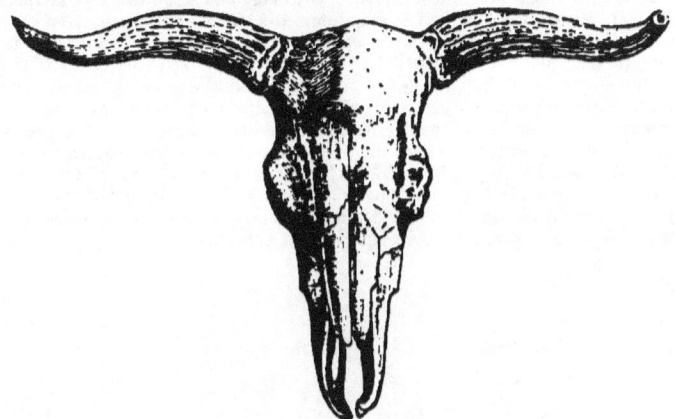

Fig. 40.—Skull of the Urus or Wild Bull, (*Bos primigenius*); Pleistocene and Recent Formations.

Thus we have in Huntingdonshire no clear line of division between the two great masses of the *Oxford* and *Kimmeridge* Clays. The St. Ives rock is also traceable at High Papworth, while a higher band was exposed in the railway cutting at Bluntisham, and a lower one at St. Neots. From the Oxford Clay, near the latter town, the remains of a large new species of reptile (*Plesiosaurus Evansi*) has been obtained.

THE LOWER GREENSAND. This interesting deposit is composed of brown or yellow sands, the tint being due to the presence of oxide of iron, which is sometimes absent, the beds then being white: it forms a range of low hills which extend from Potton and Sandy, in Beds, by Everton and Gamlingay, past Gransden, Eltisley, and Caxton; it has been worked for the valuable band of coprolites which it contains.

THE DRIFT.—The glacial deposit, called the Great Chalky boulder-clay, is of a bluish colour, and is crowded with masses of rock which often exhibit the polished and scratched appearance characteristic of glacier or ice-borne masses; it extends more or less over the whole of the county, and is often of great thickness. Masses of chalk and flint are abundant in it, with oolitic and liassic fossils, millstone-grit, sandstone, mountain limestone, &c.; it seems to have been deposited in a shallow sea by a great glacier pushing down from the north. The *valley gravels* of the Nene are of later date; near Elton they have been cemented together by the infiltration of calcareous matter so as to form indurated masses. At Overton Waterville numerous shells occur, with remains of the mammoth, &c.

The Fens are formed of silt or warp, a fine mud deposited by the sea when the land was at a rather lower level than at present, with intercalated beds of peat, due to the growth and decay of mosses (*Sphagnum*) during long ages. Some of the old Fen lakes, as Whittlesea Mere, Ramsey Mere, &c., have been drained of late years, and are now marked by patches of alluvium, so rich in shells as to constitute a very fertile soil.

Of the peat beds there are usually two, separated by marine silt. Trees of large size, as the oak, birch, &c., are occasionally dug out of the peat. The remains of animals which once roamed through these old forests are also found, as the wild ox (fig. 40), the Irish elk, the red-deer, bear, otter, beaver, wolf, fox, &c. In the silt or "buttery clay" between the peat beds, marine shells, such as the common cockle and oyster, and a shell called *Scrobicularia*, are found, with many beautiful species of foraminifera. Drainage has often caused contraction of the peat, &c. At Holme, an iron column erected on the foundation of the subjacent Oxford Clay, shows that a contraction of the superficial deposits of the fens in that place, to the extent of seven feet, has occurred since 1848.

The flat bottoms of the valleys of the Ouse and Nene are covered with a fine black loam, deposited during overflows of the river; these flats are often under water in winter, but in summer they constitute admirable grazing grounds.

Mr. John Evans records no flint or stone implements from Huntingdonshire in his great book on the subject. We have heard from him, however, that stone tools of an early type have been lately found near Peterborough. All gravel pits should be searched for them, especially those on the sides of the valleys, about 40 or 50 feet above the present streams.

No. 19.

GEOLOGY OF KENT.

NATURAL HISTORY AND SCIENTIFIC SOCIETIES.

Croydon Microscopical Club. Report.
East Kent Natural History Society; Canterbury.
Folkestone Natural History Society.
Maidstone Natural History and Philosophical Society.
West Kent Natural History and Microscopical Society.
Mid-Kent Natural History Society.

MUSEUMS.

Canterbury Museum.
Museum in Fort Pitt, Chatham.
Museum in Royal Engineers' Barracks, Chatham.
Dover Museum.
Folkestone Museum.
Greenwich Museum.
Maidstone Public Museum.
The Rotunda Museum, Woolwich.

PUBLICATIONS OF THE GEOLOGICAL SURVEY.

Maps. - Sheets, 1 : London ; 2, Sheerness ; 3, Canterbury, Margate, Ramsgate. Dover ; 4, Folkestone, Rye ; 5, Hastings, Newhaven, Hailsham ; 6, Bromley, Chatham, Maidstone, Tunbridge.

Books.—The Geology of the Country between Folkestone and Rye, by F. Drew, 1s. Geology of the Weald, by W. Topley, 28s. Geology of the London Basin, by W. Whitaker, 13s.

IMPORTANT WORKS OR PAPERS ON LOCAL GEOLOGY.

See Lists of Works on Geology of this county in the two Government Survey memoirs, Whitaker's Geology of the London Basin, and Topley's Geology of the Weald.

1874. Price, F. G. H. The Gault of Folkestone. Journ. Geol. Soc., vol. xxx. p. 342.
1874. Topley, W.—The Channel Tunnel. Popular Science Review, vol. xiii. p. 394.
1877. Lucas, J.—The Chalk Water System. Proc. Inst. Civil Eng. vol. xlvii. p. 70.
1877. Price, F. G. H. Beds between Gault and Upper Chalk near Folkestone. Journ. Geol. Soc., vol. xxxiii. p. 431.
1880. Davies, W.—The Musk-Ox at Crayford. Geol. Mag., p. 246.
1880. Ettingshausen, Baron Von.—Report on Fossil Flora of Sheppey. Geol. Mag., p. 37.
1881. Shrubsole, W. H.—Diatoms of the London Clay. Journ. Roy. Microscopical Society, p. 381.

See also General List, p. xxviii.

THE inhabitants of Kent have great facilities for the study of the rocks of their county. All along the coast admirable sections are exposed in the cliffs, and inland the absence of the surface deposits which, under the name of drift, so obscure the geology of the counties north of the Thames, permits the various strata to be traced with comparative ease.

The entire county has been mapped by the Geological Survey, and these maps, on the scale of one inch to a mile, are so coloured that anyone by their aid can determine exactly the character of the rocks of the neighbourhood in which he may happen to dwell. There are also two admirable memoirs published by the same department explaining the maps, and giving full particulars of the various sections and fossils: the first of these, by Mr. Whitaker, is entitled the "Geology of the London Basin," and describes the northern part of the county; whilst Mr. Topley's "Geology of the Weald" is a full and faithful authority for the remainder of the district. These works represent the sum of our present knowledge, towards which many hard workers in the field of geological science have contributed; among these may be mentioned Drs. Mantell and Fitton, Sir Charles Lyell, Messrs. Martin, Godwin-Austen, S. V. Wood, jun. and Professor Prestwich. The works of Mantell and Lyell are well known, but many valuable papers by other authors may be found in the publications of the Geological Society, in the pages of the "Geological Magazine," or in the Proceedings of the Geologists' Association of London, whose members have frequently visited the county.

GENERAL ARRANGEMENT OF THE STRATA.—In the account of the geology of Essex, Herts, and Middlesex, we have described how the chalk of the Chiltern Hills dips to the south-east under the London clay: the same chalk strata rise again from below the tertiary strata, and form the North Downs, which have a gentle northerly slope, while their southern sides are comparatively steep and precipitous. With the South Downs the opposite is the case; their northern sides are steep, while their south slopes, except where abruptly terminated by the sea, are gentle. Further, lying between the two chalk ridges, we find two sets of beds, corresponding generally with each other, every one on the north with a corresponding bed on the south. The probable explanation of this, that at once strikes one, is that these sets of beds were once continuous with one another over the intervening space; but that by the action of the sea, and subsequently by that of rain and rivers, the intermediate mass of rocks has been worn away. In the strata of Middlesex we have an example of what is geologically termed a basin, or *synclinal* curve, the rocks having a trough-like arrangement; but in Kent we get just the opposite of this; the rocks have been bent into a saddle-shaped or *anticlinal* curve, and once formed a dome-like mass, the top of which has been removed by the eroding agencies already alluded to. The oldest rocks, those which underlie all the others, will consequently be found about midway between the two chalk escarpments. We shall commence with the description of these, and gradually rise to newer and later-formed rocks.

NEOCOMIAN FORMATION. *The Wealden Beds.*—The strata included under this term have a maximum thickness of more than 2,000 feet: they have been subdivided as follows, the different beds being named from places where they cover much ground, and are thick and well exposed.

		Feet.
	Weald clay	400 to 1,100
Hastings Beds.	Upper Tunbridge Wells sand (with Cuckfield clay)	100 to 200
	Grinstead clay	0 to 80
	Lower Tunbridge Wells sand	30 to 130
	Wadhurst clay	100 to 180
	Ashdown sand	150
	Fairlight clays	360

It will thus be seen that there are two main divisions of the Wealden Beds, viz. (1) the Hastings Beds, and, above them (2) the Weald Clay. The entire mass from bottom to top gives evidence of having been deposited in the estuary of a great river, for the fossils are either shells, such as *Cyrena, Unio*, and *Paludina*, which inhabit fresh water, or bones of crocodiles, or of a gigantic land reptile, the *Iguanodon*.

The Hastings Beds.—These are chiefly sandy, and occupy the south of the county. A line drawn through Chiddingstone to Tunbridge, and thence by Brenchley to Kennardington, would roughly indicate their northern boundary: thus their line of strike, or surface extension, has a direction on the whole from east to west. The *dip* or slant of the beds is of course at right angles to this, or nearly due north. The scenery of this tract of Hastings Beds is varied and beautiful: the ground is hilly, Tunbridge being 88 feet above the sea, Bounds 430 feet, and Tunbridge Wells 378 feet; the hills to the north and west of this favourite watering place are considerably higher, as is also the neighbourhood of Goudhurst. The alternation of clayey beds with those of sand has been the great cause of the variety of landscape which is seen in this district. Very fine natural exposures of the sandstone rocks are to be seen round Tunbridge Wells: those known as *Harrison's Rocks*, and *Toad Rock* on Rusthall Common, are striking examples.

Fig. 41.—Section, showing the two escarpments formed by the Chalk and the Lower Greensand, 1½ miles west of Westerham.

a Gravel.   *c* Upper Greensand.   *e* Folkestone Beds.   *g* Atherfield Clay.
b Chalk.   *d* Gault.   *f* Hythe Beds.   *h* Weald Clay.
　　　　　　　　　　　* Sea level.

Iron Ore.—The ore known as clay ironstone occurs at the base of the Wadhurst Clay, and a band of calcareous iron ore occupies about the same position in the Ashdown Sand: these two beds were formerly largely worked, but owing to the want of fuel (the woods having been almost used up) the smelting and manufacture of iron came to an end in the Wealden in the first quarter of the present century. Remains of the bell-shaped pits from which the ore was mined may still be seen in many woods.

The Weald Clay.—This constitutes a low flat tract of land from four to six miles wide, which runs east and west by Tunbridge, Shipborne, Hunton, Frittenden, Pluckley, and High Halden to Romney Marsh: it is a brown or blue clay, ill drained and mostly in pasture: oaks grow well upon it. At Hythe, the top beds of the Weald clay contain a mixture of freshwater and marine fossils, showing probably a depression of the great estuary and the beginning of marine conditions. The strata dip gently to the north.

The Lower Greensand.—The strata known by this name form a picturesque hilly ridge running from the coast at Hythe by Lympne, in a north-west direction to Sevenoaks, and then due west towards Westerham. The height of this ridge on the south of Sevenoaks is 666 feet, and a little further west at Brasted Chart, 810 feet. The following subdivisions of the Lower Greensand have been mapped by the officers of the Geological Survey. (See also fig. 41.)

		Feet.
LOWER GREENSAND.	Folkestone Beds	90
	Sandgate Beds	80
	Hythe Beds	120
	Atherfield Clay	30

The lowest stratum, or Atherfield clay (named after a place in the Isle of Wight, where it is much thicker) is seen resting on the Weald clay, on the south side of the escarpment. Above it come the Hythe Beds, which are by far the most important: they contain valuable beds of limestone known as "Kentish Rag," which are largely quarried at Maidstone, more than 50,000 tons being annually sent away from the neighbourhood: many London churches are built of this material. The sandy beds which intervene between the layers of good building stone are known as "hassock." The fossil shells *Gervillia*, *Trigonia* and *Exogyra* are common here, and one quarry is known as the "Iguanodon Quarry," from a fine specimen discovered many years since by Mr. Bensted, and now in the British Museum. The Folkestone Beds are sandy, and are dug for building purposes, and for glass-sand at Aylesford, Borstead, Hollingbourn, &c.

The Gault. Between the ridge formed by the lower greensand, just described, and the chalk escarpment, there is a narrow and well-marked valley: the bottom of this valley is formed by the gault, which is a stiff blue

Fig. 41A.—Copt Point, Folkestone. Gault resting on Lower Greensand.

clay, 100 feet thick at Folkestone, where at Copt Point a fine section of this bed is seen resting on the Lower Greensand. It is highly fossiliferous, but the shells are difficult of preservation; they should be soaked in a hot solution of gelatine immediately after their extraction. *Ammonites splendens* and *Rostellaria carinata* are perhaps the most characteristic fossils. At Aylesford and Burham fine sections are exposed in the brick pits.

The Upper Greensand. This formation is very thin indeed in Kent. North of Folkestone and in Eastwear Bay we see greenish sandy beds, about 20 feet thick, lying above the gault. Inland there is no exposure except in the brick pits where the gault clay is dug.

The Chalk. The chalk escarpment, abrupt to the south, gently sloping northwards, we have already alluded to: its height, north of Westerham, is 812 feet; Knockholt, 783 feet; near Deptling, 657 feet; Hollingbourn Hill, 606 feet; Charing, 640 feet; Paddlesworth, 626 feet; Tolsford Hill, 562 feet; Folkestone Hill, 575 feet.

The total thickness of the chalk varies from 600 feet on the west of the county near Westerham, to 900 feet as shown in the cliffs on the coast. In the splendid sections afforded on the coast, the following subdivisions may be recognised, commencing on the north in the Isle of Thanet and walking towards Folkestone:—

	Feet.
Chalk with few flints (Margate)	80
Chalk with many flints (Broadstairs to Dover)	350
Chalk with few flints (Dover)	130
Chalk without flints	270
Chalk marl	70

Chalk is a white, soft, earthy, very pure limestone: its structure is homogeneous to the naked eye, but when examined microscopically it is found to be composed of countless numbers of shells of *foraminifera*. It was probably formed in a deep and open sea. Between Knockholt and Orpington, the chalk outcrop is about 5 miles wide. Passing eastwards we find it broaden to about 12 miles, running north right to the Thames at Dartford and Gravesend. From Chatham to Faversham the breadth north of the escarpment is from 4 to 8 miles, but in East Kent it is again wide, stretching north nearly to Canterbury and Sandwich. Dipping under the river Stour, the chalk rises again to form the Isle of Thanet: it is also exposed at Cliffe in the Hundred of Hoo, and along the upthrow (south) side of a line of fault which extends from Lewisham to Woolwich Dockyard. *Belemnitella mucronata* (fig. 42) is a characteristic fossil of the upper chalk at Margate and the Isle of Thanet; it is a belemnite with a narrow slit on one side.

In the valley of the Medway, north of Maidstone, the chalk without flints is largely worked, being ground up and mixed with clay, when it forms "Portland Cement."

TERTIARY PERIOD. The chalk ends a series of deposits classed as Secondary or Mesozoic (middle life), and the strata we have now to describe belong to an entirely different and later time, known as the Tertiary or Cainozoic (recent life) Period: they constitute a tract of land from six to eight miles wide, running along the south side of the estuary of the Thames.

EOCENE FORMATION. The Eocene (dawn of recent life) beds contain the first traces of creatures identical with those now existing. The lowest Eocene beds are the *Thanet Sands*: these were so named by Professor Prestwich, not because they cover much ground in the Isle of Thanet, but because no other Eocene strata occur there, and also because in that region

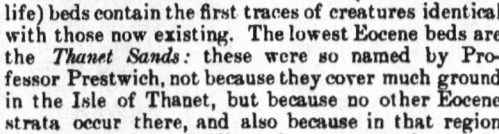

Fig. 42.—*Belemnitella mucronata*, the "guard" of a cephalopod; from the Upper Chalk.

they are fairly fossiliferous, whilst, as we follow them westwards, fossils are found to be almost or entirely wanting: they are well shown in the cliffs at Pegwell Bay, and again at Reculvers: everywhere they rest evenly upon the chalk, and are seen to be fine grey or buff sands from 20 to 60 feet thick, with a bed of green-coated flints at the base. Where the covering of sand, however, is thin, the chalk has often been worn into pipes or hollows by the action of carbonated water trickling down from above. *Cyprina Morrisii* is a characteristic shell.

Woolwich and Reading Beds. These repose upon the Thanet Sands, and with the latter are finely shown in the great pits at Lewisham, and at Charlton, near Woolwich. There is a pit in Sundridge Park noted for the numerous fossils it has yielded, and the beds are also well exposed at Cobham Park and near Canterbury, and in the cliffs east of Herne Bay: they are from 20 to 50 feet thick, and consist of alternations of mottled clays with sands and pebble beds: thick bands of oyster shells also occur. *Ostrea bellovacina* and *Melania inquinata* are the best known fossil shells. (See figs. 43 and 44.)

Oldhaven Beds.—These are beds of flint-pebbles and sand, so named by Mr. Whitaker because they are well seen at Oldhaven Gap, west of Reculvers:

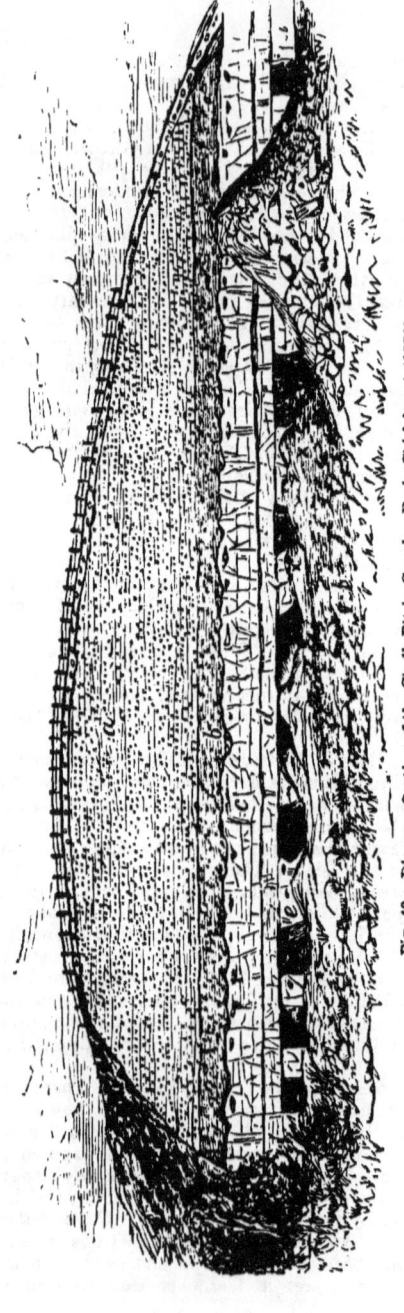

Fig. 43.—Diagram Section of the Chalk Pit in Camden Park, Chislehurst (1871).

Loamy and gravelly soil on top.

Thanet Sand. { *a* Clayey bedded sand, somewhat darker than usual, about 30 feet.
{ *b* Base bed, 5½ feet, clayey, greenish, and with green-coated flints, as usual, the lower half very dark. Many very small pipes in the chalk.

Chalk. { *c* Chalk, with flints bedded, showing a dip of about 1° N.W. along the face. At a depth of seven or eight feet more or less hard (*d*) for about forty inches down, the upper part of this marked bed being very hard and rather darker, and the bottom forming the even roof of the galleries. The bed (*e*) next below is also hard at top. The chalk has been worked by means of galleries, driven for a considerable distance under the hill.

they form the surface of Blackheath and Plumstead Common, and are from 20 to 30 feet thick.

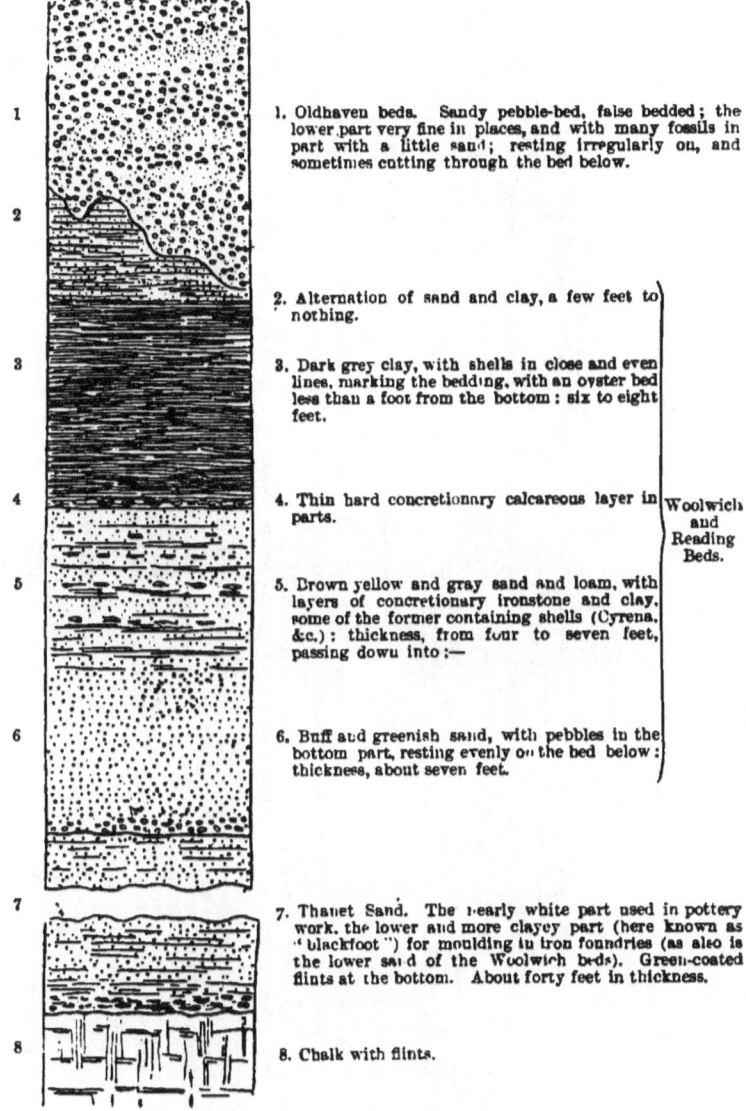

Fig. 44.—General Section at Charlton, Woolwich.—(Scale eight feet to an inch.)

Sands of the North Downs.—Capping the chalk escarpment between Folkestone and the valley of the Stour we find, mostly preserved in pipes or

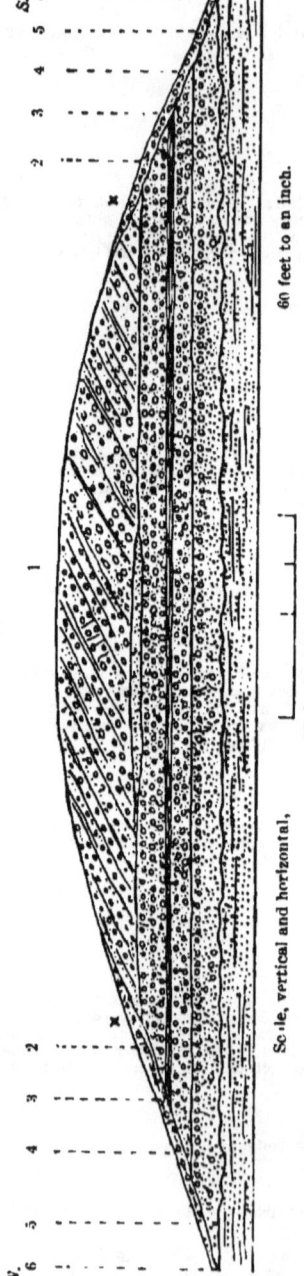

Fig. 45.—Cutting on the South-Eastern (Lewisham and Tunbridge) Railway at the North-Western corner of Camden Park (Chislehurst).

* Wash of pebbles, &c., on the slopes.

Oldhaven Beds.
1. Pebble bed, fossiliferous, false bedded, and in parts hardened into a conglomerate along the lines of false bedding; sometimes with a little greenish or light-coloured sand at bottom; about 25 feet in thickness.
2. Pebble-bed, much like the above, but more evenly bedded and not false bedded; fossiliferous, except in the lower part, the shells being mostly perished; hard blocks in parts, and at the middle a marked black pebble bed, partly hardened; about eight feet in thickness.

Woolwich and Reading Beds.
3. Green and red mottled plastic clay, mostly about a foot, with a layer of white earth at top and bottom; which may be owing to the decomposition of the shells in the pebble beds above, by the infiltration of the water and the deposition of their carbonate of lime from the water when it reached a less permeable bed, as suggested by Prof. Morris.
4. Irregular pebble bed, the upper part whitish and irregularly bounded beneath by a brownish line (bleaching by infiltration and deposition of the dissolved colouring matter at a lower level), with mostly a little white earth below; greenish at bottom, and with white earth, about five feet in thickness.
5. Bottom bed. Rather pale green sand, with a bed of pebbles, a foot or more thick, some two feet down, and scattered pebbles below. At one part a line of white earth above and some patches of the same in the pebble layer; about eight feet in thickness; resting irregularly on the sand below.
6. Fine light-coloured Thanet Sand.

(*Drawn and described by Mr. W. Whitaker*, 1865).

hollows, certain deposits which Professor Prestwich classed with the crag of Norfolk and Suffolk, but which Mr. Whitaker refers to either the Oldhaven beds or the Woolwich and Reading series. Fossils are few and imperfect, and the question of their age cannot yet be regarded as settled.

The London Clay.—This is a stiff blue or grey clay. Under London it is 400 feet thick, but 480 feet in the Isle of Sheppey: it forms the top of Shooter's Hill and the hills at Sydenham: it also constitutes the well-wooded tract lying between Canterbury, Whitstable, and Herne Bay: it is the top bed in the fine exposure at Loam-pit Hill, Lewisham, but the best place by far at which to study it is the Isle of Sheppey, provided always that the weather be dry and that the visitor has on a pair of " geological boots : " here it is seen to contain bands of *septaria*, which are nodules of impure carbonate of lime, and also much iron pyrites. The former are used in the manufacture of " Parker's cement," and the latter to make copperas. The London Clay is also dug at many points for brick and tile making. Great numbers of fossils have been collected at Sheppey, no fewer than 200 species of plants have been determined, especially fruits of palms (*Nipadites ellipticus* most common), and many species of shells, as the nautilus; also crabs, turtles, sharks, snakes and pig-like mammals. Of late years bones of some remarkable birds have been found at Sheppey, whose bills were armed with teeth! Near the base of the London Clay, Mr. W. H. Shrubsole has detected a layer of diatoms, beautifully mineralized by iron pyrites.

Bagshot Beds.- The Lower Bagshot Sands may be seen capping the London Clay in the cliffs near the East-end Coastguard station, in the Isle of Sheppey.

POST-TERTIARY DEPOSITS.—The absence of the Drift (glacial clays and gravels) allows us at once to state that all the various beds of gravel, sand, or brick-earth, which rest in an irregular manner on all or any of the rock-masses we have been describing, are due to the agency of various atmospheric agents of denudation, of which rain and rivers take the chief place. Many of the gravel beds occur far up on the sides of the valleys, and so give us a means of estimating the amount of denudation which the rivers have effected. Thus the " Bone-bed " at Folkestone is a deposit of marl and flint-gravel, which is considered to have been deposited by the stream which now flows into Folkestone Harbour, no less than a hundred feet beneath it. The beds of brick-earth and gravel at Crayford and Erith are favourite hunting-grounds of London geologists: here many species of land and fresh-water shells occur, especially *Cyrena fluminalis*, now living in the waters of the Nile. Bones of large quadrupeds—the elephant, rhinoceros, bear, &c., have also been found. Elephant bones have occurred in the gravels at Aylesford and Maidstone.

Of late years the sea has made considerable inroads on the coast-line; Warden Church in the Isle of Sheppey has recently been removed in consequence of the recession of the cliffs, due mainly to the great land-slips which took place in 1856 and again in 1859.

A deep well-boring at Crossness, near Blackwall, has lately afforded a splendid opportunity of discovering the nature of the rocks at a great depth: the beds passed through were :—

	feet.
Recent deposits	39
Woolwich Beds	47
Thanet Sands	51
Chalk	637
Upper Greensand	65
Gault	176
Red Marls (Devonian?)	60
	1,075

PREHISTORIC MAN.—Numerous and important discoveries of the flint implements, which are almost the only records we possess of the first inhabitants of these islands, have been made in the county of Kent: on the sea-shore between Herne Bay and Reculver more than a hundred of the early or *Palæolithic* type of implements have been picked up: they have undoubtedly fallen from the gravel-beds which cap the cliffs. The gravels near Canterbury, and those of the Darent and the Cray, have also yielded several specimens, and two have been found in the brick-earth of Crayford: generally they are fashioned out of flint, and are round at one end and pointed at the other. The average length is from five to eight inches. At Crayford large numbers of flint flakes and imperfect implements have lately been found by Mr. Spurrell, under 37 feet of gravel and sand.

Of the later, or *Neolithic* Stone Period, many finds of celts (axe-heads), flint flakes, arrow-heads, &c., are recorded from Bigberry Hill (Canterbury), Folkestone, High-street near Chislet, Isle of Thanet, Leeds Castle, Maidstone, Ramsgate, Ozengal, &c.: these occur either in mounds, where they were buried with the dead, or haphazard on the surface, having been lost or thrown away by their owners.

No. 20.

GEOLOGY OF LANCASHIRE.

NATURAL HISTORY AND SCIENTIFIC SOCIETIES.

Literary and Philosophical Society of Liverpool. Proceedings.
Liverpool Geological Society. Proceedings.
Historical Society of Lancashire and Cheshire; Liverpool. Transactions.
Liverpool Naturalists' Field Club.
Manchester Geological Society. Transactions.
Literary and Philosophical Society of Manchester. Proceedings and Memoirs.
Manchester Field Naturalists' Society.
Manchester Scientific Students' Association.
Manchester (Lower Mosely Street School) Natural History Society.
Barrow Naturalists' Field Club; Barrow-in-Furness. Transactions.
Lunesdale Naturalists' Field Club; Lancaster.
Lunesdale Entomological Society.
Lunesdale Geological Society.
Warrington Literary and Philosophical Society.

MUSEUMS.

Blackburn Museum.
Bolton Public Library and Museum.
Museum of the Royal Institution, Liverpool
Brown Free Library and Museum, Liverpool.
Liverpool College Museum.
Owen's College Museum, Manchester.
Salford Library and Museum, Peel Park, Manchester.
Museum of the Literary Institute, Preston.
Warrington Museum.

PUBLICATIONS OF THE GEOLOGICAL SURVEY.

Maps. Quarter Sheets: 79 N.E. Liverpool; 80 N.W. Prescot, St. Helens, Warrington; 80 N.E. Altrincham, Knutsford; 81 N.W. Stockport; 88 S.W. Manchester, Oldham; 88 N.W. Todmorden; 89 S.E. Bolton-le-Moors, Salford; 89 S.W. Wigan, Ormskirk; 89 N.W. Preston, Kirkham, Chorley; 89 N.E. Haslington, Blackburn; 90 S.E. Formby, Waterloo; 90 N.E. Lytham, Southport; 91 S.W. Blackpool, Fleetwood; 92 S.W. Clitheroe; 98 S.E. Kirkby Lonsdale; 98 S.W. Ulverston. (Survey not completed.)

Books.—Geology of the Country round Prescot, by E. Hull, 8d. Geology of the Country round Stockport, Macclesfield, Congleton, and Leek, by Hull and Green, 4s. Geology of the Country round Oldham, by E. Hull, 2s. Geology of the Country round Bolton, by E. Hull, 2s. Geology of the Country round Wigan, by E. Hull, 1s. Geology of the Country between Liverpool and Southport, by C. E. De Rance, 3d. Geology of the Country between Southport, Lytham, and South Shore, by De Rance, 6d. Geology of the Country between Blackpool and Fleetwood, 6d. Geology of Furness District, by Aveline, 6d.

The Burnley Coal-field, by E. Hull, &c., 12s. Superficial Geology of South-west Lancashire, by De Rance, 17s.

IMPORTANT WORKS OR PAPERS ON LOCAL GEOLOGY.

List (by Messrs. Whitaker and Tiddeman) of 561 works on the Geology of Lancashire in the Survey Memoir "On the Burnley Coal-field."

1863. Morton, G. H. Geology of the Country round Liverpool. Smith, Watts & Co., Liverpool.
1864. Hull, E., and Green, A. H. Millstone Grit of South-east of Lancashire. Q. Journ. Geol. Soc., vol. xx. p. 242.
1868. Hull, E. Relative Ages of the Leading Physical Features and Lines of Elevation of the Carboniferous District of Lancashire and Yorkshire. Q. Journ. Geol. Soc., vol. xxiv. p. 323.
1868. Hull, E. The Pendle Hills. Q. Journ. Geol. Soc., vol. xxiv. p. 319.
1872. Reade, T. M. Post-Glacial Geology of West Lancashire. Geol. Mag., vol. ix. p. 111.
1872. Tiddeman, R. H. Ice-Sheet in North Lancashire. Q. Journ. Geol Soc., vol. xxviii. p. 471.
1873. De Rance, C. E. Mineral Veins in Carboniferous Rocks of Lancashire. Geol. Mag., vol. x. pp. 64, 303.
1879. Wilson, E. Age of the Pennine Chain. Geol. Mag., pp. 501, 573.
1879. Mackintosh, D. Erratic Blocks of West of England, &c. Q. Journ. Geol. Soc., vol. xxxv. p. 425.
1880. Dickinson, Jos. Rock-salt at Preesal. Trans. Manch. Geol. Soc., vol. xvi. p. 26.
1880. Watts, W. The so-called Arctic Peat Bog near Oldham. Trans. Manch. Geol. Soc., vol. xvi. p. 43.
1880. Wild, Geo. Marine Fossil Shells at Ashton Moss Colliery. Trans. Manch. Geol. Soc., vol. xvi. p. 37.

See also General Lists, p. xxv.

PROBABLY in no other county is the study of the rocks more important to the interests of the population than in Lancashire. Most valuable deposits of coal and iron constitute the great wealth of the district, and have determined its importance as a mining and manufacturing region; the water-bearing properties of some of the strata are of great value to the inhabitants of the towns, while the nature of the soil, which results from the decomposition of the surface rocks, is of paramount interest to the agriculturist.

Many minds have been occupied with the problems offered by the geology of Lancashire; in the Geological Survey Memoir on the Burnley Coal-field, Messrs. Whitaker and Tiddeman give the titles of 561 books, papers, &c., published up to the year 1874, all written during the last two centuries, and by far the greater part within the last twenty years. Many of these have appeared in the Transactions of the Geological Societies of Manchester and Liverpool, in the "Geological Magazine," the Quarterly Journal of the Geological Society, &c. The chief private contributors have been Messrs. E. W. Binney, J. Aitken, D. Mackintosh, G. H. Morton, and Professors W. C. Williamson and Phillips, while the work of the Government Geological Survey has been mainly executed by Professors Hull and Hughes, and Messrs. Aveline, Tiddeman, A. H. Green, J. R. Dakyns, C. E. de Rance, and J. C. Ward.

The maps of the Geological Survey give the most valuable information as to the position of the various rock-beds, and the memoirs descriptive of these maps are thoroughly reliable and full guides to a knowledge of the country they describe.

It is not easy to give in a few words an account of the general nature, arrangement, and disposition of the rock-beds which form Lancashire. In the first place the outlying portion of Furness and Cartmel contains very old strata (of Silurian age), and properly belongs to Westmorland; the rest of the county slopes down from the Pennine Range westwards to the sea, although this uniformity of surface is broken by lines of hills which run roughly from west to east, or at right angles to the Pennine Chain. The Pennine Chain marks a line of great upheaval, along or in the neighbourhood of which the lowest Carboniferous rocks have been brought to the surface, and as we walk westwards we pass over the middle and upper Coal-measures, then over "red rocks" of Permian and Triassic age, and, lastly, find ourselves on a low plain close to the sea, composed of peat, sand, and clay, of very recent geological age.

SILURIAN FORMATION. The oldest rocks of Lancashire occupy the surface only in the extreme north-west of the county, but they doubtless extend southwards beneath the later-formed strata, and probably form the floor on which the great Carboniferous series reposes, so that a boring at Manchester would strike, at a depth of 2 or 3 miles, the same rocks which rise up on the north in the mountains of the Lake district, and on the south-west to form the corresponding heights of Wales.

The lowest Silurian rocks exposed are the *Skiddaw Slates*, which are so brought up by faults as to occupy an area of about 1 square mile immediately on the south of Ireleth in Furness.

The strata called the *Volcanic Series of Borrowdale* (Green Slates and Porphyries of Sedgwick) form the north-west corner of the Furness district, lying between the county boundary and a line drawn from Duddon Bridge to Brathay Bridge; this is a wild and mountainous tract, including Dunnerdale, Coniston Old Man (2,633 feet), Grey Friars, Dowcrags, &c. The rocks are all of a volcanic nature, consisting of old lavas and consolidated ash-beds, the latter yielding good slates. These beds contain no fossils, but include valuable mineral veins of copper and iron ore. From the copper mine near Coniston, 646 tons of ore were raised in 1877, valued at £3,373.

The Coniston Limestone Series. The distinguishing rock of this series is a hard compact blue limestone, which can be readily traced from Duddon Bridge by Broughton Mills, Appletreeworth, and Ash-ghyll to Coniston, Pull-wyke, and the head of Windermere; including the associated beds of shale, the average thickness is 300 feet. Fossils are numerous, and are best obtained where the limestone has been exposed for many years to the influence of the weather, as in the stone walls; for the freshly-quarried rock is compact and shows few traces of organic remains, but these being harder than the rest of the stone, "weather out" in course of time; they include corals, brachiopod shells, and crustacea. In the Furness district the Coniston Limestone is clearly seen to lie unconformably on various members of the Borrowdale series: it has a south-easterly dip at a high angle, and so forms a band at the surface of only 100 or 200 yards in width. East of Ireleth the Coniston Limestone appears again as a narrow band ranging north and south for a distance of 3 miles; here the shale beds are well developed, there is a remarkable band of ashy sandstone, and the top bed of all is conglomeratic. The Coniston Limestone is useless for lime-burning; it was formerly made into cement at Broughton Mills, but is now chiefly used for stone walling.

The *Stockdale Shales* are of a pale colour and rather sandy; they are of inconsiderable thickness and contain no fossils; they repose everywhere upon the Coniston Limestone.

Coniston Flags and Grits.— The thickness of these beds is about 6,000 feet. They extend from Broughton-in-Furness to Brathay and Coniston Lake, and thence southwards to Colton, Lowick Bridge, and Ulverston. The flag quarries at Brathay are well known, and both flags and slates are worked 2 miles further south in the Coldwell quarries, which is a good locality for fossils, especially (fragmentary) trilobites. Speaking generally we may say that sandstones prevail in this division, which forms a hilly and somewhat barren tract.

Next come the *Bannisdale Slates*, which lie still further to the south-east;

they undulate between Coniston Lake and Windermere, extending southwards by Newby Bridge to Cartmel, and eastwards across the Winster. They consist of flags, shales, and sandstones of great thickness (5,000 feet), with few fossils.

Owing to the general similarity of the strata, and the scarcity or ill-preserved condition of the fossils, the task of mapping accurately the subdivisions of the Silurian rocks of the Furness district proved a very difficult one; great and numerous dislocations or faults have also occurred during the various elevations and subsidences which these old beds have undergone since their original deposition and consolidation; as a consequence of this rocks of very dissimilar age and nature are often brought into juxtaposition, which led an Ireleth farmer to declare that there was "every sort of stone that God Almighty ever made, lying on the hill at High Haume."

CARBONIFEROUS FORMATION. The close of the Silurian Period was marked in Western Europe by an upheaval of the land, resulting in the formation of a continent, whose western limits probably extended further into the Atlantic than those of the British Isles. In great depressed areas on this old continent the *Old Red Sandstone* was deposited in inland fresh waters, but the Lancashire area apparently then formed dry land between a lake in Herefordshire on the south and one in Central Scotland on the north; the red beds of conglomerate which occur in the Lake District are now known to form the base of the Carboniferous Formation.

Carboniferous Limestone. The Mountain or Scar Limestone rests upon the Upper Silurian rocks; it consists of massive beds of limestone 20 to 300 feet in thickness, separated by thin layers of shale and sandstone; the total thickness is about 3,000 feet. We find this rock in the south of Furness between Ulverston, Dalton, and Baycliff; then crossing the Leven we see it extending from Cartmel to Lindal and Flookborough, and from Arnside to Burton and Warton. In all this region the Limestone has a prevailing south-easterly dip, and passing in that direction under newer rocks it rises up again, forming the axis in each case of the anticlinals of Slaidburn and Clitheroe. The limestones of Pendle Forest, both black and white, are largely quarried, and make excellent lime; the lime from the lower or black limestone is preferred where whiteness is an object, the carbonaceous matter to which the colour is due being completely removed by burning. At many sections in the neighbourhood of Clitheroe, the limestone beds are seen to be bent or contorted in a remarkable manner. The fossils of the Mountain Limestone include great numbers of crinoids, of which the broken arms or stems are alone usually met with, the " heads " being rare, brachiopod shells such as *Productus, Chonetes, Spirifera*, &c., lamellibranchs, gasteropods, and cephalopods, including above 300 species, which have been diligently collected by Mr. James Eccles, F.G.S., and other working palæontologists.

Hæmatite Iron-Ore. Enormous deposits of red hæmatite or kidney-iron ore (peroxide of iron $Fe_2 O_3$), occur in connection with the Mountain Limestone of Furness. The quantity of the ore and its suitability for the Bessemer process have led to a great development of the mining industry of this region within the last ten or fifteen years. The ore occurs in caverns or fissures, or filling irregular hollows or pockets in the rock. It has without doubt been deposited in its present position by water; fossils are often met with which have been partially or wholly converted into hæmatite, a change which we can only explain by the action of aqueous agencies; the ore is much softer than that of the Whitehaven district, being chiefly worked by the pick, while the Whitehaven ore has all to be blasted. The chief mines are at Lindal Moor (Harrison, Ainslie, and Co.), Park (Barrow Hæmatite and Steel Company), Roanhead and Askham, Stank, Elliscales, &c. In 1877 there were raised 933,000 tons of ore, value £651,170.

The Yoredale Beds. It is difficult in North Lancashire to define closely the upper boundary of the Mountain Limestone; it has been drawn by the

Geological Survey at the point where shales begin to predominate, and to certain beds of grit, shale, and limestone which lie above, the name of *Yoredale Series* is applied. These beds are from 2,000 to 3,000 feet in thickness; the lower and upper divisions are gritty, between which black shales occur, being best developed on the fell-sides of Bowland.

The Yoredale Grits form the high ground of Pendle Hill, 1,831 feet, and extend thence across the Ribble to Ribchester; further north they occur round Wennington, between Burton and Carnforth, &c. The black Yoredale shales of Bowland have been frequently mistaken for Coal-measures, and numerous remains of shafts which have been sunk in search of Coal-seams may be seen in the basins of the Ribble and the Hodder; *Posidonomya Gibsoni* is a common fossil shell here.

Fig. 46.—Feather-edge Coal in Millstone Grit, as seen in the quarries at Hill Top, north of Bury. The main mass of the grit underlies the coal, and is upwards of sixty feet in thickness. Upon this lies a bed of under-clay, with *stigmaria*, 2 ft. thick, and on this clay there rests the feather-edge coal, also about 2 ft. thick.

Millstone Grit.—The following grouping of the members of this well-known series is that generally adopted; it is by Professor A. H. Green :—

Coal-measures.

 Rough Rock.—Almost invariably a coarse and massive felspathic grit, about 100 feet thick; a band of flagstone frequently at the base; called also *First Grit*. Seam called Feather-edge Coal near top which has been largely worked north of Rochdale. (See fig. 46.)
 Shales.
 Second Grit.—A somewhat variable group of finely grained sandstones and shales.
 Shales.
 Third Grit (or Haslingden Flags).—A coarse massive gritstone between 200 and 300 feet thick, remarkable for its regular jointing and bold wall-like escarpments. Fine flaggy beds at the base, and a coal on the top frequently.
 Shales (of Sabden Valley, &c).
 Fourth, or Kinder-Scout Grit.—Coarse, often conglomeratic, very massive, with occasional beds of flagstone. Where most largely developed, is in two beds with shale between; 500 feet thick.

Yoredale Beds.

The Millstone Grit contains few fossils with the exception of rolled stems of carboniferous plants; it appears to have been deposited in a shallow sea traversed by strong currents running from the north or north-east, for the sandstones are frequently false-bedded, the layers inclining to the west or south-west.

These beds run along the eastern boundary of Lancashire from near Colne by Boulsworth Hill (1,700 feet), Black Hambledon (1,573 feet), to Blackstone Edge (1,551 feet). (See fig. 47.)

From Todmorden a line of hills formed of Millstone Grit runs westwards by Haslingden to near Chorley, forming the "Rossendale Anticlinal," which separates the Burnley coal-field from the larger Coal-measure area of South Lancashire. In the north of the county again these coarse sandstones occupy a considerable region between Lancaster and Garstang on the west and the Bleasdale Moors, Littledale Fells, and Tunstall on the east. In this northern district the Second and Third Grits are so split up by shale-beds as to be better grouped under one head and called "Middle Grits."

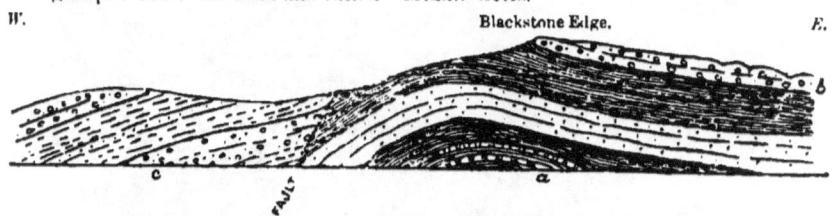

Fig. 47.—Section across Blackstone Edge, showing the "Anticlinal Fault" at its western foot.
  a Yoredale Beds.   *b* Kinder Scout Grit.   *c* Millstone Grit.

Coal-measures. We now enter on the consideration of a series of beds of shale and sandstone of great thickness, whose economic importance is due to the fact that they include numerous workable seams of coal. Professor Hull recognises the following sub-divisions:

Upper Coal-measures. The shales, sandstones, coal-seams, and limestones of Ardwick, near Manchester, 2,000 feet. These beds are distinguished by their red or purple tint.

Middle Coal-measures, including the strata from the Worsley Four-feet coal to the "Arley Mine" seam, 3,000 feet.

Lower Coal-measures or Gannister Beds, consisting of flags, shales, and thin coals (the Mountain mines), with marine fossils, 2,000 feet.

All the Carboniferous strata of Lancashire were originally deposited in a horizontal position; first the Carboniferous Limestone was accumulated in a tolerably deep sea, whose floor was formed of Silurian rocks; as this sea was shallowed by the continued deposit on its bottom of the remains of the countless myriads of shells, crinoids, &c., which now constitute thick beds of limestone, it became the recipient of quantities of mud and sand brought by rivers and marine currents from a neighbouring continent, and so the Yoredale beds were formed; then in the Millstone Grit series we find in the thin coal-seams evidence that parts of the sea-bottom became land, on which a rank vegetation grew and decayed for many generations.

During the formation of the Coal-measures, more general land conditions appear to have prevailed; all this country then probably presented the appearance of a low flat marsh very little above the sea-level, on which plants of a low degree of organization (*cryptogams*) flourished with exceeding luxuriance. But there was no fixity of level; the land at irregular intervals underwent depressions of a few feet, and the accumulated vegetable growth of many years was covered over by a layer of sand or mud, on which a new forest grew; and this must have gone on for perhaps a million of years to allow of the accumulation of the 8,000 feet of strata which form the Coal-measures of Lancashire.

At the close then of the Carboniferous Epoch, all Lancashire, together with a surrounding area of great extent, probably including the greater part of the

British Isles, was composed of horizontally spread-out carboniferous strata. At this time Professor Hull has shown that great pressure was exerted on these rocks from the north and south, by which they were thrown into folds running east and west, or rather, in Lancashire, from a little north of east to a little south of west. The tops of these folds—the anticlinals as they are geologically termed, would be much fissured and very much exposed to denudation; and in fact they have since been worn away, so that the axes of these anticlinals are now usually found at the bottoms of valleys, while in the synclinal curves which are the bottoms of the original folds, we have a structure well calculated to resist the denuding influences of the atmosphere and the sea. This is the origin of the lines of hills which run across the western part of Lancashire from south-west to north-east, of which the Pendle range is the best example. (See figs. 3 and 4).

The present Coal-fields owe their preservation to their lying in the hollows or synclinal folds; they once stretched continuously all over the county, but the exposed portions have been worn away, and so the great South Lancashire Coal-field has been separated from that of Burnley, and this again from the little coal basin of Ingleton and Black Burton in the north of the county.

The total or combined thickness of the workable coal-seams of Lancashire is estimated by Mr. Dickinson at 62 feet, divided into 18 seams; he estimates the quantity remaining to be worked (at a less depth than 4,000 feet) at five thousand millions of tons. The amount raised in 1877 from 517 collieries was 17,621,531 tons, at which rate the coal will be exhausted in three centuries.

Lancashire contains the deepest coal-mine in the British Isles, that at Ashton Moss Colliery, where the coal is got at a depth of 2,691 ft. There is another very deep shaft in the same district at Rosebridge, near Wigan, where the "Arley Mine" is worked at a depth of 2,445 feet; the temperature of the rock at this depth, as shown by thermometers deeply embedded in it, is $93\frac{1}{2}°F.$; the average temperature at the surface is $49°F.$, so that there is an increase of temperature at the rate of $1°F.$ for every 54 feet we go down. The temperature of the working galleries at the bottom is, however, only $79°F.$, owing to the ventilating air-currents, and it seems probable that we shall be able to work coal down to a depth of 4,000 feet.

Among the Lancashire seams, the Cannel coal of Wigan is the most valuable; it is 3 feet thick at Wigan, but thins in every direction from that town. Next comes the "Arley Mine" from 3 to 5 feet in thickness, and of great horizontal extent; it is called the "Little Delf," at St. Helens, the "Riley Mine" at Bolton, the "Dogshaw Mine" at Bury, and the "Fulledge Main-coal" in the Burnley basin.

We may now briefly consider the separate divisions into which the Coal-measures of Lancashire have been thrown by the great upheavals along lines running east and west to which we have already alluded, and which took place after the deposition of the Coal-measures, but before the formation of the Permian strata.

South Lancashire Coal-field. Only the northern half of this great basin rises to the surface; the southern extension of the beds is covered over by Permian and Triassic strata, whose thickness may long prevent the Coal-seams being worked between Manchester and Newton on the north, and Warrington and Stockport on the south, although there is little or no doubt but that the Coal-measures stretch underground southwards of the Mersey, rising up against a ridge of old rocks for whose probable existence between Congleton and Chester Professor Hull has given good reasons. Quite recently it has been found, by boring, that the Coal-measures lie only 340 feet deep below Warrington.

In consequence of the basin-shaped arrangement of the strata, older and older beds crop out as we go from south to north. Leaving the Upper Coal-measures, which only occur near Manchester, we pass over the valuable tract of Middle Coal-measures, ranging from Oldham to Bolton, Wigan, St. Helens, and Prescot, then the Gannister series rises up at Rochdale, Bury, and Coppull, and lastly the Millstone Grit forms high moorlands, separating the district from the Burnley coal-field further north. The boundaries on the east and west are great faults ranging from north to south, which were formed after the Permian Epoch. (See Figs. 47 and 48.)

Burnley Coal-field. — This also lies in a hollow, basin, or trough; the southern boundary of Millstone Grit which separates it from the great South Lancashire coal-field we have just mentioned; the beds similarly rise on the north, and first the Millstone Grit, then the Yoredale beds, and lastly the Mountain Limestone form the Pendle Range and the country round Clitheroe. The towns of Burnley, Padiham, Accrington, and Blackburn, are the most important on this coal-field.

Coal-field of Ingleton and Black Burton. — The work of the Geological Survey has shown this small coal-field to be of much larger area than was originally supposed, and it may include from 15 to 20 square miles, but the country is deeply covered by drift, and few sections of the beds are visible, those in Cant Beck and along the river Greeta being the best. Two coal-seams, the lower or "Deep coal" (6 feet thick), and the "Main coal" (4 feet

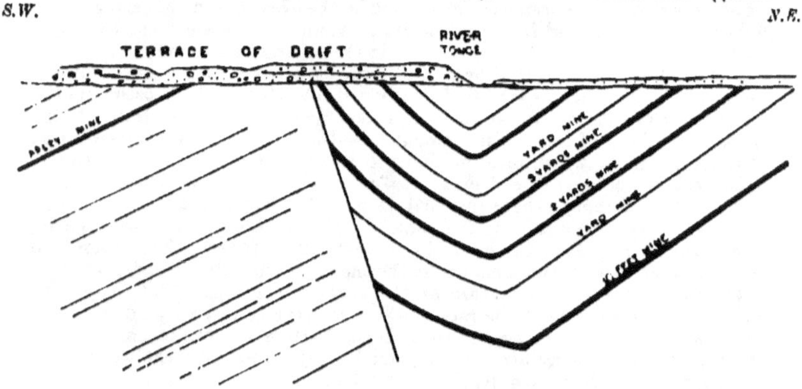

Fig. 48. — Section across the Irwell Valley Fault, as proved in the Earl of Bradford's collieries near Bolton. This fault runs from the Mersey west of Stockport, to the Millstone Grit hills, north of Bolton. The downthrow is on the eastern side, and at Farnworth is as much as 1,050 yards. It is a very clean line of fracture, so that there is little disturbed ground on either side of the fault.

thick), have been worked along the Greeta, east of Black Burton; here they dip to the north at 10 degrees, but their further extension is unknown.

Besides the great quantity of coal mentioned above as having been raised in Lancashire in 1877, there were also obtained 2,500 tons of iron pyrites (coal-brasses), value £1,250; also 111,478 tons of fire-clay, and 1,520 tons of oil-shales.

PERMIAN FORMATION. The Permian strata can be traced resting on the Coal-measures along the southern margin of the South Lancashire coal-field, from Sutton near St. Helens, by Haydock, Edge Green, Astley, and Manchester, to Stockport. The lower beds are bright-red sandstones, well seen at Collyhurst near Manchester, above which, at Worsley, are red marls, with numerous bands of fossiliferous limestone containing dwarfed specimens of the shells *Schizodus, Bakevellia, &c.*

In North Lancashire Permian beds form the southern end of the Furness promontory, and Walney Island, though at the latter place they are not visible at the surface, being deeply covered by the drift. Magnesian Limestone has been quarried at Old Holebeck, south of Stank, and Upper Permian red sandstones are largely worked at Hawcoat, for the new docks and buildings at Barrow, while in former ages they yielded the material for Furness Abbey. At Rampside a boring has been put down to the depth of 2,210 feet in search of coal; at the depth of 250 feet a spring of water was cut in the Permian Red Sandstone, which yields 13,500 gallons of water daily; the beds beneath the Permians were found to be the Yoredale series, the Coal-measures being absent. Connecting these two regions we find Permian beds (1) around Garstang and Wateby; (2) as an outlying patch of about 2 square miles round Masongill and Ireby, on the Ingleton coal-field; (3) at Roach Bridge, on the river Darwen; (4) at Waddon Hall, near Clitheroe; the two last-named are very small patches, but they suffice to show the former extension of Permian beds all over Lancashire, and have been preserved by their position on the downthrow side of great faults.

Origin of the Pennine Chain.—After the formation of the Permian strata a fresh series of great earth movements took place, accompanied necessarily by some dislocation or "faulting" of the beds; the lines of upheaval were now in a north and south direction, and the chief elevation took place along what is now the Pennine Chain. At the time of their formation the Coal-measures and Permian strata of Lancashire were one continuous sheet with those of Yorkshire, but all these were swept off the top of the great fold which was raised from the Cheviot Hills due southwards to Buxton and Matlock, and so the coal-fields of Lancashire and Yorkshire were dissevered.

One notable result of this elevation is the Anticlinal Fault, along which the Burnley and South Lancashire coal-fields end abruptly on the east, and which can be traced from Colne along the western slopes of Boulsworth Hill, and Black Hambledon, across the vale of Todmorden, along the western base of Blackstone Edge (see fig. 47), and thence southwards to Leek and Wetley in Staffordshire, a total distance of 55 miles. The rocks on the east side of this fault have been elevated about 1,000 feet above their normal position, and, being the hard sandstones of the Millstone Grit, they have well withstood denudation and so form the hills and bleak moorlands which constitute the eastern boundary of Lancashire.

THE TRIAS; OR NEW RED SANDSTONE.—The beds we have next to describe form a broad and comparatively low district on the west side of Lancashire, from Preston, by Ormskirk, to Liverpool, and thence along the valley of the Mersey eastwards to Manchester. These Triassic strata have been arranged under two heads:—

Keuper Beds	{ Keuper Marls. { Keuper Sandstone.
Bunter Sandstone	{ Upper Red and Mottled Sandstone. { Conglomerate or Pebble Beds. { Lower Red and Mottled Sandstone.

The lower or Bunter beds are noted for the quantity of water they contain, and they have been tapped at several points, especially near Liverpool, by deep borings, from which great quantities of water are pumped; at present above seven millions of gallons daily are obtained at Liverpool.

Within the last few years many borings have been executed by Messrs. Mather and Platt of Manchester in the neighbourhood of that town, and at Ormskirk, Warrington, St. Helens, &c., in the sandstones of the New Red, and almost invariably with the result of obtaining a good supply of water.

The various divisions of the Trias are traceable at the surface only with great difficulty, for the low country they form is for the most part deeply covered by drift deposits. Their aggregate thickness is about 2,000 feet. The Bunter Sandstones are largely quarried near Liverpool; the Keuper Marls occur in the country north and south of the mouth of the Ribble, being here of a greenish-grey colour, with few red layers. At Preesal, 1½ miles south of Fleetwood, a boring passed through saliferous marls for 258 feet, and then through 266 feet of rock-salt with layers of shale; this deposit is of the same geological age as the rock-salt of Cheshire.

Fossils are almost unknown in the Trias, and this fact and the occasional presence of salt in the upper beds lends probability to Professor Ramsay's view that the strata were formed in inland salt lakes of great extent.

Great movements of the earth's crust have taken place since the Triassic time; the faults caused by these are very numerous, and cross the Lancashire coal-field in a direction from north-west to south-east; one of these faults, with a great down-throw to the west, forms the western boundary of the Coal-measures, bringing the Bunter Sandstones to a level with them.

THE GLACIAL DEPOSITS, OR DRIFT.—The most puzzling geological deposits of Lancashire are the irregular but often thick masses of clay and sand which lie at random upon the edges of the solid rocks. These deposits are most fully developed in the plains of West Lancashire: the lowest is a dark leaden-coloured clay called the Till; it occurs at elevations of 200 feet and upwards only, contains angular blocks of local origin but no shells, and seems to have been formed by a sheet of land-ice. Upon it rests the *Lower Boulder Clay*, of reddish-brown colour, with many stones and boulders of Lake-district rocks; it is finely exposed in the cliffs north of Blackpool, is often stratified, and the stones, though striated, are partly rounded. Hence it would seem to tell us of a period of submergence, when ice drifting from the north and east dropped its stony cargo in a shallow sea; shells of cold-water species are not uncommon. (Figs 49 and 50.)

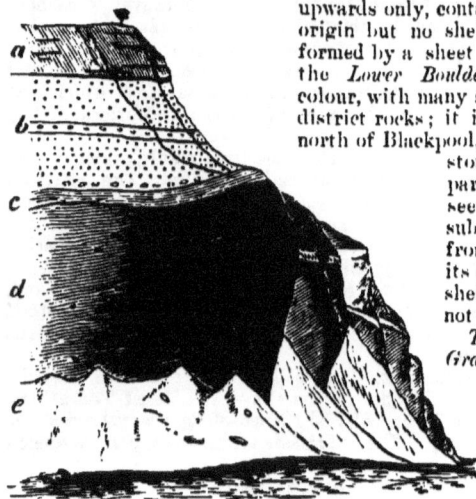

Fig. 49.—Section of Norbreck Cliff, north of Blackpool.
a Upper Boulder Clay. *b* Sand or Gravel (Middle Drift).
c Bluish Silt.
d Dark Brown Clay. } Lower Boulder Clay, 30 feet.
e Talus-heap.

The *Middle Drift Sand and Gravel* marks a mild period, when the glacial cold had decreased. It is sometimes as much as 70 feet thick, but is often absent. The molluscan remains show a mixture of northern and southern shells. It is commonly false-bedded.

The *Upper Boulder Clay* is sometimes 100 feet thick. It is of a dull red tint, weathering on the surfaces or joints which may be exposed to the air to a bluish-white. Large boulders are rare, but glaciated stones and shell fragments are common. The boulder clay is largely dug for brick-making. Geologists are now agreed in ascribing these boulder clays to the action of ice in some form or other, and we often find traces of the passage of ice over the solid rocks beneath in the shape of deep grooves or striations; many such have been noted on the Bunter Sandstone near Liverpool, at Salford, &c., and on the hard Lower Carboniferous rocks of the centre and north-east of the county. The drift is not nearly so widely spread or thick on the hills or "Fells" as on the low plains; travelled blocks can be traced up to a height of 1,300 feet on Boulsworth Hill and Black Hambledon. The movement of the glaciers or icebergs was clearly from north to south; the boulders are mostly Lake-district rocks, especially the Eskdale Granite and felspathic masses from the Borrowdale series. Other granite boulders are like the stone of Criffel in Kirkcudbrightshire; great numbers of these occur at Blackpool and Liverpool. Many of these interesting "erratics" are yearly destroyed for road-mending or building; it would be well to preserve the more remarkable specimens. At Preston a fine granite

Fig. 50.—Drift Deposits in the Banks of the Ribble, below Ribchester.

a Upper Till or Boulder Clay. { Red clay, apparently laminated, resting on the underlying sand with an even well-defined line of demarcation, 55 to 60 feet thick.
b Middle Sand. { Fine reddish sand with beds of gravel of rounded waterworn pebbles, and with a gravel bed at the base, 50 feet thick.
c Lower Till or Boulder Clay. { Dark brown and bluish stiff clay, apparently not laminated, and with angular pebbles and boulders, more than ten feet thick.

boulder measuring 5 by 4 by 3 feet has lately been removed to one of the public parks.

Ninety species of shells are recorded by Mr. De Rance from the Glacial Drift of Lancashire.

POST-GLACIAL DEPOSITS.—To these belong the terraces of alluvium or old

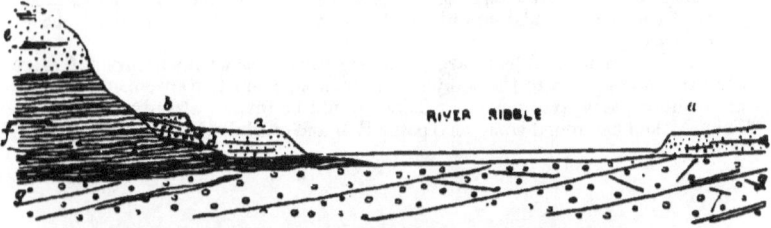

Fig. 51.—Diagram of Banks of the Ribble in Mete House Wood.

a Recent Alluvium. *b* Sand. *c* Gravel. *d* Peaty Clay. *e* Middle Drift Sand. *f* Lower Boulder Clay. *g* Pebble Beds of Bunter Sandstone.

The present bank of the stream is formed by (*a*); this deposit rests against an older bank (*b*); while the glacial clays and sands must be older still, since they lie underneath both (*a*) and (*b*).

river muds seen ranged one above the other along the valleys of the Irwell, Mersey, and Ribble, together with the later-formed muds and gravels nearly on a level with the present beds of those rivers. (See fig. 51.)

Dunes or sand-hills, formed by the wind, occur on the coast between the mouths of the Mersey and Ribble, and again at Lytham and north of

Blackpool; they are from 30 to 80 feet in height, and extend inland for 2 miles: the sands of Shirdley Hill and Aintree are probably ancient sand-dunes.

Peat.—Extensive beds of peat occur in south-west Lancashire forming what are called "mosses," as Chat-Moss, Sefton Moss, &c. This peat is chiefly the result of the growth of a kind of moss called *Sphagnum;* the peat lands are now mostly drained, and yield excellent crops of potatoes. Trunks and stumps of trees are frequently found in the peat, as at the mouth of the river Alt, and, indeed, at many points on the Lancashire coast, forming "submerged forests." A bed of peat covered by boulder-clay has been discovered at Rhodes Bank, Oldham, at a height of 570 feet above sea-level. It is considered possible, however, that this overlying boulder-clay was not formed where we now see it, but that it has slipped down upon the peat, so that the latter may after all be the newer of the two deposits.

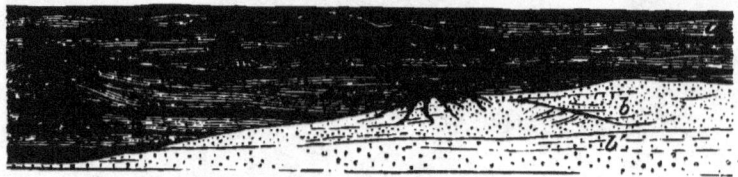

Fig. 52.—Section of Carr Moss, Halsall, exposed in a sluice.
a Peat with trunks and roots of oaks, pollard ashes, &c.
b Dark brown sand.
b¹ White and yellow sand. } Shirdley Hill Sand.

PREHISTORIC MAN.—Specimens of the stone tools which were used by the earliest human inhabitants of our islands have not occurred very plentifully in Lancashire. All which have been found belong to the Neolithic or newer stone age. Celts or stone axe-heads have been found at Toxteth (clay-slate, 4 inches long), Shaw Hall near Flixton (13 inches long by 3 wide and 2 thick), Newton (compact slate, 17 inches by 3 by 2); an axe-head of porphyritic greenstone perforated for a handle from Ayside near Newby Bridge, Windermere, is in the collection of Mr. John Evans; others have been found at Dean near Bolton, Hopwood, Claughton Hall, Garstang, Winwick near Warrington, and Haydock near Newton. Stone querns or hand-mills for grinding meal, made of the rough sandstones of the Millstone Grit, have occurred at one or two places.

The above-named objects are of a large and conspicuous nature; if more attention were given to the subject no doubt such smaller specimens, as flint arrow-heads, scrapers, knives, or flakes, would be found; already, indeed Mr. John Aitken has found some on Tooter Hill and on Bull Hill near Bacup.

No. 21.

GEOLOGY OF LEICESTERSHIRE.

NATURAL HISTORY AND SCIENTIFIC SOCIETIES.

Literary and Philosophical Society of Leicester. Report and Transactions.
Scientific Association, Leicester.
Literary and Philosophical Society of Loughborough.

MUSEUMS.

The Town Museum, Leicester.
Melton Mowbray Museum.

PUBLICATIONS OF THE GEOLOGICAL SURVEY.

Maps.—Quarter Sheets: 63, S.W. Atherstone; 63, S.E. Lutterworth, Market Harborough; 63, N.W. Ashby, Market Bosworth; 63, N.E. Mountsorrel, Leicester; 71, S.W. Castle Donington; 71, S.E. Loughborough. Sheets: 64, Melton-Mowbray, Medbourne; 70, Waltham, Belvoir Castle.

Books.—Geology of South Leicestershire, by Aveline and Howell, 8d. Geology of Rutland and East Leicestershire, by Prof. Judd, 12s. 6d. Geology of the Leicestershire Coal-field, by E. Hull, 3s. Triassic and Permian Rocks of the Midland Counties, E. Hull, 5s.

IMPORTANT WORKS OR PAPERS ON LOCAL GEOLOGY.

List of 43 works in Geology of Leicestershire by W. J. Harrison.

1834. Mammatt, E.—Coal-field of Ashby-de-la-Zouch.
1842. Jukes, J. B.—Geology in Potter's "Charnwood Forest."
1863. Sorby, H. C.—Microscopical Structure of Mountsorrel Syenite. Rept. Geol. Soc. W. Riding Yorks, p. 296.
1866. Brodie, Rev. P. B. Lias at Barrow-on-Soar. Rept. Brit. Assoc., p. 51.
1876. Harrison, W. J.—Rhætic Beds in Leicestershire. Journ. Geol. Soc., vol. xxxii. p. 212.
1877. Harrison, W. J.—Geology of Leicestershire.
1877 to 1880. Hill and Bonney, Revs.—Rocks of Charnwood Forest. Journ. Geol. Soc., vols. xxxiii., xxxiv., xxxvi.
1878. Harrison, W. J.—Digging out a Boulder. Midland Naturalist, vol. i. p. 153.
1879. Harrison, W. J.—Garnets in Charnwood Rocks. Midland Naturalist, vol. ii. p. 77.
1879. Allport, S., and Harrison, W. J.—Rocks of Brazil Wood. Midland Naturalist, vol. ii. p. 243.
1879. Harrison, W. J.—Rambles with a Hammer. Midland Naturalist, vol. ii. p. 117.
1880. Harrison, W. J.—The Rhætic Star-fish Bed. Science Gossip, p. 49.
1880. Harrison, W. J.—Scheme for Examination of Glacial Deposits of Midlands. Midland Naturalist, vol. i. p. 242.

See also General Lists, p. xxv.

THE rocks of Leicestershire are more varied in age, character, and composition than those of any of the adjoining counties. Probably no district could be chosen which would better illustrate the effects of geological structure upon the life and occupations of its inhabitants. Fine local collections of rock-specimens and fossils may be seen in the Leicester Town Museum.

If we take a brief general survey of the county, we shall find the rocks fall into five broad divisions:

    I. In the north-west rises the hilly, almost mountainous region of Charnwood Forest, composed of igneous and metamorphic rocks.

    II. Westwards of Charnwood, and extending across the western boundary of the county into Derbyshire, the Coal-measures, with their accompanying beds of shale and limestone, form the region known as the Leicestershire Coal-field. Some pebbly beds of a rather later geological period the Permian also occur here.

    III. "Red Rocks," of the Triassic age, form much of the land north, east, south, and south-west of Charnwood, covering, in fact, all the western half of the county not occupied by I. and II. The river Soar may be regarded as the eastern boundary of this division.

    IV. In the eastern half of the county, stiff bluish clays of Liassic age preponderate, with a hard bed of marlstone; whilst above them, in the extreme north-east, and in one or two outlying patcnes elsewhere, sands and limestone of Oolitic age are found.

    V. Lastly, scattered in varying thickness and with great irregularity over all the rocks already mentioned, there are beds of clay, gravel, sand, and pebbles, which we call Drift relics of the last Glacial period and submergence beneath the sea. The alluvial deposits of our rivers still in course of formation—bring our geological history down to the present day.

Let us now consider each of these great rock-divisions a little more in detail.

THE CRYSTALLINE AND SLATY ROCKS OF CHARNWOOD FOREST. These have long been regarded as a geological puzzle; no fossil remains have ever been found in them which we might have compared with those found in rocks elsewhere. In structure they somewhat resemble the Silurian beds (Borrowdale Series) of the Lake district, but discoveries made within the last few years render it almost certain that these Charnwood rocks are of very high antiquity, belonging, in fact, to the PRE-CAMBRIAN FORMATION. In appearance and composition, they present a remarkable variety, and the question of their origin has not been less debated than that of their age. Extending from Blackbrook on the north-west to Bradgate on the south-east, a distance of about 7 miles, a line can be traced, dividing the slates, &c., which can be seen to dip or slant to the north-east from those which dip to the south-west. This is, therefore, the line along which the original elevatory forces specially acted when these rocks were upheaved. Eastwards of this anticlinal line we find coarse gritty slates, "ashy" in character, at Moorley Hill, Nanpantan, &c., passing into finer-grained slates further south and east. At Whittle Hill the grain is very fine and close, and the "Charley Forest Hones," which have a high reputation, all come from a little quarry here. At Swithland there are some large and deep slate pits, from which fine slabs are obtained, but as the grain is not so fine as that of the Welsh slates, it is not so well suited for roofing purposes. Good slabs of slate are also got at Groby, and one or two other spots.

It is as a source of what is popularly called "granite," that Charnwood Forest has become especially celebrated. Mountsorrel is, however, the only spot which produces a rock entitled, mineralogically, to this name. If we pick up and closely examine an ordinary piece of Mountsorrel granite (there are two varieties, a greyish-white and a red, either will do), we shall see opaque oblong crystals of felspar, which can be readily scratched with a penknife;

other crystals, colourless and glassy, of great hardness, which are quartz, whilst a little clear green hornblende is interspersed. Here and there are a few black glistening specks of mica, and it is the presence of this last-named mineral which entitles the Mountsorrel rock to the name of granite. The "face" of rock which is being worked is distinctly visible from the Midland Railway, and a very large number of workmen are employed: the stone is broken up into cubes and kerbs for paving purposes. At Groby and Markfield, there are also large quarries. The stone here is a true syenite, being composed of pinkish felspar, and green partly decomposed hornblende. Cropping out at intervals for some nine miles south of Charnwood Forest, we find several small bosses of a very fine-grained syenite at Enderby, Croft, Stoney Stanton, Narborough, and Sapcote: quarries are being rapidly opened in this district, and the stone is very tough, and cleaves into cubes with great facility. The researches of Professor Bonney and the Rev. E. Hill have shown that the old rocks of Charnwood are of volcanic origin; most of them are volcanic ashes ejected from numerous cones and accumulated chiefly on land, but partly perhaps in small lakes. The anticlinal axis which traverses the Forest from north-west to south-east is also a line of fault, the strata on the west of it having been upheaved from 500 to 1,000 feet above those on the east. The fine-grained ashes and flinty slates of Blackbrook and the Old Reservoir are the lowest beds of all; above these come coarse ash-beds and masses of agglomerates, wonderfully developed in the region round the Monastery, and probably marking the neighbourhood of some of the cones; higher still are banded slates and the quartz-grit bed seen at Steward's Hay Spring, the Stable Quarry Bradgate, and the Brande; while the workable slates of Groby and Swithland are on top of all. The total thickness may be estimated at 10,000 feet.

The rock at Brazil Wood should be called a "micaceous schist" rather than "gneiss;" it contains numerous garnets, and can be traced passing into a coarse slate, just like that of Swithland.

The granite, syenite, &c., is clearly intrusive, breaking through and altering the beds of grit or slate.

There are many spots on the Forest where hard stone suitable for road-mending could be worked to great advantage, but much judgment is required in the selection. The comparative nearness ought to command the custom of the east and south-east of England; and with improved communication the future will, doubtless, see a great development of the special industries of this district. The rocks of Charnwood probably extend beneath the surface far to the south and east; so that endeavours to find coal due east of Leicester will certainly be futile.

II. THE LEICESTERSHIRE COAL-FIELD.—When the seams of coal, with their accompanying strata, were first deposited, they doubtless stretched continuously all over the region of Charnwood Forest, and were connected with the Nottingham Coal-field. Since the upheaval of the Forest district all the coal-beds which once covered it have been swept clean away—denuded off—together with a vast thickness of the syenites and slates themselves, until now not a patch of coal is left on that area. The lowest bed of the great Carboniferous formation is the *Mountain Limestone*, which is found cropping out at intervals to the north-west of Charnwood, at Grace Dieu, Osgathorpe, Barrow, Breedon Cloud, and Breedon. This limestone is very impure, containing much carbonate of magnesia; it is, in fact, a "dolomite," and is largely quarried to burn into lime. It also occurs at Dimminsdale, where it contains ores of lead, copper, iron, zinc, &c., but in small quantities only. Several fossil shells occur, mostly as casts, *Productus* and *Bellerophon* being the most common. The limestones are overlaid by beds of shale, which are well seen in the quarries at Grace Dieu. They are about 50 feet thick in this district, and are capped by the millstone-grit, a very hard, coarse sandstone, which is only exposed in one or two small patches near Thringstone, &c. This is the "Farewell rock" of the Welsh miners. No seams of coal occur beneath it.

The Coal-measures. About 2,500 feet of sandstones, clays, shales, and coal-seams overlie in this district the millstone-grit. Of these, however, the lower half may be regarded as unproductive, the coal-seams contained in it being too thin to admit of profitable working.

The coal-seams of the upper part, too, do not spread in uninterrupted sheets at regular depths. The same great upheaval which acted chiefly under Charnwood Forest also dislocated and disturbed all the surrounding country, elevating portions here and there above the rest, and so breaking the continuity of the coal-seams and producing what are called "faults." The surface inequalities so produced have since been worn down to a tolerably level surface, so that we only know of the existence of these "faults" when in following a coal-seam we find it suddenly stop short, abutting on a rock of quite different character. The chief "faults" run from north-west to south-east, parallel to the anticlinal axis of Charnwood Forest. They are named (1) the Coleorton, or Boundary Fault, which stretches from Ticknall to Bardon Hill; (2) from this the Heath End Fault seems to branch out, whilst further to the south-west the Boothorpe Fault and the Moira or Main Fault run nearly parallel to it. A number of minor faults run generally at right angles to these. (See fig. 53).

The effect of unequal upheaval was to elevate the district between the Heath End and Boothorpe Faults, *i.e.* all the part to the north-west and south-east of Ashby-de-la-Zouch. From this part then exposed to denudation

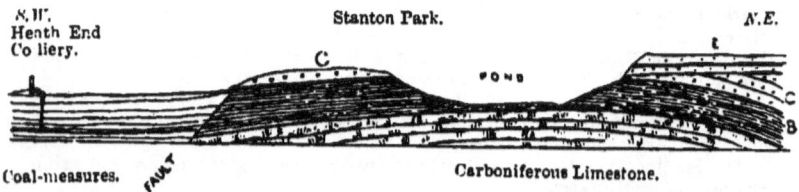

Fig. 53.—Section across Stanton Harold and Heath End; showing the Coleorton Boundary Fault near its western extremity.
B Carboniferous Limestone Shale.   *C* Millstone Grit.   *E* New Red Sandstone.

the coal-beds were speedily pared off, and we thus have the lower unproductive measures standing like a wall between the western coal-producing district of Moira and the eastern or Coleorton division. We may, therefore, suppose that the chief seams of coal of these two latter districts are identical and were once continuous, and from the general similarity in character, number, and order of succession of the beds of each, this seems a fair conclusion.

In the Moira district the main coal is 14½ feet in thickness, but it is not of equal quality throughout, and is not all extracted. In the Coleorton district it is about 6 feet thick. The boundaries of the Moira district by Stanton, Gresley, Donisthorpe, Measham, Willesley, and Woodville, are pretty clear, the coal-seams cropping out on all sides, but the Coleorton district extends southwards and eastwards beneath the new red marl to an unknown distance. At Whitwick colliery a bed of dolerite was struck at a depth of 165 feet. This is a true volcanic rock, and was probably ejected from a vent somewhere in the great "fault" at the foot of Bardon Hill: it is 60 feet thick, and has turned the bed of coal on which it rested to cinders.

Still further to the south, coal has been reached at Ibstock, Nailstone, and Bagworth; whilst recently similar beds have been proved further east at Desford, but this is close to the eastern boundary, and the seams were so faulted and disturbed as to be unworkable. About 6 miles south of Desford, at Elmesthorpe, a boring of 1,500 feet in depth found unproductive Coal-measures dipping nearly vertically; and westwards, between the Nuneaton and Ashby Coal-fields, several borings have been made, near Market Bosworth, Snarestone, &c., but so far without success. In a boring on the Spinney Hills, close to the east side of the town of Leicester, dark slaty-looking rocks of Palæozoic age were reached at a depth of 800 feet. In 1879 there were 27 collieries at work in the Leicestershire Coal-field, and 1,035,016 tons of coal were raised.

The Permian Formation.—In a few spots to the south and west of the Ashby coal-field, as at Measham, Packington, &c., there occur beds of pebbles embedded in a paste of red marl. These are almost certainly of Permian age, and as the pebbles they contain resemble the Carboniferous and Silurian rocks of the West of England, being at the same time angular and some even striated, they probably afford a proof of the recurrence of glacial periods at various geological epochs.

III. THE NEW RED SANDSTONE, OR TRIAS.—The lowest division of the "Red Rocks" is the *Bunter (variegated) Sandstone.* Pebbly beds of this age occupy the district about Netherseal and Donisthorpe. In Leicestershire, as elsewhere, there is considerable difficulty in separating the Triassic and Permian beds, from want of good sections and the absence of fossil evidence. Both are unconformable to the Carboniferous rocks and also to one another. From facts obtained elsewhere, however, geologists draw between them a great line of

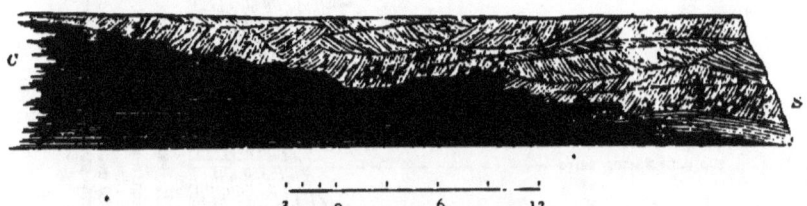

Fig. 54.—Section of the Rock-Fault in the Main Coal, Coleorton.
C Main Coal   *S* Sandstone, false bedded.
This rock-fault is not a "fault" at all in the true geological sense of the term, since there is no displacement of the strata. The coal has been washed away, and the sandstone deposited in its place.

demarcation, placing the Permian in the Palæozoic (old life) system, whilst the Trias is the first of another great system—the Mesozoic (middle or later life period).

Above the Bunter Conglomerates come the Keuper Beds (so named from their containing in Germany much copper ore). The lowest member is a thick sandstone called the "*Water-stones,*" from its everywhere affording an abundant supply of water when pierced by wells sunk through the red marl above.

The Red Marl with Upper Keuper Sandstone constitutes the greater part of Leicestershire west of the river Soar, and usually extends a mile or so east of that river. The town of Leicester is built on the uppermost beds of this division; it contains beds of sandstone seen at Orton-on-the-Hill, Diseworth, and especially at the Dane Hills, close to the west side of Leicester. Here, in a cutting of the Leicester and Burton Railway, a fine section is seen, showing the characteristic "false bedding" (produced by the action of currents) in perfection. A little crustacean, *Estheria minuta,* is found in these beds, with teeth and spines of *Hybodus.* A massive nodular band of gypsum occurs towards the top of the red marl; it is well seen in the cutting of the Midland Railway's main line near Thurmaston. It is in the red marl of Shropshire, Cheshire, and elsewhere, that thick beds of rock-salt occur; indeed the whole Triassic system seems to have been deposited in inland seas, saturated with

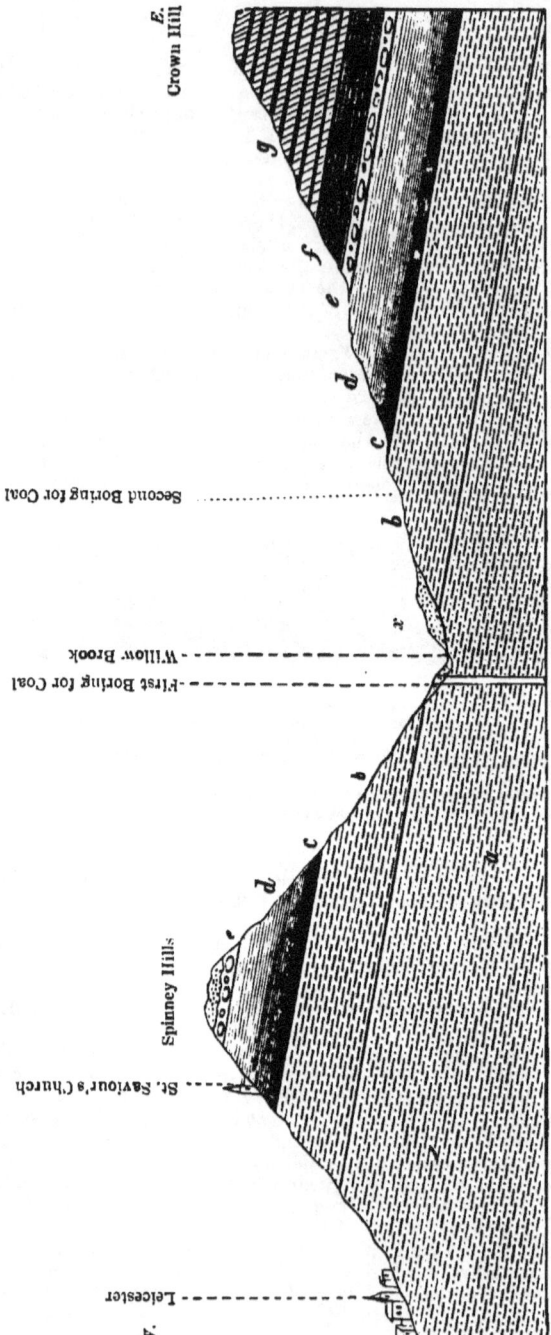

Fig. 55.—Section from the Spinney Hills, east of Leicester, to Crown Hill, near Evington.

a Upper Keuper Red Marls, with Gypsum.   *b* Rhætic Grey Marl.   *c* Bone-Bed.   *d* Avicula contorta shales.
e Nodular Limestone (Rhætic).   *f* White Lias.   *g* Lower Lias Limestones and Shales.
x Alluvial deposits.

The site of a third boring for coal was just beyond, or to the east of, Crown Hill.

salt. In Leicestershire we have traces of this in the numerous casts of salt crystals which cover the surface of the light-coloured marls in many places. Ripple-marks, sun-cracks, and rain-pittings, afford further signs of the condition of things at that period. The red colour of the marls is due to the presence of oxide of iron, and it is generally found that where this is abundant traces of life (fossils) are few.

IV. THE LIAS AND OOLITES OF EAST LEICESTERSHIRE.—*The Rhætic Formation.* Resting on the top of about 1,000 feet of red marl and sandstone, there occur certain thin bands of grey marls, black shales, and nodular limestones, whose fossil contents are of the highest interest. The only clear sections of them at present known in this county are to be seen in certain brick-pits at the northern extremity of the Spinney Hills, a low range forming the eastern boundary of the town of Leicester and of the Soar valley. The top of the Keuper red marl is there visible, and above it is a thick bed of grey marl, which in turn is surmounted by the famous Bone-Bed, a layer some 3 inches thick, composed of remains of Ichthyosaurus, Plesiosaurus, Labyrinthodon, with numerous fish remains, as Hybodus, Ceratodus, Nemacanthus, &c., with many other interesting fossils. Next come about 9 feet of black and light-coloured shales, containing the characteristic shells *Avicula contorta*, and *Cardium Rhæticum*, and the whole is capped by bands of hard nodular limestone. It is still a matter of dispute whether these beds belong to the Triassic or to the Liassic formation. Probably they are true "passage beds," having in the lower part Triassic and in the upper Liassic affinities. In these Spinney Hill beds was discovered a new species of star-fish, since named *Ophiolepis Damesii*, which has since been found in several other Rhætic sections in England. (See figs. 55 and 56.)

THE LIAS.—The Lower Lias beds have long been worked at Barrow-on-Soar and at Kilby. They consist of an alternation of bands of Limestone and dark shales. The former when burnt yield a very valuable hydraulic cement. The characteristic fossils are *Ammonites catenatus*, *Eryon Barrovensis* (resembling somewhat a cray-fish in appearance), *Dapedius orbis* (a fish with a very rounded outline), *Nautilus striatus*, *Lima gigantea*, &c. In rather higher beds *Gryphæa incurva* occurs in great numbers with a large ammonite, *A. Bucklandi*. In the northern division of the county there are probably many spots where the hydraulic limestones could be worked to advantage, but southwards they are overlaid by too great a thickness of Drift.

The total thickness of the Lower Lias beds is about 600 feet, for a boring for coal near Billesdon Coplow reached that depth entirely in this formation; had it been persevered with a little further, it would doubtless have entered the red marl. The district occupied by the stiff bluish clays of this division is mostly in the condition of pasture land, and forms the chief portion of the great hunting and stock-rearing district of East Leicestershire. The tunnels on the Nottingham and Melton railway afforded some good sections of these beds. They contained the shells *Hippopodium*, *Cardinia*, and *Pleurotomaria* in abundance. In the Geological Survey Map the boundary of the Lias near the town of Leicester has been drawn too far to the east; light-coloured clays and compact limestones cap the Rhætics of the Spinney Hills and constitute the Highfields district and (part of) Victoria Park.

The Marlstone or Middle Lias.—The "Rock-bed" is a limestone containing much iron; exposed to the air it weathers brown, and constitutes excellent corn-land. In the north-east, at Wartnaby and Holwell, it is about 25 feet thick, but decreases to about 1 foot between Keythorpe and Hallaton: its outline is very irregular, as it sends out bold spurs to the west, as at Burrow Hill and Billesdon, and indeed forms an escarpment along its whole extent. At its junction with the clay beds beneath numerous springs issue: a good example is in the picturesque little ravine of Holwell Mouth, where the river Smite issues from the base of the Rock-bed in this manner. At Tilton the marlstone is finely exposed in the new railway cutting; it here contains

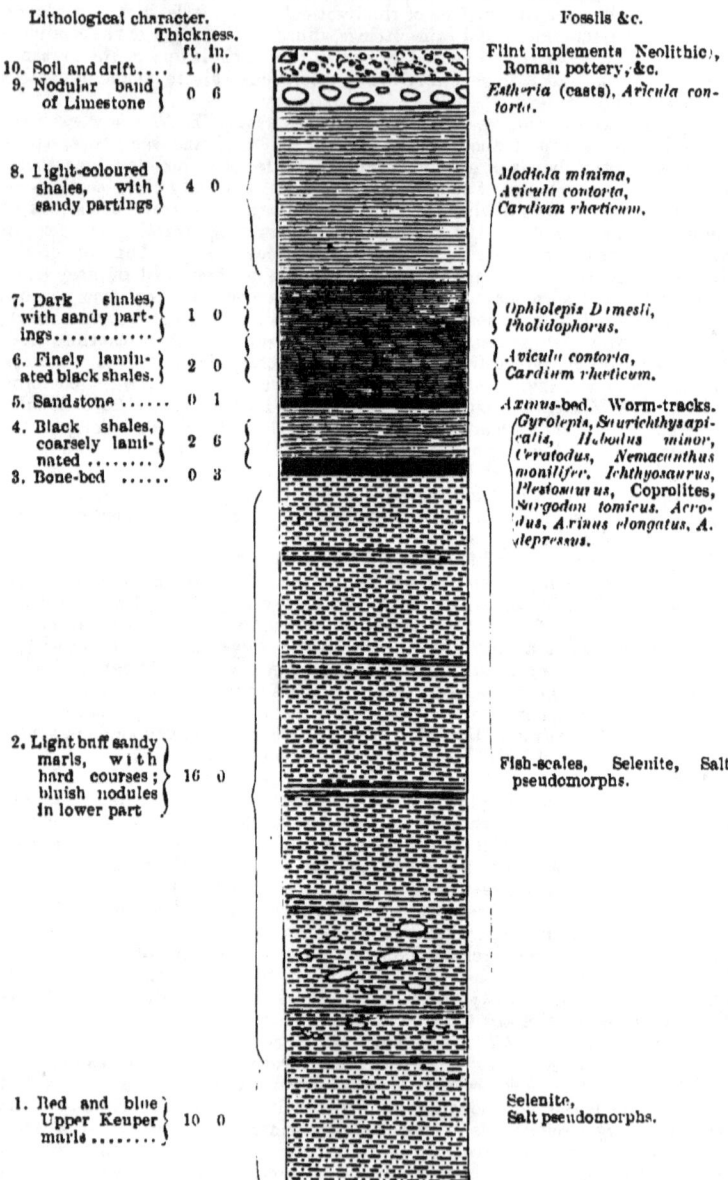

Fig. 56.—Vertical Section of Rhætic Beds, Spinney Hills, near Leicester.

countless numbers of the two brachiopod shells, *Terebratula punctata* and *Rhynchonella tetrahedra*.

The *Upper Lias Shales* are rarely seen, being covered over with Drift. The "fish and insect beds" are exposed in some pits near Keythorpe. The new lines of railway from Melton Mowbray to Market Harborough, with the branch from Tilton to Leicester, have exposed some grand sections of the different members of the Lias in the deep cuttings which the hilly nature of the ground rendered necessary. (See fig. 57.)

THE OOLITE.—Resting upon the Upper Lias shales is a stratum called the *Northampton Sand*, from its development further south, where it is extensively

Fig. 57.—Section exhibited in a pit between Keythorp and Hallaton, East Leicestershire. The Marlstone Rock-bed is here only one foot thick, and seems to have been eroded before the deposition of the Upper Lias.

	a Soil, &c.
Upper Lias	{ *b* Nodular Limestones of the Fish and Insect Beds. *c* Clays.
Middle Lias	{ *d* Marlstone Rock-bed. *e* Clays, &c.

worked for ironstone. In Leicestershire it only occurs in detached patches or outliers, although no doubt it was once continuous over the whole district. Robin-a-Tiptoes, a hill some 750 feet high, and about 11 miles east of Leicester, is capped by hard calcareous beds of Northampton Sand, as are the adjacent eminences of Whadborough Hill, Barrow Hill, &c. Near Neville Holt it is about 20 feet thick. Iron-works were commenced here, but have been discontinued.

The Lincolnshire Oolite Limestone.—The labours of Professor J.W. Judd have made it clear that this important bed belongs to the Inferior Oolite, thus being older than the noted Bath limestone, with which it was formerly considered contemporaneous. The main line of outcrop just enters the county on the north-east, near Crown Point, and the limestone extends near Croxton and Stonesby. In the south-east, at Neville Holt, this bed also occurs as an outlier, capping the Northampton Sand.

We have now considered briefly all the great stratified rocks of Leicestershire. They have, on the whole, a slight inclination or dip to the south-east. This brings the oldest rocks up in the north-west of the county, and as we walk from there eastwards we are continually passing over the edges of newer and later-formed strata.

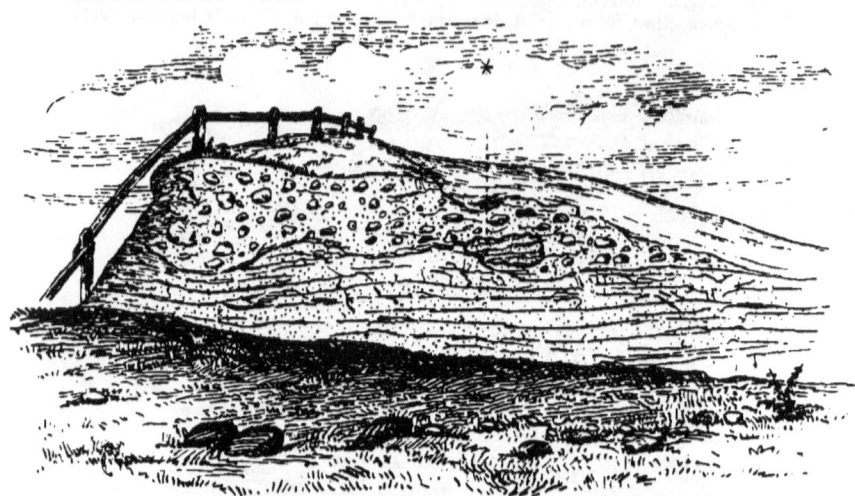

Fig. 58.—Sketch of Section in pit above Hallaton Ferns, East Leicestershire, showing a small outlying patch of the Northampton Sand, capped by Boulder Clay.
* Large boulder, 18 inches long, of the hard Siliceous Limestone (Pendle) of the Inferior (Lincolnshire) Oolite.

V. THE DRIFT; OR BOULDER CLAY AND SURFACE GRAVEL. Scattered indiscriminately and unconformably over the upturned edges of all the abovementioned beds, we find beds of clay, sand, and gravel, which are relics of the last Glacial Period, when England was altogether or in part covered by ice, and submerged once, if not twice, beneath the level of the sea.

(a) *Pre-Glacial or Mid-Glacial Beds.* —These contain rocks *similar to those of the adjoining district only.* At Rotherby brick-yard there is a section showing finely laminated, recomposed red marl. The surfaces of the laminæ bear dark carbonaceous markings: it is well stratified, and shows false bedding. About 15 feet in thickness is exposed, and it is covered by sandy clay, in which is embedded a limestone boulder about 3 feet by $1\frac{1}{2}$ feet: similar beds occur at Melton, Billesdon, Ouston, &c. Near Leicester Abbey is an interesting sand-pit containing black seams of carbonaceous matter, which show admirably the false bedding; it is overlaid by Boulder Clay. (See fig. 59.)

(b) *The Great Chalky Boulder Clay.*—This occurs principally in the central and eastern parts of the county; it consists chiefly of chalk, chalk-flints, oolite, and marlstone, embedded in a stiff bluish clay with quartzose pebbles, millstone grit, coal, mountain limestone, &c. In the upper part great boulders of Forest Rocks occur. Great sheets of glacial ice from the northeast swept down and over our Charnwood hills, rounding their outlines and tearing off great portions of them to be spread over the surrounding plains. Some of these boulders are of historic interest. Such was probably

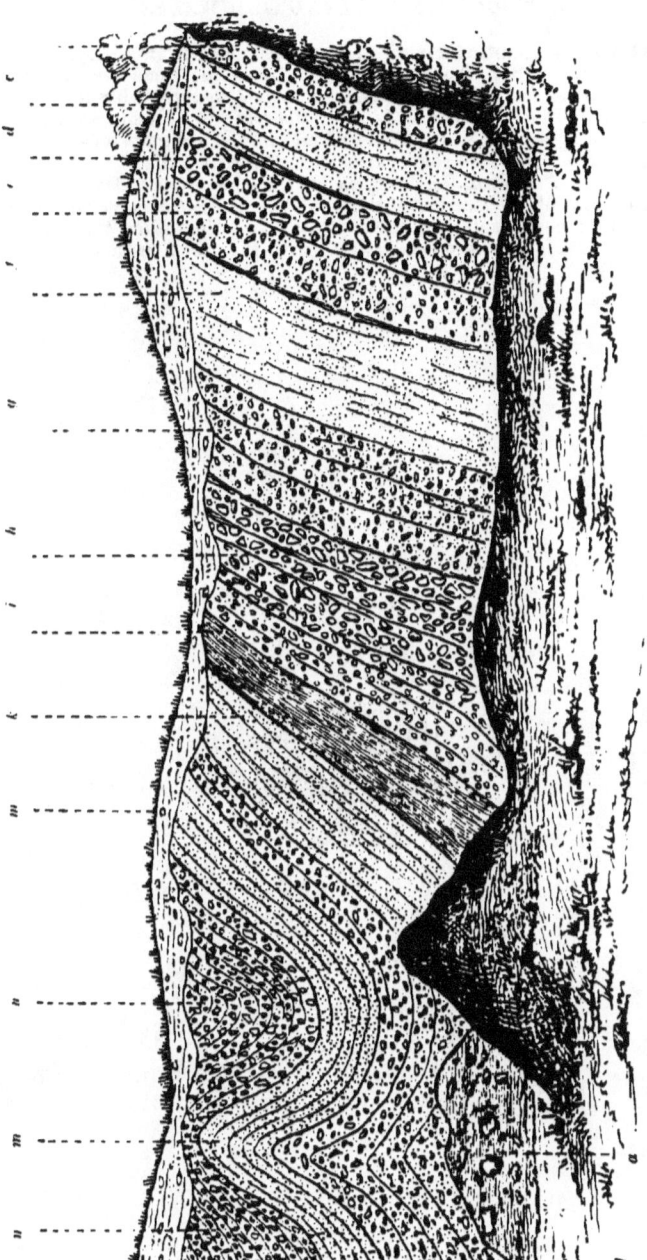

Fig. 59.—Pit in Glacial Gravels between Whadborough and Ouston, East Leicestershire; showing Contortions possibly caused by the grounding of Icebergs. (After Judd, Geol. Survey.)

a Boulder Clay.   *d h* Coarse Gravel.   *b e g l n* Fine Gravel.   *c f k m* Sands.   *l* Loam.

"John's Stone," a granitic mass which stood near Leicester Abbey, but was destroyed about 1836. The stone which occupies the same site, and which has been mistaken for it, is a hard sandstone slab, and probably formed the base of the former monolith. The *Holstone* of Humberstone, near Leicester, is a 15-ton boulder of Mountsorrel granite, and numerous blocks of the same rock can be traced southwards along the Soar Valley.

River Gravels, Peat Beds, &c., bring us down to the present day. In these deposits remains of the mammoth, rhinoceros, reindeer, &c., have been found. Flint and stone implements, too, occur, relics of the time when the early inhabitants of this island were unacquainted with the use of metals.

No. 22.

GEOLOGY OF LINCOLNSHIRE.

MUSEUMS.

Museum of the Literary and Scientific Institute, Stamford.

PUBLICATIONS OF THE GEOLOGICAL SURVEY.

Maps.—Sheet 70, Belvoir Castle, Grantham. (Survey not completed.)
Books.—The Geology of Rutland and Part of Lincoln, by J. W. Judd, 12s. 6d. Geology of the Fenland, by Skertchly, 40s.

IMPORTANT WORKS OR PAPERS ON LOCAL GEOLOGY.

1853. Morris, Prof. J.—On some Sections in the Oolite District of Lincolnshire. Journ. Geol. Soc., vol. ix., p. 317.
1867. Burton, F. M.—Rhætic Beds, near Gainsborough. Q. Journ. Geol. Soc., vol. xxiii. p. 315.
1867. Judd, J. W.—Strata which form the Base of the Lincolnshire Wolds. Journ. Geol. Soc., vol. xxiii. p. 227.
1868. Wood, S. V., and Rome, Rev. J. L.—Glacial and Post-Glacial Structure of Lincolnshire. Journ. Geol. Soc., vol. xxiv. p. 146.
1870. Judd, J. W. — Neocomian Strata of Lincolnshire. Journ. Geol. Soc., vol. xxvi. p. 326.
1875. Cross, Rev. J. E.—Lias and Oolite of N.W. Lincolnshire. Journ. Geol. Soc., vol xxxi. p. 115.
1875. Blake, Rev. J. F.—Kimmeridge Clay of England. Journ. Geol. Soc., vol. xxxi. p. 196.
1879. Jukes-Browne, A. J.—Southerly Extension of the Hessle Boulder Clay. Journ. Geol. Soc., vol. xxxv. p. 397.
1879. Wilson, E.—On the Boring for Coal at South Scarle. Journ. Geol. Soc., vol. xxxv. p. 812.
1879. Burton, F. M.—Rhætic Beds and Keuper, near Gainsborough and Retford. Rept. British Assoc. (Sheffield), p. 336.
1880. Anon.—The Frodingham Iron-field. Colliery Guardian, vol. xl. p. 1051.

See also General Lists, p. xxv.

TILL within the last few years, Lincolnshire has, with respect to its geology, been one of the most neglected counties in England. This may, perhaps, be accounted for by the supposed absence of important minerals, by the want of good sections, and by the large amount of *drift* or surface *débris* strewn over the surface—as thick masses of clay and sand—concealing the strata which lie beneath.

Of late years some important papers on the county have been published in the Quarterly Journal of the Geological Society, by the Rev. J. E. Cross, on the Lias of the north-west corner of the county, and the Rev. J. F. Blake on the Kimmeridge Clay, in vol. xxxi.; by Prof. Judd on the Chalk (vol. xxiii.), Speeton Clay (vol. xxiv.) and Neocomian beds (vol. xxvi.); by Professor Morris on the Oolites of the south-west of the county (vol. ix.); and there are some scattered notes in the "Geological Magazine." The definite mapping of the county by the Geological Survey was, however, long postponed in order that the coal-fields might be first completed. In 1875 the issue of sheet 64 gave us a thorough and complete knowledge of the extreme south of the county, from Stamford, Little Bytham, and Creeton on the west, to Bourn and Market Deeping on the east. This sheet also includes the whole of Rutland, and portions of Leicester and Northampton: it was the work of Professor J. W. Judd, F.R.S., and, with the accompanying memoir, has contributed to place that gentleman in the first rank of living geologists. Sheet 70, which lies immediately to the north of 64, has also been nearly completed by Mr. Holloway*, and will be published in 1881; it extends as far north as Newark, Tattershall, &c.

Lying on the east coast of England, Lincolnshire includes none of the hard rocks which stand up so boldly in the west, forming the mountains of Wales and Scotland. The rocks which compose the county belong to the MESOZOIC or Middle Period of geologists, and comprise a fairly complete series from the Trias to the Chalk.

As the different beds of rock extend in long bands nearly due north and south, they coincide with the main extension in length of the county: as moreover, these rocks are of varying degrees of hardness, being an alternation of beds of clay and limestone, we get longitudinal valleys scooped out in the clays, while the harder limestones stand out as bold ranges of hills running from north to south. This simplicity of structure is, however, complicated by the immense quantity of *drift* which has been deposited here and there, filling up some of the ancient valleys, and by the extent to which the eastern and south-eastern portion has been planed down by the sea, reducing it to one level plain, affording none of those cliff-sections which have so aided the study of the neighbouring county of York.

The oldest rocks found in Lincolnshire occur on its western side, and form the slopes of the Trent valley. THE TRIAS, as these rocks are called from their triple division in Germany, is represented here by the upper member only, which is known as the *Keuper*, from its containing copper ore abroad: it consists of stiff red marl, with intercalated bands of sandstone and gypsum: it constitutes the foundation of the Isle of Axholme, although only occurring at the surface round Epworth, and extends in a varying band for a short distance on the east of the Trent, broadening out, however, to the southward in Notts. The presence of the peroxide of iron, which tinges the beds red, seems to have been prejudicial to animal life, for very few fossils occur in the Keuper. Fish-scales and teeth, with bones and footprints of the *Labyrinthodon*, should, however, be looked for in any bands of sandstone which may be met with. The red clay is frequently dug for brick-making: the beds have a gentle dip or slant to the eastward.

THE LIAS is the name applied to a series of limestones, shales, and clays, which rest upon the Red Marl. At the junction between the two great formations we find a series of beds termed *Rhætics* (from their great development in the Rhætian Alps), which seem to mark a transition from the one to the other. These beds are in part exposed in some pits near Newark, and extend northwards by South Scarle, Gainsborough, and Blyton, to the point where the Trent joins the Humber, passing thence into Yorkshire. At Lea, about 2 miles south of Gainsborough, they were noted by Mr. Burton in 1866, in the cutting of the Great Northern line. The characteristic shells, *Avicula contorta* and *Cardium rhæticum*, were found here, together with a thin stratum called the *bone-bed*, from its being full of fish-teeth and scales.

* By the death of this gentleman since the writing of these words we have lost a careful and painstaking geologist.

LINCOLNSHIRE.

The Lower Lias Limestones and Shales come next in order, and the hard ands of limestone stand out as an escarpment, well seen in the Frodingham cutting of the Ancholme and Grimsby railway. These beds are very fossiliferous, containing many *Ammonites, Gryphæa incurva* in thousands, and, what is more important, a very valuable bed of ironstone, which is now largely worked. This important discovery and its development are chiefly due to Mr. Charles Winn, M.P. The ore is the hydrated peroxide, and the following quantities were raised in 1874:—

	Tons.	Value.
Appleby (Brigg)	105,109	£21,021
Frodingham (Brigg)	235,028	47,005
Frodingham	47,894	9,578
Scunthorpe	74,322	14,864

This bed of iron ore lies low down in the Lias, and is not the same as the Cleveland bed, with which it has been confounded; but the ferruginous limestone abounding in *Ammonites semicostatus*, which is so well developed at Redmile, in the Vale of Belvoir, is on the same horizon as the Frodingham ore (fig. 60).

The bed of ore is about 27 feet in thickness, and crops out at the surface, forming the plain on which the village of Scunthorpe stands, so that the workings are open and shallow. As is the case with similar ironstone beds elsewhere, we find the ore loose, and of a dark-brown tint near the surface, and passing downward into a tolerably hard rock: there is a hard limestone band at the base. The presence of limestone renders the ore a very suitable one to mix with another containing silica. Fortunately a bed of iron ore rich in silica has lately been found near Lincoln, and this is now mixed with the Scunthorpe stone in the proportion of one-eighth of the whole. The yield of metal is about 1 ton to 3¼ tons of ore, or 28 per cent.

In 1879 the total quantity of iron-ore raised in Lincolnshire was 695,326 tons, valued at £103,000.

The Middle Lias or Marlstone. — This is a very variable formation, both in thickness and mineralogical character: the *Marlstone Rock-bed* is a ferruginous limestone, which in the south-west of the county is as much as 20 or 30 feet

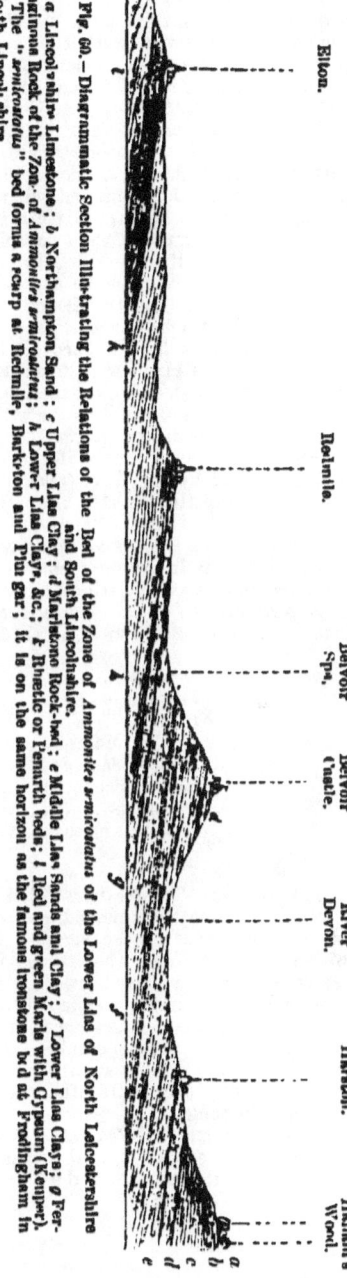

Fig. 60.—Diagrammatic Section Illustrating the Relations of the Bed of the Zone of *Ammonites semicostatus* of the Lower Lias of North Leicestershire and South Lincolnshire.

a Lincolnshire Limestone; *b* Northampton Sand; *c* Upper Lias Clay; *d* Marlstone Rock-bed; *e* Middle Lias Sands and Clay; *f* Lower Lias Clays; *g* Ferruginous Rock of the Zone of *Ammonites semicostatus*; *h* Lower Lias Clays, &c.; *k* Rhætic or Penarth beds; *l* Red and green Marls with Gypsum (Keuper). The "*semicostatus*" bed forms a scarp at Redmile, Barkston and Plungar; it is on the same horizon as the famous ironstone bed at Frodingham in North Lincolnshire.

thick, forming the bold outlier on which Belvoir Castle stands: it enters Lincolnshire near Woolsthorpe, and as it passes northwards becomes thinner, so that at Santon it is seen to be only 8 or 9 feet thick. Here in the cutting it is exposed as hard light-grey limestone, weathering brown, and full of a plaited bivalve shell, known as *Rhynchonella tetrahedra*. The rock-bed is worked for iron-ore at Caythorpe near Lincoln; it occupies the same geological position as the famous Cleveland seam.

The Upper Lias is in Lincolnshire, as generally elsewhere, a mass of thick clay. In the south-west, near Stainby, where it enters the county, it is, perhaps, as much as 200 feet thick. Passing northwards, by Grantham and Lincoln, we find it at Santon reduced to 60 feet, and composed of blue shales with casts of Ammonites. Everywhere it forms the western slope of the range known as the Cliff, and where it is thickest, there we find the hills highest, and the scenery most picturesque and diversified.

The Liassic formation thus occupies a band in the west of Lincolnshire, about 8 or 10 miles in width in the south of the county, but narrowing steadily northwards, owing to the thinning out of the various beds of limestone and clay, until on the Humber it is about a mile in breadth. Everywhere *the Cliff* marks the eastward extension of the main mass; but where the river Witham has cut through the limestone ridge at Grantham, and in the still more striking gorge formed by the same river at Lincoln, we must extend the various members of the Lias for some miles to the eastward, as *inliers*.

Good exposures of the Lias are rather rare in Lincolnshire: in the Marton cutting of the Gainsborough and Lincoln railway, Messrs. Burton and Waugh collected a number of fossils of the zone of *Ammonites angulatus*, which have been described by Mr. Tate (Quart. Journ. Geol. Soc., vol. xxii. p. 306). On the slope of Spittlegate Hill, near Grantham, there is a good section of the Upper Lias, showing the *Fish and Insect Beds* at its base, surmounted by ferruginous clays with *Ammonites communis*. The Liassic clays form a stiff clay land, mostly in pasture; but the marlstone gives rise to a rich, brownish loam, on which excellent wheat crops are grown.

Deep Boring at Scarle. A very remarkable boring for coal at South Scarle, between Lincoln and Newark, in 1876, passed through the following strata:

	Feet.
Drift	10
Lower Lias	65
Rhætics	66
Keuper Marls	573
Keuper Sanstone	244
Bunter Sandstone	542
Permians	519
Coal-measures (?)	10
	2,029

THE OOLITE.—The strata which lie next above the Lias have received the above name from the rounded, roelike structure—ώον, an egg seen in most of the limestones of the series. They consist of an alternation of beds of limestone and clay, the former standing out as lines of hills, while the latter occupy the valleys between. The commencement of the Oolitic Formation in Lincolnshire is marked by the striking ridge so well known as the "Cliff:" this is a bold escarpment facing the west, and running in a nearly straight line for about 90 miles: along the top runs the celebrated Roman road known as Ermine street. The escarpment is only broken at the two points already noted, Grantham and Lincoln, where the Witham has cut through the range.

LOWER OOLITE.—*Northampton Sand.*—This is a stratum of very varying mineralogical character, well developed in Northamptonshire, where it is 70 feet thick, but much reduced as we trace it northwards. Near Stamford it is 30 feet thick, at Grantham 48 feet, near Lincoln 30 feet, and in the north of the county some 10 or 15 feet only. Everywhere it forms the top of the Cliff, surmounting a long slope of Upper Lias Clay, which is usually tinged red by the downwash from the ferruginous beds above. The Northampton Sand forms a light soil, usually of a reddish hue, and well adapted for the growth of spring crops: in the county from which it takes its name it is largely worked for ironstone; but the ore diminishes in value as we trace it northwards.

Lincolnshire Oolite Limestone.—The researches of Prof. Judd have shown that this important bed belongs to the Inferior Oolite, and not to the Bath Oolite as was formerly supposed: immediately overlying the Northampton Sand, it forms a capping to the "Cliff," constituting the broad Oolitic plateau which stretches for 3 or 4 miles east of it. In Mid-Lincolnshire it is fully 150 feet thick, at Grantham 100 feet, at Stamford 80 feet, near Kirton 90 feet; but in South Yorkshire it is only 15 feet; and as we trace it further, both to the north and south, it at last thins out altogether and disappears.

At its base the limestone often passes into thin, sandy, fissile beds, which are the well-known *Collyweston Slates.* Fossils are numerous: here and there banks occur which seem to represent old coral reefs, and altogether we have evidence that local depression took place within an area having a diameter of about 90 miles, the amount of depression being greatest in Central Lincolnshire. As a consequence of this, there was slowly accumulated, by the growth of coral reefs, and the action of marine currents sweeping small shells and their fragments along the sea-bottom, a mass of calcareous strata, presenting many variations in its local characters, and constituting the formation to which we have applied the name of the "Lincolnshire Oolite Limestone." The western half of this great lens-shaped mass of limestone has been removed by denudation, and on the summit of the Cliff we get the exposed section of the remaining eastern half, which, dipping to the eastward, lies buried beneath the newer rocks which come on as we pass in that direction.

Excellent building stones are obtained from this formation, the quarries at Ancaster being especially famous. The commissioners appointed in 1839 to report on the building stones of England for the construction of the new Houses of Parliament, state that "many buildings constructed of a material similar to the Oolite of Ancaster, such as Newark and Grantham churches, and other edifices in various parts of Lincolnshire, have scarcely yielded to the effects of atmospheric influences."

The Great Oolite has for its base beds of sandy clay of varied tints. These are called the *Upper Estuarine Series*, for, like the Northampton sand below, they appear to have been formed near the mouth of a large river. At Little Bytham and other places, bricks of great hardness and durability are made from these clays. They constitute much of the cold unkindly land of the "Heath." At Stamford and Grantham they are 25 feet thick; but northwards they disappear, or are merged into the *Great Oolite Limestones.* This rock is from 20 to 30 feet thick, and, with the clays beneath, forms frequent outliers on the plateau of Lincolnshire limestone; above it are the *Great Oolite Clays*, from 10 to 20 feet thick. The limestones of this series are quarried for lime-burning and for road-metal, while the clays are dug for brick-making.

The Cornbrash.—This well-known rock is a ferruginous limestone, usually very fossiliferous, and in this district about 15 feet thick: its top is marked by a thick bed of a large rugose species of oyster—*Ostrea Marshii.* It forms a red soil of no great fertility, and is esteemed as furnishing hard road-metal.

The region occupied by the Great Oolites and Cornbrash forms a band down Mid-Lincolnshire, of which Bourn and Sleaford may indicate approximately the eastern limit. This region is characterized by a series of low, flat-topped hills, and some admirable sections of the beds are exposed in the cuttings of the Great Northern Railway, as at Essendine, Little Bytham, and Ponton.

The Middle Oolite is represented in Lincolnshire by the thick mass of the *Oxford Clay:* it has at its base certain hard sandy beds called Kellaways Rock, 10 to 20 feet in thickness; but the clays above are probably at least 500 feet thick. They abound in *Ammonites* and the well-known *Belemnites*—the darts or thunderbolts of the peasantry. Brigg, Wickenby, Wragby, and Stixwould may be named as indicating the eastern boundary of the Oxford Clay.

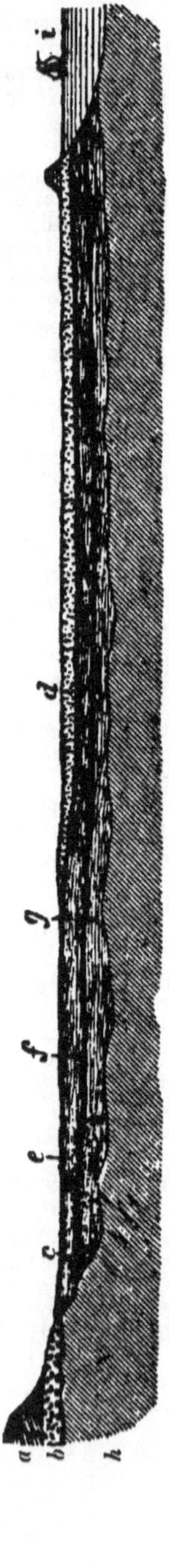

Upper Oolite.—*The Kimmeridge Clay* forms a broad band eight or nine miles wide in the south, but greatly narrowed in the extreme north by the overlap of the Cretaceous beds. It comes up to the western foot of the Wolds, and Fulletby, Tealby, Caistor, and Worlaby indicate its eastern edge: it is above 600 feet thick, and is exposed in numerous brickyards, in the cuttings of the Louth and Lincoln Railway, &c. At the top of the Kimmeridge Clay there is a very bituminous bed of shale, resembling that of the Coal-measures. Early in the present century much money was thrown away in consequence, in boring for coal in the valley of the river Bain, where these shales are well exposed. At Brinkhill there is a stratum of iron pyrites, known from its yellow hue as "Brinkhill Gold."

The Neocomian Formation.—The Chalk Wolds of Lincolnshire are well known; but on their western slopes there are a series of sandstones, iron-stones, and clays which are not so easily recognised. These constitute a sub-division, to which the term *Neocomian* is applied. We can trace them northwards into Yorkshire, where they fringe the Vale of Pickering, and thence, turning eastwards, they form the Speeton Cliffs, which overlook Filey Bay. In Lincolnshire they appear near Worlaby, and pass southwards by Caistor, Tealby, Dennington, and Scamblesby.

The West Yorkshire Coal and Iron Company have opened an iron mine in the middle Neocomian beds at Acre House, between Claxby and Nettleton. The existence of fragments of iron slag, Roman pottery, &c., in this part shows that the ore was known and worked in very early times. Professor Judd describes the bed now being worked as a rock almost entirely made up of small and beautifully polished oolitic grains of hydrated peroxide of iron. The earthy material, full of larger concretionary masses of ironstone, which was at first thought to be equally valuable, is found to yield so small an average percentage of iron that it is not worked. Fossils are numerous, especially a large shell named *Pecten cinctus.* The ore contains much limestone, and is consequently well adapted for smelting, mixed with the clay-ironstone from the Coal-measures. It is mostly sent to Leeds for that purpose. 885 tons, valued at £177, were raised in 1874; and 323 tons, value £49, in 1879.

Cretaceous Formation.—*The Hunstanton Red Chalk.* —This band, there about four feet thick, is familiar to all who have visited the Norfolk coast: it is traceable all through Lincolnshire from near Gunby, by Candlesby, Nettleton, Grasby, Barnetby, and South Ferriby.

Fig. 61.—General Section across the Fenland.

a Boulder Clay. *b* Gravel. *c* Upper Peat. *d* Warp. *e* Buttery Clay. *f* Lower Peat. *g* Boulder Clay. *h* Oxford and Kimmeridge Clays. *i* Sea.

Everywhere a curious sponge, *Spongia paradoxica*, is a common fossil, together with *Belemnites minimus;* this stratum is as much as 14 feet thick in Mid-Lincolnshire.

The Chalk.—Above the "Red Rock" there are several bands of pinkish chalk which have often been mistaken for it. One such occurs near Louth. As we ascend the chalk becomes harder, and, having resisted denudation better than the softer strata which occur both above and below it, now forms an almost continuous ridge along the western side of the Wolds, being only cut through by the Withern Eau and the tributaries of the Witham. The Chalk of Lincolnshire, being much more covered by superficial deposits than the hills of the North and South Downs, has not the smooth, rounded, and gently swelling aspect of the latter: it has mostly been brought under the plough, and with very satisfactory results. In the numerous pits the characteristic chalk fossils may readily be obtained, such as the rounded sea-urchins *Ananchytes* and *Micraster;* a furrowed shell, *Inoceramus*, is common, and *Terebratulæ* abound. Bands of flint also occur.

THE DRIFT.—In few counties are the stray deposits to which this term is applied better developed. Scattered over the district we find immense beds of clay and sand, full of pebbles and bits of older formations, which in some places are about 300 feet thick. These were formed in the *Glacial Period*, when an ice-sheet descending from the north forced its way over at least part of the county. From its seaward termination huge icebergs, breaking off, carried their loads of stones and mud, dropping them as they passed, melting, to the southwards. Some of the masses of rock so transported are of enormous size: near Stoke Tunnel the Great Northern Railway passes right through an enormous mass of Lincolnshire Oolite Limestone, which, as noted by Professor Morris, rests on undoubted Boulder Clay, and is therefore itself undoubtedly a transported mass.

FEN BEDS.—Over the district bounded on the north by Gibraltar Point, Horncastle, and Lincoln, and thence by Sleaford and Corby to the extreme south of the county, and all along the eastern coast, the whole surface as far as the sea, is covered by drift, which in turn is overlaid by beds of silt or warp with intercalated peat-beds. Digging through these, blue clay belonging to either the Oxford or Kimmeridge series is usually found beneath. The total area of the "Fenland," as the low flat country round the Wash is called, is 1,300 square miles, and about one-half of this lies in Lincolnshire. Unsuccessful attempts to obtain water by means of deep borings have given the following sections:—

BORING AT BOSTON.	Feet.	WELL AT FORSDYKE, SPALDING.	Feet.
Fen Beds	24	Fen Beds	78
Boulder Clay	166	Yellow Sandy Clay	37
Kimmeridge Clay	294	Great Chalky Boulder Clay	51½
Oxford Clay	88	Kimmeridge Clay	159½
	572		326

The silt or warp is a deposit of fine mud thrown up in past ages by the sea; the peat-beds between often contain large trunks of trees, and mark intervals during which vegetation flourished. Similar beds of silt and peat form most of the Isle of Axholme. (See fig. 61).

POST-GLACIAL PERIOD.—Much has been done by the hand of man in draining the Fen lands: in such works traces of prehistoric man have been met with. At Spalding a large and heavy stone celt was found, and in Newport, Lincoln, a bored stone, which doubtless served as a hammer-head; flint flakes or knives occurred, with burnt bones and a bronze arrow-head in some cinerary urns dug up at Broughton. A large and broad arrow-head of flint was found at Gunthorpe, and another near Manton. Bone lance-heads and pins have also been met with. A celt or axe-head of flint was found at Edenham, near Bourn, and another at Kate's Bridge, south of the latter town; a finely polished celt, 5½ inches long, occurred in peat at Digby, south of Billinghay; ancient canoes made of the hollowed-out trunks of trees have

Fig. 62.—Trees near Holbeach. The long rows of aspens bordering the dykes all incline to the north-east, being bent in that direction by the prevalent south-west winds. The trees of the buried forests lie pointing in the same direction. (After Skertchly, Geol. Survey.)

been dug up at Kyme, Billinghay, Langtoft, and Pinchbeck Bars. These relics take us back to the meeting-point between Geology and Archæology—to the Stone Age, when man was as yet unacquainted with the use of metals, and when he inhabited this country in company with the mammoth, the elk, and other animals now extinct, whose bones we find embedded with his implements in the strata of latest formation.

No. 23.

GEOLOGY OF MIDDLESEX.

Natural History and Scientific Societies (including all in London).
Quekett Microscopical Club; London. Journal.
West London Scientific Association and Field Club. Proceedings.
Old Change Microscopical Society.
South London Microscopical and Natural History Club.
South London Entomological Society.
New Cross Microscopical and Natural History Society.
Sydenham and Forest Hill Microscopical Club.
British Archæological Association, 32, Sackville Street, W.
British Association, 22, Albemarle Street, W.
Royal Society, Burlington House, Piccadilly.
Society of Antiquaries, Burlington House, Piccadilly.
Linnean Society, Burlington House, Piccadilly.
Zoological Society, 11, Hanover Square, W.
Anthropological Institute, 4, St. Martin's Place, W.C.
Entomological Society, 11, Chandos Street, Cavendish Square.
Royal Agricultural Society, 12, Hanover Square.
Royal Archæological Institute, 16, New Burlington Street, W.
Royal Geographical Society, 1, Savile Row.
Royal Microscopical Society, King's College.
Victoria Institute of Great Britain, 7, Adelphi Terrace, W.C.
Society of Arts, John Street, Adelphi.
See Guide to Natural History Clubs of London, by H. Walker. Sold at 97, Westbourne Grove; price 2d.

Museums.

British Museum, Great Russell Street, Bloomsbury.
Natural History Museum, South Kensington.
South Kensington Museum, Art and Education.
Bethnal Green Branch Museum.
Museum of Practical Geology, Jermyn street.
East India Museum, South Kensington.
Museum of Patents, South Kensington.
Christy Collection (Prehistoric Man), Victoria Street, Westminster.
United Service Museum, Whitehall.
Soane Museum, 13, Lincoln's Inn Fields.
Architectural Museum, Dean's Yard, Westminster.
Museum of the Geological Society, Burlington House, Piccadilly.
Museum of the Society of Antiquaries, Burlington House, Piccadilly.
Museum of the Royal College of Surgeons, Lincoln's Inn Fields.
Museum of the London Missionary Society, Blomfield Street, Moorfields.

PUBLICATIONS OF THE GEOLOGICAL SURVEY.

Maps.—Large Sheet: London and its Environs, price 22s.; also in Quarter Sheets 1 N.W., N.E., S.W., S.E.
Books.—The Geology of Parts of Middlesex, by W. Whitaker, 2s. Guide to the Geology of London, by W. Whitaker, 3rd Edition, 1s. Geology of the London Basin, by W. Whitaker, &c., 13s.

IMPORTANT WORKS OR PAPERS ON LOCAL GEOLOGY.

List (by W. Whitaker) of 505 works on the Geology of the London Basin, in the Survey Memoir, vol. iv.
1868. Wood, S. V., jun.—Pebble Beds of Middlesex. Journ. Geol. Soc., vol. xxiv. p. 464.
1869. Clutterbuck, Rev. J. C.—Farming of Middlesex (Geological Features and Character of the Soils). Journ. Roy. Agric. Soc., Ser. 2, vol. v., p. 1.
1870. Symons, J. G. -Increase of Temperature Downwards, in the boring at Kentish Town. Rept. Brit. Assoc,, p. 182.
1870. Lane Fox, Col.—Palæolithic Flint Implements at Acton and Ealing. Rept. Brit. Assoc., p. 130.
1878. Prestwich, Prof. J.—Artesian Well at Meux's Brewery, Tottenham Court Road. Journ. Geol. Soc., vol. xxxiv., p. 902.
1880. Harrison, W. J.—Deep Borings in the South-East of England. Midland Naturalist, vol. iii. p. 188.
1881. Smith, W. G.—Palæolithic Implements in Thames Valley. Nature, p. 308.

See also General Lists, p. xxv.

THE geological structure of the county of Middlesex is perhaps better known than that of any other English county. This arises in part from its being the metropolitan county, and consequently the residence of numerous workers in science.

The whole surface has been mapped by the Geological Survey, and described in a full and admirable "Memoir on the Geology of the London Basin," by Mr. W. Whitaker and his colleagues. Messrs. Prestwich, Searles Wood, Edwards, Col. Lane Fox, J. Evans and others have also done good work among its rocks. Many useful papers have been published by the *Geologists' Association*, whose head-quarters are at University College, Gower Street, and whose members make weekly excursions to places of geological interest in the neighbourhood of the metropolis, varied by occasional longer visits to more distant places. In the Geological Museum in Jermyn Street, the student will not only find most extensive collections of rocks and fossils, but also an admirable model of London and the neighbourhood, constructed by Mr. T. B. Jordan, showing the combined geological and topographical features of about 165 square miles.

It may be useful to give first a table of the various beds of rock which occur at the surface in Middlesex, together with their maximum thickness in this area:—

	Feet.
Alluvium (Recent river deposits)	15
Post-glacial Beds (Brick-earth, gravel, &c.)	50
Glacial drift (Boulder-clay, gravel, &c.)	80
Lower Bagshot Sands	100
London Clay	420
Woolwich and Reading Beds	90
Chalk with flints	800

We will commence with the lowest and oldest rocks, and briefly describe them in turn, in ascending order.

THE CHALK.— So little of the surface of Middlesex is occupied by this rock that at first sight it seems scarcely worth mentioning. We find this rock just entering the county on the eastern side of the Colne Valley, near Harefield, and again for a very small distance on the north-west side of South Mimms. If, however, we examine the *dip* or inclination of the chalk beds which extend northwards from the above-named places until they form the Chiltern Hills, we shall find it is towards the south and east. Then going southwards into Kent and Surrey we see the chalk of the North Downs dipping to the north and west. This leads us to think that the beds which form these two lines of hills are continuous beneath the surface, and, indeed, as a matter of fact, we know that they are, for deep wells and borings show that anywhere in Middlesex the chalk may be reached at depths never exceeding 400 or 500 feet, and usually much less. (See fig. 63.)

THE WOOLWICH AND READING BEDS.—These are plastic clays and mottled sands, which rest upon the chalk: we can trace them from a point about two miles north of Uxbridge, by Harefield, and again in the lane-cutting near South Mimms. An inlying mass is exposed at Pinner, Ruislip, and Ruislip Wood; at the latter point the junction with the chalk beneath, may be seen in several pits.

THE LONDON CLAY.—This well-known bed forms much the greater part of the surface of the county; it is a stiff brown or bluish clay, in which occur at intervals layers of septaria, which are rounded nodules of impure carbonate of lime. The colour of the clay at the surface is always brown, for the carbonate of iron, which tinges the lower beds blue, changes to brown peroxide of iron when exposed to the action of the atmosphere. Where the *London Clay* rests upon the *Reading Beds* the junction is marked by green and yellow sands, containing rounded flint pebbles. This is the "basement bed," and is only a few feet thick; it contains numerous fossils, as fish-teeth, shells, &c. The upper part of the London clay is somewhat sandy, and good bricks are made from it: in thickness it increases as we follow it eastwards, so that under London it is from 400 to 440 feet thick. It is exposed in numerous brickyards, but fossils are not to be easily found: they seem to occur in bands, certain levels being very productive, whilst above and below they are scarce or wanting. In the various railway and road cuttings on the north of London, especially in the neighbourhood of Highgate, many beautiful fossils have been found, and in the brickyards here and at Hampstead good specimens still occasionally occur.

The fossils of the London clay indicate a rather warm climate; thus numbers of fossil turtles occur, with the fruits of palms, &c., true crocodiles also, and such shells as *Nautilus, Conus, Voluta, &c.*, which now inhabit warm or even tropical waters. Mr. W. H. Shrubsole, F.G.S., has lately detected a thin but very constant layer in which diatoms mineralized by iron pyrites occur.

THE BAGSHOT BEDS.—Of the sandy strata which bear this name just a corner of the main mass enters the south-west of Middlesex in the neighbourhood of Littleton. Formerly, however, it stretched continuously over the London clay, and in proof of this we still find outlying masses forming the upper part of Harrow, Hampstead, and Highgate Hills, each of which is a marked feature of the landscape. There are several pits on the well-known gorse-covered heath at Hampstead in which we may see fine brown sands, with occasional thin layers of clay. The junction of these sandy beds with the London clay beneath is well marked by the boggy ground and by the springs which issue forth, as in the field just west of Rosslyn House. The rain falling on the surface percolates through the sands until it reaches the underlying clay, which is impervious; it then flows along the line of junction until it comes out on the hill side.

The Bagshot Sands are almost, if not quite, unfossiliferous.

GLACIAL DRIFT. -The beds we have now to describe are far more complicated and difficult to map and measure than any of the older formations. After the deposit of the Bagshot Sands, the surface of Middlesex either continued as dry land for many long ages, or if it was submerged the rocks which were deposited upon it were afterwards removed by denudation. Probably the former of these hypotheses is the more correct, for the surfaces of the older rocks show evidences of having been long exposed to the action of the atmosphere.

The deposits which we term *glacial drift* rest irregularly upon all or any of the older rocks; they contain (in Middlesex at least) no fossils except such as have been washed out of earlier deposits, and many of the included rock-fragments are polished and scratched in a manner which can be most reasonably accounted for by the action of ice. The earliest drift deposits are the pebble-gravels which occur on Stanmore Heath, south-east of Shenley, and west and north of Barnet: they resemble the far older pebble-beds of the Bagshot Sands, but are distinguished by the presence of pebbles of quartz and quartzite mingled with others of flint.

The *Middle Glacial Beds* are formed of sand and gravel, sometimes with masses of clay interspersed; they stretch southwards as far as Finchley, and may also be seen at Hendon, Southgate, Colney Hatch, Whetstone, Muswell Hill, and near Harefield. Above them comes the *Great Chalky Boulder Clay*, which is probably the deposit left by a great glacier stretching southwards as far as the northern edge of the Thames Valley: it is a dark bluish grey clay crammed with fragments of all kinds of rocks, especially lumps of hard chalk, and contains quite a collection of fossils rubbed out of older rocks. It may be studied in the brickyards at Finchley, and at Southgate, Enfield, and Potter's Bar.

POST-GLACIAL DEPOSITS.- When the cold period during which the glacial beds were deposited had begun to ameliorate, and when the glaciers retreated far to the northward, the Thames Valley would probably present the appearance of a sloping plain, the hollows having been filled up by the deposits just described. Soon, however, the river began its work of re-excavation, and, by the immense quantity of matter which it has since removed and carried downwards into the North Sea, has proved its power as an agent of erosion, although it is true that the time we must allow it for the execution of this task must be reckoned in thousands of centuries. The post-glacial deposits may also be termed "*River Drifts*," for they were all formed by the rivers when the latter ran at much higher elevations than they do at present. At this period, owing to an elevation of the land, England was certainly a part of the continent of Europe, and the Thames ran across the plain which now forms the bed of the North Sea, to join the Rhine. In the beds of gravel, sand, and brick-earth then deposited by the Thames, along its sides, we find the remains of huge animals - the mammoth, rhinoceros, hippopotamus, with the hyena, wolf, lion, bear, &c., which must have crossed over when as yet the Straits of Dover had no existence. With them, too, we find the first traces of man in the form of the flint implements rudely fashioned by him. The brick-earths of this period may be seen in the pits at West Drayton and Southall (where they are underlain by gravel), Hanwell (west of Asylum), Acton, Halliford, Highbury, Stoke Newington, Tottenham, Ponder's End, &c.

ALLUVIUM. — Under this term we include the deposits of mud and loam with occasional seams of gravel, which border the present streams, and which have evidently been formed by their overflowing in times comparatively recent ; in fact, the formation of such deposits is still going on. Such is the character of the Thames bank between Westminster and Vauxhall Bridges, and similar deposits occur along the course of the Colne and Lea.

SCENERY.—In the north and north-west of Middlesex the Woolwich and

Reading beds form a flat narrow surface, rising but little above the chalk. In turn the former are surmounted by the London clay, which forms a chain of low hills along the northern limit of the county, rising to a height of 505 feet on Stanmore Common, 440 feet at Brockley Hill, and 341 feet at Muswell Hill, and from these points slopes gently southwards. On this slope we find the outliers of Bagshot sand already named at Harrow (405 feet), Highgate (426 feet), and Hampstead (443 feet), constituting the only well-marked features of the surface of the county.

WATER SUPPLY.—Until a late period the beds of gravel and sand forming the glacial and post-glacial beds yielded a considerable amount of water, by means of shallow wells, to the towns and villages situated upon them, and many houses still derive a supply from these contaminated sources. Within the last half century, however, numerous deep wells have been sunk to the chalk, which yields an excellent supply of pure though hard water. The following list of the *depth from the surface to the chalk* at various points may be of use :—

	Feet.		Feet.
Acton	284	Haggerstone	165
Bank of England	234	Hampstead	378
Blackwall	237	Harrow	158
Bow	174	Hendon	244
Brentford	315	Isleworth	420
Chiswick	299	Pinner	60
Colney Hatch	189	Ruislip	76
Covent Garden Market	260	Tottenham	150
Fulham	250	Uxbridge	100

After reaching the chalk, however, it is usually necessary to bore into it for from 30 to 100 feet in order to ensure a good supply.

Rocks below the Chalk.—Two very deep borings in London have proved the existence of beds beneath the chalk which do not crop out to the surface in Middlesex. (See fig. 63.)

Kentish Town.		*Meux's Brewery, Tottenham Court Road.*	
	Feet.		Feet.
London Clay	236	Gravel and Clay	21
Reading Beds	61½	London Clay	63½
Thanet Sand	27	Reading Beds	51
Chalk	644¾	Thanet Sand	21
Upper Greensand	13¼	Chalk	655½
Gault	130½	Upper Greensand	28
Red Beds of doubtful age (probably Devonian)	188½	Gault	160
		Lower Greensand	64
		Devonian	70
Total	1,302	Total	1,134

The above sections show the absence in the one case and comparative thinness in the other of the Lower Greensand, a formation which it was hoped would furnish a supply of water ample enough, perhaps, for the wants of the metropolis. The presence of rocks of Devonian age is, however, of the deepest interest, and confirms the views held by Messrs. Prestwich and Godwin-Austen that a ridge of old rocks ranges below the surface in the neighbourhood of London, connecting the rocks of Belgium with those of Somerset, and affording hopes of concealed coal-fields under the south-east of England. The Devonian beds pierced in the Tottenham Court Road boring were red and green shales, dipping at an angle of 30 degrees, and containing fossils (*spiriferæ, &c.*) which show them to belong to the Upper Devonian period, resembling the Eifelian type. After penetrating about 70 feet of these shales, the boring was given up, as, after their age was once determined, it would have been useless to continue it.

PREHISTORIC MAN. - The earliest traces of man are the roughly-fashioned flint implements found in the gravel beds along the sides of the river-valleys, at heights of from 50 to 80 feet above the existing streams. Since the formation of these deposits the rivers have had time to excavate the valleys so much deeper, a fact which is alone sufficient to indicate the great antiquity of the relics we allude to.

The gravels of Hackney Down have yielded an oval flint implement, and Mr. Evans found one somewhat similar in the brick-pit at Highbury New Park, near Stoke Newington, in 1868. In the gravels at Ealing Dean and Acton, Col. Lane Fox found numerous specimens at depths of from 7 to 13 feet. These earliest of all tools are only roughly-fashioned nodules of flint, and are considered to belong to the *Palæolithic* (old stone) *age*. There is a fine specimen in the British Museum, which was found in Gray's Inn Lane, with the bones of an elephant, as long ago as the end of the seventeenth century: it is of a pointed form, 6 inches long by 4 inches wide at the butt. From the gravels, which in the north of London form a terrace about 70 feet above sea-level, Mr. W. G. Smith has with great industry collected 113 perfect palæolithic implements, besides some 1,400 flakes.

To the Newer Stone age (*Neolithic period*) belongs the polished celt (axehead) of grey flint found during the main drainage works for London, and now in the British Museum. It is 7 inches long, and tapers from 2 inches at edge to 1 inch at butt. Flint flakes, evidently

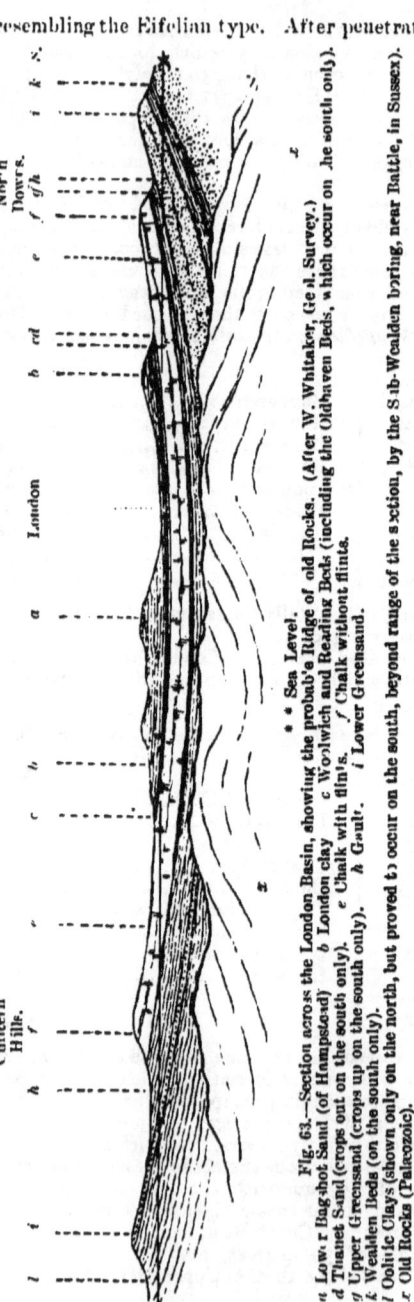

Fig. 63.—Section across the London Basin, showing the probab's Ridge of old Rocks. (After W. Whitaker, Geol. Survey.) *a* Lower Bagshot Sand (of Hampstead) *b* London clay *c* Woolwich and Reading Beds (including the Oldhaven Beds, which occur on the south only). *d* Thanet Sand (crops out on the south only). *e* Chalk with flints. *f* Chalk without flints. *g* Upper Greensand (crops up on the south only). *h* Gault. *i* Lower Greensand. *k* Wealden Beds (on the south only). *l* Oolitic Clays (shown only on the north, but proved to occur on the south, beyond range of the section, by the Sub-Wealden boring, near Battle, in Sussex). *r* Old Rocks (Palæozoic).

fabricated by man, and used as knives, have been found at Teddington, and a large and well trimmed flake, which seems to have been designed for a spearhead, was found by the Rev. J. C. Clutterbuck on Hounslow Heath. These things, in fact, are commoner than is usually supposed, and if educated people generally were to make themselves acquainted with the forms of these prehistoric implements, so as to recognise them at sight, it cannot be doubted that many fresh discoveries would be made.

No. 24.

GEOLOGY OF MONMOUTHSHIRE.

PUBLICATIONS OF THE GEOLOGICAL SURVEY.

Maps.—Sheets: 35, Forest of Dean (southern part); 36, Pontypool, Cardiff; Quarter Sheets: 42, S.E. Abergavenny; 42, N.E. Hay, Talgarth; 43, S.W. Forest of Dean to Monmouth.

Memoir, vol. i. On the Formation of the Rocks of South Wales and the South-West of England; Ramsay on the Denudation of the Rocks of South Wales and the South-West of England.

IMPORTANT WORKS OR PAPERS ON LOCAL GEOLOGY.

1861. Glass, Rev. N. Silurian Strata near Cardiff. Geologist, vol. iv. p. 169.
1872. Lee, J. E.—Notice of Veins or Fissures in the Keuper filled with Rhætic Bone Bed at Goldcliff. Rept. British Assoc., p. 116.
1879. Sollas, W. J.—The Silurian District of Rhymney and Pen-y-lan, Cardiff. Q. Journ. Geol. Soc., vol. xxxv. p. 475.
1881. Sollas, Prof. W. J.—Striated Triassic Pebbles near Portskewet. Geol. Mag. p. 79.

See also Transactions of Woolhope Club; Transactions of South Wales Institute of Mining Engineers, &c.; Murray's Guide to South Wales (G. P. Bevan), &c.

See also General Lists, p. xxv.

THE strata which form this county belong mainly to two great geological formations—the Old Red Sandstone and the Carboniferous—and though the structure of the district is in consequence simple and easily understood, it still includes points of great geological interest.

Looking first at the broad physical features, we note (1) extensive tracts of low flat marsh land, on the south along the Severn; (2) the centre and east of the county is fertile, with much rich red soil and many flat-topped hills, increasing in height northwards; (3) on the west we have a mountainous mining district, the eastern portion of the great South Wales coal-field.

All the rocks which form Monmouthshire are of aqueous origin—that is, they consist of sandstones, clays, and limestones, which were originally formed on the floor of the sea or lakes, and have since been hardened and elevated. The present irregular outline of hill and vale is due to the action of the weather—of rain and rivers, and frost and ice, which have acted on the strata, wearing them away unequally according to their relative hardness and positions.

We will now describe the various layers or beds of rock in order, commencing with those which form the foundation so far as is anywhere visible of the whole county, and which must therefore be of the earliest formation and greatest age.

SILURIAN FORMATION.—In the very centre of Monmouthshire, Silurian strata crop out, forming an oval area eight miles in length from Ffynonau on the north to Llandegfydd on the south, and four miles in width from Usk on the east to Ysgubor on the west: the river Usk divides this Silurian area into two parts, that on the south-west of the stream being the larger. Although the features are not so well marked, this *inlier* of old rocks bears much resemblance to the similar outcrop at Woolhope in Herefordshire; the lowest beds exposed are the *Wenlock Shales*, here containing numerous beds of sandstone, and occurring in the centre of the area at the Tucking Mill, Cilfigan Park, Prescoed, Cefn Mawr, Monkswood, and Kemeys Commander.

The *Wenlock Limestone* is of considerable thickness, and encircles the dome of Wenlock Shales; it may be examined at Trostrey, Radyr, Glascoed, and Tynewydd.

Then come the *Ludlow Beds*, consisting in the lower part of argillaceous shales, which become sandy in their upper division; they contain calcareous bands, which represent the *Aymestry Limestone*, as at Llancago Hill, Hill Barn, and Llanbadock; *Pentamerus galeatus* is a common fossil shell. The Ludlow Beds sweep round from Bettws Newydd by Usk, Llangibby Castle, to Llandegfydd, and thence northwards to Glasgoed Common, Pentwynn, and the Chain Bridge.

The total thickness here exposed of these Silurian beds is 1,570 feet, of which 270 feet is limestone. The district is hilly, but the hills formed of the harder beds of limestone and sandstone run generally north and south, and do not present the amphitheatre-like shape which we see at Woolhope. On the east side the beds dip to the east, and on the west side to the west, the general structure being that of an anticlinal running north and south, of which the top has been worn away. All the beds are evidently of marine origin, for they contain numerous remains of trilobites (the genus *Homalonotus* is common) and shells, which must have lived in the sea. One point of special geological interest about these Silurian rocks of Monmouthshire is that they link together to some extent the eastern and western Silurian areas, the lower sandy beds reminding us of the Grits of Denbighshire, while the thick Wenlock Limestone indicates similar conditions to those which prevailed at Wenlock Edge, Dudley, &c.

To the south of the main exposure, near Usk, there are several smaller outcrops of Silurian rocks, as at Llanfrechfa and near Rhymney; the latter area has lately been fully described by Mr. Sollas. The Wenlock and Ludlow beds here exhibit a total thickness of 954 feet of shales, sandstones, and mudstones, including only 4 feet of limestone. They form an east and west anticlinal, and are well exposed on the banks of the Rhymney below Cae Castell, and in a quarry where they are worked for road-metal. Fossils are numerous, including corals, shells, trilobites, &c.; there is a thin bone-bed just at the junction with the Old Red Sandstone above.

THE OLD RED SANDSTONE.—Rocks of this age form more than one-half of the surface of the county. The great extent of these beds in Hereford and Brecknock is well known, and they stretch into Monmouthshire in an unbroken mass on the north and north-west from the Black Mountains and Pontrilas, reaching southwards past Monmouth and Abergavenny, encircling the Silurians of Usk, and extending to the south-west past Newport. This mass of red rocks attains in the north of the county a thickness of from 8,000 to 10,000 feet, but thins southwards to about 4,000 feet between Newport and Risca; the change from the grey Silurian shales below to the red sandstones above is rapid and striking, and this change of colour is due to the presence in the latter of a little peroxide of iron. With this change of colour there is, too, a great change in the fossils; these are not so numerous in the Old Red as in the Silurian series; they belong, too, to creatures which probably inhabited lakes and rivers, so that the Old Red Sandstone is considered to have been deposited in great freshwater lakes, or perhaps, as Professor Hull has lately suggested, in the estuary of some large river.

In Monmouthshire the beds of *Old Red Sandstone* undulate in the centre and north of the county, but near the Wye they dip steadily eastwards and pass underneath the Forest of Dean Coal-field, while on the west side of the county they similarly dip westwards, disappearing beneath the South Wales Coal-field.

The following divisions have been recognised:

Upper or Conglomerate Series { Red and variegated sandstones and quartzose conglomerates, containing scales and plates of the fossil fishes *Holoptychius* and *Pterichthys*.

Middle or Brownstone Series { Marly sandstones and flagstones, red shales, and thin cornstones, with remains of *Cephalaspis* (a fossil fish).

Lower or Cornstone Series { Pale coloured sandstones, red and variegated marly beds with cornstones in lower part. Fossils: *Pterygotus* (a crustacean), *Pteraspis*, *Cephalaspis* (fishes).

There is no break or unconformability; from bottom to top stratum succeeds stratum with perfect regularity. The "cornstones" are irregular bands of red or yellow nodular or concretionary limestone; they form a rich soil; formerly they were much quarried for burning into lime and for mending roads. Tintern Abbey is built of Old Red Sandstone obtained from the Barbadoes Quarry hard by; those portions of Chepstow Castle which were built of a similar stone in the twelfth century are now much decomposed. The principal hills formed of Old Red are the Sugar Loaf, 1,852 feet above sea-level, Skyrrid Fawr 1,498 feet (both in north-west of county, the "brown stones" form their summit); Beacon Hill, which overlooks the Wye, 1,000 feet; Kymin Hill near Monmouth is capped by quartz conglomerates. Fossils are not common, but the enamelled plates, scales, and spines of a few species of fishes, remains of a large lobster-like crustacean (*Pterygotus*), and a few plants may be found by the diligent geologist.

CARBONIFEROUS FORMATION. The Carboniferous strata lie mainly in the west of Monmouthshire, the beds being those which form the eastern extremity of the great South Wales Coal-field.

Carboniferous Limestone. This is a grey or blue crystalline limestone, composed of the remains of encrinites, shells, &c.; it must have slowly accumulated on the bottom of a deep and clear sea: shales occur at the base

the *Lower Limestone Shales*; being soft, their position is often marked by a narrow valley: the uppermost bed also consists of limestones alternating with dark shales.

The Carboniferous or Mountain Limestone occurs in this county in two distinct areas, which were once connected.

1. The Wye, between Tintern and Chepstow, has cut a deep and winding way through this rock, which extends thence westwards to Itton, Lanvair, Penhow, Magor, and Portskewet; the beds, which are here about 1,000 feet in thickness, dip to the south, passing under the Bristol Channel. The scenery along the Wye at Windcliff, Piercefield Park, and Chepstow is magnificent.

2. In the west the Mountain Limestone forms a narrow fringe round the coal-field by Machen, Risca, Trevethin, and curves round the northern slopes of the Blorenge; it has a steep westerly dip of from 20 to 60 degrees. This rock contains many caverns, which are often lined with beautiful crystals of calcite and quartz; sometimes they also contain valuable deposits of iron-ore (hæmatite); lead-ore also occurs, but is not now worked. There are large quarries in which the limestone is worked to supply the great quantities annually used to mix with the iron-ore for smelting; the limestone acts as a flux, a much less heat sufficing to melt the ore when it is thus employed. The total thickness of the limestone, with its accompanying shales, is in this area from 500 to 700 feet.

The Millstone Grit. This is a hard, coarse sandstone, sometimes a conglomerate, which has been used for making cider mill-stones. Its thickness is 330 feet; miners call it the Farewell Rock, because it underlies all the workable seams of coal. We can trace it on the west of and resting upon the band of limestone just described from Risca, by Pontypool to Capel-newydd, then it caps the Blorenge, 1,720 feet in height, a grand bastion-like eminence, which forms the north-east corner of the South Wales Coal-field. Like the limestone beneath, this grit forms a poor and barren soil, chiefly used to pasture sheep. Its surface is flat, forming a table-land with a slight southerly slope, on which rise rivers, whose course takes them to the Bristol Channel.

THE COAL-MEASURES. —These include beds of shale and sandstone, 11,650 feet in thickness, between which there occur 75 seams of coal, only 25 of which, however, are more than 2 feet in thickness; the lower beds are also rich in ironstone. The minerals raised in 1877 were: coal, 4,350,785 tons, from 116 collieries; fireclay, 27,846 tons; the iron-ore returns are not separated from those of South Wales, so they cannot be given.

The following subdivisions have been recognised:—

1. *Lower Coal-measures*, with 34 coal seams (8 workable); ironstone bands with fossils, some of marine species.
2. *The Pennant Grit.*—Thick sandstones, with 15 coal seams (5 workable).
3. *Upper Coal-measures*, with 26 coal-seams (9 workable).

The coal of Monmouthshire is chiefly bituminous; it is composed, as in other districts, of compressed vegetable matter. Although in the coal itself the individual plants cannot be distinguished, owing to the manner in which they have been squeezed and pressed by the immense weight of the rocks above them, yet in the beds of shale we often find single plants, as ferns, reeds *(Calamites)*, gigantic plants allied to our club-mosses *(Lepidodendron* and *Sigillaria)*, &c.

The surface features of this coal district are remarkable; the rivers run from north to south, and have cut deep parallel valleys, which are separated from one another by terraced hills rising to heights of from 1,500 to 1,800 feet, and capped by the Pennant Grit. The valleys of the Ebwy, Sirhowy, and Rumney all converge towards the port of Newport, and each of these valleys is traversed by a railway which brings down the mineral treasures of the district, spoiling the quiet beauty and isolation of the land, but very useful for commercial purposes. Many of the coal-seams crop out on the hill-sides and are worked by adits, or galleries, which sometimes extend for miles. On the northern outcrop, near Tredegar, Ebbw Vale, Nantyglo, Beaufort, &c., what is called "patch-working" is practised; this is nothing more than quarrying the coal at its outcrop, in the open air.

It is the Lower Coal-measures which in this district are chiefly worked; the Pennant Grit above them is comparatively unproductive; of the Upper Coal-measures, only the well-known Mynydd Isslwyn seam extends thus far east. Six or seven well-marked faults run across the coal-region from north-west to south-east, ending off on the south against a cross-fault, which extends from Trevethin to Newbridge.

THE TRIAS.—Beds of this age occur only in the south-east corner of the county, between Chepstow, Portskewet, and Undy.

The Dolomitic Conglomerate occurs in six or seven patches at Undy, Lanfihangel, Mounton, and Claypit Farm south of Chepstow; it is composed of lumps of Carboniferous Limestone, cemented together by carbonates of lime and magnesia; it was, in fact, formed as a beach at the foot of the limestone hills.

The Keuper Red Marls form a narrow strip extending from Mathern to Portskewet, Caldicot, and Undy; they also occur between Llanmartin and Christchurch; no fossils are found in them.

RHÆTIC FORMATION. Grey marls and black shales form the lower part of Gold Cliff on the coast, and may also be traced east, north, and west of the Lias at Bishopston and Llanwern.

THE LIAS.—The blue limestones and shaly clays of this age form the capping of Gold Cliff: inland they occur at Langstone, Bishopston, Milton, and thence westwards to the Usk, on the other side of which there is a small patch at Maes Glas. Several lime quarries have been opened in these beds.

SURFACE DEPOSITS.—Thick beds of gravel occur in the valleys, and cover the low ground formed by the softer beds of the Old Red Sandstone. These appear to be in the main of local origin, yet glaciers may have contributed to their formation. Very few observations have yet been made in this district upon the surface deposits, and it is greatly to be desired that local observers would search for and record the existence of any masses of rocks foreign to the district, which appear to have been brought by natural agencies.

The rivers have formed deposits of mud and gravel along their courses, but more especially at their mouths. where they enter the Bristol Channel. Here the sediment brought down has been deposited until a tract of low level land has been formed, which extends for 24 miles from Portskewet to Cardiff, and is from one to three miles in width. This region is known as the Caldicot and Wentloog Levels ; it is but little above the level of the sea, and is protected by an embankment or sea-wall.

PREHISTORIC MAN.—Few discoveries have been made in this county of relics belonging to tribes to whom iron was unknown. In a barrow, or burial-place, on Penhow, Mr. Morgan found part of a whetstone, with a bronze dagger, and numerous flint flakes.

No. 25.

GEOLOGY OF NORFOLK.

NATURAL HISTORY AND SCIENTIFIC SOCIETIES.
Norfolk and Norwich Naturalists' Society; Norwich. Transactions.
Norfolk Microscopical Society.
Norwich Geological Society.
Norwich Science Gossip Club.

MUSEUMS.
King's Lynn Museum.
The Norfolk and Norwich Museum, Norwich.

PUBLICATIONS OF THE GEOLOGICAL SURVEY.
Books.—Geology of the Fenland, by Skertchly, 40s.; Manufacture of Gun-flints (at Brandon, &c.), by Skertchly, 17s. 6d. Geology of the Country Round Norwich, by H. B. Woodward.

IMPORTANT WORKS OR PAPERS ON LOCAL GEOLOGY.
1824. Taylor, R. C.—On the Crag Strata, at Bramerton, near Norwich. Trans. Geol. Soc., 2nd series, vol. i. p. 371.
1824. Taylor, R. C. –On the Alluvial Strata and Chalk of Norfolk and Suffolk. Trans. Geol. Soc., 2nd series, vol. i. p. 374.
1826. Taylor, R. C.—Notice of Fossil Timber, on the Norfolk coast. Trans. Geol. Soc., 2nd series, vol. ii. p. 327.
1829. Woodward, S. P.—On some remarkable Fossil Remains found near Cromer. Proc. Geol. Soc., vol i. p. 93.
1833. Woodward, Samuel.—Outline of the Geology of Norfolk.
1834. Taylor, Jno.—Strata in a Well at Diss. Trans. Geol. Soc., series 2, vol. v. p. 137; Proc. Geol. Soc., vol. ii. p. 93.
1836. Fitch R.—Mastodon Tooth in the Crag at Thorpe. Proc. Geol. Soc., vol. ii. p. 417.
1837. Clarke, Rev. W. B.– Physical Relations of Suffolk with Norfolk and Essex. Trans. Geol. Soc., 2nd series, vol. v. p. 359; Proc. Geol. Soc., vol. ii. p. 528.
1838. Mitchell, Dr. J.—On the Drift. Proc. Geol. Soc., vol. iii. p. 3.
1839. Gunn, Rev. J.– On Paramoudras, and on the Drift. Proc. Geol. Soc., vol. iii. p. 170.
1839. Lyell, Sir C. On the Relative Ages of the Crag Deposits. Proc. Geol. Soc., vol. iii. p. 126.
1840. Lyell, Sir C.—On the Drift and Associated Freshwater Deposits Comprising the Mud Cliffs of Eastern Norfolk. Proc. Geol. Soc., vol. iii. p. 171.
1840. Trimmer, J.—On the Drift between Lynn and Wells. Proc. Geol. Soc., vol. iii. p. 185.

1841. Lyell, Sir C. Fossil Fishes of Mundesley. Proc. Geol. Soc., vol. iii. p. 362.
1844. Trimmer, J.—On the Drift between Weybourne and Happisburgh. Journ. Geol. Soc., vol. i. p. 218.
1844. Trimmer, J. On Pipes in the Chalk. Journ. Geol. Soc., vol. i. p. 300.
1845. Munford, Rev. G. Submerged Forest at Hunstanton. Gentleman's Magazine.
1851. Trimmer, J. Generalizations respecting the Erratic Tertiaries of Norfolk. Journ. Geol. Soc., vol. vii. p. 19.
1856. Bunbury, C. J. Draining a Mere, near Wretham Hall. Journ. Geol. Soc., vol. xii. p. 355.
1858. Trimmer J. On the Boulder Clays of Gorleston Cliffs. Journ. Geol. Soc., vol. xiv. p. 171.
1860. Prestwich, Prof. J. London Clay in a Well-boring at Yarmouth. Journ. Geol. Soc., vol. xvi. p. 449.
1864. Gunn, Rev. J. Sketch of the Geology of Norfolk.
1864. Seeley, Prof. H. On the Hunstanton Red Rock. Journ. Geol. Soc., vol. xx. p. 327.
1866. Fisher, Rev. O. Relation of the Norwich Crag to the Chillesford Clay. Journ. Geol. Soc., vol. xxii. p. 19, and vol. xxiii. p. 175.
1867. Harmer, F. W. A Third Boulder Clay in Norfolk. Journ. Geol. Soc., vol. xxiii. p. 87.
1868. Dawkins, W. Boyd.—New Deer from Norwich Crag. Journ. Geol. Soc., vol. xxiv. p. 516.
1869. Grantham, R. B. The Broads of East Norfolk. Journ. Geol. Soc., vol. xxv. p. 258.
1869. Flower, J. W. Flint Implements in the Drift. Journ. Geol. Soc., vol. xxv. pp. 272, 449.
1869. Wood, S. V., and Harmer, F. W. Intraglacial Erosion, near Norwich. Journ. Geol. Soc., vol. xxv. pp. 259, 445.
1869. Wiltshire, Rev. T. On the Red Chalk of Hunstanton. Journ. Geol. Soc., vol. xxv. p. 185
1870. Gunn, Rev. J. Position of the Forest Bed and Chillesford Clay. Journ. Geol. Soc., vol. xxvi. p. 551.
1871. Prestwich, Prof. J. On the Crag. Journ. Geol. Soc., vol. xxvii. pp. 115, 325, 452.
1871. Wood, S. V., and Harmer, F. W.—Outline of the Geology of the Upper Tertiaries of East Anglia. Palæontographical Society's, vol. 1871.
1877. Wood, S. V., and Harmer, F. W. Later Tertiary Geology of East Anglia. Journ. Geol. Soc., vol. xxxiii. p. 74.
1877. Reid, C. Modern Denudation in Norfolk. Geol. Mag., vol. xiv. p. 136.
1877. Belt, Thos. First Stages of the Glacial Period in Norfolk and Suffolk. Geol. Mag., vol. xiv. p. 156.
1877. Norton, H. The Forest Bed of East Norfolk. Geol. Mag., vol. xiv. pp. 320, 335 (Rev. J. Gunn), and 432 (B. S. Breese).
1877. Reid, C. Beds between the Chalk and Lower Boulder Clay, at Cromer. Geol. Mag., vol. xiv. p. 300.
1878. Harmer, F. W. Testimony of the Rocks in Norfolk. Hamilton, Adams, & Co.
1879. Newton, E. T. Fossil Tortoise from Mundesley. Geol. Mag., vol. xvi. p. 304.
1880. Fisher, Rev. O. On the Cromer Cliffs. Geol. Mag., vol. xvii. p. 147.
1880. Reid, C. The Glacial Deposits of Cromer. Geol. Mag., vol. xvii. p. 55.
1880. Woodward, H. B. Address to the Norwich Geological Society. Geol. Mag., vol. xvii. p. 72.
1880. Reid, C. Classification of Pliocene and Pleistocene Beds. Geol. Mag., p. 548.

Fossils of Norfolk. See the publications of the Palæontographical Society, including the Crag Mollusca, by Searles V. Wood, Esq. (sen.); Cretaceous Corals, by Prof. Duncan; Echinoderms, by Prof. Forbes and Dr. Wright; Foraminifera, by Prof. Rupert Jones and Mr. Brady.

See also General Lists, p. xxv.

THE strata of Norfolk have long been the subject of diligent investigation by local workers. Mr. Samuel Woodward published an excellent "Outline of the Geology of Norfolk," in 1833, and before his time Mr. R. C. Taylor had done good work at the Crag and Cliff sections. In the Journal of the Geological Society we have since had excellent papers on the Red Chalk of Hunstanton by the Rev. T. Wiltshire and H. Seeley, on the Crag and Forest Bed by Professor Prestwich, Mr. John Gunn, the Rev. O. Fisher, &c. Mr. Searles V. Wood's account of the Crag Mollusca, published by the Palæontographical Society, in which every species is figured, was a grand contribution, whilst for the same Society Professor Duncan has described the Cretaceous Fossil Corals, Dr. Wright and Professor Forbes the Echinoderms, Mr. Sharpe the Cephalopoda, and Rupert Jones and Brady the Entomostraca and Foraminifera. Mr. S. V. Wood, jun., assisted by Mr. Harmer, has described the Glacial Deposits with a skill and care which are worthy of the highest praise, and among other workers we may mention Messrs. Trimmer, Rose, John Evans, J. W. Flower, Norton, and J. E. Taylor.

In the Norwich Museum there is a splendid collection of local fossils, and the Norwich Geological and Naturalists' Societies and the Norwich Science Gossip Club have done and are doing good work.

It is, however, to the Government Geological Survey that we look for a full and decisive account of the rocks of any part of England. The officers of the Survey, however, began their work some forty years ago in the south-west of England and in Wales, and have ever since been advancing eastwards, until quite lately they have begun to examine the counties which border on the North Sea. The geological map of Norfolk, on the scale of one inch to one mile, which they will prepare, together with the memoirs published to describe the country, will, doubtless, set at rest many vexed questions and form a noble contribution to the science of Geology. Already Mr. S. B. J. Skertchly has written an excellent account of the "Fenland," together with a most interesting memoir on the gun-flint manufacture at Brandon, whilst valuable notes have been issued by Messrs. W. Whitaker, Horace B. Woodward, W. H. Penning, J. H. Blake and Clement Reid.

As a resident of a county, two-thirds of whose outline is encompassed by the sea, the geologist dwelling in Norfolk enjoys a great advantage; indeed, but for the fine sections of the glacial deposits exposed in the cliffs it is almost certain that these puzzling beds could never have been satisfactorily examined or understood. The coast line undulates both in height and outline: King's Lynn (St. John's church) is 16 feet above the sea, Holkham (church) 94 feet, Cromer Lighthouse 248 feet, Cromer Church 68 feet, Mundesley (church) 115 feet, Trimingham (church) 195 feet, Hasborough (church) 70 feet and Winterton Lighthouse 61 feet.

The surface of the country generally is flat and undulating: along a line across the centre from west to east the Ordnance Survey levelling gave the following heights:— Walsoken (church) 13 feet, West Bilney 54 feet, Swaffham (church) 238 feet, East Dereham (church) 165 feet, Hockering (church) 163 feet, Norwich Castle 111 feet, St. Mary's Church 18 feet, and Great Yarmouth (St. Peter's Church) 23 feet.

Taking a general view we find, on the extreme west, a small portion, just the edge, of the Fen district, extending from Welney and Hilgay fen to the Wash; then comes a wooded escarpment of sandy hills (Lower Greensand) running from Snettisham to Sandringham, beyond which is a gently rising slope lending up to a line of low chalk hills, which run from Hunstanton and Burnham Westgate southwards by Swaffham to Thetford: east of this line the chalk stretches right across to the sea, but is so covered over by later deposits as to be rarely visible at the surface, except on the foreshore between Weybourne and Cromer, and in the valleys of the principal rivers. The watershed of the county lies nearly in the centre, along a line passing from Brancaster, through Litcham, Dereham, Hingham, and Attleborough, to Lopham Ford, where the Waveney and Little Ouse rise within a few yards of each other and then flow in diametrically opposite directions.

The rocks of Norfolk fall naturally into two divisions: the first includes those of Secondary and Tertiary age, which form what we may call the *solid geology* of the county; these have a general easterly dip, so that the oldest rocks crop out on the west side of the county: the second division comprises all those surface accumulations, whether of glacial, alluvial, or marine origin, which lie upon the older beds in an unconformable and irregular manner.

THE OOLITIC FORMATION.—On the east side of King's Lynn and running northwards along the Wash nearly to Hunstanton, and southwards past Watlington to near Downham Market we have a narrow strip of *Kimmeridge Clay*; this is the *Oaktree Clay* of William Smith, who noticed that oaks grew remarkably well upon it: it is a tenacious dark-coloured clay, but exposures are scanty in this neighbourhood: dipping eastwardly it passes under the chalk, &c., and was probably reached at a depth of 743 feet in a deep well-boring at Holkham; but before we reach the east coast it is absent, having thinned out or been denuded, as the deep boring at Harwich showed no trace of it. The Purbeck and Portland beds, which rest upon the Kimmeridge Clay in the south of England, are here absent.

THE NEOCOMIAN FORMATION.—The Wealden strata being also absent, the *Lower Greensand* reposes directly on the Kimmeridge Clay: it is locally termed *Carstone*, a word which may be derived from "Quernstone," as some of the hard beds were formerly used to make the querns or stone hand-mills in which our ancestors ground their corn. This *Lower Greensand* consists of alternating beds of red and white sand and sandstone about 70 feet in thickness: it forms the lower portion of Hunstanton Cliff, where it is of a yellow tint above and dark-brown below, loose and sandy and full of small pebbles: near the base is a line of nodules containing *Ammonites Deshayesii, Perna mulleti &c.* From this point the sandstone can be traced southward by Snettisham and Castle Rising to Downham Market. The hard beds called "ginger-bread stone" are used for building purposes, and the white sands in the manufacture of glass. At one place there is a seam of fullers' earth.

THE GAULT (AND RED CHALK). One of the most attractive geological sections on the coast of England is exhibited in Hunstanton Cliff: its base formed of Carstone we have just described: upon the yellow sands which constitute the top of this division there rests a band of bright "Red Chalk," four feet thick, above which, forming the top of the cliff, comes the White Chalk 40 feet thick; the extreme height of the section is only 60 feet, but as all the beds slope or dip to the north and east at an angle of about two degrees, a slightly greater thickness of strata is visible than if they were all horizontal. The bed of Red Chalk, as it has been named, has attracted much attention: it is full of fossils —*Belemnites minimus, Spongia* (or *Siphonia*) *paradoxica, Terebratula biplicata, Ammonites auritus*, and about fifty other species, of which two *Bourgueticrinus rugosus* and *Terebratula capillata* are, in England, confined to this deposit. This red bed can be readily traced for 8 miles southwards to Sandringham: a little further south, at Flitcham, a red clay was found to underlie the white chalk, and further south still a stiff blue clay the well-known *Gault* appears to have taken the place of this red clay. Thus the *Gault* and the *Red Chalk* would seem to be of the same age, and this impression is confirmed by a study of the fossils, which are to a large extent common to both formations: the colour is no objection, but rather tends to confirm this opinion: analyses by Mr. David Forbes showed that ordinary Gault contained nearly 6 per cent. of protoxide of iron, while the Red Chalk contained almost exactly the same amount of peroxide of iron: now the former salt of iron readily changes into the latter, as we see when the blue Gault Clay is burnt into red bricks; the change is brought about by the simple addition of oxygen.

It is possible that the upper portion of the Red Chalk may represent the *Upper Greensand* of the southern counties, and this is a point to which the attention of local geologists may well be directed, as its proof or disproof must rest mainly on extensive collections of fossils made with great care and with special reference to the position in the bed in which they occur. In the Holkham boring, after passing through 20 feet of gravel and 635 of chalk, the Red Chalk was found to be 8 feet thick, and to rest on 10 feet of "Blue Gault," so that here the red bed would seem to represent the upper part of the Gault only.

At Norwich the boring showed 12 feet of alluvium, then 1,042 feet of Chalk, 6 feet of "Upper Greensand" and 36 feet of Gault, in which the boring stopped: the stratum termed "Upper Greensand" may only have been an upper sandy division of the Gault.

THE CHALK.—This rock is seen on an ordinary geological map to form a larger proportion of Norfolk than of any other English county; yet the portion of which it is the actual surface is comparatively small: this arises from the fact that in such maps the *Drift*—as the glacial deposits are termed—is not shown, and the Drift lies thickly upon the chalk in this county. Every person is familiar with the appearance of ordinary white chalk: it is a soft white earthy limestone nearly pure carbonate of lime in fact, which the microscope shows to be largely composed of the minute shells of foraminifera: these are minute animals, occupying a very low place in the scale of creation—mere specks of jelly to look at which construct for themselves a calcareous covering or shell: they seem to live chiefly near the surface and at a considerable distance from land. Many soundings of late years have shown that the bed of the North Atlantic and other portions of the ocean floor is composed of a whitish ooze formed of the shells of forams which have sunk to the bottom after the death of the inhabitant: in this manner the chalk was probably formed, and the period required for the accumulation of this deposit, having a thickness in Norfolk of upwards of 1,000 feet, must have been of immense duration. Three divisions can be made in the chalk of this district.

(1) *The Chalk Marl.* In the Hunstanton Cliff section about 4 feet of a greyish-white limestone is seen to rest on the red band; this white bed is the chalk-marl: it is only about four feet in thickness, and in the lower part we find the same branching sponge which occurs in the Red Chalk; the upper portion is full of the fragments of a shell called *Inoceramus;* some large ammonites also occur in it.

(2) *The Lower or Hard Chalk.* This forms the upper portion of Hunstanton Cliff. Passing inland, we can follow it by means of quarries through Snettisham, Gayton, Marham, Narborongh, Whittington, and Stoke-ferry to Feltwell and Hockwold-cum-Wilton: it is much used in western Norfolk, together with carstone, for the construction of cottages and farm buildings; many carvings and monuments for the interior of churches have also been executed in it. No flints occur in this Lower Chalk: its thickness in the deep boring executed by Mather and Platt for Messrs. Colman, at Norwich, was found to be 102 feet, and in the Holkham boring 116 feet. Of fossils it has yielded bones of a large saurian—*Ichthyosaurus campylodon*—and two ammonites—*Ammonites peramplus* and *A. austenii*—large species about two feet in diameter.

(3) *The Upper Chalk, or Chalk with Flints.* This division comes on to the eastward of that last described and forms the great bulk of the formation: in the Norwich boring it was found to attain the great thickness of 940 feet, and at Holkham (where the upper portion has been denuded) 519 feet: it yields a large supply of good but hard water. The dip of the Chalk is to the east-

wards, about 48 feet per mile, or half a degree. The lower part of this division is exposed at Brancaster, Docking, Lexham, Litcham, Swaffham, and Thetford; the texture of it more compact, and the flints fewer than in higher beds. As we pass eastwards we find numerous chalk-pits at Wells, Holt, Coltishall, Horstead, Whitlingham, Postwick, Trowse, &c. The flints mark the stratification well, and are usually arranged in horizontal layers about five feet apart. On the origin of flint, Mr. S. Woodward makes some acute remarks in the Geology of Norfolk, which, as we have already mentioned, he published in 1833. He writes: "The fact of the regular parallelism and stratification of the flinty nodules, is, we conceive, a sufficient refutation of their animal origin; added to which there is not a greater number of organic remains to be found in the flinty stratum than is seen distributed in the adjoining bed of Chalk. . . . . In accounting for the different appearance of the Chalk formation in its upper and lower beds, we venture to advance it as our opinion, that, the whole being deposited in one homogeneous mass, in which about 12 per cent. of silex was equally distributed, the lower part . . . . . began to consolidate ere any arrangement of the siliceous particles could take place, whilst in the upper part these particles, by chemical attraction, congregated themselves into the nodular and tubular forms under which they appear." In several quarries near Norwich, as at that famous one at Horstead, figured and described by Sir Charles Lyell, and at Trowse, immense pear-shaped flints, from two to five feet in length, and half as much in breadth, are found, several often occurring one above the other; for these Dr. Buckland imported the name *Paramoudra*, an Irish word for similar masses occurring in the chalk near Belfast : they are often hollow, and seem to have been formed by the accumulation of flint round gigantic decaying sponges. Of shells, the commonest species, near Norwich, are *Terebratula carnea*, and *Rhynchonella plicatilis*. At Bishop's Bridge fine specimens of a great reptile the *Mosasaurus* have been met with, but these are usually so decayed that only the teeth can be extracted. "Remains of *Leiodon anceps*, a Lacertilian reptile allied to *Mosasaurus*, have been obtained by Mr. T. G. Bayfield, at Lollard's Pit, Norwich. But the very highest beds of chalk known in Norfolk, are those which occur at Trimingham, and which contain fossil sponges in the flints, only found elsewhere in the coarse flint-gravel which caps Mousehold Heath." (H. B. Woodward.)

In many sections the upper 5 to 20 feet of the chalk appears to be re-arranged, mixed up with clay and sand, and containing broken flints with pebbles of quartzite, &c.: the beds beneath, too, are often tilted and disturbed: these appearances are due to the passage of glaciers over the surface, as will be described further on.

TERTIARY PERIOD. An interval of time, whose duration must have been great indeed, elapsed after the formation of the chalk before any fresh deposits were laid down in this area: the best proof of this great interregnum is found in the entire change of life which took place in this interval. In the strata we are now about to describe not one fossil is identically the same as those which occur in the older beds: this point then is selected by geologists as a great line of demarcation ; the chalk is assigned to the Secondary Period, and the stratified beds which rest upon it to the Tertiary.

EOCENE FORMATION. The presence of beds of this age in Norfolk was not suspected until the cores brought up from a deep well-boring at Great Yarmouth in 1840 (for Sir E. Lacon & Co.) were examined many years later by Professor Prestwich, together with the Rev. John Gunn and Mr. Rose: the boring passed first through 50 feet of blown sand, then through 120 feet of recent estuarine deposits: at a depth of 170 feet a light-brown clay with numerous concretions was found, and proved to be 310 feet thick ; this, though no fossils were brought up, Professor Prestwich states is undoubted *London Clay ;* beneath it came 46 feet of *Woolwich and Reading Beds:* the chalk was found at a depth of 526 feet, and pierced to a depth of

57 feet. These Eocene beds are nowhere exposed at the surface in Norfolk, but they may extend some little distance inland towards Mundesley and Reedham, covered over and concealed by newer deposits.

PLIOCENE FORMATION.—In the neighbourhood of Norwich, and along a narrow strip on the north side of the river Waveney, near Bungay, we see resting on the chalk some beds of sand, clay, and shingle, not more than 30 feet in thickness, and often containing seams or banks of shells; these beds are termed the *Norwich Crag* (*crag* is derived from the Celtic *creggan*, a shell, the beds being often so shelly as to be used for improving land). They were first well described by R. C. Taylor in 1823; subsequently E. Charlesworth distinguished them from the Suffolk Crags by the name of the Mammaliferous Crag. The term *Fluvio-marine Crag* has also been applied to the lowest portion of the crag. In Suffolk and Essex beds called Red Crag and Coralline Crag also occur, which were deposited (the latter at all events) before the time of the Norwich Crag, although the Norfolk series may perhaps be coeval with the upper part of the Red Crag of Suffolk. In the large chalk-pits near Norwich, as at Trowse, Whitlingham, Thorpe Limekiln, and at Bramerton and Postwick Grove, also at Horstead and Coltishall, in the Bure Valley, the Norwich Crag is well exposed. Here, resting on the chalk, can be seen the "stone-bed," showing an eroded surface of chalk, bored by Pholas, on which lie large flints, upon and between which are many bones of large mammals, as *Mastodon arvernensis, Elephas meridionalis,* the horse, stag, ox, beaver, &c. Above comes the Norwich Crag proper laminated clay, sand, and shingle, with here and there patches of shells; the commonest species are *Tellina obliqua, Mactra ovalis, Purpura lapillus,* &c. Resting on the Norwich Crag, of which they may be regarded as a sub-division, we find the *Chillesford Beds*—sands and clays about 5 or 10 feet thick, containing such shells as *Astarte borealis, Tellina lata,* &c. A fine section of the Chillesford Beds is exposed in a brick-pit near Aldeby and about four miles from Beccles. This pit has been carefully searched by Messrs. Dowson and Crowfoot, who have obtained more than 70 species of mollusks from it. Altogether 111 species have been obtained from the Norwich Crag proper, 24 of which are land or fresh water forms, and there are 18 extinct species. From the Chillesford Beds 87 species have been obtained, 14 of which are not known as living. The uppermost portion of the Crag, sometimes called the Bure Valley Beds, contains *Tellina Balthica*. It is seen at Belaugh and Weybourne.

All the above beds were probably deposited in a shallow sea, the coast line of which lay some few miles west of Norwich, and near or in the estuary of a large river. The shells show a gradual change of climate, the Mediterranean forms met with in the lower strata gradually disappearing, and those of colder seas becoming more abundant.

In a few places on the coast we see other PRE-GLACIAL BEDS exposed, as at Cromer, Runton, and Weybourne; these include the Forest Bed, which can be traced from Runton by Cromer and Hasborough to Kessingland, in Suffolk, a distance of about 40 miles. At low water, or after heavy storms, the stumps of trees may be seen imbedded in laminated clays and sands; cones of Scotch fir and spruce are common, with bones of such large mammals as *Elephas antiquus, Hippopotamus, Rhinoceros,* many species of deer, &c. It is now generally considered that these bones were washed out of some older deposit, and it is even doubtful whether the trees grew on the spot where we now see them. The *Weybourne Sands* and *Bure Valley Beds* are of the same age as the Forest Bed; they contain *Tellina balthica*, a characteristic glacial shell.

QUATERNARY PERIOD.—The PLEISTOCENE FORMATION includes all the deposits connected with the last glacial period; of these Norfolk presents a series more complete than is to be met with in any other part of the British Isles. Fossilized remains of large mammals have been obtained in great numbers by the fishermen who with their nets sweep the shallow floor of the German Ocean. Of these the late Rev. James Layton formed a magnificent collection which is now in the British Museum; it is stated that he had at one time as many as 600 grinders of the elephant in his possession. Fine collections have also been made by the late Miss Gurney, of Northrepps, the Rev. John Gunn, and Messrs. C. B. Rose, Owles, Steward, and Nash. Many fine tusks of the mammoth, measuring from six to nine feet along the curve, have been dredged up from the Knole Sand, off Hasborough. Miss Gurney's collection is now in the Norwich Museum.

LOWER GLACIAL BEDS. Next we have a series, about 200 feet thick, commencing in the *Lower Boulder Clay*, or *Cromer Till*, which shows an undoubted glacial origin; it is a greyish clay, containing rounded and scratched stones, such as chalk, flint, basalt, pink granite, &c.; then we have the *Contorted Drift*, a yellowish, loamy and gravelly deposit, remarkable for the way in which it is seen to be bent or "contorted" by some force acting after its deposition. It contains enormous masses of chalk, some of which measure hundreds of feet in length. The contortions and included chalk masses are believed by some authorities to be due to the grounding of great icebergs floating slowly southwards; by others they are considered as the effects of land ice. A few marine shells occur, among which *Tellina balthica* is recognizable. These Lower Glacial Beds stretch inland and cover the surface of a considerable area in the north-east of Norfolk, as far south as the course of the Yare and Wensum.

MIDDLE GLACIAL BEDS. These are sands which appear to mark a comparatively warm or "inter-glacial" period; they are from 15 to 70 feet thick and occupy large areas, cropping out from beneath the chalky boulder clay. Shells have been found at Billockby, eight miles north-west of Yarmouth, in a thin shelly seam, about four or five feet below the top of the sand. Great interest attaches to these beds, for near Brandon, brick earths of this period have yielded flint implements -the undoubted work of man.

UPPER GLACIAL BEDS. The well-known *Great Chalky Boulder Clay* spreads over the centre, south, and south-west of this county; its extreme thickness here is above 100 feet. It is a stiff bluish clay, quite unstratified, and containing fragments of chalk of all sizes, together with a heterogeneous collection of rocks and fossils, derived from almost every formation. It appears to have been formed beneath or pushed before a great glacier advancing from the north. The *broads* are due to fluviatile and in part, perhaps, estuarine action the *meres* occupy hollows in the boulder clay, and are most likely due to dissolution of the chalk beneath and subsidence of the boulder clay.

Hill, Plateau, or Flood Gravels. These are coarse gravels, which seem to have been formed by floods resulting from the melting of great glaciers. The "cannon-shot" gravel of Mousehold Heath is a good example; this is mainly composed of large much battered flints, with quartz and quartzite pebbles.

POST-GLACIAL DEPOSITS:

Valley Gravels. Numerous beds, of very different ages, are included under this term; some in West Norfolk belong to a system of rivers, which flowed nearly at right angles to those of the present day. These beds have yielded many flint implements.

In the cliffs at Mundesley there is an old filled-up river bed, which has lately yielded remains of a tortoise (*Emys lutaria*).

Blown Sand forms the "Marram Hills," between Winterton and Hasborough: these form a natural barrier against the sea, and are kept up with great care.

The Fen-beds. The gravelly beds between Brandon and Hockwold-cum-Wilton, and the Nar Valley Clays, which extend from Watlington to Narford, seem to be the oldest of the Fen deposits: the peat of Hilgay Fen comes next; and, as we follow this northwards, we find it becomes interstratified with a marine deposit of silt, which forms the land between Downham Market and Walsoken northwards to the Wash. This Fen district is the "Marshland" of Norfolk; and, in its flat surface, its remarkable drainage system, buried forests, &c., has many points of interest all its own. Of the former condition of this region, Mr. Skertchly has given us a striking picture.—" Great meres existed which received the surplus waters; and, surrounded with reed-brakes. such as even now the country produces with surpassing beauty, afforded shelter to myriads of wild birds, which found abundant food in the waters. Dank morasses, covered with sedge and rush and flags, abounded on the peat lands, and the cushion-clumps of sedge (*Carex paniculata*) afforded a hazardous foothold to the nimble wayfarer. On these morasses, and on the firmer, or rather drier soil, grass attained a rank luxuriance; and here the cattle grazed, and throve wondrously. But in winter, nearly all the peat-land was drowned, or as the old fen-men say, " surrounded," and then the hardy inhabitants went from island to island in small boats, or travelled quickly over the smooth ice. The border-land was clothed with forest-growth; and, seawards of the timber-trees, clumps of willow and sallow gave shelter to the wild-boar and the wolf. On the silt-lands the lower portions were surrounded in winter, and often far into the summer; and East and West Fens (and especially the former) almost always presented a lake-like appearance. The soil was fertile; the waters, the woods, and the air were tenanted with game. Famine could never be known, for the land literally overflowed with food, and, as a consequence, the people degenerated into a thriftless race, whose only strong passion was a love of freedom."

Now all this is changed; the great works which have been carried on, almost without intermission, from the time of Charles I. to the present day have so drained the land, that in dry years (as in 1874) there is a positive lack of water.

Denudation of the Coast.—The coast of Norfolk, especially where it rises into cliffs of any height, has suffered greatly from the action of the sea, combined with that of landsprings; the sea, by dashing against the base of the cliffs, using as missiles the fallen stones, rapidly undermines them, when the upper part falls in and is swept away by the waves: the springs flowing along the junction of pervious beds (sands) with impervious ones (clays) loosen the adhesion of the beds and the upper part slides down on to the beach: in this way the whole coast is receding at the rate of perhaps one yard per annum. "Since the Conquest the villages of Shipden, Keswick, Clare, Wimpwell, Eccles, and Ness, or the greater part of them, have been washed away. The remains of Eccles Church are still to be seen, buried as it were within the Marram Hills." (S. Taylor.) A letter by Mr. C. Reid in the " Geological Magazine," thus describes the damage done by the storm of January 30th, 1877: " I have examined the coast from Hasborough to beyond Sherringham, and the damage done is marvellous. Probably the loss of land along the whole line of coast mentioned may be estimated at a yard. At the life-boat gap at Bacton the amount that has gone is 15 yards, and a strip of about that width is missing as far as the Walcot Gap (three furlongs). . . . . The most serious loss is at Lower Sherringham, there Mr. Upcher has lost two acres; nearly all the sea-wall has been swept away none of the gangways are left; a cottage and a shed have fallen into the sea, the inn on the cliffs has had the windows broken, and is in a very unsafe condition, and should another gale occur now, much of the village will go. Mr. Upcher informs me that he reckons his loss of land during the past sixty years to be thirty acres at the very least."

PREHISTORIC MAN.—Of the comparatively rough flint implements which belong to the older or *Palæolithic* division of the Stone Age, many specimens have been found in the old river-gravels along the valley of the Little Ouse, at Lopham Ford, Shrub Hill near Feltwell, White Hill and Red Hill near Thetford, Bromehill near Weeting, &c.: these were chipped into shape but never polished; they are either pointed or oval in form, and the average size is from 5 to 7 inches long and 2 to 4 broad. Of the later or *Neolithic* Stone Age numerous specimens of celts (axe-heads), arrow-heads, flakes, cores, &c.,

have been found on or near the surface, or in barrows at Attleborough, Aylsham, Cromer, Necton, North Walsham, Pentney, Sporle near Swaffham, Norwich, &c. In the meres at Wretham, 5 miles north of Thetford, traces of undoubted lake-dwellings were met with; these have been described by Professor A. Newton and Sir C. Bunbury. The Neolithic tools are of delicate and varied shapes, and often bear marks of rubbing or polishing. In 1831 a polished flint celt was found embedded in a tree-trunk in the "submerged forest" near Hunstanton. The flints, of which these tools are almost invariably composed, were of course obtained from the chalk, and at a place called "Grimes' Graves" in the parish of Weeting, about 3 miles north-east of Brandon, it seems probable that we have pits sunk by these old Neolithic workers for the purpose of obtaining this material. One of the pits here was explored by Canon Greenwell in 1870; he found some of the old tools- picks foremd of deer horn which in one place had been buried underneath a falling

Fig. 64.—Front view of Core, with Flakes replaced, showing the points of percussion. The central mass of flint, from which no more flakes can be struck, is called the core.

of the chalk. "The day's work over, the men had laid down each his tool, ready for the next day's work; meanwhile the roof had fallen in, and the picks had never been recovered. . . . It was a most impressive sight, and one never to be forgotten, to look, after a lapse, it may be, of 3,000 years, upon a piece of work unfinished, with the tools of the workmen still lying where they had been placed so many centuries before." These picks still retained upon their chalky incrustation, the impressions of the workmen's fingers.

There is no passage from the Palæolithic to the Neolithic period; the two stand distinctly apart with no links between them: this great gap is now accounted for by Dr. Geikie, Mr. Skertchly, and other eminent geologists, by the supposition that a period of glaciation intervened and drove man out of this country to warmer climes: the men who made and used the Palæolithic implements were Pre-glacial or Inter-glacial; the Neoliths were Post-glacial.

No. 26.

GEOLOGY OF NORTHAMPTONSHIRE.

NATURAL HISTORY AND SCIENTIFIC SOCIETIES.
Peterborough Natural History and Scientific Society.
Northampton Natural History Society and Field Club.

MUSEUMS.
Northampton Museum.
Peterborough Museum.

PUBLICATIONS OF THE GEOLOGICAL SURVEY.

Maps. Sheet 64, Peterborough, Stamford, Uppingham. Quarter Sheets: 45 N.W. Banbury, Deddington, Chipping Norton; 45 N.E. Buckingham, Brackley; 46 N.W. Newport Pagnel, Woburn; 52 N.W. Kettering, Wellingborough; 52 N.E. Huntingdon; 52 S.W. Northampton, Olney; 53 N.E. Clipston, Crick, Braunston; 53 S.W. Southam, Kingston; 53 S.E. Towcester, Daventry, Weedon; 63 S.E. Lutterworth, Market Harborough.
Books. Geology of Part of Northamptonshire, by Aveline & French, 8d. Geology of Parts of Northamptonshire and Warwickshire, by Aveline, 8d. Geology of Rutland and Part of Northamptonshire, by Professor Judd, 12s. 6d. Geology of the Fen-land, by Skertchly, 40s.

IMPORTANT WORKS OR PAPERS ON LOCAL GEOLOGY.

1870. Sharp, S.—Oolites of Northamptonshire. Journ. Geol. Soc., vol. xxvi. p. 354; xxix. p. 225.
1871. Herman, W. D.—Allophane and an Allied Mineral found at Northampton. Journ. Geol. Soc., vol. xxvii. p. 234.
1881. Thompson, B.—Notes on Local Geology. Journ. Northampton Nat. History Society.
See also General Lists, p. xxv.

GEOLOGICALLY speaking Northamptonshire cannot be called a neglected county. Maps of the southern and central divisions, coloured so as to show where each rock occupies the surface, were issued by the Geological Survey between 1859 and 1864, and these, together with the accompanying descriptive memoirs by Messrs. Green, Aveline, and Hull, should be obtained by all who wish to have a thorough knowledge of the district. Of the northern portion Prof. Judd's admirable Map and Memoir, published in 1872, have given us an exhaustive account, while Mr. Sharp has also worked indefatigably and has formed a grand collection of local fossils.

The rocks that compose this county are in age and character midway between those older beds which lie to the west, forming the coal-fields of Warwickshire, &c., and the newer chalk deposits which occur to the east. The strata run in long bands of varying width from north-east to south-west, coinciding almost exactly with the longest axis of the county, which is much elongated in this direction. The dip or slant of the rock-masses is towards the south-east, so that they overlie one another in this direction. Thus if we were to make a very deep boring, say at Stony Stratford or Peterborough, we should find at a considerable depth, the same blue clays (lias) which occupy the surface on the west, near Market Harborough and Rugby.

The coincidence of the form of the county with the strike or direction of the strata, is thus favourable to the display of any particular bed or beds, as we can trace them over great distances and compare them at many points, but it is not favourable to the occurrence of any considerable series of beds. In fact Northamptonshire may be designated an Oolitic county, *i.e.* one in which the Oolitic strata are most interestingly shown, but to the almost entire exclusion of anything else. Owing to the tilting of the rock-beds the oldest or lowest come to the surface in the extreme west of the county, and we shall commence with them, as they form the foundation or base upon which the others are deposited. It should be remembered that the term rock is applied geologically to a mass of any substance which enters into the composition of the earth's crust, whether it be soft or shifting as clay or sand, or of the hardness of granite and slate.

THE LIAS. This term is applied to a thick series of clays, shales, and limestones, which stretch across England from Dorset to the Yorkshire coast. The total thickness is about 800 feet, but this is divided into three parts by variations in the character of the rocks and by the different fossils we find, according as we examine the lower, middle, or upper part of the series.

Lower Lias Clays.—These are bluish clays, in thickness not less than 500 or 600 feet, which form the western side of the county, stretching in from Warwickshire and Leicestershire, where their lowest beds occur. From Banbury northwards by Cropredy, Claydon, Priors Hardwick, Braunston, Kilsby, Claycoton, and on to Market Harborough we find these blue shaly clays forming a flat and uninteresting country, sparsely habited and chiefly in pasture. In all the upper part of the valley of the Nen too, from Rislingbury to Badby, the bed of the river is formed by an inlier of the Lower Lias, the stream having cut its way down through the overlying formations.

The Marlstone (or Middle Lias).—The hard "Rock-bed" as it is termed, is very thin in the north, near Harborough, not more than a foot or two thick, but it thickens southwards to about 20 feet where it crosses the Cherwell; the junction with the Lower Lias is easily traced, as being harder it has better withstood denudation, and so forms a low escarpment facing to the west. The rock-bed is a ferruginous limestone, reddish brown where exposed to the weather, but green or blue when dug at any depth. Fossils are very numerous but of a few species only, *Rhynchonella tetrahedra* and *Terebratula punctata* occurring in beds or masses. From a well 168 feet deep the marlstone furnishes to the town of Northampton a supply of pure water. Long Buckby, Daventry, Brockhall, Bugbrook, Middleton Cheney and Thenford may be indicated as standing on the marlstone: it is frequently quarried for building purposes, when its rich brown hue forms a pleasing contrast with the white oolitic limestone generally used in conjunction with it.

Upper Lias Clays.—These are between 150 and 200 feet thick: owing to their softness they have been much denuded, so that their western outcrop is very irregular, and they form numerous outliers scattered over the marlstone plateau. Many of the streams which drain to the east have cut down to these tenacious clays, which thus form the lower parts of the valleys of the Tove and its numerous tributaries, the valley of the Nene from below Northampton to Thrapston, &c.

The clays are worked in many places for brick-making. Fossils are not uncommon, chiefly remains of large saurians, with Ammonites, Belemnites, Ostrea, &c.

THE OOLITE.—This formation is composed of alternating beds of sandstone, limestone, and clay, of considerable thickness; it is named from the resemblance of the texture of many of the limestones to the roe of a fish (ὠον—an egg).

INFERIOR OOLITE.—*The Northampton Sand.*—In comparatively recent geographies we were told, that the mineral wealth of England lay west of a line drawn from the mouth of the Tees through Chesterfield to the mouth of the Exe. But the extensive iron workings opened in the formation we are about to describe have necessitated a modification of this statement. Since 1853 a valuable bed of iron ore has been worked in Northamptonshire, the amount raised in 1860 being 95,664 tons; in 1870 this had increased to 887,020 tons and to 1,412,256 tons in 1873; the next year, however, saw a decrease, consequent on the depression of trade generally, and only 1,056,478 tons were got in 1874, of an estimated value of £211,295; lastly, in 1879, there were raised 1,211,406 tons of iron ore, valued at £175,558. Much of the ore is sent away to Merthyr in South Wales, to Staffordshire, and to the north, because being of a siliceous character it acts as a flux when mixed with the clay ironstone of the coal-measures. There were however 14 blast furnaces in work in 1874, which produced 53,760 tons of pig-iron; in 1879 there were 17 furnaces in blast, and 165,317 tons of pig-iron were made.

Examining in detail this valuable bed we find it to be a sandstone, more or less compact, impregnated with varying amounts of oxide of iron and yielding in a few spots good building stone, whilst at Duston the limestone parts into such thin regular laminæ as to be denominated and used as "slates." We find it comparatively thin in the south-west, where it enters the county near Aynho. Then ranging by Newbottle and Farthingho it is cut back to Brackley by the Ouse Valley. North of the Tove the Northampton Sand spreads over a wide tract from Eydon (an outlier), Adston, and Maidford to Stowe Nine Churches, Gayton and Blisworth; it is in this region more developed, and is here largely worked for iron ore. North of Northampton it attains its full thickness of 80 feet, and at Duston it yields good building stone. Spreading out widely we find outliers on the west round Spratton, Hazlebeech, and Cold Ashby, whilst a peninsula of singularly serrated form stretches past Church Brampton to East and West Haddon and Guilsborough. Eastwards we find it on both sides of the Nen Valley at Wellingborough, Irchester, Irthlingborough, and Thrapston, and passing round to the north-west by Kettering, Rushton, and Desborough. The most northerly point where the iron-ore is worked is at Burleigh Park, near Stamford.

Fossils are not common in the *Northampton Sand*, but plant-markings and both marine and fresh-water shells occur. These facts would lead us to infer that it is an estuarine formation deposited by a great river probably flowing from the north-west, for as we trace the deposit to the south-east it gradually thins out until it disappears altogether, as near Grafton Regis, Grendon, and east of Oundle.

It is not probable that the iron was present with the sand at its first formation, but rather that this important ingredient was afterwards deposited from water highly charged with carbonate of iron slowly percolating through the rock. In fact when the ironstone is followed to any depth we find it as the blue carbonate still; but in the surface portions, which have been exposed for long ages to the action of the weather, the carbonate has been converted into the hydrated peroxide, which is of a rich brown tint and occurs in cakes and layers. The Northampton Sand forms a rich but rather light soil.

The Lincolnshire Oolite Limestone, with the Collyweston Slate.—North of Harrington and Maidwell, and west of Kettering and Water Newton, we find an important bed of limestone, which gradually thickens as we follow it northwards, until in the county whence it takes its name, it is as much as 200 feet thick.

At the base the beds of limestone are thin and fissile, splitting under the hammer (after having been exposed to frost) into thin layers, which are used for roofing purposes. These lower beds are called the *Collyweston Slates*, from the name of the village where they have long been worked; they are of local occurrence only, but are also met with at Kirby. They are of a yellowish white tint and have been used by Sir Gilbert Scott and other eminent architects for the roofs of churches and Gothic buildings generally. As to fossils the surfaces of the "slates" often bear ripple-markings, worm-tracks and plant remains, whilst shells of the genera *Pinna, Gervillia, Lucina*, &c. occur.

The Lincolnshire Limestone proper has very commonly an oolitic structure, and where the little rounded grains constitute the mass of the rock as at Ketton and Weldon, it furnishes a first-class freestone: it is easily worked, is of a rich creamy tint, and hardens on exposure to the air. The famous quarries at Barnack are now quite worked out, but in the Middle Ages they furnished the stone for Burleigh House, Peterborough Cathedral, and indeed most of the churches in Lincolnshire and Cambridge. This bed of limestone, together with the underlying Northampton Sand, was formerly assigned to the period of the *Great Oolite*, an error which was confirmed by the Geological Survey in their examination of the southern part of the county. In fact the relations of the strata were very difficult to make out from an examination of the Southern Oolites only, but when Mr. (now Professor) Judd came to map the northern portion he had had much previous experience of the same beds in Lincolnshire, and he has clearly proved that both the above-named divisions belong to the *Inferior Oolite*, the lowest member of the Oolitic system.

THE GREAT OOLITE. This division is composed of four members, two bands of limestone separated by two beds of clay; the lowest of these four sub-divisions is the—

Upper Estuarine Series.—This is a sandy clay, and where the Lincolnshire Limestone is absent is seen resting upon the Northampton Sand. The junction is usually very irregular, and is always marked by a band of ironstone nodules, indicating a line of unconformity. This is well seen in the pit on the Racecourse at Northampton, and in similar sections at Weekley and Old Head Wood. Wood is not uncommon, and bones of that huge reptile the *Cetiosaurus* also occur. The clay is dug for brickmaking.

Great Oolite Limestone. This bed occupies much ground in the south about Radston and Biddleden, and thence by Whittle Wood to Stony Stratford. We meet with it again round Tiffield, and it forms the ridge to the north-west of Irthlingborough and that from Twywell by Brigstock and Great Oakley. The marly beds are much valued and largely dug for lime burning, and here and there hard blue shelly bands occur, which take a fair polish and are called "Alwalton Marble." It forms a blackish soil, which produces excellent crops. Notwithstanding its name it is seldom of an oolitic texture, and it frequently contains beds of small smooth oysters—*Ostrea Sowerbyi* and *O. subrugulosa*.

Great Oolite Clays.—These are variegated in colour and vary in thickness from one or two to 20 or 30 feet. They represent the "Forest Marble" of the Cotteswold Hills. In South Northamptonshire they appear to be absent. Sections are rare, but Lilford Park, Oundle and Peterborough afford fair exposures in cuttings or brickyards. They form a cold and wet soil, often constituting the slope of the hills capped with Cornbrash, which are dotted over the Great Oolite Limestone plateau. Fossils are rare, but mostly such as occur in the limestone below. *Placunopsis socialis* is a characteristic shell.

The Cornbrash.—This is the last bed of the lower oolites and occurs only in the north-east of the county: it is an irony limestone, weathering red, very fossiliferous and not more than 15 feet thick: it contains a band of the thick rugose oyster *Ostrea Marshii* at its summit, by which it can usually be recognised, as well as by the wall-like appearance it presents in sections; *Ammonites Herveyi* and *A. macrocephalus* are also common. The outcrop may be traced from near Wimmington by Raunds, Lilford, Warmington and Elton to Peterborough. North and west of this town it occupies a considerable tract, and it occurs on both sides of the valleys of the numerous brooks which run from the west into the Nene between Wansford and Oundle: it is very hard and is quarried for road-metal, but furnishes a poor soil.

MIDDLE OOLITE.—*The Oxford Clay.*—Overlying the rock last described we find thick blue clays, which at their base however are frequently hard and sandy. These lowest beds, representing the *Kellaway's Rock* of Wiltshire, are worked in the brickpits at Oundle, Southwick, Benefield, &c. The dark-blue clays above stretch eastward, right out of the county, from Luddington and Barnwell to Lutton and again by Glinton, Werrington and Eye. There is also an extensive outlier west of Oundle, and a small one in Brigstock Park. Good exposures except in the brickyards are rare, as the Oxford Clay is usually overlaid by a considerable thickness of drift. As fossils, bones of large saurians, wood (near Peterborough) and spines of fishes, together with *Belemnites Owenii* and *hastatus* are most common. (See fig. 65.)

THE DRIFT.—The reader must not suppose that all the preceding rocks

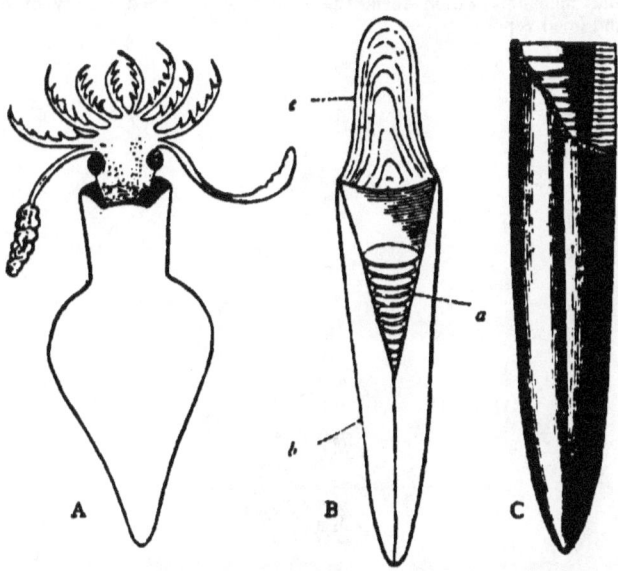

Fig. 65.—*A*, Restoration of the Animal of the Belemnite; *B*, complete Skeleton of a Belemnite, showing (*a*) the chambered phragmacone, (*b*) the guard, and (*c*) the horny pen; *C*, *Belemnites canaliculatus*, from the Inferior Oolite.

occupy the surface in the regions over which they are said to extend. They are frequently covered over by irregular masses of sand and clay which we term drift, because they are supposed to have been brought from a distance, chiefly by the agency of glaciers and icebergs, and strewn without order over the surface of the country at the termination or during the retreat of the glaciers, or on the melting of the icebergs. The lowest beds are gravels or sands, perhaps mid-glacial, and marking a cessation in the period of intense cold to which this country was then subjected: such deposits are to be seen near Upper Benefield and Moulton: above these is the *chalky boulder clay,* full of flints, pebbles of limestone, &c., and often containing well-preserved Liassic or Oolitic fossils.

RIVER GRAVELS. -Lastly in the deposits made by existing rivers we come upon traces of early man. Such is the tapering celt or stone axe-head found near Towcester, also a somewhat similar one from Gilsborough, now in the

Northampton Museum, with another 7 inches long and 2¾ inches broad, found in that town. At Oundle flint scrapers have been found by Mr. John Evans, and leaf-shaped arrow-heads and spear-heads are recorded from the same place.

In an old proverb Northamptonshire is said to be famous for "springs, spires, and squires;" now all these owe their importance more or less directly to the geological structure of the county. The excellent springs result from the alternation of porous sands and limestones with impermeable clays. Everywhere we find the junction of two such formations marked by a line of springs, which have plainly determined the sites of the villages and towns.

The beauty of the churches is largely owing to the excellent building-stones which abound, together with the fertility of the soil, which enabled the district to support a large and prosperous population.

To this last cause and also to the fine scenery and pleasant sites are due the many old mansions which form the homes of a landed gentry of ancient name and good repute.

No. 27.

GEOLOGY OF NORTHUMBERLAND.

NATURAL HISTORY AND SCIENTIFIC SOCIETIES.

Berwickshire Naturalists Club; Alnwick. Annual Report and Proceedings.
Natural History Society of Northumberland and Durham; Newcastle-on-Tyne. Transactions.
North of England Institute of Mining and Mechanical Engineers; Newcastle-on-Tyne. Transactions.
Tyneside Naturalists' Field Club. Transactions.
Newcastle-on-Tyne Literary and Philosophical Society.
Newcastle-on-Tyne Antiquarian Society.
Newcastle-on-Tyne Entomological Society.

MUSEUMS.

Alnwick Castle Museum.
Barnard Castle Museum.
Museum of the Literary and Philosophical Society, Newcastle-on-Tyne.
Museum of the Antiquarian Society, Newcastle-on-Tyne.

PUBLICATIONS OF THE GEOLOGICAL SURVEY.

Maps--Quarter Sheets: 105, S.E. South Shields, Sunderland; 105 S.W. Gateshead; 105 N.W. Morpeth, Bedlington, South Gosforth; 105 N.E. Tynemouth, Blyth; 106 S.E. Allendale; 106 N.E. Bellingham; 109 S.E. Newbiggin; 109 S.W. Rothbury. (Survey not completed.)

IMPORTANT WORKS OR PAPERS ON LOCAL GEOLOGY.

1868. Tate, George—Geology of Northumberland. Nat. Hist. Trans. of Northumberland and Durham, vol. ii. p. 1.
1877. Lebour, Prof. G. A.—Geology of Northumberland.
1877. Lebour, G. A. and Topley, W.—On the Intrusive Character of the Whin Sill. Journ. Geol. Soc., vol. xxxiii. p. 406.
1877. Lebour, G. A.—On the terms "Bernician" and "Tuedian." Geol. Mag., p. 19.
1880. Miller, Hugh.—Tynedale Escarpments. Nat. Hist. Trans. Northumberland and Durham, vol. vii.

See also General Lists, p. xxv.

THE rocks of this county have not hitherto attracted so much attention as they deserve: this however will soon be remedied, for the officers of the Government Geological Survey have been for some time minutely mapping the strata. Of the important coal-field in the south-east of the county such maps have already been published, together with horizontal and vertical sections, which show exactly the position of each stratum, and Messrs. Howell, Topley, and Hugh Miller are now engaged in completing the survey of the rest of the county.

Of early geological observations we find papers in the Transactions, &c., of the Geological Society of London by Messrs. N. J. Winch, Hon. H. G. Bennett, W. C. Trevelyan, W. Hutton, and G. C. Atkinson, the nature of the Whin Sill being the chief point of discussion. In the Transactions of the North of England Institute of Mining Engineers there are several valuable papers by Prof. Lebour, Messrs. D. Burns, E. F. Boyd, Bewick, and others; and much information may be gained from the publications of the Tyneside Naturalists' Field Club, the old Northumberland Natural History Society, &c. The late Mr. G. Tate's History of Alnwick, with several guide-books to Rothbury (W. Topley), Morpeth (Rev. W. Howchin), &c., have also yielded us much information. Prof. G. A. Lebour has recently published an excellent little volume containing a general account of the rocks; and the same gentleman has edited a series of drawings of fossil plants, made by Hutton and Lindley in 1840, together with a catalogue of their collections now in the museum of the Natural History Society at Newcastle, which also contains a very fine series of the rocks and fossils of the county. A good account of the Cheviot district was given by Dr. James Geikie, in *Good Words* for 1876.

The general arrangement of the rock-beds which form Northumberland is that of a series of layers whose exposed edges run or *strike* from north-east to south-west, while they incline or *dip* towards the south-east. The oldest or bottom rocks occur in the north-west as a consequence of this arrangement; and if we walked from Cheviot to Tynemouth we should pass across the edges of all the beds in turn, coming continually to newer and newer strata: lastly, if at Tynemouth we were to make a deep boring, we should pass through in turn the continuations of all the rocks over which we had been walking, until at a depth of perhaps 5,000 or 6,000 feet we should reach the beds which form the surface of the north-west of the county.

The highest ground lies also in the north-west, and from thence the surface slopes on the whole to the sea. Cheviot is 2,676 feet above the sea, Hedge-hope Hill 2,348 feet, Carter Fell Toll-house 1,292 feet; in the Pennine Range on the west we have Caplestone Fell 1,555 feet, and Blacklaw Hill 1,942 feet: on the eastern slope Cambo church is 682 feet; Belford (Church) 209 feet; Alnwick (St. Mary's Church) 180 feet; Hebburn (Church) 329 feet; Morpeth (Parish Church) 168 feet; Haltwhistle (Parish Church) 400 feet; Hexham 200 feet; Corbridge (Parish Church) 138 feet; and Newcastle (St. Nicholas Church) 100 feet.

IGNEOUS ROCKS. *The Cheviots.*—These hills are mainly composed of a rock called porphyrite, which has a red felspathic matrix, in which are embedded crystals of the same mineral: it is quarried near Biddlestone Hall, and contains fine agates, specimens of which may often be picked up in the gravels of the upper parts of the rivers Coquet and Rede. At Yevering and Cheviot the rock is a syenite, which can be traced passing into granite. On the other hand we have dark-coloured rocks called Dolerites, which may be of later date, for Professor Lebour states that "at Puncherton Burn (running into the Alwyn), the Dolerites are seen to pass into a brecciated conglomerate in which blocks of rock, apparently of Lower Carboniferous origin, besides others of earlier age, and mixed with fragments of Porphyrite, are embedded in a matrix of Dolerite." The rocks which form the Cheviot Hills have however as yet been little studied, and there can be no doubt but that their thorough examination by competent geologists would bring to light new facts of the highest interest: they are believed to be of later date than the Silurian strata next to be described, for in the Upper Coquet, between Philip and Makendon, the Silurian beds are seen to have been bent up by protrusion of the Cheviot rocks. Probably a part at least of the whole series will prove to be ash beds ejected from sub-marine or sub-aërial volcanoes, or both, and since metamorphosed: this would account for the appearance of bedding and dip which they often present, as in the banks of Ridlees Burn.

The Great Whin Sill.—This remarkable bed of basalt appears to have its head-quarters in Teesdale, where it underlies the Tyne-bottom Limestone. From this point it runs northwards, along the western side of the Pennine Chain, until, following the general strike of the beds, it turns to the north-east, and enters Northumberland near Greenhead: here it rests on the "Great Limestone," a bed about 700 feet above the Tyne-bottom Limestone. This fact at once proves the Whin Sill to be an intrusive sheet of igneous rock, forced in a melted condition between the other strata. The Roman or Picts' Wall runs for some distance along its edge, and we can clearly trace it by Swinburn and Sweethope to Farney Law. At the Elf Hills Quarry, Sir W. C. Trevelyan has described the Whin Sill as sending up strings of melted rock into the limestone above it, and everywhere it alters and hardens the rocks, especially the shales, with which it comes into contact. From the last-named point the Whin Sill can no longer be traced continuously along the surface, but only comes up in detached bosses here and there: such a mass is found south-east of Alnwick: then it forms the cliffs on the coast at Dunstanborough and Bamborough, whence, on the one hand, it runs inland as far as Kyloe Crags, and on the other forms the Farne Islands. At several points along its course the Whin Sill is quarried for road-metal. The period at which it was injected between the other strata cannot be positively determined, but it was probably about the close of the Carboniferous Epoch: its usual thickness is 100 feet, but it varies from 20 feet to 150 feet.

Basaltic Dykes.—Several other sheets of igneous rock cut across the sedimentary strata of Northumberland. These differ from the great Whin Sill, chiefly in the fact that they are more or less vertical, cutting through the strata at a considerable angle, whilst the Whin Sill, as we have seen, preserves its parallelism with the stratified rocks for considerable distances. We can distinguish two sets of these dykes, viz. those which run from south-west to north-east, and others running from south-east to north-west, and therefore crossing the first set at right-angles. Of the former division we know the *Brunton Dyke*, between Whitfield and St. Oswald's Chapel, of this the *Bavington Dyke* may be a continuation; the *Lewis Burn Dyke*, between Short Cleugh and Billsmoor, is perhaps continued as the *Black Burn Dykes*; the *Plashetts Dyke*, the *Boulmer Dyke* (100 feet thick on the coast), the *Hampeth Dyke* near Shilbottle, the *Howick Dyke* and the *Cornhill Dyke*, between Coldstream and Mattalees.

Of the south-east and north-west series, the longest is perhaps the *Acklington Dyke*, extending from Bondicar on the coast to Clennell, a distance of 20 miles, and thence into Scotland. The *Hebburn Dyke* crosses the Tyne at Walker, and passes thence along Newcastle Town Moor to Slatyford: it is continued as the *Coley Hill Dyke*. Others in this series are the *Cramlington*, *Bedlington* and *Lower Wansbeck Dykes*, the *Trobe's Dene Dyke*, the *Beadnell Dyke* and the *Holy Island* or *Lindisfarne Dyke*, of which last Mr. G. Tate has given an excellent description: he says, "It is one of the largest in the county, and indeed, has been erroneously described as part of the Whin Sill, to which it has some resemblance, as it rises in Lindisfarne or Holy Island in high craggy hills of columnar basalt. It crosses the southern part of the island nearly from west to east, and is seen two miles seaward, forming the Plough and Goldstone rocks on which the "Pegasus" was wrecked. The Castle crowns a high, craggy, basaltic hill, and on the west side of the island the dyke is exposed in a high cliff, and is there 120 feet wide, with a slope 85 degrees southward: large blocks of limestone, highly metamorphosed, are enveloped in the basalt, and the strata, broken through, have been relatively altered in position, those on the south side having been considerably upcast. A calcareous shale, very fossiliferous, and a limestone beneath it, abut against the dyke and are metamorphosed, and near to the Castle a vein of basalt penetrates the shale. This dyke is seen near Fenham on the mainland, and further westward, near Kyloe church, where its width is from 20 to 60 feet, for it widens as it descends; in one part it is covered by shattered beds of shale: it cuts through the Lowick coal beds, and is traceable further westward to Leitham, the whole ascertained course being about 14 miles." Where these dykes come into contact with the coal, the latter is often completely coked. If we assign the dykes to the same age as those of the south of Scotland, with which they agree in direction, composition &c. we shall place them in the Tertiary Period, when those great Hebridean volcanoes were in full activity whose remains Professor Judd has traced with such skill. The Northumbrian dykes are almost invariably composed of a close-grained blue basalt, weathering brown and red, and yielding fair road-metal.

SILURIAN FORMATION. A very small area of these old rocks is exposed in the banks of the river Rede and Whitelee Burn, and also in the valley of the Upper Coquet between Philip and Makendon. They consist of shales or clay-slates (formerly termed greywacke), and are tilted at a high angle against the igneous rocks of the Cheviots. They probably belong to the great spread of Silurian strata, which on the other side of the Cheviots forms so much of the south of Scotland, and may belong to the Riccarton Series, which forms the base of the upper Silurian division. No fossils have yet been found in these beds in Northumberland, but they are traversed by dykes of diorite.

CARBONIFEROUS FORMATION. To this great division nine-tenths of the rocks of this county belong. The lowest beds rest on the denuded and contorted edges of Silurian rocks, for of the *Old Red Sandstone*, which should lie between, there is no trace in this region.

The Tuedian Beds. This term was proposed by Mr. G. Tate, in 1856, for the lowest beds of the Carboniferous Formation: it is derived from the river Tweed, in whose basin they are largely developed. Their base consists of red sandstone conglomerates, as may be well seen in Roddam Dene and Biddlestone Burn, where they abut on the Cheviots, and contain rolled masses of porphyry of all sizes, from that of a pea to a man's head. These conglomerates are probably the remains of old beaches, which gathered round the Cheviots when the latter stood as islands in the early Carboniferous sea. Beds of red sandstone overlying the conglomerates show distinct ripple-marks, such as are forming to-day on every sandy shore. Higher still we get beds of purple shale, greenish sandy clays, coarse sandstones and compact cream-coloured limestones. The upper boundary of the Tuedian division is formed by a bed called the Dun or Lamberton limestone, and in the region of the Upper Coquet, where that limestone cannot be traced, by the base of the Harbottle Grits. Thus the Tuedian beds form a narrow band extending along the Tweed and encircling the Cheviots, passing round by Wooler, Branton and Alnham to near Carter Fell. Their total thickness is estimated by Mr. Tate at 1,500 feet, of which 500 feet belong to the red conglomerates. Fossils are not common in them. In the sandstones plant remains, especially of the genus *Ulodendron*, occur.

Bernician Beds.—In Northumberland it is impossible to draw any lines of division in the beds which lie above the *Tuedian* and below the Millstone Grit. Further south, in Yorkshire and Derbyshire, a series of shales and sandstones, to which the term *Yoredale Series* was applied by Professor Phillips, underlies the Millstone Grit, and in turn rests upon thick limestones the *Carboniferous* or *Scar* Limestone. But as we follow these beds northwards we find that in place of the Scar limestone and Yoredale Series we have a great mass of sandstones, shales and limestones which constitute one natural whole, and to which Professor Lebour has given the name of *Bernician Beds*, from Bernicia, the ancient name of Northumberland. They are the "Lead-measures" of Westgarth Forster, and include the Carbonaceous and Calcareous Limestones of Mr. G. Tate; they occupy a considerable breadth of the western side of the county, extending northwards from Haltwistle and Hexham nearly to the Cheviots, and north-eastwards by Rothbury and Alnwick to the sea, where they form the coast line from Alnmouth to Berwick; in the

South Tyne district their thickness is 2,500 feet, and it is about the same in the north-east of the county near Alnwick and Scremerston, but in the region intermediate between these points it has been ascertained by the Geological Survey that the Bernician Beds thicken to 8,000 feet, a remarkable and previously unsuspected fact.

In the Bernician Series there are three great beds of grit or coarse sandstone. The lowest of these—the Harbottle Grits—forms the hilly craggy district between the Upper Coquet, the Redewater, and part of Carter Fells. The Simonside Grits form hills 1,600 feet high near Rothbury, and lastly the Inghoe Grits occur between the Tyne and Wansbeck. There are also numerous beds of bluish limestone, six at least of which can be traced continuously for great distances, and several of them are underlain by thin seams of coal; thick beds of shale (or "plate") occur more or less between all the above-named strata. The Beadnell, Three, or Five-yard Limestone is the lowest continuous band, the others are the Six-yard or Acre Limestone, the Four-fathom or Eight-yard, the Great Limestone, the Little Limestone, and the Fell-top Limestone. These are quarried at many places to burn into lime, and they are traversed by mineral veins which in some localities are rich in lead ore. In 1877 the lead mines of Durham and Northumberland raised 25,086 tons of lead-ore, which yielded 18,984 tons of lead and 75,686 ounces of silver; the greater part of this, however, belongs to Durham. Carbonate of Barytes, to the amount of 4,345 tons (valued at £8,688), was raised at Fallowfield and Settlingstones. Of the thin coal-seams, that under the Little Limestone is worked at several points for local use; it is 2 or 3 feet in thickness.

Some good bands of iron-ore (clay ironstone) occur between the Great and Little Limestones, and are worked near Bellingham, &c. In 1879 the quantity of lead ore raised in the two counties had decreased to 14,187 tons, valued at £150,000.

Fossils—chiefly marine shells—are numerous, especially in the limestones. They are the ordinary mountain limestone species, *Productus giganteus*, *Strophomena depressa*, *Chonetes Hardrensis*, &c., &c. Professor Lebour enumerates 16 species from the Fell-top limestone at Harlow Hill; the Rev. E. Jenkinson has obtained 131 species from the Lowick quarries, chiefly from the "Great" and "Four-fathom" limestones, and 91 species have been obtained from the Ridsdale ironstone beds. The shell called *Posidonomya Becheri* occurs at Budle and near Alnwick, and the genus *Agelacrinus* is also noteworthy, as it is not usually found save in Silurian rocks. Foraminifers abound, and one large species—*Saccamina Carteri*—is very characteristic in the Four-fathom limestone. It gives the stone a spotted appearance, and in weathered masses stands out like beads on the surface of the rock.

The Millstone Grit.—This division of the Carboniferous Formation is not well represented in Northumberland, we may group with it the strata called *Gannister Beds*, and assign to it the whole of the beds lying between the Fell-top limestone and the Brockwell coal seam. Their total thickness is about 300 feet, and they consist chiefly of thin sandstones and shales, with some thin and usually worthless seams of coal. A seam called the *Crankey Seam* has, however, been worked on the south of Harlow Hill. Between the Derwent and the Tyne the Millstone Grit forms the surface over a considerable area; it thence runs north-east by Morpeth to the coast near Warkworth, getting thinner in this direction.

Gannister is a well known name for siliceous under-clays, capable of standing great heat, which occur in most coal-fields in the lowest part of the coal-bearing series. Beds of this kind are seen at Saltwick Stocksfield, &c., and in the same beds, at Whittonstall, Professor Lebour, in 1878, found a few marine fossils, including *Aviculopecten papyraceus*, an *Orthoceras*, and some *encrinite* stems. "For many years a bed of ironstone was wrought in the upper portion of the gannister series in the Derwent district. It was known as the "German Band," a grotesque name, due, not to any covert allusion to itinerant musicians, but to the small colony of German sword-makers who in former days worked this ironstone and plied their trade at Derwentcote." *(Lebour).*

The Coal-measures. —Taking the Tyne as a base-line, the coal-bearing strata extend from the mouth of that river westwards to near Wylam, a distance of 14 miles; a line drawn from Wylam to the mouth of the Coquet will mark the western boundary, whilst to the east the beds dip under the sea, beneath which they doubtless rise up, or crop out at a distance of several miles say from 5 to 10 from the shore. Thus in this county we have the northern and smaller portion of a great oval basin-shaped coal-field, which is tilted to the eastward, so that its eastern part is covered by the waters of the sea. On the land the Coal-measures form tame scenery—low undulating ground with a clayey, heavy soil; their thickness is 2,000 feet, composed of about equal proportions of sandstone and shale, with 16 seams of workable coal of an aggregate thickness of 46 feet. The quantity of coal remaining still to be got was estimated by Mr. Forster, before the Royal Coal Commission of 1871, at 2,576,000,000 tons, and it was considered possible to work seams extending beneath the sea, for a distance of about 2 miles from the coast, giving a further quantity of 403,000,000 tons.

In 1877 there were 173 collieries at work in Northumberland and North Durham, and these raised 11,975.250 tons of coal; in 1879 there were 183 collieries at work, and 12,245,597 tons of coal were raised.

Of the coal-seams the *Brockwell* or *Splint Coal* (1 to 4 feet thick) is the lowest workable one; about 350 feet above it is the *Low Main* or *Hutton Seam*, from 2 to 6 feet thick: it is continuous throughout the whole extent of the coal-field, from Warkworth in the north, to Haswell and Hetton in the south, and it also yields the best description of three different varieties of coal, suitable for purposes not at all similar to each other, viz., the best household (in Durham), the best gas and the best steam coal (in Northumberland). The Bensham seam is 75 feet above the last named, and 200 feet higher still lies the famous High Main or old "Wallsend" coal: once the most valuable seam in the coal-field, it is now almost entirely worked or burned out. The *Hebburn Fell* or *Monkton Seam* lies 600 feet above the *High Main*, and 450 feet higher is the *Closing Hill Seam*, the highest known, only seen in a quarry on the north side of the Ninety Fathom Dyke, about half-a-mile from Killingworth House.

Formation of Coal. During the Carboniferous Period changes in the level of land and sea appear to have been numerous and extensive. The lower deposits from the Tuedian to the top of the Bernician indicate a moderately deep and tranquil sea; by elevation of the sea-bottom the millstone grit was formed in shallower and current traversed waters; lastly, during the formation of the Coal-measures, the land was on the whole above, but not much above, the level of the water, and rank forests grew in swamps and marshes over great areas near the sea. Every coal-seam consists of the compressed vegetation formed by the growth and decay of generation after generation of plants in one of these great forests; a thick mass of vegetable matter having thus been formed, the land on which the forest grew must have suffered slow depression, so as to permit the waters to encroach on it and deposit layers of mud and sand; alterations of this kind, going on for some hundreds of thousands of years, resulting at last in the formation of the mass of strata some 2,000 feet in thickness which constitutes the Great Northern Coal-field. The plants whose remains we find in the coal-measures are ferns, or reeds *(Calamites)*, or gigantic club-mosses *(Sigillaria* and *Lepidodendron)*. Freshwater or estuarine shells called *Anthracosia* form seams called mussel-bands, of which six at least can be traced. The *fire-clay* which occurs under nearly every coal-seam is the soil in which the ancient forest grew: it furnishes a valuable material for the manufacture of "seggars," or vessels which have to stand great heat. In 1875 there were raised 102,433 tons of fire-clay in this county. Some of the shales which forms the "roofs" of coal-seams contain great numbers of the scales and teeth of fishes, saurians, &c. From the roof of the Low Main Coal Mr. Thomas Atthey, of Gosforth, has obtained thousands of such specimens, which are now in the museum of the Natural History Society. Mr. Barkas has also diligently worked up this subject.

Of the sandstones the bed called the Grindstone Sill or Post yields the famous Newcastle grindstones; it forms the chief heights near Newcastle as Byker, Benwell Hill, &c.

Faults.—The Ninety-fathom Dyke is a very well known break or dislocation of the strata which extends from Cullercoats by Denton and Whittonstall to Minsteracres. The beds on the north or downthrow side of it are from 500 to 1,200 feet higher in the geological series than those on the south side. The Stublick Dyke is a similar fault or dislocation, which commences a little south of Corbridge, and runs, westwards, parallel to the Tyne into Cumberland: it also has a downthrow on the northern side exceeding 1,000 feet in amount. In consequence of this, several little detached semi-basins of coal are brought in along its northern side near Whitfield and Lambley. There are numerous other "faults" of smaller amount, and some greater ones, which are supposed to effect the Bernician Rocks only, but which have not yet been clearly traced.

THE PERMIAN FORMATION.—Of beds of this age, often called the *Magnesian Limestone*, only three or four outlying patches occur in Northumberland. These are seen overlying the Coal-measures unconformably at Tynemouth, Cullercoats, and Whitley, and at Seaton Sluice near Hartley, all near the mouth of the Tyne. The bottom beds are *red sandstones*, which contain plants of Coal-measure species; above these come unfossiliferous *yellow sands;* next we find hard shaly beds, about 3 feet thick, called the *marl-slate*, which contain finely preserved remains of fishes of the genera *Palæoniscus, Platysomus,* &c. Cullercoats is the only point at which these are obtainable. Then comes the *magnesian limestone* proper, of which the lower and middle divisions only are here present. The lower beds are 200 feet thick; they may be seen in the Whitley quarries, now used as reservoirs. The middle beds have been mostly denuded off; they have a cellular structure. At Tynemouth, though not easy of access, they have yielded many shells, such as *Productus horridus, Axinus dubius,* &c.

GLACIAL DEPOSITS.—The Permian strata are the latest—the newest —of the stratified rocks which form the solid mass of Northumberland. The great series of Secondary and Tertiary deposits are entirely wanting here, and the beds of which we have now to speak are, geologically, but of the most recent formation creatures of yesterday and, indeed, were hardly noticed by geologists of the last generation. They consist of irregular deposits of clay and gravel, scattered over nearly the entire area of the coal-field, and rising westwards to the height of about 1,000 feet. The commonest deposit is a stiff brown or blue clay—the boulder clay—so called because it is full of rounded and scratched blocks of stone of all kinds, but mainly such as would have come from the Cheviots or the Pennine Chain. The surface of the rocks beneath this *boulder clay* is often seen to be smoothed and scratched, as in the Farne Islands, and on the limestones at Belford, Swinhoe, Belsay, &c. We now think that this Boulder clay is the result of the action of glaciers—great masses of ice—which some quarter-of-a-million years ago occupied the Cheviots and the Pennine Range, and slowly ground their way down to the sea. Beds of gravel and sand rest on the clay, and appear to have been derived from it, whilst the moraine heaps which run in a crescent form across the small valleys in the hills (as near East Woodburn in the Lisle Burn) mark the small glaciers of the close of this Glacial Period. Winding mounds of gravel, called *kaims*, are well seen between Alnwick and Berwick and along the Tweed Valley.

RECENT DEPOSITS.—Of raised beaches Professor Lebour records two; one, half-a-mile north of the mouth of the Wansbeck, and the other on Holy Island, near the Beacon. River-gravels and sands, brick-earths and clays, may be seen along the lower courses of most of the existing rivers; the loam near Acomb is worked to make pottery, at Redheugh Bridge for brick making. From the extensive gravels which occur in a valley between the Tyne and the Irthing, west of Haltwhistle, Mr. D. Burns, F.G.S., has shown that formerly the latter river ran eastward into the Tyne, instead of westward into the Eden. Peat mosses, or turbaries, still cover much ground in Western Northumberland, and form the "moss hags" which run up to the summits of the Cheviot Hills. Peat is chiefly composed of a moss called *Sphagnum*, and requires a great amount of moisture for its continued growth; hence drainage is an effectual means of destroying these bogs.

Fig. 66.—Finely-chipped Flint Blade, found in a cist, with the remains of a burnt body, on Ford Common, Northumberland.

PREHISTORIC MAN.—Of the earlier, or Palæolithic stone age, no indications have been found in Northumberland; but of the latter or Neolithic Period we have good evidence in the form of many stone implements, some buried with the dead in the mounds, or barrows, which are the places of interment of the early races who inhabited this district; others found in peat-bogs, or on the surface of the moors, &c. These stone tools carry us back to a period whose date may be roughly stated as lasting in these islands up to about 10,000 years ago, when man was unacquainted with the use of metals, and when stone, especially flint, was the material out of which he constructed such rough implements and weapons as he required.

Stone celts, or axe-heads, have been found near Harbottle, Burradon, Ponteland, Doddington, Ilderton, Branton, Bellingham, and Throckley Fell; they are usually from five to seven inches long and two or three broad; they were fastened in handles of wood or bone, but these have perished. Stone axe-heads, with a hole through the centre to admit a handle, have been found at Alnwick (12 inches long), Thirstone, Shilbottle, and Hipsburn,: a fine specimen of this kind, fashioned out of an oval quartzite pebble, and hollowed on both faces, was found in a cist at Seghill in 1866, and is now in the collection of Canon Greenwell at Durham, who also has one made of basalt, which was found at Twisel. A grooved stone hammer-head, found at Percy's Leap, is in the Alnwick Castle museum. The many stone balls which have been found at Corbridge Fell are considered by Canon Greenwell to have been used in some game. Flint flakes, or knives, were found in graves at Amble, and on Ford Common, &c. Flint arrow-heads have been found in Kielder Burn. Of a later date are the four cists found at Tosson, near Rothbury, which contained skeletons, two of them accompanied by an urn. In one cist were three buttons, apparently made of jet or cannel coal, and two inches in diameter, and in another was an iron javelin head. But with the discovery of metals we enter on historic times, and the geologist gives place to the historian and the geographer.

No. 28.

GEOLOGY OF NOTTINGHAMSHIRE.

NATURAL HISTORY AND SCIENTIFIC SOCIETIES.
Nottingham Literary and Philosophical Society. Annual Report.
Nottingham Naturalists' Society.
Nottingham High School Natural History Society. Magazine, "The Forester."
Nottingham Working Men's Naturalists' Society.

MUSEUMS.
School of Art and Museum, Nottingham.
The Castle Museum, Nottingham.

PUBLICATIONS OF THE GEOLOGICAL SURVEY.
Maps.—Sheets: 70, Grantham; Quarter-sheets: 71 S.E. Loughborough; 71 S.W. Castle Donnington; 71 N.E. Nottingham, Southwell; 71 N.W. Belper; 82 S.E. Mansfield, Ollerton; 82 S.W. Chesterfield; 82 N.E. Tickhill, Worksop; 82, N.W. Sheffield. Survey not completed.
Books.—Geology of the Country round Nottingham, by Aveline, 2nd edition, 1s. Geology of Parts of Notts and Derbyshire, by Aveline, 1s. Geology of Parts of Notts, Derby, and Yorkshire, by Aveline, 1s.

IMPORTANT WORKS OR PAPERS ON LOCAL GEOLOGY.
1829. Sedgwick, Professor A.—On the Magnesian Limestone, &c. Trans. Geol. Soc., Ser. 2, vol. iii. p. 37, and Proc., vol. 1. p. 63.
1841. Bailey, Thomas.—On the Gravels near Basford. Proc. Geol. Soc., vol. iii. p. 411.
1859. Lancaster, J., and Wright, C. C.—On the Sinking of the Shireoak Colliery, Worksop. Journ. Geol. Soc., vol. xvi. p. 137.
1876. Irving, Rev. A.—Some Recent Sections near Nottingham. Journ. Geol. Soc., vol. xxxii. p. 513.
1876. Wilson, E.—Permians of North-East of England. Journ. Geol. Soc., vol. xxxii. p. 533.
1877. Shipman, J.—Conglomerate at Base of Lower Keuper. Geol. Mag., p. 497.
1878. Shipman, J.—Geology of East Nottingham. Midland Naturalist, vol. i. pp. 18, 30.
1879. Wilson, E.—Age of the Pennine Chain. Geol. Mag., p. 500.
1879. Wilson, E., and Shipman, J.—Keuper Basement Beds near Nottingham. Geol. Mag., p. 532.
1880. Boot, J. T.—Sections of New Collieries and Boreholes in the Midland Coal-field. Trans. Manchester Geol. Soc., vol. xvi. p. 83.
1881. Wilson, E.—Permian Formation in N.E. England. Midland Naturalist, pp. 97, &c.

See also General Lists, p. xxv.

GEOLOGY OF ENGLAND AND WALES.

THE stratified rocks of England occupy the surface, broadly speaking, in long lines stretching from north-east to south-west; the oldest rocks being in the west, in Wales and Cumberland, whilst those of latest deposition are found in the eastern counties. As Notts is a long and somewhat narrow county, with its greatest length in a north-east and south-west direction, it will be seen that its position is on the whole unfavourable to the display of a *variety* of rocks, although it would afford long and continuous sections of those formations which happen to pass through it.

Glancing at the surface of the county as a whole, we shall note, as a striking feature, that long escarpments lines of low, tolerably well-marked hills — abrupt to the west, gently sloping to the east, run in the direction of its greatest length. In the valleys between these, the two chief rivers, the Trent and the Idle, flow to the north-east, whilst the little tributaries of these streams enter them nearly at right angles.

This structure of the district is entirely dependent on the varying hardness of the different strata. The hard limestones and sandstones, having better withstood the wear and tear of ages than the softer sands and marls, now rise above them, forming bold and distinct features in the landscape.

We will now examine the geological structure of the district more in detail.

THE COAL-MEASURES. These are the oldest rocks in the county: they occupy the surface of a small portion in the south-west, from near Nottingham to Kirkby-in-Ashfield, and thence westwards into Derbyshire, forming the southern portion of the York, Derby, and Nottingham Coal-field, the most extensive in England; but this area only represents that part where the coal strata actually constitute the surface rocks. At many places further east the coal is reached by pits sunk through superincumbent, later-formed rocks: indeed, the Coal Commission which sat in 1871 considered that coal was to be got at a less depth than 4,000 feet as far east as Lincoln. The boring at South Scarle (see Lincolnshire) has justified this prediction; another boring at Owthorpe (South Notts) reached Coal-measures at a depth of 1,066 feet.

There are many seams of coal, but only six or seven of these are of such quality and thickness as to admit of profitable working. Nearest the surface is the "*Top Hard*," 8 feet thick. This celebrated coal was reached at Cinderhill Colliery at a depth of 655 feet; at Hucknall Torkard at 1,236 feet; at Kirkby Woodhouse 750 feet; Clifton 207 feet; Pleasley near Mansfield 1,542 feet; Langwith near Bolsover 1,617 feet; and under Newstead Abbey 1,600 feet.

The "*Deep Soft*" seam lies about 500 feet lower down, and about 20 feet lower still, there is the "*Deep Hard*" coal. The *Kilburn Coal* is probably 800 feet beneath this.

In 1854 the Duke of Newcastle sank shafts at Shireoaks at once, without preliminary boring, to a depth of 1,527 feet, at which the Top Hard was struck: of this depth the first 227 feet were through Permian Beds. New pits have been sunk within the last few years at Clifton, Annesley, Bestwood, and other places, so that in 1874 there were 45 collieries at work in Nottinghamshire; these raised in the year 3,127,750 tons of coal. In 1879 there were 41 collieries, and the coal raised amounted to 4,249,242 tons; of iron ore 20,000 tons, valued at 10s. per ton, were obtained.

In sinking the Shireoaks pit, what appeared to be a valuable bed of ironstone was met with: it is a red ore, and crops out at Anston Stones: attempts to work it proved unprofitable, as the band varied much in character: it yielded about 40 per cent. of metallic iron.

THE PERMIAN FORMATION. These rocks occupy a long narrow strip on the west of the county. By far the most important member is the Magnesian Limestone: commencing near Bulwell it runs northward, Sutton-in-Ashfield, Mansfield, Warsop, and Shireoaks being situated nearly on its eastern limit, while westwards it extends into Derbyshire: its thickness decreases from above 100 feet at Shireoaks to about 30 feet at Bulwell. In fact, as we go southwards we are approaching its ancient limit or shore-line—the elevated ground round Charnwood Forest. At Bulwell, Kimberley, Linby, Grives Wood, &c., the Magnesian Limestone is largely worked to burn into lime, the lower beds, however, being used for paving-stone: going northwards its quality improves.

At Sutton-in-Ashfield there are large quarries, where the stone is got for mending roads, for building, and for lime. The eastward dip is here well seen to be about two or three degrees, and it is nowhere on an average more than five degrees. This gentle dip, together with the slope of the ground being in the same direction, accounts for a rock of such little vertical thickness covering so much surface.

At Mansfield there are some quarries of an excellent white freestone, south of the town, and others at Rock Valley and on the Chesterfield road, where the stone is of a red tint: this is probably the same bed, coloured in the latter case by an infiltration of iron. Stone from both these quarries was used for the terrace in Trafalgar Square, and Sir Gilbert Scott speaks of it as "one of the best building stones in the kingdom." The thin beds are much used for paving, cisterns and troughs being made from the thick beds. At Mansfield Woodhouse the Magnesian Limestone is described by the Geological Surveyor, Mr. Aveline, as a massive, irregular bedded, crystalline limestone, yellow, speckled with black; this quarry furnished the stone for Southwell Cathedral, the foundation and lower part of the Houses of Parliament, &c. Unfortunately all these beds of stone are very local and limited; at a comparatively short distance they alter completely in character.

The *Middle Permian Marls and Sandstones* overlie the limestone: they are worked for bricks at Bulwell, Hucknall, &c., and occupy a strip some nine miles long by one to two broad, north and south of Worksop. Above these marls is another bed of limestone, largely quarried in the north-west, near Carlton, near which place some beds of *Upper Permian Marls* also occur.

THE TRIAS, OR NEW RED SANDSTONE.—This formation constitutes the greater part of the county; its lowest division is the *Bunter Sandstone*. The bold cliff on which Nottingham Castle stands is composed of the massive, yellowish sandstone, full of pebbles, which is characteristic of the Bunter: it has been largely excavated to form caves, cellars, &c., in and near the town of Nottingham. From Nottingham it extends northwards, widening out so as to include the whole of Sherwood Forest, and reaching from Warsop and Worksop on the west to Ollerton and Retford on the east; its lower beds furnish in many places valuable moulding sand, but as a rule it forms a light sterile soil, and sections of it are rare. The pebbly beds are evidently a shore formation, as they contain "clay boulders," lumps of clay perhaps from the Permian Marls, which having been rolled about on an old beach contain embedded fragments of various rocks. Good examples of oblique lamination, or false-bedding, produced by currents, are noticeable.

The next division of the Trias in this county is the *Lower Keuper Sandstone*, or *Waterstones*: this occupies a strip on the east, by Kirklington, Wellow, Walesby, West Drayton, and East Retford. It is well seen in the brick-yards south of Ollerton, where certain blue clays and sandstones which constitute its base are exposed: footprints of *Labyrinthodon*, ripple-marks, &c., occur in the sandstones (figs. 67 and 67A).

The *Upper Keuper Marls and Sandstones*. These form a ridge some 200 or 300 feet high on the east, separating the basin of the Idle from that of the Trent. As the various little tributaries of the Idle, the Maun, the Meden, &c., reach this escarpment, they turn northwards, not having been able to penetrate the stiff red clays. On the east slope numerous small streams running into the Trent have cut little valleys. These Keuper Marls form a good rich corn-land. South of the Trent they spread from Thrumpton and Loughborough by Bingham, to Newark. They contain numerous bands of gypsum, which are worked at Thrumpton, Gotham, Newark, &c. Casts of cubical salt crystals, ripple marks, &c., are common; but, except a few fish-scales in the higher beds, the red marls have yielded no trace of life. This

Fig. 67.—Footprints of the *Labyrinthodon*, on a Slab of Bunter Sandstone; the slab is also traversed by sun-cracks.

Fig. 67A.—Single Footprint of *Labyrinthodon*, a reptile formerly called the *Chirotherium*; from the Bunter Beds of the Trias.

is probably due to the salt, bitter water of the great inland seas in which they were deposited. Some of the beds of marl are covered with hemispherical protuberances, having corresponding pittings in the next layer, and these have usually been described as rain-pittings. It is more probable in many cases that they are the result of a peculiar efflorescence of salt. There are several bands of interstratified sandstone of no great thickness.

A boring at Owthorpe, near the southern boundary of the county, passed through the entire thickness of the Trias, giving the following section:—

	Feet.
Lower Lias and Rhætic Beds	66
New Red Sandstone (Trias)	1,000
Coal-measures	——

THE RHÆTIC BEDS.—Of insignificant thickness, and of no practical importance, these beds are, nevertheless, of the highest interest to the geologist. It is in them, in Germany, at Watchet in Somersetshire, and in America, that teeth of the oldest known mammal have been found. Everywhere these Rhaetic beds are found between the Trias and the Lias when the junction of those formations is exposed. Till lately sections of them have been very rare except in the West of England; but in 1874 they were detected near Leicester, and at Elton, in Nottinghamshire. At this latter spot about 16 feet of light marl constitutes the bottom stratum, with some 9 or 10 feet of black shales above it. Near Stanton-on-the-Wolds the new cutting on the Nottingham and Melton railway has exposed these beds, and the Rev. A. Irving, F.G.S., has recorded their occurrence at Leadenham. Fish teeth and scales, with the characteristic shells, *Avicula contorta* and *Cardium Rhæticum*, occur in abundance.

THE LIAS.—This well-known formation scarcely enters into Nottinghamshire; it occupies a little ground in the south-east, along the Vale of Belvoir, and to the east of Newark. Only the lower beds occur, and these contain bands of limestone, which when burnt yield a cement valuable for its property of setting under water. This is probably due to its rather argillaceous character, and to the fact that it contains a little carbonate of magnesia. It is quarried for this purpose near Barnston, &c. Bones of the large extinct saurians, *Ichthyosaurus* and *Plesiosaurus*, occur plentifully near Cortlingstock; *Ammonites*, too, abound. When seen in quarries the face of the rock has a banded appearance, owing to the alternation of beds of dark shale and light-coloured limestone.

THE DRIFT.—Where the Bunter Sandstone occupies the surface it is difficult to distinguish between its pebbly beds and those of the latest glacial epoch. The tops of the hills and the low grounds usually have a covering of Drift gravel or clay, but this has been washed off the slopes and carried down into the lower part of the river valleys. From Robin Hood's Hills southwards, over Annesley Park, the Drift lies thick, and contains numerous fragments of foreign rocks. The so-called Druidical remains near Blidworth are cemented masses of Drift gravel, which now stand out boldly, the loose unconsolidated surrounding portions having been carried away by denudation.

ALLUVIUM.—The river gravels of the Trent stretch from one to two miles on either side of that river, marking the extent to which it formerly wandered to and fro when unchecked by the devices of man. These gravels are mostly derived from the Drift, and are composed of a great variety of rocks from all parts of the area drained by the Trent. The river Idle has also deposited much mud on the low flat lands on either side of its present bed.

No. 29.
GEOLOGY OF OXFORDSHIRE.

NATURAL HISTORY AND SCIENTIFIC SOCIETIES.
Ashmolean Society ; Oxford. Transactions and Proceedings.
Oxford Natural History Society.
Banbury Natural History Society and Field Club.

MUSEUMS.
The Ashmolean Museum, Oxford.
The Bodleian Library and Museum, Oxford.
The New University Museum, Oxford.
Museum at Wadham College, Oxford.

PUBLICATIONS OF THE GEOLOGICAL SURVEY.
Maps.—Sheets: 7, Wendover, High Wycombe; 13, Oxford, Reading, Wantage; 34, Fairford; 44, Burford. Quarter Sheets: 45 N.W. Banbury, Deddington, Chipping Norton; 45 S.W. Woodstock, Witney; 45 N.E. Buckingham, Brackley; 45 S.E. Bicester, Brill; 53 S.E. Towcester, Daventry, Weedon; 53 S.W. Southam, Kingston.
Books.—The Geology of Parts of Oxfordshire and Berkshire, by Hull and Whitaker, 3s. Geology of Country round Banbury, Woodstock, Bicester, and Buckingham, A. H. Green, 2s. Geology of the country round Woodstock, by E. Hull, 1s. Geology of the London Basin, by W. Whitaker, 13s.

IMPORTANT WORKS OR PAPERS ON LOCAL GEOLOGY.
1859. Hull, E.— South-easterly Attenuation of the Lower Secondary Rocks, and Probable Depth of the Coal-measures under Oxfordshire. Journ. Geol. Soc., vol. xvi. p. 63.
1859. Phillips, Professor J.- Sections of the Strata near Oxford. Journ. Geol. Soc., vol. xvi. pp. 115, 307.
1861. Whitaker, W.— On the Chalk Rock in Oxfordshire. Journ. Geol. Soc., vol. xvii. p. 166.
1871. Phillips, Professor J.— Geology of Oxford and the Thames Valley, 21s.
See also *General Lists, p.* xxv.

THE rocks of this county have been well studied, both by those men who laid the foundations of the science of geology in the first half of the present century; and by their successors of to-day. Professor Buckland, who was rector of Islip (where he is buried) and Professor of Geology in Oxford University, wrote several papers on the subject. With him may be named Fitton, Strickland, and Ogilby. The late Professor J. Phillips, of Oxford, gave us a capital work in his "Geology of Oxford and the Thames Valley." The Government geological maps of the county were published between 1859 and 1863, the work having been chiefly done by Messrs. Hull and Whitaker, and descriptive memoirs by these gentlemen and by Mr. Green have since been issued. Large collections of fossils were made by Mr. Whiteaves, and Professor Owen has frequently described the huge reptiles whose bones are not uncommon in the oolitic strata. The scenery of the central portion of Oxfordshire, where clays predominate, is somewhat dull and flat, but in the north-west, and in the chalk ridge which occupies the south-east corner, we get some respectable elevations.

The oldest rocks are found in the extreme north-west, and as we walk thence towards the chalk, we are continually passing over newer and newer beds which lie one upon the other, all having a general slant or dip towards the south-east. If on the contrary we walk towards the north-east or south-west, we can often manage to keep on one bed of rock all the way, because the different formations extend or strike in long irregular lines in that direction.

By a deep boring lately made near Burford, the interesting fact was ascertained that Coal-measures underlie Oxfordshire, at a depth, in this spot, of 1,184 feet; the cores brought up were full of fossil ferns.

THE LIAS.—Commencing in the north-west we find the eastern portion of the Vale of Moreton to be composed of bluish clays—the *Lower Lias*—which also reach along the valley of the Cherwell as far as Steeple Aston, and along either side of the river Evenlode to near Charlbury: these Lower Lias clays contain a bed of hard shelly limestone called Banbury Marble, which takes a fair polish, and is worked into chimney-pieces, &c.

The Middle Lias or *Marlstone* forms a prominent ridge overlooking this flat country, and ascending it we find ourselves on a plateau of the rock, which is a sandy limestone tinged red by oxide of iron, extending for 5 or 6 miles north-west and south of Banbury: this hard top band is termed the Rock-bed. The fossils found in the beds beneath, which are usually sandy clays 40 or 50 feet thick, show that they belong to the Marlstone, but from want of sections it is almost impossible to separate them from the great mass of Lower Lias clays beneath. The Rock-bed is largely worked for building and for road-metal; when dug at any depth, or under a covering of clay, it is a hard bluish limestone; where it has been much exposed to atmospheric influences, it has weathered brown, and has become comparatively friable. In this state it yields brown hematite—a valuable iron ore which has been largely worked at Adderbury, north of Deddington: here 36,808 tons were dug in 1874, valued at £7,721; but in 1879 only 1,233 tons, valued at £216, were obtained. At Fawler, Swalcliffe, and several other points the ore is also of good quality, and will doubtless be worked at a future day. From Burford to Charlbury the thickness of the Rock bed varies from 10 to 20 feet. Fossils are common, such shells as *Rhynchonella* and *Terebratula* forming masses called "jacks" by the workmen.

Upper Lias Clays.—These are bluish clays which once everywhere covered up the Marlstone; they have been so denuded, however, as now only to exist in narrow strips, and in detached patches called outliers. As we follow them along a narrow strip on each side of the Valley of the Evenlode they become thinner and thinner, until between Charlbury and Stonesfield they entirely disappear. It results from this that a boring, say near Oxford, might possibly reach coal at a depth of less than 2,000 feet. In sections near Banbury the beds are very fossiliferous, containing numerous *Ammonites*, &c. This district has been well worked by Messrs. T. Beesley and E. A. Walford.

THE OOLITE.— In North Oxfordshire we find, resting upon the Upper Lias Clays, some sandy beds, here changing into a white oolitic freestone, as at Brailes Hill, or into ironstone, as at Mine Hill. Shenlow Hill, which rises to a height of 836 feet, is capped with these sands, as is also Epwell Hill, a little south of it. South of these points the sands spread out considerably, covering the Rollwright Ridge, and extending southwards to Chipping Norton, and eastwards to Dunstew and Steeple Aston. The term *Northampton Sand* has been applied to these beds, but it is very important to understand that here they really include two formations, the lower part containing marine shells and belonging to the *Inferior Oolite*, and the upper portion almost destitute of fossils, except wood and plant markings, being of *Great Oolite* age. In fact in North Northamptonshire and Lincolnshire we have a thick bed of limestone between these two sandy beds which have here run together, and are almost undistinguishable. *Ammonites Murchisonæ* is a very characteristic fossil of the lower division.

Great or Bath Oolite. At the base of this well-known series occurs the "Stonesfield Slate," which is a laminated sandstone, splitting readily along the bedding planes into slabs thin enough to be used for roofing. The working of the slates is very expensive, and they are now only used for architectural effect. They have produced a large number of fossils, indicating their formation in shallow water. *Trigonia impressa* is the commonest shell, but the most remarkable fossils are the lower jaws of four small species of mammals, which are the oldest known except those in the Upper Trias. In similar beds at Sarsden, the Earl of Ducie obtained the lower jaw of a *pterodactyle* - an extinct species of flying reptile. Many sections of beds belonging to the lower part of the Great Oolite are exposed on both sides of the Valleys of the Evenlode and the Dorne. At Tainton, near Burford, these beds have been extensively quarried, and the oldest buildings in Oxford are of stone obtained from this spot, which also furnished the material for Blenheim Palace, and for the inside of St. Paul's Cathedral.

The *upper* part of the Great Oolite is in Oxfordshire a thick-bedded white limestone, compact or marly, and about 20 or 30 feet in thickness. Entering the county near Westwell it passes by Burford and Wychwood Forest and spreads out into a flat plateau on which stand Ditchley Park, Kiddington, Wootton, &c. Crossing the Cherwell the outcrop is narrowed by that stream; good sections are exposed in the quarries at Enslow Bridge, where bones of saurians, especially *Teleosaurus*, are not uncommon; thence it passes northwards by Stoke Lyne, Cottisford, and Mixbury. The total thickness of the Great Oolite beds in this District is about 200 feet. Fossils are numerous, about 150 species being recorded, thick beds of the shell known as *Terebratula maxillata* being frequently exposed in quarries, as at Enslow Bridge.

Forest Marble. This term is applied to the hard flaggy limestones, much ripple-marked and false-bedded, which compose the larger portion of Wychwood Forest. They are quarried for wall building and flags, and are often composed of broken oyster-shells cemented together by carbonate of lime. This formation is very fragmentary and occupies more ground as outliers than along the outcrop; in the Forest it is 25 feet thick. In the large outlier at Blenheim Park it has become reduced to 15 feet, and then south of Tackley seems to die out altogether. It again appears, however, on the left bank of the Cherwell, opposite Enslow Bridge: here it is clayey and is worked for brick-making. At Middleton, Stoney, and Bucknell the Forest Marble can also be recognised, but then thins away rapidly; it forms a poor soil.

Cornbrash. This is a rubbly cream-coloured or brown band of limestone about 10 or 15 feet thick, which is as constant in its occurrence as the Forest Marble is inconstant. From the village of Broughton Poggs in the south-west we can trace it by Broadwell and Kencott to Norton Bridge, where there are good sections. Passing round Witney the outcrop narrows from above a mile to about 200 yards or so. At Bladon it broadens out to Woodstock and Shipton; hereabouts the Cornbrash is very fossiliferous. From Middleton Stoney to Bicester this limestone forms a broad plateau more than two miles wide, and then from Fringford and Coddington it circles round to Buckingham: it forms a "brashy" loose soil of a reddish-brown hue, and well adapted for the growth of wheat. A curious row of inliers occur at Islip, Charlton, Merton, and BlackthornHill: these are dome-shaped masses, rising out of the dull plain of Oxford Clay, and it is worth noticing how they have all been seized upon as sites for villages. *Avicula echinata* is a common fossil.

The Oxford Clay. The clunch, as William Smith called it, is a very thick blue clay which weathers yellow at the surface, and covers a broad tract of country, the scenery of which is remarkably dull and uninteresting. Entering near Lechlade it forms all the district between Bampton, Ducklington, Church Handborough and Bletchington on the north, and the Thames on the south. Then extending from Kidlington to Oxford we can follow it due north-east over the dreary ground of Ot Moor, past Launton and Piddington. There are some outliers on Wychwood Forest and one at Combe: at Oxford it is 600 feet thick. Sections are rare and fossils scarce. Near Bicester it is dug for bricks. *Gryphæa dilatata* is a characteristic fossil. Everywhere the Oxford Clay forms cold stiff land, difficult to cultivate and usually in pasture.

Coral Rag.—The term "rag" is applied by workmen to any stone which is largely composed of shells, corals, &c., bound together by a calcareo-siliceous or other cement. We can trace such beds from Sandford-on-the-Thames passing east of Cowley to Headington, Elsfield, Stanton St. John, and thence to Wheatley; from this point the sands and limestones die out and appear to be replaced by beds of clay. At Headington Hill, near Oxford, there are large quarries in this Coralline Oolite, and the stone has been much used in that city. It is not very durable, however, and requires to be laid as in the quarry, so that the bedding-planes are not exposed; Wadham College is perhaps the best example of a building constructed of this stone. The lower division, or Calcareous Grit, is about 70 feet thick, and the upper part, or Coral Rag proper, about 50 feet; the latter is largely formed of corals, *Thecosmilia* and *Isastrea* being very numerous.

Kimmeridge Clay.—This is a very stiff unctuous shaly clay, with occasional bands of limestone nodules. It is well seen in the brick-pits near Headington, and its thickness in Oxfordshire is about 100 feet; it forms the vales of pasture land between Sandford, Toot Baldon, Cuddesdon, and Waterperry. Crystals of selenite, called "fossil water" by the workmen, are of frequent occurrence. *Ostrea deltoidea* is very common. At Cumnor Hurst, near Oxford, the bones of a large extinct species of reptile called the *Iguanodon*, were found in 1879 in a sandy layer near the top of the Kimmeridge Clay.

Portland Stone and Sand.—The lower part is formed of brownish sands which contain masses of grit of strange forms, such as those seen on the old London road in going from Oxford up Shotover Hill. These beds make no great show in Oxfordshire, although they are from 50 to 90 feet thick, being frequently concealed by the overlap of the Cretaceous series. The Portland stone is a white limestone, which is worked at Garsington. At Haseley the stone is about 8 feet thick. Similar beds occur at Cuddesden, Great and Little Milton, and east of Thame.

Purbeck Beds 4 feet thick are said to occur on Shotover Hill.

THE CRETACEOUS FORMATION.—Of the *Lower Greensand* a considerable patch extends from Culham to Burcott, and thence north to Nuneham Courtney and east to Chiselhampton; it is there covered by the overlap of the Gault, but reappears at Great Haseley and Albury. Outliers cap Shotover Hill. These beds are variegated sands containing much siliceous iron ore towards the top and about 80 feet thick. Freshwater shells, as *Unio* and *Paludina*, occur here.

The Gault is a pale blue clay about 200 feet thick. We can trace it as a level tract on the left bank of the Thames at Dorchester and Warborough, and past Studhampton, Chalgrove, Rycote, Tetsworth, and Sydenham, the outcrop being from 1½ to 4 miles wide; it is dug for brick and tile making.

Upper Greensand.—This overlies the Gault, and forms a band about a mile wide from Crowmarsh and Bensington to Cuxham, South Weston, Chinnor, and Henton; the lower part is whitish with dark clayey greensand above.

The Chalk.—The junction of the Lower Chalk with the Greensand is well marked by a line of springs, which have determined the sites of numerous villages, as Mongewell, Pyrton, Shirburn, Lewknor, Aston Rowant, Crowel, Chinnor, and Bledlow. The *Lower Chalk* is a white earthy limestone, almost destitute of flints: the *Chalk Rock* is the term applied to a hard band at the top which constitutes the crest of the Chiltern Hills. Beyond this, to the east, as far as the county extends, we have the *Upper Chalk* with flints. There are some good sections near Great Marlow.

TERTIARY PERIOD. —A few outlying fragments of EOCENE age are found scattered over the chalk downs, proving that they once covered them all over. At Binfield Heath and Emmir Green the vestiges of the *Reading Beds* cover more than a square mile, and in the latter place a little *London Clay* is to be seen in the brickyard. Similar outliers are found on Turville Common, Maidengrove Common, Nettlebed, Woodcot Common, &c. These are mostly sandy, but at Nettlebed there is about 40 feet of greenish-white sandy clay, very valuable for making drain-pipes, &c.

THE DRIFT.—The south-west of Oxfordshire is tolerably free from those surface masses of clay and sand, which interfere greatly in the northern and eastern divisions with the work of the geological surveyor. In gravel-pits we see that the drift, as it is termed, is mainly composed of fragments of Liassic and Oolitic rocks, broken up and mixed together, probably by the action of glaciers or floating icebergs. Through this *débris*, which partly filled up the ancient valleys, the present rivers have often cut their way, and the present river-gravels are usually composed of re-assorted drift pebbles.

Alluvial Deposits. - The rivers have also deposited much mud, or alluvium, on either side of their banks, forming the rich green water-meadows of the Thames and the Cherwell. These river accumulations are from a mile to a mile and a half broad in the neighbourhood of Oxford. The Windrush has also formed similar beds, averaging three quarters of a mile wide, from Witney to its junction with the Isis.

PREHISTORIC MAN. Implements fashioned out of stone are of not infrequent occurrence; they are met with in the surface soil or in ancient interments or encampments. Flint arrow-heads have been found at Stanlake, Abingdon, and by Colonel Lane Fox in an old earthwork on Callow Hill. A stone celt or axe-head found at Eynsham is in the collection of Mr. John Evans; he describes it as a short thick specimen, 4½ inches long, which is also the length of a similar specimen from Abingdon; they are both polished, and the latter has a facet on the edge, having been reground to sharpen it. Another, made of greenstone, is recorded from Stanlake. A celt is said to have been found with some Roman remains at Alchester, but the association was probably accidental. These stone axes or adzes, or chisels, seem to have been set in sockets, which, however, are rarely met with. One such socket or handle formed of the horn of the red deer is said to have been found with human remains and pottery of an early character at Cockshott Hill, in Wychwood Forest. A hammer-stone formed of a quartzite pebble was obtained by Colonel Lane Fox within an ancient earthwork at Dorchester, and a sharpening or grinding stone was found at Burcott in the same parish in 1835, together with a stone celt. Flint flakes or knives are abundant on Callow Hill, and one, of which the edges were serrated so as to act as a saw, was found at Brighthampton. From Callow Hill a flint scraper for dressing hides is also to be mentioned, and a fragment of a spear-head from Dorchester Dykes. A diligent look-out would lead to many more discoveries of these old stone tools, which are the only remaining traces of savage tribes who inhabited this country in pre-historic times.

No. 30.

GEOLOGY OF RUTLAND.

PUBLICATIONS OF THE GEOLOGICAL SURVEY.

Map.—Sheet 64.
Memoir.—On the Geology of Rutland, &c., by Professor J. W. Judd, 12s. 6d.

IMPORTANT WORKS OR PAPERS ON LOCAL GEOLOGY.

List (by W. Whitaker) of 110 works on the Geology of Rutland, and Parts of the adjoining Counties, in the Survey Memoir by Professor Judd.
1836. Fitton, Dr. W. H.—Strata between the Chalk and Oxford Oolite in South-east of England (Chalk at Ridlington).
1839. Delabeche, H. T. and Smith.—Report on Selection of Stone for New Houses of Parliament. Blue Book.
1873. Sharpe S.--Oolites of Morcot, Ketton, &c. Journ. Geol. Soc., vol. xxix. p. 238.
1876. Harrison, W. J.—Geology of Rutland.
    See also General Lists, p. xxv.

WITH the exception of a few papers by Professor Morris, the Rev. P. B. Brodie, and Dr. Porter, but little had been published concerning the geology of this county till the year 1872, when an admirable geological map of the district, by Professor J. W. Judd, was issued by the Government Geological Survey. This was followed by a descriptive memoir by the same author, who has treated the subject so thoroughly and fully as to leave future workers in the district but little employment except the filling in of details. To Mr. Judd's work we must refer all who wish for full and minute particulars, only professing here to give a brief sketch of the structure of the county. In studying the rocks of any district, we must be careful not to confuse the irregular beds of drift-clay and gravel, which may or may not be scattered over the surface, with the true subjacent beds. In the geological maps the Drift has not hitherto been coloured, but in a separate series now being prepared it is specially marked. To agriculturists its presence is very important, as of course where present it determines the character of the surface soils. Leaving this "Drift" for after consideration, we shall examine first the stratified rocks of the district. No igneous or metamorphic rocks, such as slate or granite, are found in Rutland. The whole county is composed of Liassic and Oolitic strata, two series which are often coupled, and described as the Jurassic system, from their development and exposure in the Jura mountains of Switzerland. They form a remarkably regular alternation of clays and limestones, and stretch across England from the Yorkshire coast to Dorsetshire. The total thickness of the various beds exposed in Rutlandshire is about 450 feet, and were they now horizontal as when deposited on the sea-bottom, it is evident we could know little or nothing about the lower members. The same tilting, however, as described in the case of Leicestershire, has caused the strata to dip eastwards about 1 in 120, or 44 feet per mile.

As might be expected, the harder strata, mostly limestones containing iron, form hills and escarpments, whilst the clay beds form the slopes of valleys. The name of the county, Rud, or red-land, is derived from the colour of the ferruginous limestones, which also tinge the slopes with their down-wash. Traces of former workings for iron are frequent, the masses of slag which often cover the surface of newly-cleared fields being very striking. This industry died away with the destruction of the forests and consequent lack of charcoal, but is once again being revived.

THE LIASSIC FORMATION.—The oldest rocks found in Rutland belong to the

Lower Lias.—These occur in the north-west of the county, near Whissendine, where the banks of the river Eye, the railway cuttings, and new brickyard, afford sections of clays with nodules containing ammonites of this age.

Middle Lias.—The Marlstone Rock-bed forms productive corn land about Oakham and elsewhere, the clays above and below it being in pasture. It is here 8 or 9 feet thick, and contains numerous fossils, as *Rhynchonella, Terebratula,* and *Belemnites.* The railway from Luffenham to Melton runs through the fertile vale of Catmos, whose bottom is formed of this marlstone, while the sides are long slopes of Upper Lias Clay, capped on the east by oolites. About Teigh the Rock-bed forms a bold escarpment. Eastwards it is exposed in the valley of the Gwash at Braunston, and in the valley of the Chater near Withcote, projecting like a shelf on the sides of the valleys.

The Upper Lias is in this district a bed of clay about 200 feet thick, which is often dug for bricks. It occurs chiefly in the west of the county from Edmundthorpe to Caldecot, and, owing to its great local thickness, the scenery about here is much bolder and more picturesque than where the same beds occur further south, in Northamptonshire. It is indeed comparable in many parts to that of the Cottesswolds. The name Lias is a corruption of the word "layers," as pronounced by workmen in describing the alternate bands of limestone and shale which so frequently occur in this formation.

THE OOLITIC FORMATION.—The word oolite is from the Greek "*oon,*" an egg, in allusion to the rounded grains of which so many rocks of this age appear to be composed. Its lowest member in Rutland is the *Northampton Sand.* In some places this is merely a white shelly sand, but the lower part has usually been converted into a rich ironstone rock, which when dug in deep wells is of a blue or green colour, but where it is near the surface and has been exposed to the action of the weather it is of a rich brownish hue. In the former case the iron is in the condition of a carbonate, but in the latter it is in the state of an oxide. The upper white sands contain plant-markings, and have been named the *Lower Estuarine Series.*

About Uppingham the Northampton Sand is 30 feet thick. It thins out eastwards to 2 or 3 feet at Barrowden, and then disappears altogether. From the valley of the Welland near Harringworth due north to Grantham, and on to the Humber, the hard beds of ironstone of the Northampton Sand cap the steep slope of Upper Lias Clay already mentioned. The result is the bold escarpment facing westwards, known in Lincolnshire as the "Cliff," which can be traced for fully 90 miles, and on the top of which runs the celebrated Roman road known as "Ermine Street." Numerous springs issue from the junction of the Northampton Sand, with the impervious Lias Clays. It forms a rich light soil, especially adapted for the growth of spring crops. It is probably an estuarine deposit formed in the delta of a great river flowing from the west. The question of the origin of the iron in the Northampton Sand is of great interest. The ore was probably not there when the beds were first deposited. But during long ages after, water containing carbonate of iron in solution percolated slowly through the beds, depositing the carbonate of iron on the grains of sand as it did so. In that part which has since been exposed to the weather, the carbonate has been changed to an oxide of iron, and it is only where this has been effected that the ore is worth working. The surface too, has been broken up, and the lighter non-ferriferous portions carried away by rain-water, so that the ore is usually richest in iron near the top. (See fig.68.)

The Lincolnshire Oolite Limestone, which comes next above the Northampton Sand, is so named because it attains its maximum thickness and development in the county whence it takes its name. At Stamford it is 80 feet thick; southwards, it thins out, and disappears altogether near Harrington, and

Soil.
Clay.

Fine White Sand
(with thin bands of clay)

gradually passing into

Brown Sand,

Sand Rock,

Cellular Ironstone,

and finally into

Hard, Ferruginous Rock.

Clay (Upper Lias).

Fig. 68.—Stone Pit near Uppingham, on the road to Stockerston. This section shows the Northampton Sand (Inferior Oolite) in its unweathered and in its weathered condition.

eastwards, near Water Newton. Northwards it thickens towards Lincolnshire; whilst westwards it has been so denuded that we are ignorant of its original extent, though it probably stretched nearly to the river Soar. At its base, in certain localities, there are found beds of a thin sandy limestone, which, after being exposed to the action of frost, split into thin flags suitable for roofing purposes. These are the well-known Collyweston Slates; they have been especially favoured by Sir Gilbert Scott and other architects as a material for church roofs. They contain some interesting shells, as *Pinna cuneata, Pterocera Bentleyi*, &c. During their deposition the sea-bed was gradually sinking, for the bed now to be described—the Lincolnshire Limestone proper—is largely constituted of coral reefs and shell banks. The main mass stretches in a band, some three or four miles broad, across the east of the county, from Barrowden, Tixover, and Stamford in the south, to Thistleton and Stretton on the north. In this tract it furnishes a light soil of a reddish tint, owing to the occurrence of a band of ironstone nodules, which lies at the top. The line of outcrop is remarkably distinct, but there are outliers of this limestone at Seaton and Manton, and between Morcot and Luffenham.

In many places the "Lincolnshire Limestone" is dug to burn into lime, but it is still more largely quarried for building purposes. The Ketton quarries exhibit very fine sections. Here the upper beds of stone called "Crash," and "Grit" or "Rag," are too hard to be easily dressed, but underlying them comes the celebrated Ketton "Freestone." This is one of the best building stones in England, and was highly eulogized by the Commission appointed in 1839 to report on the best stone for the new Houses of Parliament. It consists entirely of beautifully rounded grains, cemented together by carbonate of lime. Its durability is very great, and its uniformity of structure is such that it "can be placed in any position in buildings without exhibiting any tendency to weathering." The thickness of the bed, however, is only about 3 feet. Ketton stone has been used for many of the colleges of Cambridge, at St. Dunstan's church, in Fleet Street, London, in the restoration of Peterborough and Ely cathedrals, and in most of the neighbouring towns. The new municipal buildings at Leicester are faced with it. About one hundred workmen are employed in the quarries. There are also large quarries in this limestone in the east of the county, at Clipsham and Little Casterton, whilst it is got for local use at many other spots. Specimens from the most important quarries are exhibited in the Geological Museum, Jermyn Street, London.

The Great Oolite.—In Rutland the "Lincolnshire Limestone" may be considered as forming a plateau, on which, in the east, stand detached patches of *The Great Oolite*. This, the next formation, is an alternating series of clays and limestones, of which the lowest bed, resting on the Lincolnshire Limestone, is a light-blue clay, about 30 feet thick. It is named the *Upper Estuarine Series*, from its containing bands of both fresh-water and marine fossils. It forms a cold unkindly soil, as at Luffenham Heath, the woods near Pickworth, Stretton, Empingham Wood, &c. In the Lincolnshire Limestone quarries above mentioned these clayey beds are seen on the top, and by their presence have preserved the stone beneath from the percolation of water, and so have much improved the quality. Above these clays come certain beds of marly limestone, called the *Great Oolite Limestone*. They give rise to an excellent black soil, and contain numerous fossil oysters, mostly small and smooth. They occupy a very small area in Rutland, whilst the remaining members of the Great Oolite Series, viz. the *Great Oolite Clays* and the *Cornbrash*, are still more restricted in extent, occurring, indeed, only about Luffenham Heath and Clipsham Wood. The top bed of the *Cornbrash* is a thick band composed of large rugose oysters, named *Ostrea Marshii*.

The Drift of Rutland consists of boulder clay, containing chalk-pebbles and flints in great numbers, together with masses of local rock of such size that they might readily be mistaken for outliers. Indeed, it was stated in the "Philosophical Transactions" for 1821, that chalk had been discovered at Ridlington. It was certainly a large mass of that rock transported by some huge iceberg from the eastward. This clayey drift is the "Great Chalky Boulder Clay" of Mr. Searles Wood, and was probably formed by a glacier descending from the Lincolnshire Wolds; we can trace this boulder clay southwards to the valley of the Thames; in Rutland it not unfrequently includes irregular beds of gravel and sand.

No. 31.

GEOLOGY OF SHROPSHIRE.

NATURAL HISTORY AND SCIENTIFIC SOCIETIES.
Caradoc Field Club; Shrewsbury. Transactions.
Oswestry and Welshpool Naturalists' Field Club.
Shropshire Archæological and Natural History Society.
Ludlow Natural History Society.
Severn Valley Naturalists' Field Club; Bridgnorth.

MUSEUMS.
Ludlow Museum.
The Museum of Natural History, Shrewsbury.

PUBLICATIONS OF THE GEOLOGICAL SURVEY.
Maps.—Quarter Sheets: 55 N.W. Clee Hill, Ludlow; 55 N.E. Cleobury Mortimer, Bewdley, Stourport, Titterstone, Clee Hill; 56 N.E. Clun, Knighton; 60 N.E. Oswestry; 60 S.E. Montgomery, Bishop's Castle; 61 N.E. Wellington, Shifnal; 61 S.E. Much Wenlock, Bridgnorth; 61 N.W. Shrewsbury; 61 S.W. Church Stretton; 62 N.W. Cannock Chase; 73 N.W. Whitchurch, Malpas; 73 S.W. Wem; 73 S.E. Market Drayton, Eccleshall; 73 N.E. Nantwich; 74 N.E. Llangollen, Wrexham; 74 S.E. Oswestry.

IMPORTANT WORKS OR PAPERS ON LOCAL GEOLOGY.

1862. Egerton, Sir P. de M. G.—New Species of *Pterichthys* from Farlow. Journ. Geol. Soc., vol. xviii. p. 103.
1862. Morris, Prof. J., and Roberts, G.E.—Carboniferous Limestone of Oreton and Farlow. Journ. Geol. Soc., vol. xviii. p. 94.
1863. Salter, J.W.—Tracks of Lower Silurian Crustacea. Journ. Geol. Soc., vol. xix. p. 92.
1863. Roberts, G.E., and Randall, J.—Upper Silurian Passage Beds at Linley. Journ. Geol. Soc., vol. xix. p. 229.
1865. Woodward, H.—New Genus of Eurypterida from Leintwardine. Journ. Geol. Soc., vol. xxi. p. 490.
1877. Allport, S.—Silurian Pitchstones and Perlites. Journ. Geol. Soc., vol. xxxiii. p. 449.
1877. Callaway C.—Upper Cambrian Rocks in South Shropshire. Journ. Geol. Soc., vol. xxxiii. p. 652.
1878. Callaway, C.—Quartzites of Shropshire. Journ. Geol. Soc., vol. xxxiv. p. 754.
1879. Callaway, C.—Pre-Cambrian Rocks of Shropshire. Journ. Geol. Soc., vol. xxxv. p. 643.
1881. Davidson and Maw.—Upper Silurian Rocks of Shropshire. Geol. Mag., p. 100.
1881. Smith, J., & Jones, Prof. T. R.—Silurian Entomostraca. Geol. Mag., p. 70.

See also General Lists, p. xxv.

The county of Salop is formed of a large variety of rocks, differing greatly in composition and of very dissimilar ages. These rocks have attracted the attention of many distinguished geologists, who have published books or written papers in elucidation of the structure of the district, so that our knowledge of its strata is tolerably full and complete; ample scope still remains, however, for local work, and it is especially desirable that old facts and discoveries should be re-examined with the aid of modern ideas, and appliances, as thereby many necessary corrections may be applied and additions made to the work, the main outlines of which have been so ably filled in by earlier workers in the science of geology.

In briefly reviewing the state of our knowledge of the strata of this county, it may first be mentioned that the Government Geological Survey has issued maps, coloured to show the position of each rock bed, and also horizontal sections showing the dips of the beds and vertical sections showing their thickness; these may be obtained through any bookseller, and are necessary to every one who wishes to study the subject. The early volumes of the Transactions, Journal, &c., of the Geological Society of London contain several papers, mostly on the older rocks of the county, by the Rev. J. Yates, Messrs. Murchison, Aikin, Wright and Lyell, while later volumes contain papers by Professors Ramsay and Morris, and Messrs. Roberts, Randall, Salter, Aveline, Woodward, &c. The publication of Sir R. I. Murchison's great work "Siluria," in 1839, marks an epoch in the history of geological discovery; in it the order of succession of the older rocks was first made known; the last (5th) edition of this book was published in 1872; the "Records of the Rocks," by the Rev. W. S. Symonds, tells the same story in a more popular manner. Quite lately Dr. Callaway, of Wellington, has thrown much new light on the structure of the Wrekin and the adjoining country, and Mr. Allport, of Birmingham, has done much in the same matter (Quart. Jour. Geol. Soc., vols. xxxiii., xxxiv.); Messrs. D. Jones and Scott have written on the coal-fields, which are ably described by Professor Hull in his standard work "The Coal-fields of Great Britain;" and for what we know of the Drift we are chiefly indebted to Messrs. Mackintosh, D. C. Davies, and G. Maw.

Viewing the structure of the county broadly, we see that the Severn forms a very fair boundary line between the older and the newer rocks. All the region to the south and west of that river is composed of old hard rocks, which, however, send two promontories across the stream, one near Shrewsbury, ending in Haughmond Hill, and the second and more extensive one running past the Wrekin to Lilleshall: a strip in the north-west, too, beyond Oswestry, must be classed with the older rocks: this difference in rock structure is accompanied by corresponding differences in scenery, cultivation, population and industrial occupations: the south-west region is hilly, and in parts even of a wild character, with, on the whole, a poor soil: the mines of lead, iron and coal, and the abundance of limestone, compensate for this, however, to some extent: the north and east are comparatively flat or undulating, with a richer soil, much more arable land and an agricultural population.

PRE-CAMBRIAN BEDS.—Commencing with the oldest or first formed rocks, we find these in the very centre of the county, forming the Wrekin range, and further to the south-west the hills called the Lawley and Caer Caradoc: Mr. Allport has proved by microscopical examination that the rocks which constitute these elevations are a series of beds of lava and volcanic ash, whilst Dr. Callaway has shown that they *underlie* rocks which themselves are older than any others in the county, and he consequently claims for the Wrekin the title of the oldest mountain in England (Popular Science Review, Jan., 1879): these old rocks with others which rest upon them, have been brought to the surface between great *faults* or dislocations of the strata, which can be traced running along their edges from north-east to south-west.

Pre-Cambrian Quartzite.—This is a bed of altered sandstone which rests on the volcanic rocks just described, and contains rounded fragments of their material: it is about 200 feet thick, and may be traced on the flanks of the Wrekin, Charlton Hill and Caer Caradoc: the only evidence of life which it has yielded is a worm burrow.

CAMBRIAN FORMATION.—Upon the quartzite last described we find resting a series of thin-bedded micaceous green sandstones, which by their position and fossil remains may be shown to be of the same age as the *Hollybush Sandstone* of Malvern, and the *Lingula Flags* of North Wales: upon these again are blue shales—the *Shineton Shales* of Dr. Callaway, containing numerous new fossils, such as trilobites (*Asaphellus Homfrayi*) &c.; these beds may be correlated with those at Tremadoc in North Wales: all these beds were till recently considered to be of Silurian age, and are so coloured on the maps of the Geological Survey.

The well-known Longmynd Rocks, west of Church Stretton, are also of Cambrian age: they rise to a height of 1,674 feet, and if we ascend any of the brook courses which run down their eastern sides we find they are mainly composed of laminated sandstones and grits of a purple or grey tint: they will split, though not readily, along the lines of bedding, and specimens showing ripple marks (like those to be seen any day on a sandy beach) and suncracks may readily be obtained. The late Mr. Salter discovered in these rocks many worm-burrows—little double tubes such as are made by the marine worms (*Arenicolæ*) on sandy sea shores; to the ancient ones Mr. Salter gave the name of *Arenicolites sparsus* and *A. didymus*, distinguishing in them two species: on Callow Hill the same eminent palæontologist found fragmentary remains of some crustacean: the total thickness of these Longmynd beds is immense—over 20,000 feet—and they incline at a steep angle (60°) to the west-northwest: they extend as a narrow strip northwards to beyond Shrewsbury, where they terminate in Haughmond Hill: here, as at two or three other points, small quantities of bitumen or mineral pitch are of frequent occurrence, especially in the neighbourhood of intrusive masses of greenstone.

SILURIAN FORMATION.—With the strata of this age, the name of Sir R. I. Murchison will ever be connected. In 1831 he began the study of these old rocks, then termed "grauwacke," and in Shropshire he spent much time between that year and 1839, when his great work on "The Silurian System" was published.

The Lower Llandeilo Beds or Arenig Rocks.—Going westwards from the Longmynds, we cross a valley and ascend a ridge, on whose summit there crop out some remarkable masses of quartzite, an altered sandstone; this is the Stiper Stones Ridge: it extends from Pontesbury, near Shrewsbury, in a S.S.W. direction to Snead, near Bishop's Castle, and attains a height of 1,500 feet. The rock is quarried for road-metal at Nils Hill near Pontesbury, and here ripple-marks and worm-tubes may be noted in it: it also contains a shell called *Lingula plumbea*, with trilobites. Beds of the same age form the Arenig Mountains in Merionethshire, and they are also well exposed near the town of Llandeilo in Carmarthenshire, hence their names. (Fig, 68A.)

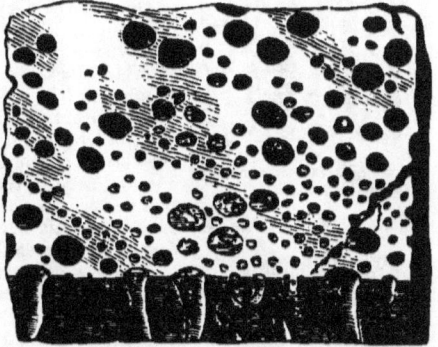

Fig. 68A.—Surface of sandy bed, with the double burrows of Annelides (Marine Worms). From the Stiper Stones and other Silurian rocks. The surface also represents, on a larger scale, the marks on the strata in the Longmynd.

The Upper Llandeilo Beds or Llandeilo Flags lie west of the Stiper Stones and form the country around and west of Shelve. Throughout the Llandeilo series, the many interstratified beds of lava and ashes tell of great volcanic activity during the period of their formation. Lead ore occurs in veins which intersect the rocks; in 1877 there were raised 6,634 tons of lead ore, which yielded 5,166 tons of lead and 5,874 ounces of silver, worth altogether about £84,086; of zinc ore, 561 tons were raised, valued at £2,192: the Snailbeach mine also yielded 142 tons of fluor spar, valued at £21: several mines also yielded barytes.

In 1879 of lead ore 6,297 tons were obtained, valued at £64,270; and of zinc ore 439 tons, value £1,662.

The Llandeilo Flags are chiefly composed of grey and black flag-stones and shales. Their thickness is about 3,000 feet. They contain some fine trilobites such as *Asaphus tyrannus* and *Ogygia Buchii*.

The Caradoc (or Bala) Beds.—These beds are found almost in the centre of Shropshire, having been brought up between the great north-east and south-west faults to which we have before alluded; we can trace them from the Craven Arms up by Ragleath, Caer Caradoc &c. to near the Wrekin; they consist of alternations of shales and sandstones. At Horderley, thick beds of freestone are seen, and Hoar Edge is formed of Caradoc grits. Good sections are exposed on the banks of the Onny, near Craven Arms Station. Fossils are common, indeed some of the sandstones contain so many that they are sufficiently calcareous to be burnt for lime; these are called "Jacob's Stones." *Strophomena grandis* is a characteristic shell. Shelly sandstones of Caradoc age also form a very small part of the north-west of the county near Llan-y-Blodwell, &c.

Upper Llandovery Beds (or May Hill Sandstone).—These consist of conglomerates and sandstones surmounted by limestone bands and purple shales, altogether about 800 feet thick. The limestones are known as the "Pentamerus beds," from the frequent presence of the shells, *Pentamerus oblongus*, *P. lens*, *P. Knightii*, &c. Rocks of this age lie very unconformably on Caradoc sandstone, between Cardington and Coalbrookdale: they fringe the south-east side of the Longmynds, and also occur north-west of Shelve.

Fig. 69.—*Syringopora verticillata*, a Silurian coral.

There is a decided break at the base of the Upper Llandovery Beds, *i.e.* it is probable that a great interval of time (represented by no deposits in this area) intervenes between them and all the rocks upon which they rest. Here then the line of division between the Upper and the Lower Silurian Rocks is drawn: at this point too the followers of Prof. Sedgwick would draw the base line of the entire Silurian system, whilst Murchison and the Geological Survey would carry it down to the bottom of the Lingula Flags; the classification adopted here is that of Sir Charles Lyell, which is a compromise between the other two.

Fig. 70.—*Syringopora compacta*, a Silurian coral.

Wenlock Beds.—The lower part of this division consists of thick shales, containing irregular concretions of limestone (Woolhope limestone) near the base. These Wenlock shales form the western slope of Wenlock Edge and the valley called Apes Dale: they are well exposed along the banks of the Severn, near Coalbrookdale and Iron Bridge, where they are known as "Die Earth," because they underlie the Coal-measures. The famous ridge called Wenlock Edge runs for 20 miles from N.E. to S.W.: it is composed of Wenlock limestone, the mass of which is rather impure and earthy, but here and there great concretionary masses of good limestone called "ball-stones" or "wool-packs" occur: these are sometimes of the diameter of 80 feet and are quarried out, leaving large cavities. At Benthall Edge, overlooking the Severn, fossils are very numerous, chiefly encrinites, corals and trilobites. Near Much Wenlock, the limestone is quarried to use as a flux for smelting iron. Of fossils, the "chain-coral," *Halysites catenularius,* the trilobite *Calymene Blumenbachii,* and the shell *Strophomena depressa* are especially common.

The Ludlow Beds.—The Lower Ludlow Beds are grey shales locally called "mudstones;" they rest on the Wenlock Limestone; their thickness is considerable, at some points as much as 1,000 feet: they have yielded a remarkable fossil—the remains of a fish, named *Pteraspis Banksii,* which was found at Church Hill near Leintwardine, by Mr. J. E. Lee, of Caerleon, in 1859: this is the oldest known vertebrate. Above these shales we get a thin bed of *Aymestrey Limestone,* sometimes 50 feet thick: it contains a brachiopod shell, *Pentamerus Knightii,* in great abundance.

The Upper Ludlow Beds are well exposed in the valley of the Teme, west of Ludlow: they contain a well-known "bone bed," a layer one or two inches thick, formed of a matted mass of bones, teeth and scales of fishes; from its appearance it is also known as "gingerbread;" it is well seen in Ludford Lane on the slopes of Whitcliffe, and at Norton, all near Ludlow. Shales and flag-stones of Ludlow age occupy a considerable area in the south-west corner of Shropshire, circling round Clun and Clun Forest.

THE OLD RED SANDSTONE.—The junction of Silurian rocks with the Old Red Sandstone is well shown near Ludlow and also in the Linley Brook, four miles north-west of Bridgnorth. Beds of sandstone called *Downton Sandstones* form the top of the Silurian formation, and are overlaid by a great thickness of red and green shales, flagstones, cornstones, and sandstones, constituting the Lower Old Red Sandstone; these beds encircle the Clee Hills, and outlying masses occupy about 100 square miles in the district of Clun; further north a small outlier forms the Forest of Hayes on the Long Mountain: the "cornstones" are concretionary limestones, and these with the red shales (coloured by oxide of iron) form a rich and fertile soil: above these come yellow and red sandstones, which are quarried at Bouldon, Bitterley Court, Abdon, Ditton, and other places; fossils are few, chiefly scales of such fishes as *Cephalaspis*: it is considered that the strata of this age were deposited in great freshwater lakes, the water of which was so charged with oxide of iron as to prevent, or at all events greatly check, the prevalence of animal life; hence the scarcity of fossils.

CARBONIFEROUS FORMATION.—As a whole, the beds of this age form an irregular disconnected band curving across the county from the north-west corner by Oswestry, Shrewsbury, and Coalbrookdale to the Forest of Wyre, and are divisible into four coal-fields corresponding with these four localities.

The Mountain or Carboniferous Limestone, which forms the natural base of the whole formation, is well developed in the north-west only; here it forms a fine ridge running south from Selattyn Hill for about eight miles, terminating in Llan-y-mynech Hill: it is here about 1,000 feet thick (containing veins of lead ore), and is overlaid by the representatives of the *Yoredale Rocks* and *Millstone Grit,* which together attain about the same thickness: then come *Coal-measures* between 2,000 and 3,000 feet thick, with seven workable seams of coal (total thickness 30 feet); there are also several bands of ironstone: a thin limestone band called *Spirorbis* limestone, because it contains a small coiled-up shell of that name, is found nearly at the top.

The Shrewsbury Coal-field lies between Haughmond Hill and Alberbury, curving southwards; another portion runs across from near Shrewsbury to Caer Caradoc. Here Coal-measures rest on Cambrian and Silurian rocks; the two or three coal-seams which occur are of little value: the *Spirorbis* limestone is found here; it is seldom more than one foot in thickness, but is very persistent.

The Coalbrookdale Coal-field has a triangular form, with its base on the Severn and its apex at Newport: the Carboniferous limestone and Millstone grit are here thin; the Coal-measures are 1,300 feet thick, with the *Spirorbis* limestone again at the top: there are six coal-seams above two feet in thickness, giving together 27 feet of coal: the western part of the coal-field is now nearly worked out, as is evidenced by the dismantled engine-houses and huge pit-banks which everywhere meet the eye: in 1870 there were raised 1,343,300 tons of coal.

The Forest of Wyre Coal Field is partly in Shropshire, and is connected with that of Coalbrookdale by a narrow band of Coal-measures running west of and parallel to the Severn: in the main mass Upper Coal-measures only occur, and these rest on the Old Red Sandstone: the *Spirorbis* limestone is present, but the coal-seams are not of great value or importance.

Two small outlying tracts of Coal-measures are perched on the summits of the Titterstone and Brown Clee Hills at a height, in the latter case, of 1,780 feet above the sea: they are each rather more than a mile in diameter, and are capped by a bed of basalt, to whose hardness they owe their preservation; this basalt issued in a melted condition from a vent on the Titterstone Clee Hill. There are two or three small collieries here whose shafts pass through the basalt to the Coal-measures below. Thin representatives of the Millstone Grit and Carboniferous Limestone crop out on the eastern side of the Titterstone Hill. Here, near Farlow, the passage beds from the Old Red Sandstone to the Carboniferous Limestone are seen in quarries; they consist of yellow sandstones and pebble-beds, with fish teeth.

The amount of coal raised in Shropshire in 1875 was 1,229,785 tons, from 64 collieries; and in 1877, 927,580 tons from 61 collieries; of ironstone (argillaceous carbonate) 270,733 tons were raised, valued at £162,439; of oil shales 2,000 tons were obtained. A bed of red marl is used in the manufacture of encaustic tiles. In 1879 the returns show a further falling off, only 854,380 tons of coal being raised, but of iron ore 300,391 tons were obtained.

THE PERMIAN FORMATION. The beds known by this name were classed as a separate formation, and named by Sir Roderick Murchison in 1841: they rest upon the Coal-measures, and form a similar band about two miles in width, which would be continuous but for the faults by which it is intersected; the best sections occur in the Enville and Bridgnorth districts; they consist of purple and red sandstones and marls, about 1,500 feet thick, containing in the middle portion some remarkable beds of breccia — angular fragments of carboniferous limestone and Silurian rocks embedded in a calcareous matrix: at Alberbury, west of Shrewsbury, these breccia beds are 400 feet thick. Professor Ramsay considers that they represent glacial deposits, and so prove the existence of a Glacial Epoch in the Permian Period. Mr. D. C. Davies has described some red Permian sandstones near Ifton, which contain thin seams of coal. With the exception of a few plant-remains, fossils are wanting.

THE TRIAS. The north and east of Shropshire is a comparatively flat country formed of soft rocks: these belong to the Triassic series, which extend northwards to form the plain of Cheshire. Two divisions are distinguishable in England in the Triassic series:—

1. *The Bunter Beds.* These consist of mottled or variegated sandstones about 600 feet thick, with thick pebble beds in the central part: such rocks are finely exposed between Bridgnorth and Shifnal, where they form the picturesque ridges of Apley Terrace, Pendlestone Rock, Abbot's Castle Hill and Kinver Edge: sweeping westward from Shifnal they run north by Market Drayton, and curve round north and west to Ellesmere. The pebbles consist of rolled quartzites.

2. *The Keuper Beds.* These have beds of breccia and conglomerate at their base: then come sandstones, which are quarried for building purposes a

Grinshill, near Shrewsbury; they also form the Hawkstone Hills: the top of the Keuper consists of red marls of considerable thickness; they extend from Whitchurch to Wem and thence round to Market Drayton.

RHÆTIC BEDS AND LIAS.—In the extreme north of Shropshire there is a well-known outlying patch of strata of this age: it is about 10 miles in length by 4 in breadth, and extends from Prees north-east into Cheshire: good sections are scarce, for the country is thickly covered with drift. The Rhætic Beds are seen in a lane cutting and brook-course near Audlem mill; they consist of buff marls surmounted by black paper-shales, containing *Avicula contorta* and other characteristic shells. The Liassic beds above are composed of blue clays with thin bands of limestone, containing *Ammonites*, *Gryphæa incurva*, and numerous other fossils.

THE DRIFT.—This is the term applied to the beds of clay, gravel, and sand, often of considerable thickness, which rest indiscriminately on the edges of the older rocks, and are, compared with them, of comparatively recent date: the north and east of Shropshire are thickly covered with drift of this nature: where best seen a triple division can be made out, viz., lower and upper stiff brown or blue clays full of angular and rounded blocks of stone of all sizes (boulder clay), and an intermediate bed of sand and gravel. The upper and lower clays are believed to have been formed by ice during the last glacial period, whose probable date Dr. Croll fixes at a quarter-of-a-million years ago: icebergs coming from the north and north-east brought granite from Scotland and the lake district, with many other hard rocks and much mud, and dropped them, by the ice melting, in the localities where they are now found. The middle gravels and sands appear to have been formed during a mild interglacial period; some deposits of white clay found in pockets in the Carboniferous limestone of the north-west may be of pre-glacial age. Our knowledge of these drift deposits is yet very incomplete: in Shropshire they contain shells and fragments of shells, and these should be diligently sought for. Old Oswestry Gravel-pit, Fox Hill field, near Lilleshall, Strethill, and indeed almost any gravel-pit or brickyard, will afford drift sections. Of the rock fragments in the drift many are of local origin; some of chalk and flint appear to have come from the eastward, whilst others must be referred to the hard Silurian and Carboniferous strata which lie on the Welsh border: their presence tells us of a former subsidence of the land in this locality, when the sea washed through the "Straits of Malvern," and the estuary of the Dee was connected with that of the Severn.

Later in formation than these glacial beds are the deposits of mud and gravel, which form the river-meadows or "hags" of the Severn and other streams.

The Wealdmoors, between the Wrekin and Lilleshall, are, according to Miss Eyton, the filled-up bed of an old lake.

PREHISTORIC MAN IN SHROPSHIRE.—Few traces of the early men to whom metals were unknown have been met with in this county: at Hardwick, near Bishop's Castle, an axe-head was found made of basalt, and perforated for a handle; its length is 10¼ inches, breadth 4¼ inches, weight 8½ pounds: a bronze celt or axe-head of fine workmanship has been found in the peat beds of Wealdmoors.

No. 32.

GEOLOGY OF SOMERSETSHIRE.

NATURAL HISTORY AND SCIENTIFIC SOCIETIES.

Bath and West of England Agricultural, &c., Society; Bath. Journal.
Bath Natural History and Antiquarian Field Club. Proceedings.
Somersetshire Archæological and Natural History Society; Taunton. Proceedings.
Royal Literary and Scientific Institution; Bath.

MUSEUMS.

Athenæum Museum, Bath.
Museum of the Royal Literary and Scientific Institution, Bath.
Frome Museum.
Langport Museum.
Museum of the Somerset Archæological and Natural History Society, Taunton.
Bristol (see Gloucestershire).

PUBLICATIONS OF THE GEOLOGICAL SURVEY.

Maps.—Sheets: 18, Crewkerne, Langport, Ilchester; 19, Wells, Glastonbury; 20, Weston-super-Mare, Bridgwater; 21, Ilminster, Taunton; 27, The Foreland; 35, Bristol Coal-field.
Books. Geology of Cornwall, Devon, and West Somerset, by De la Beche, 14s. Palæozoic Fossils of Cornwall, Devon, and West Somerset, by Professor Phillips. Geology of East Somerset and the Bristol Coal-fields, by H. B. Woodward, 18s.

IMPORTANT WORKS OR PAPERS ON LOCAL GEOLOGY.

List of 750 works on the Geology of Gloucester and Somerset, by W. Whitaker and H. B. Woodward, in the Survey Memoir on the Bristol Coal-field.
1870. Moore, C.—Mineral Veins in Carboniferous Limestone and their Fossil Contents. Report Brit. Assoc. for 1869, p. 360.
1873. Anstie, J.—The Coal-field of Somerset. 8vo. London.
1875. Tate, R. -Lias about Radstock. Journ. Geol. Soc., vol. xxxi. p. 493.
1876. Ussher, W. A. E.—Subdivisions of the Trias. Journ. Geol. Soc., vol. xxxii. p. 376.
1879. Buckman, Prof. J.—On the so-called Midford Sands. Journ. Geol. Soc., vol. xxxv. p. 736.
1879. Champernowne, A., and Ussher, W. A. E.— Palæozoic Districts of West Somerset. Journ. Geol. Soc., vol. xxxv. p. 532.
1879. Ussher, W. A. E.—Geology of Parts of Devon and West Somerset. Proc. Somerset Arch. and Nat. Hist. Soc., p. 1.
1881. Moore, C. –Abnormal Geological Deposits in the Bristol District. Q. Journ. Geol. Soc., vol. xxxvii. p. 67.

See also *General Lists, p. xxv.*

The diversified surface of the county of Somerset, its hills and vales, plains and escarpments, winding ravines and bleak moorlands, are but the expression and the result of the varied nature of the rock-masses which underlie the soil, and which have determined, not only the scenery, but also the habits and employments of the people, and the position of the towns and villages.

It was the clear, definite, and compact arrangement of the rocks in the east of the county that led William Smith, the "Father of English Geology," as long ago as 1795, to see that the strata were arranged in a definite order, bed upon bed, which was continuous over great distances; and it was while engaged in tracing their extent, relations, and contents that he came to the conclusion that "strata could be identified by their fossil remains"—*i.e.* that certain species of fossils are peculiar to certain rock-beds, so that by the aid of the former we can identify the latter.

In the long period that has since elapsed, the work of many men has given us minute and definite information about the rocks of Somersetshire; the maps of the Geological Survey, wherein the extent of each formation is indicated by a separate colour, are most useful; those of the centre and east of the county have recently been revised and improved, but the western district originally mapped by De la Beche, working as an amateur before the establishment of the Survey of which he was the worthy chief, has been left until the publication of more accurate maps by the Ordnance Survey on the scale of six inches to a mile, shall permit the puzzling and intricate rocks of this region to be mapped in a thorough and decisive manner. Of other workers we can mention but a few names; Mr. H. B. Woodward's Memoir on East Somerset and the Bristol Coal-field, gives in a masterly manner all that is known of that division, while as to the west of the county, the papers by Mr. Etheridge (Quart. Journ. Geol. Soc., vol. xxiii.) on the older rocks, and by Mr. Ussher (vol. xxxii.) on the Triassic strata, will be most useful.

In taking a brief review of the structure of Somersetshire, we note that the old or Palæozoic rocks rise to the surface both in the north-east (Mendip Hills and Bristol Coal-field), and in the west (Quantock Hills, Exmoor, &c.). The rocks of these two areas are no doubt connected underground, though as to their precise relations there may be some doubt, and the great hollow so formed is overspread and filled up by newer rocks belonging to the Secondary or Mesozoic Period, which as a rule are much softer than the old rocks they lie upon, and so form lower ground; they are not laid horizontally one upon the other, but have a general dip or inclination to the south and east, so that the lower or first-formed beds come to the surface or crop out along the shores of the Bristol Channel, and as we pass, say from Bridgwater to Yeovil, we walk across the edges of beds of rock, continually advancing to upper and newer strata.

Of altogether later date, in fact comparatively recent, are the beds of gravel, sand, peat, and mud, which at some points occupy considerable areas, and cover over and conceal the solid rocks lying beneath.

In describing the rocks which form Somersetshire, we will speak first of those which can be seen to underlie or pass under all the others. These must be the lowest or first-formed, yet by reason of subsequent elevation and superior hardness, they may now constitute the highest ground.

THE OLD RED SANDSTONE.—The most sterile portions of the Mendip Hills have a reddish stony soil, often wet and boggy. Where sections are visible we see coarse red and brown sandstones, often containing flakes of mica, mottled shales and reddish slaty clays, with beds of conglomerate formed of pebbles of quartz, jasper, and slate embedded in a red sandy matrix; the harder beds are worked for road metal and for building. In the Mendips these beds are arranged in an anticlinal curve, that is, bent like an inverted V, and dipping on one side to the north and on the other to the south, just like an ordinary roof, while the extension of the top or ridge is from west to east. From Little Elm, near Whatley, the Old Red Sandstone extends by Downhead

Common and Beacon Hill (where there are quarries) to Masberry Castle (height 979 feet); then after passing under carboniferous rocks it reappears at Pen Hill, North Hill, Egar Hill, and along the top of Black Down (1,067 feet); there are small exposures near Ebber, Cheddar, Emborrow, and Winscombe. We know of no organic remains from this formation in the Mendips.

The Old Red is also exposed in the banks of the Avon about 2 miles north-west of Clifton, whence it stretches along Sandy Lane to Windmill Hill; it also forms Portishead Down. Scales of fishes named *Coccosteus* and *Holoptychius* have been found here. It has been usual to consider the Old Red Sandstone as formed in large inland seas or lakes, whose waters were charged with the oxide of iron, which has coloured the beds red. Its thickness in this region cannot be accurately ascertained, as its base is not reached, but it may be estimated at from 4,000 to 5,000 feet.

DEVONIAN FORMATION.—We have next to describe rocks which receive their name from their great development in the county of Devon, but which extend across the boundary, and form the north-west corner of Somersetshire, and the Quantock Hills. They are considered to occupy the same position in the geological scale as the Old Red Sandstone of which we have just been speaking, that is to say, they rest upon *Silurian* strata, while above them are *Carboniferous* rocks. But these Devonian beds differ greatly from the Old Red Sandstone; they are mainly composed of slates, grits, and limestones, and their fossils, such as corals, crinoids, shells, trilobites, &c., conclusively prove them to have been deposited in the open sea. Hence the geologists of the last generation—Murchison, Sedgwick, Lonsdale, De la Beche, Phillips, and others—came to the conclusion (about 1840) that in these two great rock formations, we have a series of strata which were formed, roughly speaking, at or about the same time, but that the Old Red was deposited in great lakes or inland seas, while the Devonian rocks were being laid down in an open ocean. Hence in the Old Red we find remains of plants, fishes, and crustaceans such as would inhabit fresh water, while the Devonian fossils are all of a marine nature. In later years objections have been raised to this view, but before considering these we shall briefly describe the strata now generally included under the name Devonian.

The Palæozoic strata of North Devon and West Somerset form well-defined bands, running across the country from west to east. Looked at broadly they consist of alternations of sandstones and slates, the latter sometimes including beds of limestone. They have a southerly dip, and pass underneath the Culm-measures of the centre of Devon, rising to the surface again in the south of that county. The lowest beds are the *Foreland Sandstones*, coarse in texture and red or grey in colour; they form elevated land at Porlock, Oare, and Culborne Hills, Grabbist Hill, and North Hill; their general dip is seawards or to the north, but they roll over to the south, in which direction they are overlaid by the *Lynton Slates*, grey and gritty, and about 1,500 feet thick; these are exposed along the valley of the Lynn, and contain brachiopod shells, corals, crinoids, &c. Next come the *Hangman Grits*, red and grey in colour, and 1,500 feet thick; they form Croydon Hill, Lucott Hill, Dunkery Beacon (1,707 feet), and extend thence westwards by Black Barrow Down to the coast at Martinhoe and Trentishoe; above these we have the *Ilfracombe Beds*, satiny slates with beds of limestone, altogether 4,000 feet thick; these extend from Withycombe southwards to Nettlecombe, and then westwards by Luxborough and Exford to Ilfracombe; the limestones are quarried for limeburning, and are rich in corals, polyzoa, and brachiopods. Then we have the *Morte Slates*, quarried at Oakhampton, north of Wiveliscombe, and traceable thence by Exton to Morte Point; they are grey fissile slates, traversed by a curious series of quartz veins, unfossiliferous, and about 3,000 feet thick. In the Brendon Hills these slaty beds contain valuable deposits of spathose iron ore (chalybite, or carbonate of iron); in 1877 the amount raised was 46,894 tons, valued at £32,825. Over these lie the *Pickwell Down Sandstones*, which resemble those of the Foreland, and are perhaps 3,000 feet thick; they contain hematite (peroxide of iron), but no fossils, and form a bold chain of hills running east and west from Wiveliscombe to Dulverton and thence to the coast; they are overlaid by the *Baggy and Marwood Beds*, chiefly slates, with some sandstones; the latter contain plant-remains, but a large shell (*Cucullea*) is common in the slates. Lastly, the *Pilton Beds* are grey slates, shales, and grits, with some irregular limestone beds; from Bathealton, Stawley, &c., they stretch westwards past Brushford Anstey and Molland; they have yielded trilobites *(Phacops latifrons)*, shells, &c.

The Devonian rocks of the Quantock Hills are separated from the main mass which forms Exmoor by a valley running from Taunton to Sampford Brett, in which Triassic beds hide the relations of the old rocks: a great fault or contortion must run along this valley, for the strike of the Quantock Devonians is nearly north and south, or at right angles to that of the Exmoor rocks. Red grits, probably of the Hangman series, form the north-west portion of the Quantocks, and these are overlaid to the east by the Ilfracombe series, whose limestone beds are worked at the Ashholt, Adscombe, and Stowey quarries; they take a high polish, and have been used by Lord Taunton in the construction of his mansion at Adscombe.

The late Professor Beete Jukes believed that all the above rocks which we have been describing as the DEVONIAN FORMATION were not really the equivalents in time of the Old Red Sandstone (to which, however, he believed the Foreland sandstones to belong), but were the base and part of the Carboniferous Formation: he thought, moreover, that the Pickwell Down sandstones were a repetition by means of a fault, with a downthrow to the north, of the Foreland sandstones. His views at the time they were promulgated (1866) met with little acceptance, but lately there appears to have been a disposition to consider the question as *sub judice*. When the publication of the six-inch maps allows a minute examination of the country by the Geological Survey, the dispute will probably be finally settled.

CARBONIFEROUS FORMATION.—The succession of rocks above the Old Red Sandstone is finely shown in the gorge of the Avon at Clifton. First we see the *Lower Limestone Shales*, 500 feet thick, with many fossils, especially in the "Bone-bed," a reddish conglomerate, 6 inches in thickness, which occurs near the base. Next comes the *Carboniferous, or Mountain Limestone*, 2,600 feet, a tough, semi-crystalline greyish limestone, containing chert (impure flint) and veins filled with calc-spar; remains of crinoids (sea-lilies), corals, and brachiopod shells are numerous: this rock burns to an excellent white lime, and is also used for building and as road-metal; the upper 600 feet is shaly. Nearer still to Bristol we find the *Millstone Grit*, usually a hard, close-grained grit, or quartzite, known to miners as the "Farewell Rock," because it underlies the *Coal-measures*. It has been largely quarried for building purposes in the neighbourhood of Clifton and Bristol.

The *Mountain Limestone*, with its Upper and Lower Shales, passes south and west from Clifton by Leigh Down to Clevedon, and thence north-east to Portishead. Dipping to the south, the limestone passes under later-formed rocks, and rises up again to form the main mass of the Mendip Hills, extending from Swallow Cliff, Worle Hill, and Brean Down on the coast nearly due east for 30 miles to Frome. The mass which forms Broadfield Down connects the Mendips with the Clifton area: the two little exposures at Luckington and Vobster have been brought to the surface by a combination of "nips" and faults. In the Mendips the Carboniferous Limestone rests conformably on the Old Red Sandstone, dipping with it on one side to the north and on the other to the south, thus forming an anticlinal curve, the top of which however has been worn away by the denuding agencies of rain, frost, ice, &c. The scenery of this district is well known; it is bare, bleak, and rugged, with a thin red soil, on which turnips are sometimes grown, but which is mostly in pasturage for sheep. Caverns are numerous, as at Cheddar,

Banwell, and Wookey, and have been formed by the percolation of rain-water. The numerous and picturesque ravines, of which those at Cheddar and Ebbor are so famous, were probably formed by the falling in of the roofs of such caverns. (See fig. 71.)

Cannington Park.—There has been much dispute as to the age of the small outcrop of limestone at this point, 4 miles north-west of Bridgewater: Mr. Etheridge considers it to belong to the Devonian series, but most other observers, including Messrs. H. B. Woodward, Bristow, Ussher, Tawney, Champernowne, &c., would class it with the Carboniferous Limestone: it is of a greyish-blue tint, and is frequently oolitic; it is cut off by a fault from Devonian grits, which lie south of it.

The islets called the Steep Holme and the Flat Holme, which lie in the British Channel, are entirely composed of the Mountain Limestone.

Ores of lead and zinc are of frequent occurrence in the Carboniferous Limestone of the Mendips, and have been worked from, and even before, the time of the Romans. At present the operations are almost confined to extracting lead from the refuse left by the earlier miners: in 1877 there were obtained

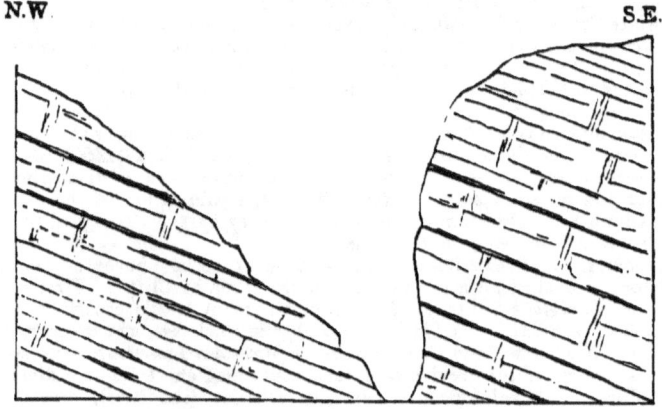

Fig. 71.—Diagram Section of the Cheddar Cliffs. a narrow winding gorge in the Carboniferous Limestone, on the south side of the Mendip Hills. (After H. B. Woodward, Geol. Survey.)

452 tons of lead ore, yielding 253 tons of lead and 1,555 ozs. of silver, of the total value of £3,514.

The Coal-measures of Somerset lie in three distinct tracts between Bristol and the Mendip Hills: two of them are of very small extent, and may be dismissed with a few words. (1.) The coal-tract of *Clapton-in-Gordano* passes northwards past Portishead, and has lately been found to extend beneath the bed of the Severn from Denny Island to Old Passage and Berkeley; (2.) the *Nailsea Basin* is a similar area, lying south-east of Clevedon.

(3.) *The Rudstock Basin.*—The Coal-measures of this district are as much as 6,000 or (according to Mr. McMurtrie) 8,000 feet in thickness: of this the Lower Coal-measures, consisting of shales, sandstones, and 26 seams of coal, include 2,500 feet; the middle member is the *Pennant Grit*, a reddish sandstone, 2,000 feet thick; while the Upper Coal-measures resemble the lower series, with 16 coal-seams, and a thickness of 2,000 feet. All the beds contain plant remains, such as ferns, gigantic club-mosses called *Sigillaria* and *Lepidodendron*, a great reed (the *Calamite*), &c., with some bivalve shells probably of fresh-water species, called *Anthracosia*, all indicating that the Coal-measures were deposited in and upon swamps and marshes, lying near and little above the level of the sea, in an area which was undergoing slow depression.

The Coal-measures only occupy the surface over a small area in the centre of the basin round Compton Dando, Publow, Pensford, Clutton, and Hallatrow; in the south they crop out in the Nettlebridge Valley east and west of Coleford, and to the north they also rise to the surface in the neighbourhood of Brislington. Everywhere else they are overlaid and concealed by newer (secondary) rocks, but the extent of the coal-seams underground is pretty well known: a line drawn from Bath to Frome would be near the eastern limit, for borings at Combe Hay and at Foxcote came upon coal-measures dipping *westwards:* in the other direction it is possible that the beds extend past Chew Stoke and south of Wrington, connecting with the Nailsea coal-field. The district just on the north side of the Mendips is greatly disturbed and faulted; some of the beds are bent into such curves that the same seam is passed through two, or even three, times by the colliery shafts: there is a "slide fault," too, by which the beds have been displaced laterally.

There appears every probability that a buried coal-field exists south of the Mendips, at a depth of from 1,000 to 1,200 feet. The various borings which have been made at Compton Dundon (depth 519 feet), Chard, High Ham near Langport, &c., have all been discontinued at depths which gave no chance of solving the question.

In 1877 there were 41 collieries in Somersetshire: of these 8 were in the Nettlebridge district, 13 round Radstock, 16 near Paulton, 3 at Nailsea, and 1 at Twerton, near Bath. The amount of coal raised was 666,500 tons; of iron-ore from the Coal-measures 1,522 tons were raised, valued at £913.

The coals of the Nettlebridge Valley are especially good for smiths' or coking coal: the Upper Coal-measures at Radstock and Farrington furnish good household coals, seams as thin as 1 foot being worked; the coal of Ashton Vale, near Bristol, is very valuable for iron-smelting. The deepest workings are at Bitton, in the Golden Valley (1,920 feet), and at Braysdown, near Radstock (1,740 feet).

PERMIAN FORMATION.—No strata of this age have been certainly identified in Somersetshire. In fact, this region was probably dry land when the Permian beds of the central and northern parts of England were being deposited.

IGNEOUS ROCKS.—As these are in this district entirely confined to the older or Palæozoic rocks, they may fittingly be described here at the close of that period.

At Downhead Common, on the Mendips, near Shepton Mallet, Mr. C. Moore detected, in 1867, an outcrop of a grey rock, which Mr. Rutley pronounces to be a *felstone*; at Stoke Lane, not far off, there is some *Pitchstone Porphyry*. In the railway cutting at Uphill a pinkish-brown earthy rock is exposed, which is a dolerite. At Wrington Warren, 8 miles south-west of Bristol, there are small exposures in Cross Comb and Cleve Comb of a *volcanic breccia*, and of decomposed *basalt*. In the Devonian beds near Adscombe, on the Quantock Hills, there is a large mass of pale-green porphyritic felstone. All these igneous rocks were forced up from below in a melted state; they have disturbed and altered (baked) the sedimentary strata in which they intrude.

SECONDARY OR MESOZOIC FORMATIONS.—All the rocks we have hitherto described rest conformably (*i.e.* are parallel) one upon the other, and they appear gradually to pass one into the other. But a great time-interval elapsed after the deposition of the Coal-measures before the formation of the newer rocks we have next to describe. In this interval great earth-movements took place; the Mendip hills, for instance, were upheaved, and the Coal-measures to a depth of several thousands of feet worn off their tops, before the first of the secondary strata were deposited on the irregular surface of denudation which resulted.

THE TRIAS.—The Triassic strata form "red ground," being coloured by peroxide of iron; they contain deposits of gypsum, celestine, and other minerals, but very few fossils, from which it has been inferred that they were deposited in extensive salt lakes. Commencing in the west, Mr. Ussher has made out the following divisions in the country between Wiveliscombe and Taunton: (1) Lower Sandstone and Breccia 1,000 feet; (2) Lower Marls, 600 feet; (3) Conglomerates, 100 feet; (4) Upper Sandstone, 530 feet; (5) Upper Marls, 1,350 feet; total, 3,580 feet. This is possibly an overestimate, but if we take off 1,000 feet we still have left a very considerable thickness. These red beds occupy the valley east of Porlock, and thence between the Brendon and Quantock Hills; then they stretch widely between Milverton, Wellington, and Taunton. As we follow the Trias eastwards we find its thickness diminish. In Mid-Somerset, from East Quantockshead by Stowey to Bridgwater and Langport, the lower beds are wanting; then north of the Polden Hills between Shepton Mallet, Wells, and Cheddar, we have the Upper Conglomerates and Marls only, while north of the Mendips even these become very thin. All this points to the fact that deposition first took place in the west, the area there being below sea-level, while the region north of the Mendips still formed dry land.

The lower sandstone beds of the west may represent the *Bunter Beds*, and certain calcareous bands at their top, well seen near Watchet, have been supposed by Mr. Blake to be of the age of the *Muschelkalk*, a limestone which is well known in Germany, but has not hitherto been detected in England. All the Upper Marls belong to the division of the Trias called the *Keuper*.

The Dolomitic Conglomerate.—Clinging to the flanks of the Mendip Hills, of Broadfield Down, and of the Downs west of Bristol, we find an old beach deposit of Keuper age consisting mainly of angular pebbles of Carboniferous Limestone cemented together by the carbonates of lime and magnesia; its thickness is seldom above 30 feet. Many towns and villages stand on this rock, as Croscombe, Wells, Westbury, Cheddar, Shipham, Blagdon, East Harptree, Wrington, and Winford. Ores of lead (galena) and zinc (calamine) occur in the dolomitic conglomerate, and have been largely worked in the past: agates are of common occurrence, and "potato-stones," which are hollow masses lined inside with crystals of either calcite or quartz.

The Keuper Red Marls are only some 200 feet thick north of the Mendips, but south of that range they rapidly thicken; near Glastonbury they may be 400 feet, at Compton Dundon 800 feet, while on the borders of Devon we have seen that Mr. Ussher estimates their thickness at no less than 1,350 feet. These Marls are red, green, and grey in colour; many orchards stand on them, and teazles, turnips, and potatoes do well; but there is also much pastureland. Formerly these beds were largely dug and used as manure, but this custom has now ceased, except where the soil is very sandy; the numerous large pits or "marly-holes" attest its former prevalence.

THE RHÆTIC OR PENARTH BEDS.—These occupy a larger surface area in Somerset than in any other county. Everywhere they are seen to form the connecting link between the Triassic and Liassic Formations. They consist of (1) buff, or tea-green marls, (2) black shales, and (3) white earthy limestones known as the White Lias. They usually contain one or two thin layers of sandstone crowded with the remains of fishes and reptiles, and known as bone-beds; their extreme thickness is only 150 feet, and often does not exceed one-third of this amount; the top bed of all is a hard, fine-grained, whitish limestone 6 to 18 inches in thickness, called the "Sun-bed," or "Jew-stone." The Cotham Marble, or Landscape Limestone is a hard bed of limestone which occurs at the top of the black shales; its tree-like markings are due to an infiltration of oxides of manganese and iron.

The Rhætic beds are finely exposed on the coast at Watchet, where in the grey marls Professor Boyd Dawkins found a tooth of *Microlestes Rhæticus*, the oldest known mammal; it was apparently a small insect-eating marsupial.

The outcrop of the Rhætic beds is so irregular, and the sections so numerous, that we can only enumerate here a few of the best exposures, as those between Newton St. Loe and Saltford, near Winford, Stanton Drew, Camerton, and Midsummer Norton, all north of the Mendips; in the railway cutting one mile west of Shepton Mallet, near Croscombe, Wells, Yarley (where the White Lias is quarried), Wedmore, along the south flank of the Polden Hills, and then south-west by High Ham and Langport towards the Blackdown Hills by whose beds the Rhætics are overlapped; 150 species of fossils have been obtained from this formation, *Avicula contorta* and *Pecten Valoniensis* are the most characteristic shells. From fissures in the Mountain Limestone at Holwell, near Frome, Mr. C. Moore has obtained a great number of fossils apparently of Rhætic and Triassic age; of one small fish alone (*Lophodus*) Mr. Moore obtained 70,000 teeth. These specimens now form part of the Moore collection in the Bath Museum.

THE LIAS.—This is in the main a clay formation. In Somerset the *Lower Lias* does not exceed 300 feet in thickness: it contains numerous bands of limestone at its base. The *Middle Lias* consists of sands and clays, with the "Rock-bed," or Marlstone (a tough irony limestone) on top; it varies from 2 or 3 to 100 feet in thickness: the *Upper Lias* is a clay deposit not more than 8 or 10 feet thick.

The Lower Lias occupies much ground in the centre of the county, from the wooded district of Ashill Forest, west of Ilminster, eastwards to Somerton, Sparkford, Evercreech, and Shepton Mallet; also round Dundry Hill, and

Fig. 72.—Section of Brent Knoll, Somerset (by H. B. Woodward, Geol. Survey).
a Cephalopoda Bed.   *b* Midford Sand.   *c* Upper Lias.   *d* Middle Lias.   *e* Lower Lias.

between Bath and the Mendips. The limestone is largely worked at Street and Keinton Mandeville; at the former place some magnificent specimens of those great extinct reptiles, the *Ichthyosaurus* and the *Plesiosaurus*, have been obtained. At Chewton Mendip and East Harptree peculiar cherty and sandy beds of Lower Lias and Rhætics occur; they are about 30 feet thick, and contains numerous fossils.

The *Middle Lias* occurs round Ilminster, South Petherton, north of Yeovil, and at Trent. From this point its outcrop is very narrow, and north of Batcomb it is so thin (a few feet only) as not to be shown on the Survey maps. Mr. Moore has obtained 183 species of fossils from the Marlstone of Ilminster, including wood, crustaceans, shells, reptiles, and fishes.

The *Upper Lias* is very thin and scarcely traceable. It includes a nodular bed called the "Fish and Insect Limestone," which has yielded many rare and perfect fossils.

THE OOLITE.—The oolitic strata form a ridge, escarpment, or line of hills bounding Somersetshire on the south-east and east. The term oolite (*oon*, an egg; *lithos*, stone) is derived from the structure of some of the limestones which are composed of rounded grains, like the roe of a fish. The lowest member is the *Midford Sands:* these are named after the little village of Midford, 3 miles south of Bath: they can be traced from this point southwards to Dunkerton, but are wanting from this point to Doulting, whence they are traced past Castle Cary, Yeovil, and Hinton St. George, to Cricket Malherbe; they cap Glastonbury Tor (500 feet above sea-level) and Brent Knoll; their thickness is from 50 to 150 feet, and they form a fertile soil, whilst many springs issue at their junction with the clayey beds of the lias beneath. The fossils of the Midford Sands include both liassic and oolitic forms, and are especially numerous in a band of brown iron-shot marly limestone called the "Cephalopoda Bed," which lies on the top of the yellow sands. Among other fossils, *Rhynchonella cynocephala* and *Ammonites Brocchii* may be named. (See fig. 72).

Inferior Oolite.—This is a buff-coloured oolitic limestone. It is exposed on the hill-sides in the neighbourhood of Bath, and an outlying mass forms the top of Dundry Hill (768 feet), which, indeed, owes its eminence to the presence of this hard, wear-resisting stone; the quarries here yield great numbers of well-preserved fossils, including some beautiful corals: its thickness here is 50 feet. Near Frome the Inferior Oolite rests on the Mountain Limestone; south of the Mendips it forms hilly ground, and yields excellent building stone, as in the well-known quarries at Doulting, Bruton, Ham Hill, Milborne Port, &c.; near Crewkerne it is broken up by many faults. Wells Cathedral and Glastonbury Abbey were built of the beautiful freestone from Doulting.

Fullers' Earth.—This is a blue and yellow clay, with a nodular band of limestone - the Fullers' Earth Rock—about the middle. Formerly it was

Fig. 73.—Section near the Bridge at Murdercombe, south of Frome. This section shows finely the unconformability of the Oolitic to the Carboniferous strata. (Delabeche; Geol. Survey.)
a Inferior Oolite.   *b* Arenaceous parting.   *c* Carboniferous Limestone.

largely dug and used for fulling at the Cloth Mills at Frome. We can trace it round the hills near Bath, and southwards by Frome, and east of Batcombe and Bruton to Bratton and Purse Caundle. Its widest outcrop is east of Crewkerne and Haselbury. It has yielded 93 species of fossils, including *Ostrea acuminata*.

The Great or Bath Oolite is a well-known, yellowish-white oolitic or shelly limestone, which forms the top of the hills round Bath, as Lansdown (813 feet above sea-level), Bathampton, Claverton Down, Odd Down, &c.; it cannot be traced more than 5 miles south of Bath. It is largely quarried for building purposes, the stone from Combe Down being considered to possess the best weathering qualities, while Farley Down yields a finer quality, better adapted for interior work.

The Bradford Clay is a pale-grey clay (30 or 40 feet thick), containing a well-known fossil, the "Pear encrinite" (*Apiocrinites rotundus*); at Charterhouse Hinton, south-east of Bath, it is seen to underlie the *Forest Marble*. This latter deposit consist of flaggy limestones, with beds of clay and grit.

It extends southward from Norton St. Philip past Frome, Wanstrow, Holton, and further south forms the ridge called Birls Hill and Abbot's Hill, which form part of the county boundary. The soil is poor, but capable of great improvement by drainage and cultivation (Buckman).

The Cornbrash enters the county at Road and Woolverton, thence it passes south by Marston Bigot, Upton Noble, Wincanton, and Horsington to Henstridge; it terminates at Closeworth and Hardington. It is an earthy limestone, about 40 feet thick, on which corn grows well.

The Oxford Clay, which comes next, forms a tract of low damp ground, 2 or 3 miles wide, from Berkley to Longleat Woods, Witham Park, Brewham, and along the valley of the Cale. It is 300 feet thick, and contains bituminous shales, which have led to trials for coal at Brewham and other places.

Coral Rag.—This only crops out at Stoke Trister and Cucklington east of Wincanton; it is surmounted by the *Kimmeridge Clay*, which occupies a narrow strip along the hill-sides west of Maiden Bradley and Alfred's Tower.

CRETACEOUS FORMATION.—In the east of Somerset the *Upper Greensand* (150 feet thick) forms the line of hills on which stands Alfred's Tower (800 feet above sea-level); it is a green or grey sand, with beds of sandstone and masses of chert. At Penzlewood are great numbers of disused pits, where the Greensand was worked for scythe-stones. *The Chalk* rises at Long Knoll, near Maiden Bradley, to a height of 948 feet. The Cretaceous beds of this tract are cut off abruptly on the south, by an east and west fault through Mere.

In the south of Somerset the Upper Greensand forms the Blackdown Hills (height from 600 to 850 feet), and it extends eastward to Combe St. Nicholas, Chard, and Cricket St. Thomas. The bottom beds here may represent the *Gault*. Outlying masses of *White Chalk* occur above the three last-named places. At Snowdown, near Chard, there is a fine section of the Chloritic Marl, a bed of yellow chalky marl with green grains and phosphatic nodules; it has yielded many fossils, including *Holaster sub-globosus* (a sea-urchin), and *Ammonites varians*.

POST-PLIOCENE OR SURFACE DEPOSITS.—These include the beds of gravel, brick-earth, silt, and peat, which occupy a large area in the centre and north of the county, as in the North and South Marshes, Burnham and Huntspill Levels, Sedgemoor, and the vale of Ilchester. The waters of the Bristol Channel extended over these districts at no very remote era—geologically speaking; but it is calculated that at any given moment the waters of the estuary of the Severn contain 700,000 tons of mud held in suspension, and the sediment so deposited formed banks, partly shutting out the sea, while rivers—the Yeo, Axe, Brue, Parrett, &c.—brought down more mud from the interior: mosses and sedges grew so thickly in the marshes as to form beds of peat, in places 15 feet thick. At Burtle there are beds of sand containing marine shells; thus about the commencement of the Christian era such places as Glastonbury, Brent Knoll, Wedmore, and Pawlet stood as islands in the middle of great bogs and marshes, presenting, indeed, a great similarity to the fen country of Lincolnshire: the Romans built sea-banks, which have been kept up ever since, and by drainage and cultivation a great tract of land has been rendered habitable.

Raised beaches are seen at Bearn Cove, north of Weston-super-Mare, and at Woodspring, indicating an elevation of land there to an extent of 20 or 30 feet; the submerged forest seen in Bridgewater Bay tells, on the other hand, a tale of depression. In the gravels which border the existing rivers, bones of mammals are of frequent occurrence. Hills of blown sand occur at several points on the coast. From the slime of the River Parret, near Bridgewater, the well-known Bath bricks are made; they are named after the person who discovered how to make them, a certain Mr. Bath of Bridgewater.

PREHISTORIC MAN IN SOMERSET.—Perhaps the earliest relics of man in this county are the tools of flint, chert, and bone found in the so-called Hyæna

Den at Wookey Hole, near Wells, which was ably explored by Professor Boyd Dawkins, between 1859 and 1863; these were associated with numerous bones of the hyæna, cave-bear, mammoth, stag, &c.

Little Solsbury Hill, near Bath, seems to have been the site of an ancient encampment of the early races to whom metals were unknown. Here Mr. Evans found flint flakes, scrapers, arrow-heads, and cores, with quartz pebbles which had been used as hammers; a flint saw, with numerous flakes and some scrapers, was found by Mr. Flower in a barrow (burial-place) at West Cranmore; Professor Buckman has an arrow-head found at Barwick, and one is also recorded from Ham Hill, near Ilchester.

No. 33.

GEOLOGY OF STAFFORDSHIRE.

Natural History and Scientific Societies.
North Staffordshire Field Club; Hanley. Annual Addresses, Papers, &c.
Dudley and Midland Geological and Scientific Society and Field Club. Proceedings.
Burton-on-Trent Natural History and Archæological Society. Annual Report.
Tamworth Natural History, Geological, and Antiquarian Society.

Museums.
Free Library and Museum, Lichfield.
Burslem Free Library and Museum.
Stoke-upon-Trent Library and Museum.
Hanley Mechanics' Institute and Museum.

Publications of the Geological Survey.
Maps. - Quarter-sheets: 54 N.W. Kidderminster, Bromsgrove; 55 N.E. Stourport, Bewdley, Forest of Wyre; 61 N.E. Wellington, Shifnal, Coalbrookdale; 61 S.E. Much Wenlock, Bridgenorth; 62 N.E. Lichfield, Tamworth; 62 N.W. Cannock Chase; 62 S.E. Sutton Coldfield, Birmingham, Coleshill; 62 S.W. Wolverhampton, Walsall, Dudley; 63 N.W. Ashby-de-la-Zouch; 71 S.W. Castle Donington, Derby; 72 N.W. Hanley, Stoke-upon-Trent; 72 N.E. Ashbourn; 72 S.W. Stafford, Stone; 72 S.E. Burton-on-Trent. Tutbury; 73 N.E. Nantwich; 73 S.E. Market Drayton. Eccleshall; 81 S.E. Buxton, Bakewell, Winster; 81 S.W. Macclesfield.

Books.—Geology of the Country round Stockport, Macclesfield, Congleton, and Leek, by Hull and Green, 4s.; South Staffordshire Coal-field, by J. B. Jukes, 3s. 6d.; Triassic and Permian Rocks of the Midland Counties. E. Hull, 5s.

Important Works or Papers on Local Geology.

1858. Hawkes, W.—Experiment on Melting and Cooling the Rowley Rag. Journ. Geol. Soc., vol. xv. p. 105.
1862. Lister, Rev. W. Drift containing Recent Shells, near Wolverhampton. Journ. Geol. Soc., vol. xviii. p. 159.
1864. Green, A. H., and E. Hull.—Millstone Grit of North Staffordshire. Journ. Geol. Soc., vol. xx. p. 242.
879. Twigg, G. H., and Beale, C.—Drift near Walsall. Midland Naturalist, vol. ii. pp. 201, 226.
1880. Bonney, Prof. T. G.— Pebbles in Bunter Beds. Geol. Mag., p. 404.
1881. Scudder, S. H.—British Carboniferous Insects. Geol. Mag., p. 293.
See also General Lists, p. xxv.

The rocks of Staffordshire are very varied in age and mineralogical composition. The district, in fact, forms a passage ground between the wild and picturesque hills found further to the west in the old rocks of Wales, and the softer newer formations which constitute the plains and gentle slopes of the eastern and southern counties of England: it contains strata representative of both these regions, but in the Coal-measures, which lie between, it has a source of mineral wealth which is peculiarly its own.

The Geological Survey officers have carefully examined the whole of the county, and their published maps, sections, and memoirs contain a fund of information relative to it which will be found of the highest value to every

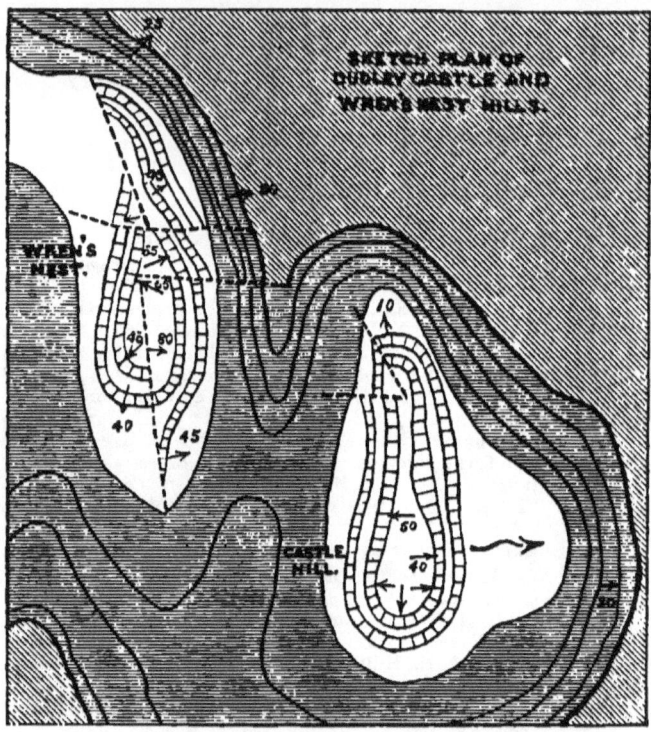

Fig. 74.—Dudley Castle and the Wren's Nest.

Dark part is Coal-measure ground, with dark lines for coal crops; the part obliquely shaded being *above* the thick coal, that horizontally shaded being *below* it.

Light part Silurian ground, with the two beds of Wenlock limestone. Dotted lines are faults. The arrows indicate the *direction* of the dip, and the figures the *amount* of dip in degrees.—*Scale, two inches to a mile.*

one interested in the subject. The Geological Society of Dudley, and the North Staffordshire Natural History Society, at Hanley, have also done good local work. There are abstracts of some good papers in the Report of the British Association Meeting at Birmingham, in 1865, and in the Journal of the Geological Society; but the contributions to the latter are hardly so important as might have been expected from a district which contains so many "practical men."

In describing the rock-beds we shall begin with those of oldest date, and gradually work our way upwards, reserving to the last the beds of clay, gravel, &c., which are scattered higgledy-piggledy over the edges of underlying rock-masses, and which are known as the Drift.

SILURIAN ROCKS.—The oldest strata in the county are shales and limestones of Upper Silurian age: these form the basis on which the Coal-measures rest, and they crop out and occupy the surface in five localities within the county, and at two points just outside its southern border.

The lowest beds are to be seen on the south-east of Hay Head, near Great Barr, where the *Llandovery Sandstone* crops out: it is here very fossiliferous; but in the Lickey Hills, between Birmingham and Bromsgrove, the same bed has been changed by heat into a compact quartzite, in which few fossils are discernible. This bed is known as the Llandovery Sandstone, because it is best seen near the little Welsh town of that name.

The Barr (or Hay Head) Limestone is the same bed as that at Woolhope, in Herefordshire: it is a valuable band, yielding an excellent hydraulic cement, and, from the small percentage of phosphorus which it contains, is especially adapted as a flux for ironstone. At the outcrop it has been so long worked as to be exhausted, but there seems no reason why it should not be largely got by underground workings. This limestone dips to the west about 10 degrees, in which direction it is overlaid by a mass of shales, perhaps 800 feet thick, which occupy the surface for a distance of 2 miles to the westward, and are known, geologically, as the *Wenlock Shales*; they are finely exposed in the new railway cutting at Walsall. At one or two localities, as the "Five Lanes," near Walsall, fossils are rather plentiful, chiefly trilobites, as *Phacops caudatus*, and such brachiopod shells as *Obolus* and *Strophomena*.

The Dudley (or Wenlock) Limestone overlies the shales just mentioned: it consists of two well-marked bands, separated by about 100 feet of shales. These run from Daw End, past Walsall, and then, dipping under the Coal-measures, they continue by Darlaston and Wednesbury, where they are worked by shafts, in a south-westerly direction, till they are again brought up, forming the dome-shaped masses known as Dudley Castle and the Wren's Nest, whilst further west they again rise to the surface at Hurst Hill. The upper or thin bed is mostly got for fluxing purposes, the lower thick band being burnt for agricultural and building operations. Perhaps no other district has such a variety of fine Upper Silurian fossils as the neighbourhood of Dudley; shells, corals, encrinites, and trilobites are found in a state of perfection such as no other locality in Britain exhibits. *Calymene Blumenbachii*, the "Dudley Locust" as it is locally called, is the commonest trilobite, but altogether several hundred species have been found, illustrated by thousands of specimens. (See fig. 74.)

The Sedgley (or Aymestry) Limestone.—This is divided from the Dudley limestone by a great thickness of shales, perhaps as much as 1,000 feet. The limestone is a dark brown nodular band, some 25 feet thick, which crops out at Sedgley Beacon, at Turner's Hill, near Lower Gornal, and at the Hayes, near Lye Waste, about 2 miles east of Stourbridge. When burnt it produces a dark-coloured lime, locally called the "Black Lime," and makes excellent mortar. Shells are frequent, a large bivalve, *Pentamerus Knightii*, being especially characteristic. Altogether, then, the Silurian rocks occupy a few detached patches only at the surface in this county, the tops of the anticlinals standing up amid the Coal-measures by reason of their superior hardness, which has enabled them to resist denudation better. But underground they form the foundation on which the Coal-measures rest, and southwards, about Wassell Grove and Manor Farm, they rise, forming an underground ridge against which the coal-seams abut, thus terminating the coal-field in that direction. This ridge probably formed part of an ancient land barrier, which stretched across Central England in Devonian and early Carboniferous times, and which consequently accounts for the absence of rocks of those ages in South Staffordshire.

THE CARBONIFEROUS FORMATION.—It is only in the north-east corner of the

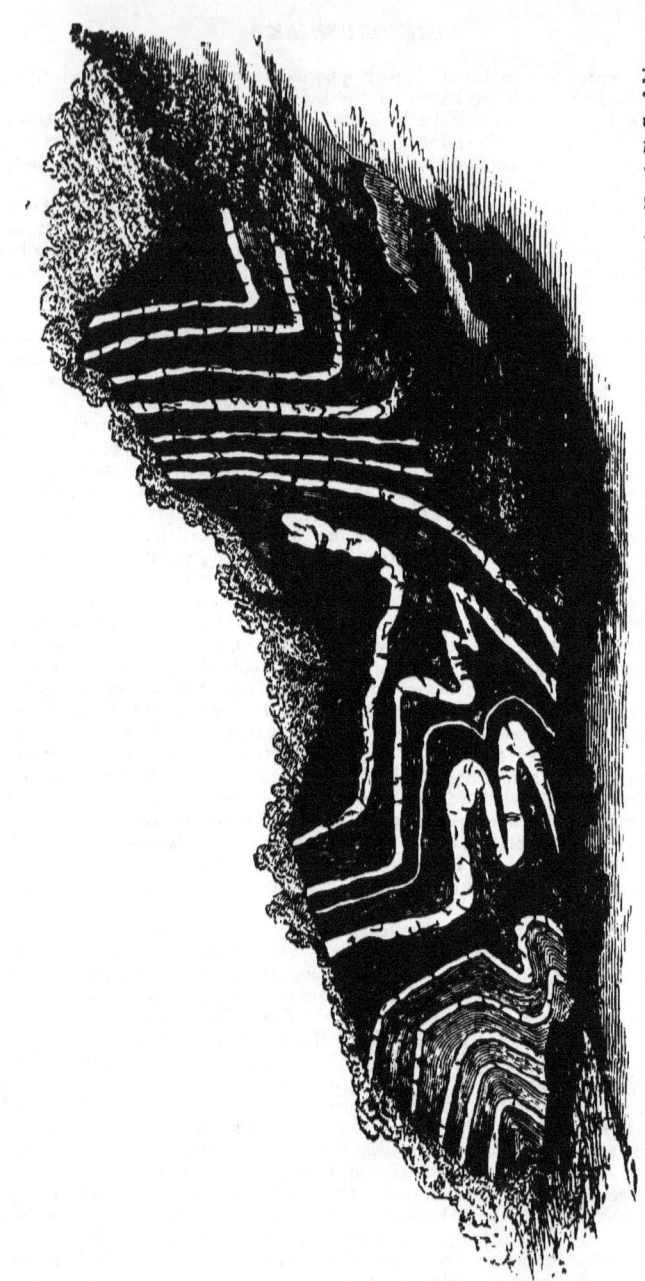

Fig 75.—Contortions in Hard Yoredale Sandstones and Shales, at Badger's Clough, between the East Mixon and West Mixon Faults, North Staffordshire. The white portions are hard sandstone beds.

county that we find the *Mountain Limestone*, which usually forms the base of the Carboniferous system: it is, in fact, an extension westwards, across the Dove, of that thick mass of rock which occupies so much of Derbyshire; and it retains all its characteristics—smoothly-topped hills covered with greensward, deep ravines, and underground watercourses, such as those through which the little river Manifold winds its way—altogether forming a comparatively wild and unfrequented region. Veins of copper and lead occur at Ecton, near Warslow, which in 1874 yielded 55 cwt. of lead ore, valued at £32, and 10 cwt. of copper ore, valued at £2; but these are not now worked. There is a small inlier of Mountain Limestone at Mixon, and the same bed is again brought to the surface just outside the county at Astbury in Cheshire, where it is got for lime. (See Fig. 15, p. 31.)

The Yoredale Rocks and Millstone Grit.—These consist of an alternation of shales with massive thick-bedded sandstones, which form bold escarpments running north and south, from the Dove on the east to Macclesfield and Congleton on the west. These beds are admirably described by Professors Hull and Green, in vol. xx. of the Geological Society's Journal. The strata are thrown into long folds, the central anticlinal line, which is traversed by a fault, stretching from Leek far to the north by Marple and Staleybridge. In the south this has on its western side one synclinal termed the Rudyerd Basin, and another on the east, the Goyt Trough. In North Yorkshire there are five beds of massive gritstone, but as we trace these southwards they gradually thin out and disappear, until in North Staffordshire only the Rough Rock (the uppermost bed) and the Third Grit remain.

THE COAL-MEASURES.—In Staffordshire, as elsewhere, the beds which contain the valuable mineral called coal consist of an alternation of shales and sandstones with certain beds of coal and ironstone: the latter, though all-important to us, constitute but a very small proportion of the total thickness, perhaps not more than one-fortieth of the whole mass. Coal is formed of compressed vegetable matter. In swamps and low flat grounds of great extent near the sea-level we must picture to ourselves great forests growing uninterruptedly for long ages, until, by the decay of many successive generations of plants, a great thickness of carbonaceous material, perhaps from 20 to 100 feet, was accumulated. Then, the land slowly sinking, the sea deposited mud or sand on this bed, and during long after-ages, by pressure and loss of gaseous material, it was compressed into an ordinary coal-seam of variable thickness. Underneath every coal-seam we find a bed of fireclay, the soil on which the ancient forest grew; and in an open work at Parkfield Colliery, near Wolverhampton, the stumps of seventy-three trees, with their roots attached, were found in the space of a quarter of an acre. The plants were mostly of low types, like our ferns, club-mosses, and horse-tails, but were, comparatively, of gigantic size. *Sigillaria*, known by the vertical seal-like markings or leaf-scars on its bark; its root, once thought to be a different plant, and named *Stigmaria;* another spirally marked tree, the *Lepidodendron*, with great reeds called *Calamites*—these, with numerous impressions of ferns, are of frequent occurrence. There are few traces of animal life, but remains of insects and reptiles have been found in the nodules of ironstone. The growth of the forests which produced coal seems to have been general and uninterrupted over a large part of England. But since that time disturbances have taken place, the strata have been upheaved along certain lines, and from those lines the Coal-measures have been washed off, so that now we get the coal-fields separated one from the other, and each lying more or less in a basin shape: thus we have in Staffordshire four detached areas or coal-fields, of which we will now speak separately.

(1.) *The Goldsitch Coal-field.*—This is a little valley or trough of Lower Coal-measures, lying in the extreme north-east of the county: it has an area of 90 acres, and contains about 117,000 tons of coal.

(2.) *The Cheadle Coal-field.*—This lies in the valley of the Churnet, and is bounded on the north by Millstone Grit, and on the south by New Red Sandstone. There are six seams of coal, containing, according to the estimate made for the Coal Commission in 1871, about 104,000,000 tons available for future use. At Froghall a valuable bed of iron ore (hydrated oxide) is largely worked.

(3.) *The North Staffordshire Coal-field.*—This fine coal-field is triangular in shape, extending from the apex near Biddulph in the north, to near Longton on the east and Madeley on the west: in the latter direction it is bounded by the great Red Rock fault, which throws down the strata on the west; so that near Astbury we have the Mountain Limestone brought on a level with New Red Sandstone, implying a displacement of about 8,000 feet. Eastwards the Millstone Grit rises in due order from beneath the Lower Coal-measures; but in the south the Upper Coal-measures themselves dip southwards, and pass

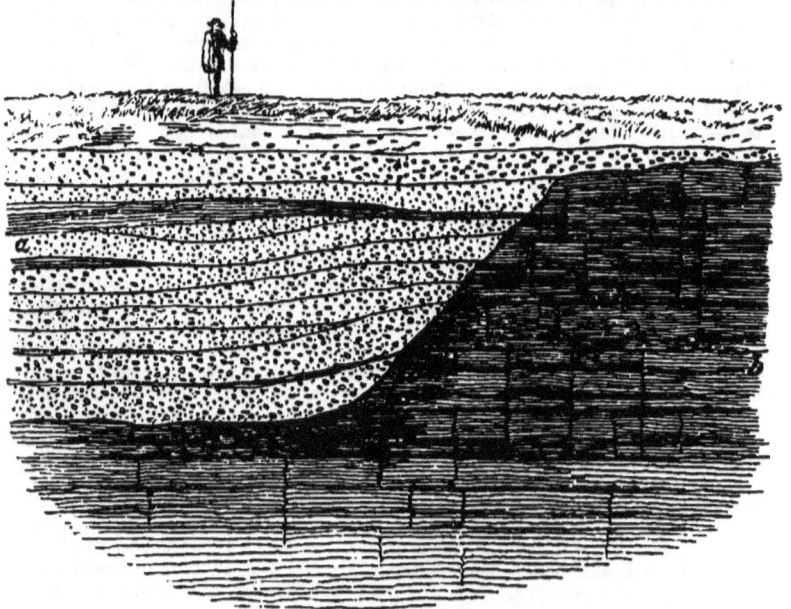

Fig. 76.—Section in railway cutting near Trindle Gate, east of Dudley, showing Lower Coal-measure Sandstones, resting unconformably on Silurian Shales.

a Coal-measures. *b* Silurian Shales.

under rocks of Permian age. Altogether, the Coal-measures of North Staffordshire are 5,000 feet thick. In the upper division, about 1,000 feet thick, there is at Fenton a band of limestone, containing a tiny curled-up shell—*Microconchus carbonarius*—which everywhere characterises these upper beds. Under the Upper Coal-measures we get about 3,000 feet of sandstones, shales, and ironstones, with about 40 seams of coal, together above 150 feet in thickness, and estimated as able to yield about 3,720 millions of tons of coal. Now, in 1874, there were 4,313,000 tons of coal raised by 156 collieries, and in 1879 there were 4,025,535 tons produced by 144 collieries; so that it is evident we have here a district of great resources; and, when we consider the possibility of its future extension towards the south, by mining through the newer overlying rocks, it is clear that there is a great future before this coal-field.

Of Lower Coal-measures we have here about 1,000 feet, containing the same band of red ironstone which is worked in the Churnet Valley. Fossils are rather numerous; scales, teeth, and spines of about twenty genera of fishes have been found. Numerous molluscan remains, too, have been collected from a bed called the "Bay-coal Bass,' and from the "Ten-foot Seam," at Hanley.

(4.) *The South Staffordshire Coal-field.*— Going southwards, we find a great reduction in the thickness of the strata. Examining the region from Brereton on the north to the Clent Hills on the south, and from Wolverhampton on the west to Walsall on the east, we find it to be an oval tract about 21 miles by 7, composed of about 800 feet of Upper Coal-measures, mostly reddish sands and clays, underneath which come 500 feet of Middle Coal-measures, resting on Silurian rocks, all the inferior carboniferous strata being absent. Of the seams worked the Brooch Coal is the highest and most persistent; it is an excellent house coal, but the supply is very limited: below this, we get about 150 feet of clay, sandstone, &c., and then we come to the famous Thick or Ten-yard Coal, composed of from ten to thirteen well-marked seams, which in the district south of Dudley rest immediately upon one another, forming a mass unequalled elsewhere in the British Isles. As we follow the Ten-yard Coal, however, to the west and north, we find it split up into separate seams, divided by beds of shale and sandstone from each other, until about Essington

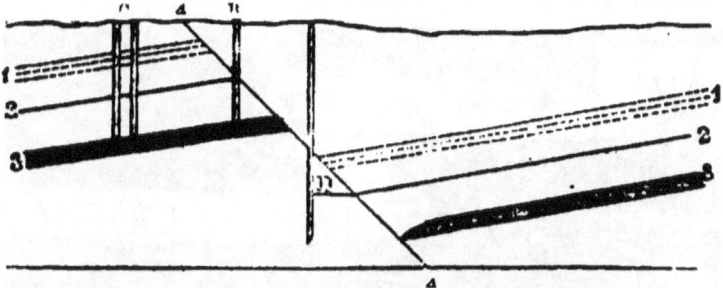

Fig. 77.—Diagram of the Corbyn's Hall Fault, South Staffordshire. (After Jukes, Geol. Survey
A B C Colliery Shafts.   1 Beds containing water.   3 Thick Coal.
D Gate Road.   2 Brooch Coal.   4 The Fault.

and Wyrley there is a total thickness of 300 feet. To explain this, we must imagine that in the Coal Period the level of the land in the south remained unaltered, allowing of the accumulation of a vast thickness of vegetable matter in one bed. Towards the north, however, depression after depression took place, with intervals of rest between. During each depression mud and sand would be deposited, and we should thus get the seams separated more widely the further we went to the north, while, following them back southwards, we should find them coalesce into one thick mass, where plant-growth had been continuous, and where the level had remained unaltered. Twenty feet below the Thick Coal we have the Heathen Coal, and below it the Sulphur Coal, the New Mine Coal, the Fire Clay Coal, and the Bottom Coal, with other variable seams.

The South Staffordshire coal-field is crossed by many *faults*, which have displaced the strata to a greater or less extent (fig. 77). In tracing and discovering the effects of these faults, geological knowledge finds one of its most useful applications.

There are several valuable bands of clay ironstone (argillaceous carbonate), which mostly occur at a little distance under the principal coal-seams. About one million tons of iron ore are raised yearly, but large quantities are imported from Northamptonshire and other districts, so that in 1874 there were 2,073 puddling furnaces and 320 rolling mills at work. In the same year 469 collieries are enumerated in the "Mineral Statistics," which raised 8,389,343 tons of coal. Of fireclay 258,792 tons were raised, mostly from Stourbridge and Tipton: in 1879 there were 425 collieries at work, producing 9,350,000 tons of coal. Thus the mineral treasures with which the county is so richly endowed are being sought for with a degree of enterprise and success which has never been surpassed. It is to be regretted, though, that in former times the task was not pursued with a due regard to economy of material, or to the interests of the future. Thus much of the "Thick Coal" which remains is "drowned out" and ungettable, though this is an evil which may yet be conquered by combination. Professor Hull calculates ("Coal Fields of Great Britain," p. 165) that there remain 205 millions of tons of coal in the southern division, and 768 millions of tons in the northern or Cannock Chase division of the South Staffordshire coal-field available for future use.

Recent discoveries, however, have shown an extension to the eastward, across the fault which bounds the coal-field in that direction. At Sandwell Park, near West Bromwich, the Brooch Coal was struck at 380 yards. Below this came the Herring Coal and the Thick Coal, the latter 20½ feet thick. The dip was east, about 8 degrees. The Heathen Coal was next proved, and at 5 yards below it was found the white ironstone, of excellent quality and large yield; a sinking at Hamstead has proved similar beds there.

Fig. 78.—"White Rock Trap," or Basalt, traversing the Thick Coal near Rowley, South Staffordshire. The igneous rock is about one foot thick; it has altered the coal—made it dull, friable, and earthy—to a distance of twelve inches.—*Scale, five feet to one inch.*

IGNEOUS ROCKS.—Eruptive masses of basalt or dolerite occupy a large area about Rowley Regis, Barrow Hill at Pensnett, Pouk Hill near Bentley, at Netherton, and other spots. Many thousands of tons per annum are got for paving purposes. There can be no doubt that these masses were injected in a fluid state from several vents, at some period after the formation of the Coal-measures. The main masses send off sheets of "greenstone," from which again veins of white-rock trap proceed, baking and altering the aqueous rocks with which they come in contact: a curious radiating and columnar structure is often observed in these trap rocks, due to their cooling under pressure. At Lye Cross, on the Rowley Hills, in 1874, a sinking for coal penetrated 60 yards of Rowley Rag, and struck all the coal-seams beneath from the Brooch to the Bottom Coal, including the Thick Coal, here 29 feet thick. (Fig. 78).

THE PERMIAN FORMATION.—Stretching irregularly round the southern terminations of both the North and South Staffordshire coal-fields, we find certain reddish sandstones, conglomerates, and breccias, of Permian age. Further south they form the Clent Hills; and the remarkable beds of polished and striated pebbles of trap, greenstone, slate, limestone, &c., are believed by Professor Ramsay to have come from the westward, perhaps from the Longmynd hills in Shropshire, and to afford evidence of glacial action in those times. Altogether the Permians are from 600 to 700 feet thick; they run, too, in a long, narrow strip on the westward, from north to south for about 10 miles, Wolverhampton being nearly in the centre. Again, in the north, we find them occupying the surface for several miles to the south of Newcastle-under-Lyme and Madeley. Many fine sections are here shown by the canals and railway cuttings.

THE TRIAS.—Red sandstones and pebble beds of Triassic age occupy the whole of the centre of the county from east to west. They are very difficult to distinguish from the Permian rocks, and great credit is due to Professor Hull for the manner in which he made clear the separate divisions. The lowest beds are called the *Bunter Sandstone*. These contain thick pebble beds, well seen in the country north of Cannock Chase, at Sutton Park, Great Barr, &c., and they form the heathy land about Swynnerton, Oulton, and south of Cheadle; the pebbles are mostly quartzites, many of which resemble the rocks round Loch Maree in the north-west of Scotland, others appear to have come from the Lickey Hills. The Bunter is an excellent source of water supply to many large towns situated upon it.

Keuper Marl and Sandstone.—This is the upper division of the Trias: it yields good building stone at Oreton Hill, Colwich, Hollington, and several other spots: its thickness is about 1,000 feet. Round Alton the scenery is very picturesque; the beds are much faulted, and a hard stratum at the base of the Keuper has resisted denudation, forming a succession of escarpments. Beds of gypsum occur near Uttoxeter, and brine springs near Stafford and Weston-on-Trent indicate the presence of salt. Ripple-markings, sun-cracks, rain-pittings, &c., often seen on the surface of the Marls, indicate that the Keuper beds were probably formed in large inland salt lakes. (Fig. 79.)

THE LIAS.—The presence of the outliers of Rhætic beds and Lower Lias at Needwood Forest, and north of Abbots Bromley, probably indicates the former extension of this formation over the greater part of the county.

THE DRIFT.—When we find irregular beds of sand or clay at the surface, containing fragments of granite, lias, oolite, and other rocks, which must have travelled from a distance, and which are often striated and grooved, we assign them to the last Glacial Period, when ice-masses from the north and west seem either to have descended as glaciers or sailed as icebergs over the whole of Central England. There is a great accumulation of travelled blocks, often of great size, in the neighbourhood of Wolverhampton; these include masses of felstone which must have come from the Arenig Mountains of North Wales, and granites like those of Eskdale in Cumberland and Criffel in the south of Scotland; boulders of the Welsh rocks lie thickly, too, between Hagley and Bromsgrove.

PRE-HISTORIC MAN.—Lastly, in the "barrows"

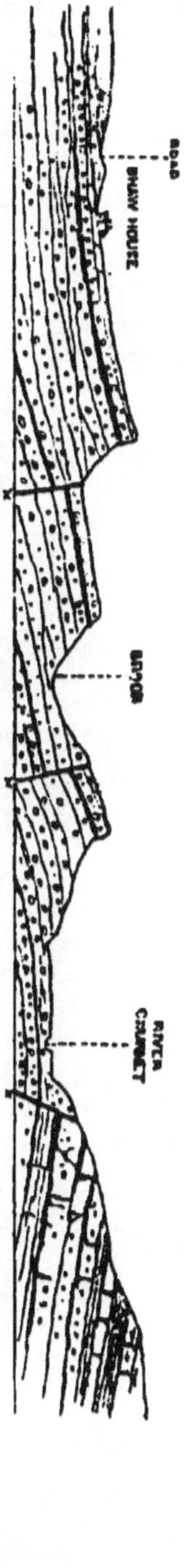

Fig. 79.—Section near Alton, Staffordshire, showing the succession of the escarpments. (After Hull, Geol. Survey.)

opened by Mr. Bateman ("Ten Years' Diggings,") we have evidence, in the flint arrow-heads, celts, and stone hammers, of the appearance of men upon the scene; men to whom the use of metals was unknown, but who had acquired considerable skill in the manufacture of their rude weapons, so that we class them as of *Neolithic* age. Of the *Palæolithic* tribes, whose implements were of the rudest description, mere chipped pebbles, no traces have yet been found in Staffordshire.

No. 34.

GEOLOGY OF SUFFOLK.

NATURAL HISTORY AND SCIENTIFIC SOCIETY.
Suffolk Institute of Archæology and Natural History; Bury St. Edmunds.

MUSEUMS.
Bury St. Edmunds Museum.
Ipswich Museum.

PUBLICATIONS OF THE GEOLOGICAL SURVEY.

Books.—Geology of the Fenland, by Skertchly, 40s. Manufacture of Gun Flints (at Brandon, &c.), by Skertchly, 17s. 6d.

IMPORTANT WORKS OR PAPERS ON LOCAL GEOLOGY.

1869. Flower, J. W.—Flint Implements in the Drift of Norfolk and Suffolk. Journ. Geol. Soc., vol. xxv. p. 449.
1870. Lankester, E. R.—Newer Tertiaries of Suffolk and their Fauna. Journ. Geol. Soc., vol. xxvi. p. 493.
1870. Gunn, Rev. J.—Relative Position of the Forest-Bed and Chillesford Clay. Journ. Geol. Soc., vol. xxvi. p. 551.
1871. Prestwich, Prof. J. -Crag Beds of Norfolk and Suffolk. Journ. Geol. Soc., vol. xxvii. pp. 115, 325, 452.
1872. Dawkins, Prof. W. B. Cervidæ of the Forest Bed. Journ. Geol. Soc., vol. xxviii. p. 405.
1874. Whitaker, W.—Thanet Beds and Crag at Sudbury. Journ. Geol. Soc., vol. xxx. p. 401.
1874. Flower, Prof. W. H.- -Skull of *Halitherium* from the Red Crag. Journ. Geol. Soc., vol. xxx. p. 1.
1876. Gunn, J.— Forest-Bed Series at Kessingland and Pakefield. Journ. Geol. Soc., vol. xxxii. p. 123.
1877. Harmer, F. W. The Kessingland Cliff Section. Journ. Geol. Soc., vol. xxxiii. p. 134.
1877. Whitaker, W.—Note on the Red Crag. Journ. Geol. Soc., vol. xxxiii. p. 122.

See also General Lists, p. xxv.

THE extensive and accurate knowledge of the rocks of this county which we now possess is the result of a long series of observations made almost entirely during the present century by the numerous talented geologists who have devoted their attention to the study of the strata in this district. Although Dale (1730), Parkinson (1811), William Smith (1816), and R. C. Taylor (1824) had specially noted the shelly deposits east of Ipswich, commonly called the "Crag," it was Mr. E. Charlesworth who, in 1835, published an important paper in the "Philosophical Transactions," classifying them in a manner which has never since been materially altered; the fossils of the Crag have been excellently described and figured by Messrs. Searles V. Wood, sen. (*Mollusca*);

Professor Busk (*Polyzoa*); Professor Duncan (*Corals*); Professor Forbes (*Echinoderms*); Rupert Jones and Parker (*Foraminifera*); Rupert Jones and Brady (*Entomostraca*), in the volumes of the Palæontographical Society. Professor Prestwich has given a minute account of the Crag deposits in the Quarterly Journal of the Geological Society (vol. xxvii.), in which also numerous papers on Suffolk geology by Messrs. G. Maw, S. V. Wood, jun., F. W. Harmer, W. Whitaker, W. H. Flower, J. Gunn, W. H. Penning, E. Ray Lankester, W. Boyd Dawkins, Rev. W. B. Clarke, Captain Alexander, Professor Henslow, Sir Charles Lyell, &c., have appeared. Excellent work has been done by Dr. J. E. Taylor, the well-known curator of the Ipswich Museum and editor of "Science Gossip," Messrs. A. and R. Bell, and others.

Fine collections of fossils have been made by Dr. Reed, of York, the Rev. H. Canham, of Waldringfield, Mr. Cavell, and many more energetic workers. The Ipswich Museum has a magnificent series, including the Canham collection, lately purchased and presented by Sir Richard Wallace.

The Government Geological Survey officers (Messrs. Whitaker, H. B. Woodward, Dalton, J. H. Blake, S. B. J. Skertchly, &c.), have now nearly completed their examination of the rocks of this county, and their maps and memoirs, which will shortly appear, will furnish a full and reliable guide to the minute geological details. A most interesting memoir, on the Gun-flint trade of Brandon, by Mr. Skertchly, has lately been published, and another, chiefly by Mr. Whitaker, giving an account of the country between Sudbury and Haverhill.

General Structure of the District.—Beginning in the extreme north-west corner of the county, we find near Mildenhall, just a corner of the Fen-land. From this level tract, which is only from 25 to 30 feet above sea-level, there rises a low range of chalk hills, the northward continuation of the Royston Downs, extending from Haverhill on the south, by Newmarket and Bury St. Edmunds to Thetford, and decreasing in height northwards. In the south-west corner near Haverhill, the chalk attains a height of 352 feet, and Haverhill Church is 225 feet above the sea. These hills have their steep slope to the west, whilst they dip gently to the south-east. In the south-east corner of Suffolk, between the sea and a line drawn from Sudbury to Aldborough, we find beds of clay and sand of the Tertiary period resting on the chalk: the height of some points (taken from the Ordnance Survey) in this region are Copdock Church, 133 feet; Belstead Church, 142 feet; Ipswich (St Mary Elms Church), 27 feet; Woodbridge Town Hall, 77 feet; Wickham Market Church, 109 feet; Saxmundham Church, 57 feet; and further north, Wangford Church, 39 feet; Lowestoft Town Hall, 73 feet; Corton Church, 63 feet.

These strata (the Chalk and Tertiaries) form what is termed the *solid geology* of the county, but they are frequently overlaid and over extensive tracts entirely concealed from observation—by the *Drift*, under which term we include all the deposits formed during the last glacial period, these being beds of boulder clay, gravel, and sand, of comparatively late geological age.

We shall now describe the different rock-beds in turn, commencing with the oldest which occurs here, viz., the chalk.

THE CRETACEOUS FORMATION.—The chalk of Suffolk forms part of the great band which extends from Salisbury Plain to the Norfolk coast. In a deep boring executed at Harwich in 1854-7, we have a key to the thickness and succession of the various divisions of the chalk: the boring was close to the harbour and just west of the Great Eastern Hotel: the beds passed through were, Earth, 10 feet; Gravel, 15 feet; London Clay, 23 feet; Reading beds, 30 feet; Chalk with Flints, 690 feet; Chalk without flints, 162 feet; Chalk marl, 38 feet; Upper Greensand, 22 feet; Gault, 39 feet; Lower Carboniferous (a hard, dark, bluish-grey, slaty rock, containing the fossil shell *Posidonomya*), penetrated to a depth of 69 feet.

Another boring, at Coombs, near Stowmarket, gave—Drift, 57 feet; Chalk, 817 feet; Upper Greensand, 10 feet; Gault, 11 feet. Other deep borings in

Herts and Middlesex show that Palæozoic rocks probably underlie the greater part of the Eastern Counties, and at depths not much exceeding 1,000 feet.

The Lower Chalk, or Chalk without Flints.—This bed forms the western slope of the low chalk escarpment; the beds below it, the chalk-marl, &c., occurring still further west, in Cambridgeshire. It is hardly correct to call it the "chalk-without-flints," as by diligent search flints may almost always be found, but in these lower beds they are comparatively scarce. Mr. Whitaker thus describes a hard bed which forms the top of this division:— "The lower chalk ends upwards in a hard crystalline bed, frequently yellow, and broken into lumps 2 or 3 inches across, with a marly substance between them; but the latter is probably a wash into the joints from the chalk above. This bed, or rather deposit divided into two crystalline beds with soft chalk between, has been described under the name of *chalk-rock*, and is tolerably persistent from Wiltshire up to, and probably far beyond, this district, though, from the scarcity of sections, the outcrop cannot be accurately traced. It contains many fossils, *Holaster planus* being somewhat characteristic, and its upper limit is sharply defined: it indicates a change in the conditions of deposit which, brief as it may or may not have been, was certainly considerable." Local geologists should look for this bed between Newmarket and Thetford; it may generally be told by its cream-colour and hardness.

The Upper Chalk.—This division includes nearly the whole of the white, soft, earthy limestone which is perhaps the best known and most easily recognised of any of the British rocks. If we draw a line from Sudbury through Ipswich to the coast near Dunwich, we shall have the south-eastern surface limit of the chalk. North and west of this line we have numerous exposures of the rock in the many pits opened to get the chalk, which is burnt for lime, and the flints, which are used for building purposes. These pits are mostly on the sides of the valleys where the streams have cut through the surface accumulations of clay, sand, &c., exposing the chalk itself; there are large openings near Bury St. Edmunds, Claydon, Cockfield, Bradfield, Sudbury, Coddenham, Welnetham, &c. In almost every pit we see the face crossed by black lines of flint nodules, and find that although there are several minor undulations, yet the beds, as a whole, incline or dip very gently to the south-east. The chalk is an excellent water-bearing formation, yielding an ample and steady supply of clear but hard water. Many wells have been sunk or bored to the chalk to get water, and these vary much in depth in the same neighbourhood, showing that the surface of the chalk is uneven; in some cases where the discrepancy is very great, it would seem that a "sand pipe" has been hit upon—this is a cylindrical hollow where the chalk has been dissolved by running water, and its place filled by the sinking in of the beds above; they may be seen on a small scale in any chalk pit.

The common chalk fossils are sea-urchins, such as *Micraster, Ananchytes, Galerites,* &c.; the shells *Inoceramus, Spondylus, Rhynchonella, Terebratula,* &c., with *Ammonites,* sponges, and occasionally teeth and scales of fishes. Chalk appears to have been formed at the bottom of a deep sea, and microscopic examination shows that the minute shells of foraminifers form the bulk of the deposit; there are also still smaller rounded bodies (coccospheres), and fragments of such (coccoliths), now known to have been calcareous sea-weeds.

TERTIARY PERIOD.—The chalk closes the list of Secondary Formations, and a great interval of time elapsed, of which we can get no information from any rocks existing in England, for there are none to be found of any age intermediate between the chalk and the tertiary strata; probably this region was elevated and remained dry land for a time.

EOCENE FORMATION.—With strata of this age, we enter on an epoch when the life existing on the earth began to be more closely allied to the animals and plants now living. Large mammals appear, and a few species of the fossil shells are identical with some now living. The lowest Eocene strata are—

The Thanet Beds. —These are only well exposed round Sudbury; there we see about 2 feet of clayey green sand resting on the chalk, with a few green-coated flints at the bottom, and above this about 12 feet of grey sand, the lower part pinkish; no fossils occur. These beds can be traced with difficulty a little north of Hadleigh, and at Ipswich. (See Fig. 80.)

Reading Beds. These are mottled clays and sands, of no great thickness. They are worked for bricks at the St. Helens pits, Ipswich, Higham Bridge, Bramford, Great Cornard, Copdock, and several other points between Ipswich

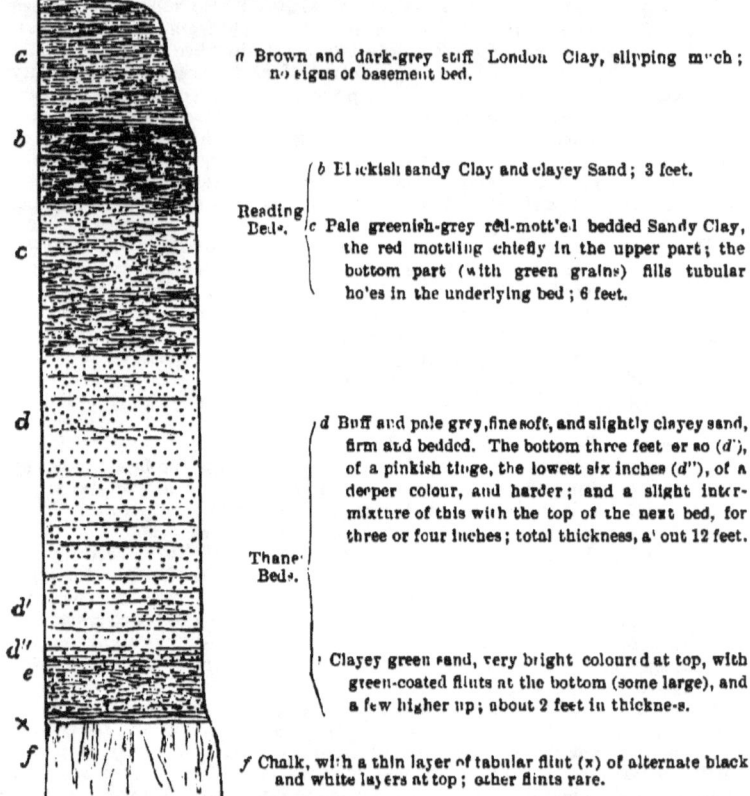

a Brown and dark-grey stiff London Clay, slipping much; no signs of basement bed.

Reading Beds.
{ *b* Blackish sandy Clay and clayey Sand; 3 feet.
{ *c* Pale greenish-grey red-mottled bedded Sandy Clay, the red mottling chiefly in the upper part; the bottom part (with green grains) fills tubular holes in the underlying bed; 6 feet.

Thanet Beds.
{ *d* Buff and pale grey, fine soft, and slightly clayey sand, firm and bedded. The bottom three feet or so (*d'*), of a pinkish tinge, the lowest six inches (*d''*), of a deeper colour, and harder; and a slight intermixture of this with the top of the next bed, for three or four inches; total thickness, about 12 feet.

e Clayey green sand, very bright coloured at top, with green-coated flints at the bottom (some large), and a few higher up; about 2 feet in thickness.

f Chalk, with a thin layer of tabular flint (*x*) of alternate black and white layers at top; other flints rare.

Fig. 80.— General Section of the Great Pit south of Balingdon (Sudbury), as shown in 1873. *Scale, eight feet to an inch.*
The beds are even and flat, or with a S.S.W. dip (up to 3°).

and Sudbury. They rest on the Thanet Sand, or in some places where that is absent the Reading beds repose directly upon the chalk. In the clay at Ipswich, Dr. Taylor states that numerous casts of a shell, probably *Cardium Plumsteadiense*, have been found.

The London Clay. —This bed takes its name from the fact that it underlies the metropolis. In Suffolk it is seen in the tract of land between the estuaries of the Stour and Orwell, around Neyland, Leyham, Stoke, Stratford, &c.; it is also exposed on each side of the estuary of the Deben, and from thence northwards forms the lower part of the cliffs to a little beyond Orfordness. It is a bluish-grey clay weathering to a brown colour at the surface, and with included rounded masses of impure limestone, called cement-stones, from the fact that they have been largely used for making Roman cement. Sometimes these nodules are traversed by cracks which have been filled up with pure crystallized carbonate of lime; they are then known as septaria. Rounded lumps of iron pyrites, brown without, yellow inside, and layers of brown phosphatic nodules, the so-called coprolites, are also common. Beautiful rhomboidal crystals of *selenite*, which the workmen call "congealed water," are abundant at Felixstowe, Bawdsey, &c.; this selenite is simply crystallized sulphate of lime, the same thing, essentially, as gypsum.

The lower part of the London Clay is called the *basement bed*; it is composed of buff-coloured sand and brown clays with some pebbles, altogether from 12 to 15 feet thick, which are well exposed on the north of Hadleigh. The London Clay is dug at several points for brickmaking, and also for spreading over or "marling" light, poor lands; it frequently contains much fossil wood, as in a pit close to Ipswich (on the Norwich road) and in the cliffs at Bawdsey. These plant-remains are always converted into pyrites except when they occur in septaria, in which latter case the structure of the wood is generally replaced by calcite (crystallized carbonate of lime). In the Bawdsey cliffs, we see the uneven surface of the London Clay on which the Crag beds lie, showing that the former formation had suffered much from denudation before the deposition of the Crag; many small faults or dislocations can be traced here by noticing the broken character of the yellow bands of pyrites which run along the lines of stratification. Many springs burst out at the junction of the London Clay with the Crag or Drift beds; the water percolating through the latter cannot make its way into the impervious clay, but flows along the junction until it finds an exit: on the coast these springs occupy hollows or recesses in the cliffs; the upper beds have fallen or slipped over, as a consequence of the erosion of the line of junction. The commonest London Clay fossils are such shells as *Modiola, Cytherea obliqua, Nautilus*, &c.; sharks' teeth abound, and the fossil carapaces of turtles, some 2 or 3 feet in length, have been obtained from the West Rocks, a shoal some few miles off the coast, which used formerly to be worked for cement-stones. One of the most interesting discoveries in Suffolk geology was made by W. Colchester, Esq., F.G.S., in 1839. At Kyson, near Woodbridge, he found in a brick-pit a layer of white and yellow sand, underlying 12 feet of London Clay; from the sandy bed were obtained many sharks' teeth, scutes of crocodiles, scales of ganoid fishes, bones of a large serpent, teeth of bats and opossums, &c., also the teeth and part of a jaw of a mammal, which Professor Owen at first considered to be a species of monkey, but which further discoveries in other parts of England and in France have shown to be an extinct animal related to the *Hyrax* (South African coney); it has been named the *Hyracotherium*.

PLIOCENE FORMATION.—In this district no beds occur of Upper Eocene, or of Miocene age; thus the Pliocene strata rest upon the London Clay, or, where they overlap that formation, upon the chalk. These Pliocene beds only occur (in England) in the counties of Suffolk, Norfolk, and Essex, where they are known as the *Crag* (Celtic *creggan*, a shell), being shelly sands containing a large per-centage of mollusca of still living species.

The Suffolk Bone-Bed, or Box-Stone Deposit.—At the base of the Crag and resting upon the London Clay, there is a very remarkable stratum, in thickness from 1 to 3 feet, containing coprolites, rolled and polished teeth and bones of mastodon, rhinoceros, tapir, hyæna, halitherium, &c., rounded masses of dark-brown sandstone (the "box-stones"), and large boulders of quartz, granite, &c.; one of dark-red porphyry found at Sutton weighed a quarter of a ton. The "box-stones" have been so named by workmen because about one in a score of them when broken open is found to contain some shell or other organic body: these correspond with the fossils of the *Diestien* or Black Crag of Antwerp. This Suffolk bone-bed appears, in fact, to have been formed by the breaking-up of some deposits of Miocene and early Pliocene age which may have occupied part of the area now covered by the German Ocean. The box-stones are most plentiful in the district between Foxhall (near Ipswich) and Felixstowe. At Waldringfield and Falkenham they are used as road-metal.

The Coralline Crag (also called the Suffolk Crag, White Crag, and Bryozoan Crag).—It has been proposed to alter the name of this division to Bryozoan Crag, as the branching forms, formerly thought to be corals, now turn out to be bryozoa or polyzoa. It is well exposed at Orford, Aldborough, Woodbridge, Sutton, Ramsholt, Gedgrave, and there is an outlying patch as far south as Tattingstone. Its extreme thickness is about 60 feet; at the base are shelly sands, above is a rock-bed yielding a soft building stone. The pits in Sudbourne Park have yielded great numbers of fine and well-preserved shells, as *Cardita senilis*, *Cyprina rustica*, *Astarte Omalii*, &c. Altogether Mr. Wood enumerates 396 species of shells from the Coralline Crag, of which 144 are now extinct (36 per cent.); many of these are of Mediterranean species, and the general conditions of the deposit indicate that it was formed in a warm sea of moderate depth—perhaps 300 to 400 feet.

The Red Crag.— This deposit is composed of red shelly sands having an extreme thickness of 25 feet. It is finely exposed in the cliffs at Felixstowe and Bawdsey, and over a large area between Sutton, Ramsholt, Trimley, Butley, &c.: at Tattingstone Park (4 miles south-west of Ipswich), at Sudbourne, and near Aldborough the Red Crag may be seen in section superimposed upon the Coralline Crag. The lower portion of the Red Crag is remarkable for its oblique lamination, which Mr. Wood believes to be due to its having been heaped up as a beach in channels and around islands formed of Coralline Crag. In examining the fossils care has to be taken to distinguish such as are derivative (*i.e.* washed out of older deposits) from those proper to the bed; excluding the former we have 248 species, of which 69 are extinct (28 per cent. nearly). The common shells are *Trophon antiquum*, (the red whelk), *Pecten opercularis*, *Pectunculus glycimeris*, *Cardium edule* (the common cockle), *Purpura lapillus* (the dog whelk), &c. Phosphatic nodules (coprolites) occur, which have probably been washed out of the Coralline Crag and London Clay: the coprolite bed or bone-bed at the base of the Red

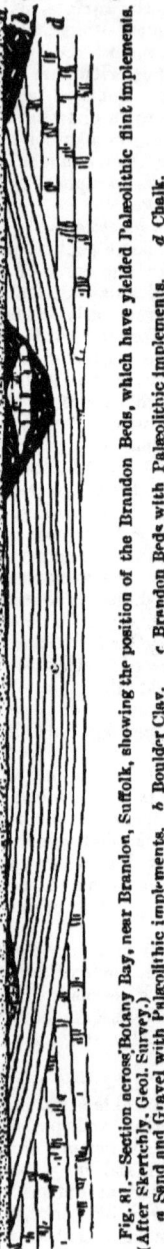

Fig. 81.—Section across Botany Bay, near Brandon, Suffolk, showing the position of the Brandon Beds, which have yielded Palæolithic flint implements. (After Skertchly, Geol. Survey.)
a Sand and Gravel with Palæolithic implements. *b* Boulder Clay. *c* Brandon Beds with Palæolithic implements. *d* Chalk. *e, g, i, k*, Clay Pits.
f No oblithic flint-pits at Grime's Graves. *h* Botany Bay Brickyard, whence Palæolithic implements were obtained in bed *c*.

Crag has been largely worked for these nodules, whose value as a manure was first pointed out by the late Professor Henslow; in 1877 there were raised in Suffolk 10,000 tons of coprolites, valued at about £3 per ton; but in 1879 the amount had fallen to 4,000 tons, valued at £2 per ton.

In many sections ordinary Red Crag full of shells is seen to be surmounted by a few feet of unfossiliferous sand. The latter, Mr. Whitaker has lately shown to be part of the Red Crag, from which the fossils have been dissolved by the percolation of water containing carbonic acid.

The Norwich Crag is found at Thorpe, near Aldborough, at Bulchamp, and at Wangford: it is only about 10 feet thick and contains fresh-water and estuarine as well as marine shells: 112 species of molluscs have been found, of which 18 are extinct (16 per cent.).

The Chillesford Beds are called after a village of that name near Aldborough: they are about 20 feet thick; the lower part sandy, with a shell-bed, above which come laminated micaceous clays. Their fossils include bones of whales and shells of Northern species, as *Astarte borealis;* 86 species are known, of which 14 are extinct (16 per cent.).

QUATERNARY PERIOD.—The Pliocene deposits tell us of a climate gradually getting colder, and a sea gradually decreasing in depth. At last the bed of

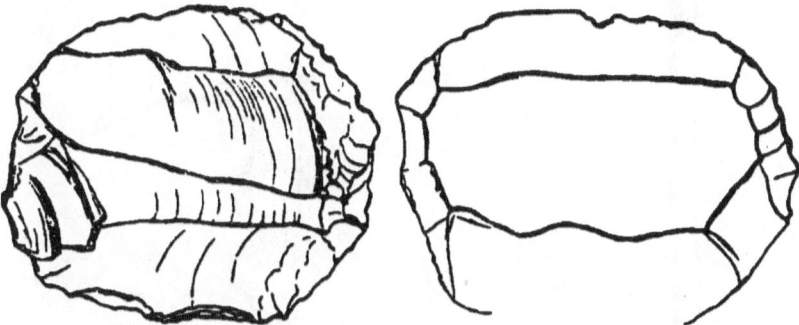

Fig. 82.—Neolithic Oval Strike-a-Light.   Fig. 83.—Modern Oval Strike-a-Light.

the German Ocean was laid dry, and England became a part of the continent, vegetation covered the surface and large mammals strayed over it. To this period belongs the *Forest Bed* of Kessingland, which can also be traced in the cliffs at Corton and Hopton, and round the coast of Norfolk. This bed has yielded many fine bones and teeth of *Rhinoceros etruscus,* the mammoth, stag, Irish elk, &c. Then subsidence took place, and in a shallow sea were formed the pebbly sands named *Bure Valley Beds* by Mr. Wood, and *Westleton Shingle* by Professor Prestwich: these reach southwards to Halesworth, Westleton, and Southwold. They contain *Tellina balthica,* a shell very characteristic of the glacial deposits. Next, Arctic conditions began to prevail, and the first or *Lower Boulder Clay* (Cromer Till) was formed; it is surmounted by the *Contorted Drift,* consisting of clays and loams seen in the Hopton cliffs, and of which there are outlying masses at Kesgrave, Blaxhall, and Boxford. Then come the *Middle Glacial Sands and Gravels,* with beds of brick earth (the Brandon Beds; see Fig. 81), formed during a milder " inter-glacial " period : these occupy much ground south of Gorleston and along the coast, but spread most widely south of Woodbridge and Ipswich: they crop out along the sides of the valleys from underneath the *Great Chalky Boulder Clay.* The latter deposit spreads everywhere over the higher grounds and marks the extreme period of cold: it is full of rock-masses (boulders), often polished and grooved by ice-action, and has an extreme thickness of 160 feet. The various subdivisions of the Drift are finely exposed in many large pits in the neighbourhood of Sudbury.

Hill or Plateau Gravels formed of large flints are found at some places (1 mile east of Bromeswell, again west of Orford, &c.) resting on the Chalky Boulder Clay; they appear to have been formed by strong currents resulting from the melting of a great ice-sheet.

RECENT DEPOSITS.—These include the river gravels, both those which lie on the slopes of the present valleys and those which are on a level with the stream; also the fen-beds near Mildenhall, and the extensive deposits of sand, shingle, and mud which form low flat tracts along the coast. These are the result of the action of the sea, which wears away the land in one place to add to it in another. Of Dunwich, once the chief sea-port of the east coast, only

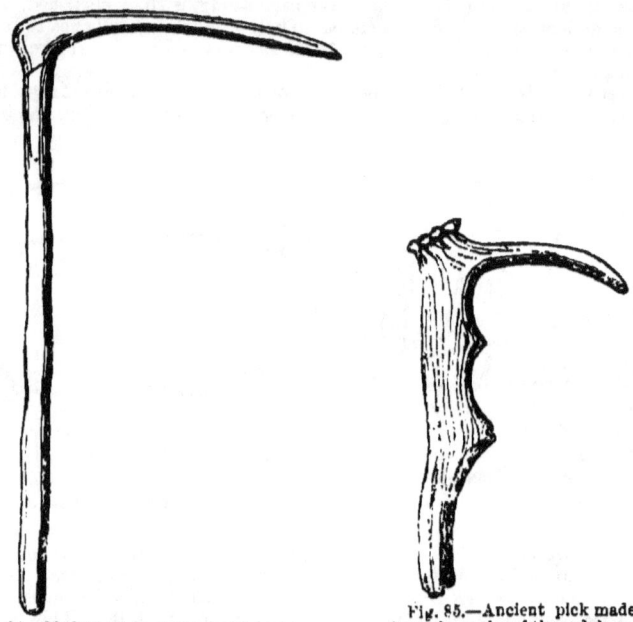

Fig. 84.—Modern pick of wood and iron; used at Brandon to extract flints from the chalk.

Fig. 85.—Ancient pick made from the antler of the red deer; used by the Neolithic flint-workers at Brandon.

a ruined church remains, and the old tower of Aldborough stood a quarter-mile east of the present shore, where the waves now roll over its site.

PREHISTORIC MAN.—During the last year or two satisfactory evidence has been obtained in Suffolk of the existence of man prior to the last Glacial Epoch, and the date of this cold period may be roughly estimated at some quarter of a million years ago. In beds of loam and brick-earth at Botany Bay (near Brandon), Mildenhall Brickyard, High Lodge, Mildenhall, Bury St. Edmunds, West Stow, and Culford, which *underlie undoubted Chalky Boulder Clay*, flint implements have been found which are, certainly, the work of man. This discovery is due to Mr. S. B. J. Skertchly, of the Geological Survey, who has been engaged for some time at Brandon in the study of the gun flint manufacture there. Rude Palæolithic implements of later date have also been found in old river-gravels at Brandon, Hoxne,

Icklingham, Lakenheath, Santon Downham, Wangford, &c. Of the Neolithic or newer stone age, numerous celts, arrow-heads, flakes, &c., beautifully fashioned out of flint and often polished by rubbing, have been met with at Botesdale, Icklingham, Lakenheath, Mildenhall, &c., but there is nothing in form or in workmanship to connect these with the roughly-chipped Palæolithic tools; they were made by a different and later set of men, and the Glacial Epoch comes between them.

The flints in the chalk of the north-west of Suffolk, at Brandon, &c., are so well suited for manufacturing purposes that Mr. Skertchly believes that the art of working flints has been continuously carried on there from prehistoric times to the present day. Thus the flint "strike-a-lights," made by thousands at Brandon during the present century, are just after the pattern of the Neolithic scrapers, many of which were not used as scrapers but for obtaining fire. (See Figs. 82 and 83). So, too, the one-sided picks used by the Brandon workmen of to-day for digging pits in the chalk in order to extract the flints, is singularly like the deer-horn picks used by their predecessors some thousands of years back. (See Figs. 84 and 85).

No. 35.

GEOLOGY OF SURREY.

NATURAL HISTORY AND SCIENTIFIC SOCIETY.
Holmesdale Natural History Club; Reigate.

MUSEUM.
Kew Museum.

PUBLICATIONS OF THE GEOLOGICAL SURVEY.

Maps.—Sheets: 1, South-west of London; 6, Bromley, Chatham, East Grinstead; 7, West of London, Windsor, Staines, Uxbridge; 8, Croydon, Farnham, Guildford, Dorking; 9, Brighton, Chichester, Midhurst, Horsham.
Books.—Geology of the Weald, by W. Topley, 28s. Geology of the London Basin, by W. Whitaker, 13s.

IMPORTANT WORKS OR PAPERS ON LOCAL GEOLOGY.

List (by W. Whitaker) of 505 works on the Geology of the London Basin, in the Survey Memoir, vol iv.
1858. Austen, R. A. C. Godwin.—A Granite Boulder in the White Chalk near Croydon. Journ. Geol. Soc., vol. xiv. p. 252.
1870. Evans, Caleb.- Sections of Chalk between Croydon and Oxtead. Proceedings Geologists' Association.
See also *General Lists, p.* xxv.

FEW districts of equal size offer such varied geological attractions as the county of Surrey. From the old Wealden clays and sands in the south we pass to the chalk ridge in the centre, on whose northern slope lie various Tertiary strata: to this alternation of rock-masses of varying composition and hardness is due the charming scenery for which many parts of Surrey are famous, and which those who understand the geological features of the district are best fitted to appreciate, for they have a knowledge of the *cause* as well as of the *effect*.

The whole surface of the county has been geologically mapped by the Government Survey, and these maps, together with Mr. Whitaker's "Geology of the London Basin," and Mr. Topley's "Geology of the Weald," form a complete guide to the nature and position of the strata. The chalk south of Croydon has been carefully studied by Mr. Caleb Evans, whose paper, with several accounts of excursions to the same district, is published in the Proceedings of the Geologists' Association of London. Other papers by Messrs. Meyer, Ransome, S. V. Wood, Godwin Austen, Professor Prestwich, and, of an earlier date, Drs. Mantell and Fitton and Mr. Martin, have afforded valuable aid in the preparation of this article.

The oldest rocks of Surrey lie in its south-east corner, and as we pass northwards from this point we continually pass over the edges of newer and newer strata. All the beds dip or incline to the north, and rest one upon the other; and consequently the latest-formed rocks are those of Bagshot Heath in the north-west. The most natural order of description seems to be to commence with the *oldest* deposits, and then to take the others in ascending order.

NEOCOMIAN FORMATION.—THE WEALD SERIES. — The lower Wealden strata are called the *Hastings Beds*, and occupy very little of Surrey, just in the south-east corner. We find them to be beds of sand and clay, lying between East Grinstead on the south and Lingfield on the north, and reaching westwards to Copthorn Common not far from Burstow.

The Weald Clay.— This blue or brown shaly clay occupies a tolerably large area; it forms the tract of comparatively flat land which extends from the neighbourhood of Chiddingfold on the west, by Cranley, Ewhurst, Charlwood, Leigh, and Crowhurst, eastwards. The Weald Clay attains its maximum thickness near Leith Hill, where it may be from 900 to 1,000 feet thick, and from this point it thins to the eastward: it contains occasional layers of shelly limestone, known as Sussex marble, full of a species of *Paludina*, a fresh-water mollusc. In the cuttings of the Dorking and Horsham Railway some fair sections of the Weald Clay were exposed, especially at Brockham Common: the beds were found to dip 3 or 4 degrees to the north, and occasional sandy beds were present. Of fossils the scales of the fish called *Lepidotus*, and the small crustaceans known as *Cyprides*, were almost the only examples. The Weald Clay forms stiff land, difficult to drain, and with bad roads.

The Wealden beds generally, offer an example of a great fresh-water deposit, formed probably in the estuary of a great river.

THE LOWER GREENSAND.—Beds bearing this name form the well-marked line of hills which, with one or two gaps, bounds the plain of Weald Clay on the north and west. Hindhead (894 feet), Holmbury Hill (857 feet), Leith Hill (967 feet), and Tilburstow Hill (591 feet), are the highest points. Several subdivisions have been traced in the Lower Greensand. Thus at the base, resting on the Weald Clay, we have a band called the *Atherfield Clay* about 50 feet thick; upon this we find the *Hythe Beds* composed of sands and a limestone called "Kentish Rag:" it attains a thickness of 300 feet at the Devil's Punch Bowl or Hind Head, and forms all the hill-tops whose heights are given above.

The "Bargate Stone," a calcareous grit, appears at Godalming, Nutfield, and Reigate; it contains a small species of *Avicula*.

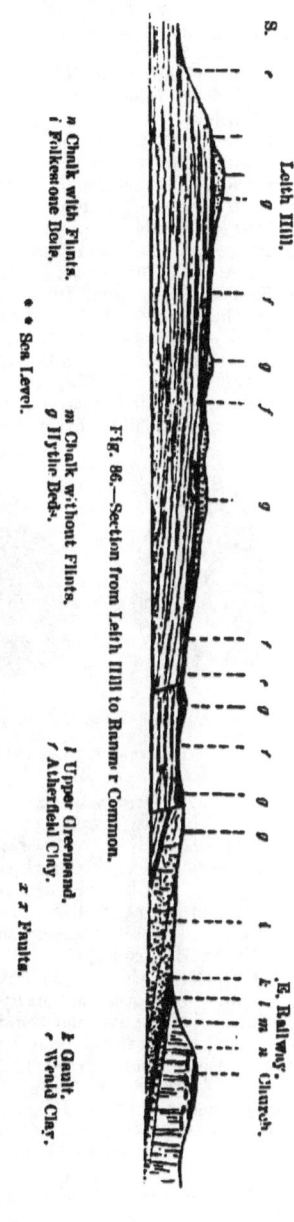

Fig. 86.—Section from Leith Hill to Banstead Common.

The Sandgate Beds are irregular and of local occurrence: they are clayey sands about 30 feet thick, which near Nutfield contain excellent beds of fullers' earth: this substance is an earthy hydrous silicate of alumina: it has a great capacity for absorbing oil or grease, and was once largely used by cloth manufacturers. The mineral known as sulphate of barytes is of frequent occurrence in the fullers' earth of the neighbourhood of Reigate and Nutfield.

The Folkestone Beds form the bottom of the northern slope of the Lower Greensand escarpment: they are light-coloured sands, well developed at Frensham, Farnham, and Thursley Commons, from whence they run eastwards by Dorking, Betchworth, and Westerham. They are often dug for glass-making and for building purposes, and contain numerous marine shells, such as *Exogyra sinuata* and *Terebratula sella*. (See fig. 41, p. 131.)

CRETACEOUS FORMATION. — *The Gault.* — This is a stiff blue clay from 30 to 80 feet thick, which forms the very bottom of the long narrow valley between the two ridges of the Lower Greensand and the Chalk.

The Upper Greensand is about 20 feet thick, and consists of green sandy beds and light-coloured calcareous sandstones, locally known as "Firestone" and "Malm Rock." We can trace it from Farnham eastwards to Godstone and Merstham: it is quarried for hearth-stones, &c. Near Farnham it contains numerous phosphatic nodules (the so-called *coprolites*), for which it

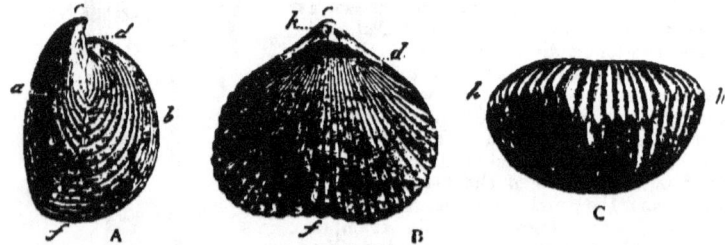

Fig. 87. — A brachiopod shell, *Rhynchonella sulcata*, from the Neocomian, or Lower Cretaceous Formation. *A*, side view; *B*, dorsal valve; *C*, view of base. *a*, ventral valve; *b*, dorsal valve; *f*, base; *c*, beak; *k*, foramen; *h*, line of junction of the two valves.

has there been largely worked. It yields a fertile soil, specially fitted for the growth of hops. (See fig. 34, p. 101.)

The Chalk. Every one is familiar with the smooth rounded outlines of the chalk downs, and with the white soft earthy limestone which composes them. The lowest chalk beds are marly, whilst the highest contain numerous nodules and layers of flints, of which the middle portion is nearly destitute. The total thickness in Surrey may be 500 feet. Beginning on the west we find the chalk forming the well-known "Hog's-back" ridge between Farnham and Guildford: this is only half-a-mile wide, the narrowness of the outcrop being due to the high northerly dip of the beds. East of Guildford the dip decreases in amount, and the outcrop widens, first to 4 miles between Box Hill and Leatherhead, and then to 7 miles between Croydon and Caterham. Along the escarpment the following heights may be noted: The Hog's-back, 504 feet; Netley Heath, 664 feet; White Downs, west of Dorking, 744 feet; north of Reigate, 767 feet; Willy Farm, south-west of Caterham, 735 feet; Tupwood Common, 760 feet; Tandridge Hill, 800 feet; Botley Hill, near Tutsfield, 866 feet. The various railway cuttings which cross the chalk hills give excellent exposures of the beds, and numerous fossils may with little trouble be obtained. Sea-urchins of the genus *Micraster* characterize the upper beds, whilst *Ammonites* abound lower down.

THE TERTIARY EPOCH.—EOCENE FORMATION.—The northern half of Surrey is composed of beds altogether of later date than those we have hitherto been describing. The Wealden Beds we have already noted are probably estuarine deposits; then a depression ensued, and in the Lower Greensand, Gault, and Upper Greensand, we get full evidence of marine conditions, while the Chalk is an oceanic deposit: then occurred a long interval of which no trace remains, and with the Eocene beds we enter on a fresh geological epoch — the *Tertiary*, or third in order, the chalk closing the series of *Secondary* strata. The Eocene Tertiary strata of North Surrey form part of the beds which constitute the southern half of the "London Basin," so called because the beds have a dip towards London both on the south and the north. The chalk underlies them all, passing under London at a depth of about 200 or 300 feet, to rise again and form the North Downs on the one hand and the Chiltern Hills, &c., on the other. (See fig. 63, p. 176.)

The Thanet Beds rest on the chalk at Addington, Croydon, Sutton, Ewell, and so on to East Horsley, beyond which point they are either absent, or concealed by the overlap of higher beds: they are composed of fine sand, and contain no fossils here. They are 40 feet thick at Croydon, and at the base there invariably occurs a bed of green-coated flints, derived from the chalk.

The Woolwich and Reading Beds.—These consist of mottled clays and sands, altogether about 30 feet thick: they form a narrow band running from Farnham Park to Croydon. The best exposures are at Peckham, and in the Brockwell Hall brickyard at Dulwich: here numerous fossils of both fresh and salt-water species have been obtained, such as *Cyrena, Ostrea,* &c., together with remains of turtles, crocodiles, and a tapir-like animal known as *Coryphodon.* Several outlying patches occur on the chalk, as to the east of Wallingham, south of Banstead, &c.

The Oldhaven Beds.—These are sandy with flint pebble beds, and are less than 20 feet thick: they are exposed on Thornton Heath, at Shirley, near Croydon, and north-west of Addington.

The London Clay.—This occupies a much larger spread of surface, running northwards from Leatherhead and Croydon up to the Thames, and extending westwards by Stoke past Aldershot: it is a stiff brownish clay, and few towns of any size stand on it, as there is a great scarcity of water. Its thickness is about 440 feet, and it is exposed in many brickyards and tile-works, as on Epsom Common, Kingston, Forest Hill, Sutton, &c. It contains layers of septaria or cement-stones, which are nodules of impure carbonate of lime. Fossils on the whole are not common, yet in certain bands they abound. The nautilus, crabs, fish, turtles, &c., with univalve shells as *Voluta, Fusus,* &c., seem to indicate the partly marine, partly estuarine nature of the deposit.

Bagshot Beds.—In their lower part these consist of 150 feet of fine light-coloured sands. They occur west of Wimbledon (an outlier) and between Cobham, Chertsey, Chobham, and Woking: upon them rest the Middle Bagshot or Bracklesham Beds which are clayey and only some 10 feet thick. On these again we find the Upper Bagshot Sands, 20 feet thick, forming the Chobham Ridges and the Fox Hills. Masses of sandstone (grey wethers) occur in these upper sands, and are discovered by sounding with long iron rods.

ALLUVIAL DEPOSITS. These are the beds of gravel and brick-earth which have been deposited by the rivers during flood-time, or in shifting their course. In the gravel along the course of the Mole, bones of elephant and rhinoceros have been found. There are also extensive gravel-beds along the rising ground which borders the Thames, as at Peckham, Camberwell, Kennington and Vauxhall, Battersea, &c.

PREHISTORIC MAN. Of the older or *Palæolithic* stone implements the only record is of one of grey flint found in gravel at Peasemarsh, between Guildford and Godalming. Of the later, *Neolithic*, or Polished Stone age, specimens of celts or axe-heads have been obtained from the Thames at Battersea, from Coway Stakes, near Egham, Kingston-on-Thames, and Titsey. One fashioned out of greenstone was found deep in the clay whilst digging for the Chelsea water-works at Kingston. A perforated piece of granite, intended probably for use as a hammer-head, was found at Titsey Bark, and one of quartzite near Reigate. Red Hill, near the latter town, is a famous place for flint flakes, or knives; some thousands were collected here by Mr. Shelley. Flint arrow-heads have been found at Lingfield, Mark Camp, and at Chart Park, Dorking. Such relics of the time when man was ignorant of the use of metals should be diligently sought for and carefully preserved.

No. 36.

GEOLOGY OF SUSSEX.

NATURAL HISTORY AND SCIENTIFIC SOCIETIES.
Brighton and Sussex Natural History Society. Annual Report.
Eastbourne Natural History Society. Papers, 4to.
Lewes and East Surrey Natural History Society.

MUSEUMS.
Brighton Museum.
Museum of the Philosophical Society, Chichester.
Museum of the Sussex Archæological Society, Lewes.

PUBLICATIONS OF THE GEOLOGICAL SURVEY.
Maps.—Sheets: 4, Folkestone to Rye; 5, Hastings, Newhaven, Hailsham: 6. Bromley, Chatham, Maidstone, Tunbridge, East Grinstead; 8, Wokingham; Farnham, Guildford, Dorking, Reigate; 9, Horsham, Midhurst, Chichester; 11, Petersfield.
Books.—Geology of the Weald, by W. Topley, 28s.

IMPORTANT WORKS OR PAPERS ON LOCAL GEOLOGY.
List (by Messrs. Topley and Whitaker) of 564 works on the Geology of the South-east of England in the "Geology of the Weald," Government Survey Memoir, by W. Topley.

1822. Mantell, Dr. G. A. —Fossils of the South Downs. London.
1871. Wood, S. V., jun. —Denudation of the Weald Valley. Journ. Geol. Soc., vol. xxvii. p. 3.
1871. Whitaker, W. Chalk at Seaford and Eastbourne. Geol. Mag., vol. viii. p. 198.
1872. Topley, W.—Agricultural Geology of the Weald. Journ. Roy. Agricultural Soc., series 2, vol. viii. p. 241.
1872-75. Messrs. Topley, Willett, Woodward, &c.—Account of Progress of Sub-Wealden Boring at Netherfield, near Brighton. Reports of British Association.
1878. Dixon, —.—Geology of Sussex, 2nd edition, edited by Professor Rupert-Jones, 42s. Brighton, W. J. Smith.
See also General Lists, p. xxv.

GEOLOGICALLY speaking, Sussex is a highly interesting and favoured county : it is composed of a considerable variety of rocks, and these are well exposed in many fine sections along the coast, and inland in numerous railway cuttings and in excavations which have been made for industrial purposes, such as chalk-pits, brick-yards, &c. Much has been written on the geology of the district: the southern portion was described by Dr. Mantell in his "Fossils of the South Downs," published in 1822, and by Mr. Dixon in "The Tertiary and Cretaceous Formations of Sussex," 1850; of this valuable book a second edition has lately been published, edited by Professor Rupert-Jones. Of the more northern portion, the principal early explorers were Sir R. Murchison, Dr. Fitton, and Mr. Martin. In later times the whole county has been geologically mapped by the Government Survey on the Ordnance maps (scale one inch to a mile), and in these and in Mr. Topley's splendid work, "The Geology of the Weald," published in 1875, we have such minute and accurate information with respect to the rock-masses, that only details are left for the future observer to fill in. Mr. Whitaker has described the chalk between Seaford and Eastbourne ("Geological Magazine," vol. viii.), Mr. Searles V. Wood the Eocene shells found so plentifully at Bracklesham (Palæontographical Society's publications), whilst valuable work has been done by Professor Prestwich, Messrs. Godwin-Austen, H. Willett, Rev. E. S. Dewick, J. E. H. Peyton, S. H. Beckles, and numerous other energetic and clever geologists.

In describing the strata which constitute the county of Sussex we shall begin with the oldest rocks those which underlie all the rest and then gradually pass upwards, noting the area occupied by each rock-bed, its nature, the fossils it contains, its economic uses, &c.

The OOLITE.—In Sussex we find just the highest Oolitic beds; they are known as—

The Purbeck Beds. On the north-west of Battle we find strata of this age occupying the surface along a line about 10 miles in length from east to west, between Whatlington and Heathfield, but only about half-a-mile in breadth from north to south. As at Purbeck, in Dorsetshire, we find two sets of limestone beds, with a considerable thickness of shales (about 130 feet) between. The upper limestones are called the *greys*, and the lower the *blues;* these limestones have been worked at various points, in Rounden Wood, Archer Wood, Poundsford, &c.; they burn into good lime, and contain numerous fossils, mostly of such fresh-water species as *Cyrena, Cypris*, &c.

The Sub-Wealden Boring. At the meeting of the British Association at Brighton in 1872 it was resolved to make a deep boring in the neighbourhood, in order to ascertain the nature of the rocks which lie far beneath the surface in the south-east of England. It was considered possible that seams of coal might be found, for Messrs. Prestwich and Godwin-Austen had pointed out that the coal-fields of Belgium and the north of France are probably connected with those of Somersetshire and Gloucestershire. The spot selected by Messrs. H. Willett and Boyd Dawkins was in the Purbeck Beds in Limekiln Wood, near Netherfield. The boring was commenced and continued to a depth of 312 feet, by Mr. J. A. Bosworth. The Diamond Boring Company then took up the work, and carried the hole down to a depth of nearly 2,000 feet, displaying great energy and perseverance in the matter. The following strata were met with:—

	feet.
Purbeck Beds	180
Portland Beds	110
Kimmeridge Clay	1,480
Coral Rag	17
Oxford Clay	54 not bottomed.

In the Purbeck Beds, at a depth of 121 feet, valuable deposits of gypsum were found, about 35 feet in thickness. The remarkable thing, however, disclosed by this boring was the enormous thickness here of the Kimmeridge Clay, which effectually destroyed all hopes of reaching the Palæozoic strata at a moderate depth. The Kimmeridge Clay, Coral Rag, &c., come up to the surface, or crop out, on the west in Wiltshire, Bucks, &c., and on the east round Calais; but at its outcrop the former bed is only about one-half the thickness which it was found to be below Sussex. As local workers, great praise is due to Messrs. H. Willett and J. E. H. Peyton for the amount of time and trouble which they took in connection with this matter. The gypsum beds discovered have since been worked, and 5,775 tons (value 6s. per ton) were raised in 1879.

NEOCOMIAN FORMATION. Strata of this age form the country included between the North and South Downs: in this area there are two well-marked series of deposits: the lower is known as the

Hastings Beds. The clays at Fairlight Cliffs near Hastings belong to the lower part of this series: they are overlaid by the Ashdown Sands, about 150 feet thick, and these are again surmounted by the Wadhurst Clay. At the base of the last named deposit is a bed of light grey stone, rich in iron, which was formerly got in large quantities and smelted by means of charcoal: Ashburnham was especially noted for the excellence of the iron produced in this manner. Owing to the destruction of the forests, and the use of coal elsewhere, the Wealden iron-works gradually decayed, and the last furnace at Ashburnham was put out in 1828. The top subdivision of the Hastings Beds is called the Tunbridge Wells Sand, because it is well seen near that popular watering-place: near Grinstead it includes a subordinate bed of clay. Thus the Hastings Beds consist of alternations of clays and sands, the latter preponderating, altogether more than 1,000 feet in thickness, and these occupy the surface between Horsham on the west and the north side of Pevensey level on the east, thus forming all the north-east of Sussex, and extending over the boundary line into Kent: this is a hilly and picturesque region; its highest point, Crowborough Beacon, is 803 feet above the level of the sea, Brightling Down being 636 and Fairlight Down 583 feet respectively.

The Weald Clay. Resting upon the Hastings Beds we find a mass of clayey strata of considerable thickness: it is blue, brown, or yellow in colour, and in the west of the county is not less than 1,000 feet thick, but thins rather rapidly as we follow it eastwards. It may be traced between Chiddingfold and Petworth, and thence eastwards to the sea by Itchingfield, Shermanbury, and Hailsham: it thus forms the low flat tract of stiff land between the South Downs and the hilly region formed by the Hastings Beds which we have just been describing. About 120 feet above the base of the Weald clay there occurs a bed of sand stone, known as Horsham stone, which is worked for building, paving, and roofing purposes. There are also one or two bands of fine-grained hard lime stone, known as "Sussex marble." These are full of the shells of a species of *Paludina*, and were formerly in request for mantelpieces; the polished altar-stairs in Canterbury Cathedral are of this stone: it is now only got for road-mending.

The Wealden Beds contain fresh-water fossils chiefly, together with bones of huge reptiles, such as the Iguanodon: they were probably deposited in the estuary of a mighty river, which drained a continent then existing on the north and west, but since submerged and gone.

The Lower Greensand. Several minor beds are included under this term. Of these the lowest (resting on the Weald Clay) is called the (1) *Atherfield Clay:* this enters the county near Haslemere, and runs along the base of the hills by Farnhurst, curving round and passing by Petworth to near Thakeham, beyond which point it has not been traced: it is a sandy clay, about 50 feet thick, but is chiefly distinguished from the Weald Clay below by its fossils, which are of marine forms, including such shells as *Panopea*, &c.

(2). *The Hythe Beds* form the hills south of Haslemere, and curve round thence by Lodsworth, Egdean, &c., but, like the preceding stratum, they rapidly thin in this direction, and are not recognisable east of the valley of the Adur. They are sands and soft sandstones, with a little fullers' earth.

(3). *The Sandgate Beds*, which succeed, are more clayey; they run from Petersfield eastwards by Pulborough.

(4). *The Folkestone Beds* are white and buff sands about 80 feet thick, whose outcrop is about a mile in width; they form a barren, unproductive soil, much of which is waste land or heather-clad common. From Midhurst they run to

Cold Waltham, then, curving by Henfield, we follow them through Hurstpierpoint, Barcombe, and past Folegate Station to the sea. Hard beds, known as *Carstone*, are quarried at Pulborough, Fittleworth, Trotton Common, &c.

CRETACEOUS FORMATION: *The Gault.*—Everywhere this is a stiff blue clay about 100 feet thick. At the base where it rests on the Folkestone Beds we find a layer of phosphatic nodules, which are a valuable ingredient in artificial manures.

The Upper Greensand.—We now begin to ascend the northern slope of the South Downs. At the base we find beds of greenish sandstone about 40 feet thick, seen at Steyning, Amberley, Bury, Barlavington, and in the cliffs north-east of Beachy Head.

The Chalk.—The South Downs have a somewhat steep, abrupt northern slope, but on the south they descend more gently towards the sea. In the west, between Bepton and Lavant, the chalk occupies a tract 7 miles in width, but it narrows to about 5 miles as we follow it eastwards, until it terminates in the magnificent range of cliffs which extends from Brighton to Beachy Head. The range is crossed by several transverse valleys, in which flow the rivers Arun, Adur, Ouse, and Cuckmere. The highest points are Butser Hill (in Hants), 882 feet; Chanctonbury Ring, 804 feet; Ditchling Beacon, 814 feet; Firle Beacon, 810 feet; Beachy Head, 532 feet. Commencing on the north side of the chalk escarpment, we find resting on the Upper Greensand a few feet of the *Chloritic Marl*, whitish, with green grains, and containing a few coprolites. Then comes the *Chalk Marl*, looking like soft greyish, rather clayey chalk, about 50 feet thick; this is overlaid by 200 feet of *Lower White Chalk* without flints, which forms all the middle part of the escarpment; and, lastly, on the top, and forming the southern slopes, we find the *Upper White Chalk* with flints, about 300 feet in thickness. From the layers of flints we can easily determine the inclination or dip of the strata, which is to the south at angles of from 2 to 5 degrees. The summit and northern face of the chalk hills are almost everywhere covered with a short close turf, the slope being fortunately too steep to admit of culture by the plough; their beautifully rounded outlines at once reveal their geological structure to the experienced eye. Here and there are straggling patches of wood, which add to the beauty of the landscape. The tree which above all others especially characterizes the chalk is the beech; but the box, the yew, and the juniper also grow well here. The lower slopes are everywhere under the plough, and form an excellent soil. The best places to seek for chalk fossils are the large pits where the rock is got for burning into lime, as at Burpham, Houghton, Balcombe, Steyning, &c. Here may be found beautifully marked rounded masses, rather smaller than a cricket ball; these are the sea urchins which crawled about on the floor of the ocean in which the chalk was deposited; when alive they were covered with spines, and these are sometimes still to be detected; the genera *Galerites*, *Ananchytes*, *Micraster*, and *Marsupites* are most common. Remains of fishes, too, are not rare, with sponges and numerous shells. Rubbing up a little of the white chalk in water with a tooth-brush, pouring off the milky liquid, and then examining the residue with a microscope, we can perceive countless numbers of minute shells known as *Foraminifera*, of which indeed the mass of the rock is composed. We therefore believe that the chalk was formed in some ocean in whose waters these tiny beings lived. As they died their hard parts sank slowly to the bottom, and in time formed the thick bed of white rock, whose cliffs standing out boldly on our southern coasts have earned for our country the name of Albion.

THE TERTIARY EPOCH.—Westwards of Brighton, by Worthing, Littlehampton, and so on to Chichester and beyond, we find a low plain lying at the foot of the chalk hills. The western portion of this plain, near and south of Chichester, is composed of beds which repose upon the chalk, and were formed at a much later period; these belong to the EOCENE FORMATION.

The lowest Eocene deposits are certain beds of mottled clay, formerly known as the *Plastic Clay*, but now denominated the *Woolwich and Reading Beds*. These run through Chichester, and then between Yapton and Arundel to a point a little north-east of Highdown Hill; they are worked for brick-making, but contain few fossils. Indications of the former eastward extension of these beds are to be found in the outlying patches which still remain between Brighton and Hove, and on the cliffs above Newhaven and Seaford.

The Bognor Beds (London Clay).—Emsworth and Chichester are built upon strata which were formed at about the same time as the London Clay, but as they differ in several particulars, it is well to have a separate name for them. They also extend from Hayling through West Wittering to Bognor. They consist of clays and sands, with three or four pebble-beds. At Bognor numerous and beautiful shells of the genera *Pectunculus*, &c., may be obtained from the rocks exposed at low water. Mr. Whitaker also detected a patch of the London Clay resting on the Woolwich Beds at Newhaven.

The Bracklesham Beds.—These form the peninsula which extends south of Earnley and Pagham to Selsea Bill. They consist of green clayey sands, and are best seen at low water in Bracklesham Bay, when thousands of fine fossil shells, such as *Cardita planicosta* and *Cerithium giganteum*, may be seen exposed. Sometimes, however, the whole is covered over by fine sand, when disappointment awaits the collector. From vertebræ found here Professor Owen has described an Eocene sea-serpent—*Palæophis typhæus*—about 20 feet in length.

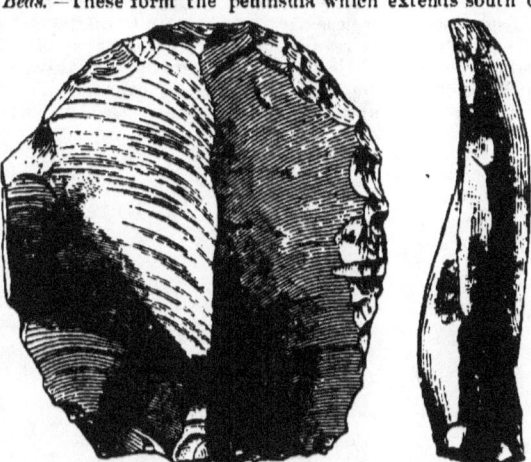

Fig. 88.—Horse-shoe-shaped Flint Scraper—front and side views; natural size. Found, with many others, on the South Downs at Berling Gap, west of Eastbourne, by Mr. John Evans.

POST-PLIOCENE DEPOSITS.—There is no boulder clay or other glacial deposit spread over the rocks of Sussex, such as we find in the counties north of the Thames. At Pagham, near Selsea Bill, however, erratic blocks of granite, syenite, and greenstone, as well as of Devonian and Silurian rocks, are found. One of these granite masses which Sir C. Lyell measured was 21 feet in circumference. These blocks were probably carried by icebergs and dropped in their present situation.

The various rivers have formed gravelly deposits along their valleys, which extend for half-a-mile or so on either side of their present courses. On the coast the sea has in many places wrought great destruction, but the material removed from one point is cast up again at another, and so have been formed the extensive flats of Romney Marsh and Pevensey Level.

PREHISTORIC MAN IN SUSSEX.—Traces of the stone age are plentiful in this county. The implements found all belong, however, to the *Neolithic*, or newer stone period, when man had learned to delicately chip, grind, and polish his flint and other stone tools, although as yet metals were unknown. Colonel Lane Fox has shown that Cissbury, near Worthing, was a regular manufactory of flint implements. In excavations in the chalk at this spot flint weapons in various stages of manufacture have been found, together with hammer-stones used in their fabrication. Quite recently Mr. Park Harrison announces the discovery of a skeleton, apparently of one of the early workers.

Along the South Downs an earnest seeker may be almost certain of daily finding some trace of prehistoric man. Berling Gap and Beltout, both near the western side of Beachy Head, have yielded many flint scrapers, flints, and knives, (see fig. 88.) Celts, or flint axe-heads, have been found at Telscombe and Bolmer, near Falmer. At Hove, near Brighton, an oaken coffin was found in a barrow, or raised mound of earth; this contained a perforated axe-head, apparently formed of some kind of ironstone, an amber cup, a whetstone, and a small bronze dagger. An oval perforated stone hammer was found in a tumulus at Cliffe, near Lewes, and a beautifully chipped flint knife near Eastbourne Flint arrow-heads have been found at Hastings; and several other places, such as Oving, Hardham, Pulborough, &c., might be named, which have yielded specimens similar to one or other of those above mentioned. To discover such specimens, it is necessary first to learn to distinguish *worked* from natural flints, which may be readily done by examining specimens in museums and local collections, and then to watch narrowly all excavations, especially those which remove the top foot or so of soil over considerable areas, as these Neolithic stone weapons usually occur on or near the surface.

No. 37.
GEOLOGY OF WARWICKSHIRE.

NATURAL HISTORY AND SCIENTIFIC SOCIETIES.

Birmingham Natural History and Microscopical Society. Annual Report.
Birmingham Microscopists' and Naturalists' Union.
Birmingham Philosophical Society.
Birmingham and Midland Institute Scientific Society.
Birmingham School Natural History Society.
Smallheath (Birmingham) Literary and Scientific Society.
Warwickshire Natural History and Archæological Society; Warwick. Annual Report.
Warwick Naturalists' and Archæologists Field Club; Warwick. Proceedings.
Rugby School Natural History Society.
Tamworth Natural History, Geological, and Antiquarian Society.
Leamington Philosophical Society.

MUSEUMS.

Birmingham and Midland Institute Museum.
Aston Hall Museum, Birmingham.
Queen's College Museum, Birmingham.
Warwick Museum.
Mason's College Museum, Birmingham.

PUBLICATIONS OF THE GEOLOGICAL SURVEY.

Maps. Sheet 44, Evesham, Worcester, Cheltenham, Burford. Quarter Sheets: 45 N.W. Banbury, Deddington, Chipping Norton; 53 N.W. Coventry, Rugby, Leamington; 53 S.W. Southam, Kingston; 53 N.E. Clipston, Crick, Braunston; 53 S.E. Towcester, Daventry, Weedon; 54 N.E. Henley-in-Arden, Solihull; 54 S.E. Stratford-on-Avon; 54 N.W. Kidderminster, Bromsgrove; 54 S.W. Droitwich, Worcester; 62 N.E. Lichfield, Tamworth; 62 S.W. Wolverhampton, Walsall, Dudley; 62 S.E. Sutton Coldfield, Birmingham, Coleshill; 63 N.W. Market Bosworth; 63 S.W. Atherstone, Nuneaton; 63 S.E. Lutterworth.

Books. Geology of the Warwickshire Coal-field, by Howell, 1s. 6d.
Triassic and Permian Rocks of the Midland Counties, by E. Hull, 5s.

IMPORTANT WORKS OR PAPERS ON LOCAL GEOLOGY.

List by Whitaker of 132 works, in Report Rugby School Nat. Hist. Soc., 1873.
1856. Brodie, Rev. P. B. Upper Keuper Sandstone of Warwickshire. Journ. Geol. Soc., vol. xii. p. 374.
1865. Brodie, Rev. P. B. Lias Outliers at Knowle and Wootton Wawen. Journ. Geol. Soc., vol xxi. p. 159.
1869. Huxley, Prof.—On Hyperodapedon. Journ. Geol. Soc., vol. xxv. p. 138.
1874. Miall, L. C.—Labyrinthodonts from the Keuper Sandstone in the Warwick Museum. Journ. Geol. Soc., vol. xxx. p. 417.
1878. Tomes, R. F. Corals of the Lias. Journ. Geol. Soc., vol xxxiv. p. 179.

1879. Allport, S. —Diorites of the Warwickshire Coal-field. Journ. Geol. Soc., vol. xxxv. p. 637.
1879 Ingram, Rev. A. H. W. —Superficial Deposits near Evesham. Journ. Geol. Soc., vol. xxxv. p. 678.
1880. Andrews, W. —Superficial Deposits near Coventry. Trans. Warwick Field Club, p. 1.
1880. Andrews, W. —Boreholes in Permian Rocks at Coventry. Trans. Warwick Field Club, p. 29.

See also General Lists, p. xxv.

In the Report of the Rugby School Natural History Society for 1873 there is a list of no fewer than 132 papers which have been written on the geology of this county: of these 21 are by the Rev. P. B. Brodie, of Rowington, who has been indefatigable in his efforts to promote the study of his favourite science. The district was mapped geologically by the officers of the Survey between 1852 and 1859, and if the reader obtains these maps, and also Mr. Howell's memoir on the Warwickshire coal-field, he will find them to contain full and accurate observations on the subject.

In giving a brief outline of the rock structure of Warwickshire we shall commence with the coal-field, because the strata composing it are older than any other in the county, and form the foundation upon which newer beds repose.

THE CARBONIFEROUS FORMATION, then, occupies the surface over a comparatively small area east and south-east of Tamworth; and Shuttington, Kingsbury, and Baddesley Ensor are respectively near the boundary line to the north and south. From these points a long narrow strip extends past Atherstone, Stockingford, Nuneaton, and Bedworth, which marks the north-eastern edge of a basin, the remainder of which is concealed by beds of Permian age, which cover it up to the south and west: a patch, however, peeps out near Arley, having been brought up by a fault.

The Millstone Grit is the lowest rock exposed, and it forms a high ridge between Nuneaton and Atherstone: it is traversed by intrusive beds of a volcanic rock, commonly termed greenstone, but whose proper scientific name is *diorite*, consisting of the minerals felspar and hornblende: these have so heated the grit that it is changed into a hard quartz rock. The junction may be seen in quarries near Tuttle Hill and Hartshill: here the millstone grit is about 500 feet thick, and is extensively quarried for road metal. At one time it was worked for manganese, the presence of which imparts a general pinkish tint to the beds. The dip or inclination of the strata is to the westward, at a considerable angle, so that they soon pass under and are overlaid by

The Coal-measures, of which there is a tolerably complete series, although the total thickness of the beds, some 3,000 feet, is much less than in the great northern and western coal-fields. At the base we find 1,500 feet of reddish shales, containing no coal-seams, and, like the Millstone Grit, traversed by dykes of intrusive greenstone. Above these is about an equal thickness of shales, clays, and sandstones, containing five workable seams of coal, having an aggregate thickness of from 26 to 30 feet. Lastly, at the top we find 50 feet of sandstones and shales, with a bed of limestone at their base, called erroneously the "fresh-water limestone:" it contains a small rolled-up serpula, *Spirorbis (Microconchus) carbonarius*. The outcrop of this bed is obscure, but it runs from Sybil Hill, near Kingsbury, round by Baddesley Ensor, Whiteford, Stockingford, and so on past Bedworth. In the collieries east of this it would of course not be met with, but we should expect to find it in the shafts of those on the west and south, as was indeed the case at Exhall: it also crops up in the inlier at Arley. Its thickness is from 2 to 3 feet: the presence here of this limestone is important, as it is an indication that the Coal-measures certainly extend to the westward under the Permian sandstones, and at a depth probably not exceeding 2,500 feet. In fact, if we draw a line past Bedworth, through Wyken, Coventry, and Warwick, and thence round on the west by Kenilworth, Berkswell, and Whitacre, we shall mark out an area beneath which at some future day coal will almost certainly be proved to extend.

Commencing in the south we find, in the Wyken and Hawksbury pits, a mass of coal about 26 feet thick, comparable with the "Ten-yard seam" of South Staffordshire. There are, however, some "partings" of fireclay, and other indications which enable us to separate this mass into the Two-yard, Bare, Rider, Ell, and Slate Coals. About 40 feet below the last-mentioned we get the "Seven-feet" Coal. At Bedworth the section is much the same; but going northwards, we find, just as in Staffordshire, that the divisions of the thick coal begin to be separated one from another by beds of sandstone and blue clay, or "bind." Passing Griff and Haunchwood, we see at Polesworth about 100 feet of such beds between the Ell and Slate Coals, while the Rider and Bare Coals have thinned out and are altogether wanting. For an explanation of this phenomenon we can only repeat what has been already said about South Staffordshire. Our coal-seams were doubtless formed by the long-continued growth and decay of vast forests, the under-clay which we find beneath every seam of coal constituting the soil in which these ancient forests grew. Now about Bedworth the growth and decay were uninterrupted, but towards the north the land at intervals slowly sank beneath the sea, allowing beds of sand and clay to become interstratified with the vegetable matter. Finally, the whole area sank beneath the level of the ocean, and under heat and pressure the masses of ferns, reeds, palms, mosses, &c., were compressed, forced to part with much of their substance in the form of gases, and became the black combustible substance now known as *coal.*

The "Seven-feet" Coal is a very continuous seam, and about 120 feet below it is the Bench Coal, which at Baddesley Colliery is 17 feet thick, with a parting of fire-clay 2 feet thick. There are also some good bands of ironstone. As in the Millstone Grit, there are several intrusive beds of basalt: large masses of this igneous rock occur at Merevale, Atherstone, and at Dosthill, about 2 miles north of Kingsbury. Here the nearest coal-seam has been so burnt as to be rendered useless. In the railway cutting near Chilvers Coton there is a good section of four dykes of *diorite,* one of which can be traced southwards to Marston Jabet, near Bedworth, where it is quarried for paving purposes. (See fig. 89).

In the "Mineral Statistics" for 1874, there are 29 collieries enumerated in the Warwickshire coal-field, and the amount of coal raised in that year was 851,500 tons. Of ironstone (argillaceous carbonate, and black band), 92,214 tons were raised, valued at £39,528. In 1879 there were 31 collieries, yielding 1,060,016 tons of coal.

The estimate of the resources of this coal-field, made by Mr. J. T. Woodhouse in 1871 for the Coal Commission, is a favourable one. He takes the area at 30 square miles, and the amount of coal unworked at 810 millions of tons, of which there can be got, say, 455 millions of tons. Considering that this district is the nearest coal-field to London, it is rather surprising that it has not been more energetically worked, but the future will doubtless see the annual output largely increased.

THE PERMIAN FORMATION.—Covering the Coal-measures from Baddesley Ensor on the north to within 2 miles of Warwick and Leamington on the south, there occur beds of red, brown, and purple sandstones and red marl, with here and there a bed of calcareous breccia or conglomerate. The total thickness of these beds is about 2,000 feet, and good sections are exposed in the large quarry near Kenilworth. It is to Professor Ramsay that we owe the determination of the true age of these beds; he also found indications in the breccias of glacial action. The Permian beds rise near Corley to a height of 625 feet above the sea. Few fossils are found in them, but remains of a large reptile, *Dasyceps Bucklandi*, turned up near Kenilworth. In a quarry formerly worked near Exhall obscure casts of a shell allied to *Strophalosia* were found, together with large fragments of silicified wood, now in the Warwick Museum.

THE TRIAS. — Wrapping all round the region we have been hitherto describing we find a considerable thickness of red and mottled marls, with here and there bands of sandstone, and pebble beds near the base. The latter belong to the *Bunter* division, as it is termed; they form the high ground between Birmingham and Lichfield, and also crop out near Polesworth.

The *Lower Keuper Sandstones*, or *Waterstones*, come next in ascending order; they are about 200 feet thick, and above them we find 600 or 700 feet of red Keuper Marls, with one or two irregular bands of white

Fig. 89.—Quarry at Marston Jabet, North Warwickshire, showing the unconformability of the Triassic Strata and the Carboniferous Beds.

A Lower Keuper White Sandstone; a fine conglomerate at the base; above the Sandstone is a bed of Red Marl, 3 feet thick. These beds lie horizontally on the edges of the Carboniferous Shales and Greenstones.
B Lower Coal-measure Shales, dipping east, 15 degrees.
C Intrusive beds of greenstone (*diorite*).

sandstone. These red beds were probably deposited in immense inland salt lakes; frequently they contain casts of salt crystals; gypsum is common; whilst ripple marks and false-bedding tell us of shallow, disturbed waters. Footprints of strange reptiles are not uncommon, together with the tiny shells of a small crustacean, *Estheria minuta*, and teeth and dorsal spines of fishes. Fossils are, however, very scarce in these beds, but the Warwick Museum possesses probably the best collection of these Triassic remains to be seen in England; it is especially rich in remains of the *Labyrinthodon*, mostly obtained from quarries in the waterstones at Coton End and near Warwick; these have lately been figured and described by Professor Miall, in vol. xxx. of the Geological Society's Journal.

On the west the Triassic and Permian beds occupy all the ground between the Warwickshire and South Staffordshire coal-fields, from each of which they are cut off by faults, this central region, in which Birmingham, Coleshill, and Sutton Coldfield are the principal towns, being let down as it were between the two coal-fields. Thus, though it is likely enough that coal may exist beneath this part, yet it probably lies at such a depth (3,000 to 5,000 feet) as to prevent its being profitably worked.

THE RHÆTIC BEDS.—Towards the top of the Triassic strata we find the red marls frequently alternating with bands of grey and blue, the difference of colour being due to the iron in the latter existing in the state of a carbonate or protoxide, while in the red marl it exists as a peroxide. At last we come to 10 or 15 feet of black shales, which mark the commencement of the Rhætic series, and of a new state of things. The famous Rhætic "bone-bed" is a thin layer of rolled and broken bones, scales, teeth, and spines of fishes, &c., which is usually found at the junction of the black shales with the grey marls beneath. Above the shales we find 30 or 40 feet of light and dark coloured clays, with nodular bands of limestone, one of which contains *Estheriæ*. The molluscs named *Avicula contorta* and *Cardium Rhæticum* are the characteristic shells, and altogether the Rhætic series seem to be beds of passage between the Trias below and the Liassic beds above. Owing to the softness of the strata and the fact that the rocks have no commercial value, there are few good sections of these beds to be seen. They have been noted at Church Lench, near Evesham, at Wotton Park, near Alcester, and in the railway cuttings near Stratford-on-Avon. The upper division, sometimes called the White Lias, has been proved at Wilmcote and Binton, and Mr. Brodie has detected outliers at Wootton Wawen and at Copt Heath, near Knowle.

THE LIAS.—Running in a broad band along the southern and eastern edge of Warwickshire, we find thick beds of shale, with limestone towards the base, to which from the regularity of their bedding, as seen in quarries, the term "layers" or "lyers" was applied by the workmen, a term which has been adopted by geologists as a name for the formation. If we pass from Stratford by Loxley, Upton, and Harborough Magna, we shall have just skirted the western edge of these beds. *The Lower Lias* is characterized by valuable beds of limestone, noted for their property of setting under water. These are largely quarried at Stockton, Harbury, Wilmcote, Binton, Grafton, and at several places near Rugby. They are highly fossiliferous, containing many *Ammonites*, numerous bivalve shells, as *Ostrea, Lima*, &c., with remains of insects, plants, fishes, and notably bones and even entire skeletons of the *Plesiosaurus* and *Ichthyosaurus*, gigantic saurians or lizard-like reptiles which inhabited the bays and estuaries of the Liassic sea. The total thickness of the Lower Lias is here about 600 feet. In the boring for water at Rugby 400 feet of this was passed through, below which came 7 feet of beds assigned to the Rhætic series, and then nearly 700 feet of Keuper marls, the waterstones being pierced at 1,140 feet: a rush of water followed, but it proved to be so impregnated with salt and gypsum as to be useless for drinking purposes.

The Middle Lias, or Marlstone, is found capping Edge Hill: it is also present at Fenny Compton, where it has been preserved by a fault running in a north-west and south-east direction. There are quarries on the Avon Dasset Hills, where it is got for building stone; it is a hard, ferruginous limestone, of a reddish tint where weathered, but naturally of a bluish-green colour. *Terebratula punctata* is the commonest fossil.

The Upper Lias Clays present few features of interest; they form the slopes of the Oolitic escarpment.

THE OOLITE. In the extreme south-west we have the critical region where it was formerly thought that the Inferior Oolite of the Cotteswolds died out and came to an end, on the west side of the Vale of Moreton. But Professor J. W. Judd has shown in a masterly manner that this is not the case, but that the *Northampton Sand* is its equivalent. At Brailes Hill and all along the ridge of Long Compton we find beds identical with those of Ilmington Downs and Ebrington Hill, on the other side of the Vale.

Near Long Compton, which is within the county, and in an outlier at Cherrington, we find beds of white freestone, underlaid by sands containing *Ammonites Murchisoniæ* and other characteristic fossils.

THE DRIFT.—Scattered irregularly over the rocks we have been describing there occur beds of sand, gravel, and clay, these being relics of a much later time, when this country was submerged beneath an icy sea, a time which we term the Glacial Period. Great masses of syenite, evidently torn off the rocks of Charnwood Forest, have been found near Rugby, but generally the boulders are small, not exceeding a foot or two in diameter. Pebbles of Silurian rocks are common, with chalk-flints, quartz pebbles, &c. The locality and rock from which the extremely hard, well-rounded quartzite pebbles were derived, is a very interesting question; almost everywhere in this county they are used for road-mending, and may be seen lying in heaps by the road-side. About one pebble in ten thousand contains a fossil, usually a shell of some kind or other, and it is only by collecting a large number of these fossils that the question as to whence the pebbles have come from can be solved.

Of still later age are the river gravels of the Avon, which have yielded at Rugby a fine jaw of *rhinoceros*, and at Warwick and Leamington remains of the mammoth, an extinct species of elephant.

PREHISTORIC MAN.—None of the roughly-chipped flint implements assigned to the Palæolithic age have been found yet, but of the newer, or Neolithic Stone age, when man had learned to polish his rude weapons, we have several examples. In Barlett's History of Mancetter there is an engraving of a perforated stone axe, found on Hartshill Common in 1770, in a small tumulus, and one of similar character was found in draining at Walsgrave-upon-Sowe, near Coventry. A quern, or stone mill, used for grinding corn, which was found near Rugby, is probably of a much later date, as it has an iron pin.

But with the advent of man Geology merges into Physical Geography, Archæology, and History.

No. 38.

GEOLOGY OF WESTMORLAND.

NATURAL HISTORY AND SCIENTIFIC SOCIETIES.

MUSEUMS.

Museum of the Natural History Society, Kendal.

PUBLICATIONS OF THE GEOLOGICAL SURVEY.

Maps. —98 S.E. Kirkby Lonsdale; 98 N.E. Kendal, Sedbergh; 102 S.W Ulleswater. (Survey not completed.)

Books.—Geology of the Northern Part of the English Lake District, by J. C. Ward, 9s. Geology of Kendal, Windermere, Sedbergh, and Tebay, by Aveline and Hughes, 6d. Geology of Kirkby Lonsdale and Kendal, by Aveline, &c. 2s.

IMPORTANT WORKS OR PAPERS ON LOCAL GEOLOGY.

1862. Harkness and Salter.—The Skiddaw Slates. Quart. Journ. Geol. Soc., vol. xix. p. 113.
1864. Murchison and Harkness. -Permian Rocks of North-west of England. Quart. Journ. Geol. Soc., vol. xx. p. 144.
1865. Harkness. Lower Silurian Rocks at base of Pennine Chain between Melmerby and Hilton. Quart. Journ. Geol. Soc., vol. xxi. p. 235.
1867. Hughes, McKenny.—Break between Upper and Lower Silurian Rocks of Lake District. Geol. Mag., vol. iv. p. 346.
1868. Nicholson.—Geology of Cumberland and Westmorland. Hardwicke.
1869. De Rance, C. E.—On the Surface Geology of the Lake District. Geol. Mag., vol. vi. p. 489.
1869. Nicholson, H.A.—Plants in Skiddaw Slates. Geol. Mag., vol. vi. p. 494.
1870. Harkness. Distribution of Wasdale Crag Blocks. Quart. Journ. Geol. Soc., vol. xxvi. p. 517.
1872. Tiddeman.—The Ice-sheet in Westmorland. Quart. Journ. Geol. Soc., vol. xxviii. p. 471.
1872. Aveline.— On Continuity and Breaks in Silurian Strata of Lake District. Geol. Mag., vol. ix. p. 441.
1874. Mackintosh. Traces of a Great Ice-sheet in Southern Part of Lake District. Quart. Journ. Geol. Soc., vol. xxx. p. 174.
1874. Goodchild, J. G.—Carboniferous Conglomerates of Eastern Part of Eden Basin. Quart. Journ. Geol. Soc., vol. xxx. p. 394.
1875. Goodchild, J. G.—Glacial Phenomena of Eden Valley and West Yorkshire Dales. Journ. Geol. Soc., vol. xxxi. p. 55.
1875. Ward, J. C.—Glaciation of Southern Part of Lake District. Quart. Journ. Geol. Soc., vol. xxxi. p. 152.
1875. Ward, J. C.—Granitic and Metamorphic Rocks of Lake District. Quart. Journ. Geol. Soc., vol. xxxi. p. 568, and vol. xxxii. p. 1.
1877. Harkness and Nicholson.—Strata between Borrowdale Series and Coniston Flags. Quart. Journ. Geol. Soc., vol. xxxiii. p. 461.

1877. Topley and Lebour.—On Intrusive Character of Whin Sill. Quart. Journ. Geol. Soc., vol. xxxiii. p. 406.
1878. Crofton, Rev. A.—Notes on Geology of Shap District. Trans. Manchester Geol. Soc., vol. xv. p. 234.
1878. Marr, J. E.—Life Zones in the Silurian Rocks of Lake District. Quart. Journ. Geol. Soc., vol. xxxiv. p. 871.
1879. Bonney and Houghton.—Mica Traps from Kendal and Sedbergh Districts. Quart. Journ. Geol. Soc., vol. xxxv. p. 165.

See also General Lists, p. xxv.

The rocks of this county must be studied in connection with those of Cumberland; the Furness district too is geologically a part of Westmorland.

All the strata belong to the Palæozoic group—the oldest division of the aqueous rocks; their great age, and the heat and disturbances to which they have been subjected, have so hardened and altered them as to enable them to resist the leveling agencies of the weather, so that we now find in this corner of England a miniature mountain group.

The general structure of the region may be roughly described as consisting of a central mass of Silurian Rocks, round which circle beds of Carboniferous and Permian age; the northern and more perfect half of this area forms the county of Cumberland; the southern half constitutes Westmorland. The foundation of all our geological knowledge of this county rests on the work done by Professor Sedgwick between 1830 and 1845; Professor Buckland and Messrs. Hopkins and D. Sharpe added minor details at this time. Within the last ten years numerous papers by Professors Harkness and Nicholson have appeared in the Reports of the British Association, and in the Journal of the Geological Society. Mr. J. G. Marshall has advanced the theory of the metamorphic origin of the granites, and Messrs. J. Clifton Ward, Marr, Tiddeman, Mackintosh, Bolton, and others have done useful private work. For the practical geologist, however, there is no guide like the coloured one-inch maps of the Geological Survey, executed by Professor McKenny Hughes, Messrs. Tiddeman, Goodchild, C. T. Clough, &c., under the direction of that experienced Silurian geologist, Mr. W. T. Aveline.

We shall begin our description of the rocks of Westmorland with those beds which are known to be the oldest, from the fact that they are seen to be at the bottom, or lie under all the other beds.

LOWER SILURIAN FORMATION.—The strata known as the *Skiddaw Slates*, which occupy so much of the surface of Cumberland, only crop out in three small isolated areas in Westmorland: the first of these is in the course of Eggbeck, on the south shore of Ulleswater; the second in Rossgill Beck and Thornship Beck, west of Shap; and the third on the east side of the Pennine Fault, in a narrow strip between Hilton and Melmerby. Here the Skiddaw Slates are finely exposed in Ellergill Beck, where they consist of dark shales and flags containing graptolites. Professor Nicholson obtained plant remains from the upper beds of the Skiddaw Slates in Thornship Beck, near Shap.

Volcanic Series of Borrowdale.—These are the "Green Slates and Porphyries" of Sedgwick. They consist of matter ejected from the craters of volcanoes which must formerly have been in active eruption in this region, but whose cones have long since been swept away. Their total thickness is about 12,000 feet, composed of layer upon layer of ashes and breccias which accumulated upon the land surrounding the old volcanoes. Some of the beds are of very coarse material, consisting of lumps up to several feet in diameter; these must mark more violent eruptions, or proximity to the volcanic sources. Other beds are of the finest powder, so that, having been since subjected to great pressure deep down in the earth, they now yield good slates. No traces of life, no fossils, have been found throughout this great series of stratified volcanic rocks. (See fig. 90).

Entering the county all along its northern and western boundaries from

Ulleswater to Windermere, the Borrowdale Series extends southwards as far as a line drawn from Ambleside to Wasdale Crag near Shap, while a line from the latter point to the east end of Ulleswater marks its eastern boundary. Within these limits the rocks rise in bold jagged and serrated elevations, forming the heights of Helvellyn (3,055 feet), Fairfield (2,950 feet), Bow Fell (2,914 feet), Rydal Head (2,910 feet), High Street (2,700 feet), Harrison Stickle (2,400 feet), Stony Cove (2,502 feet), Ill Bell, Froswick, Tarn Crag, &c.

This region forms the watershed from which the stream called Troutbeck and the rivers Kent and Sprint run southwards to the sea through the valleys of Troutbeck, Kentdale, and Long Sleddale. "These deep, well-cultivated, and often flat-bottomed valleys, with their small farm-houses, form a striking and pleasing contrast to the rocky hill-sides and high, uncultivated fells covered with heather and peat." The Borrowdale Series also occurs in the inlier of old rocks north of Hilton, where it forms Dufton Pike (1,578 feet) and Knock Pike (1,306 feet).

The strike or "run" of the Borrowdale Series is from south-west to north-east, and the general dip or slant or inclination of the strata is, of course, at right angles to this and to the south-east, in which direction they pass *under* the Coniston Limestone. But their dip is by no means constant in this direction: as we cross the line of strike, by walking for instance from Keswick to Ambleside, we shall notice continual undulations of the beds, and the high

Fig. 90.—Slate, from Shap, the particles of mica lying in all directions; cleavage imperfect. (After Sorby.)

Fig. 90A.—Slate, from Llanberis, in North Wales; the particles of mica all lie parallel to each other; cleavage perfect. (After Sorby.)

central ridge is caused by the strata lying in the form of an inverted arch or synclinal curve, which has enabled them to well resist denudation. Many quarries have been opened in the slates, but few are now worked. Several mineral veins or lodes, too, traverse the rocks; the lead-mine of Greenside in Patterdale is one of the most valuable in the North of England; in 1877 it yielded 1,600 tons of ore, from which 1,133 tons of lead and 15,726 ozs. of silver were extracted; total value, £20,811.

In Wales we find in the Lower Silurian rocks, traces of *two* well-marked periods of volcanic activity, one in the Llandeilo Flags, the other in the Bala Beds, the slaty beds between representing ordinary marine muds and marking an interval of quiescence. In the Lake District it would appear that volcanic action went on uninterruptedly, so that the Borrowdale Beds probably represent *both* the above-named divisions of rocks of North Wales.

Coniston Limestone Series.—At last the volcanic agencies became exhausted, and a depression of the land ensued, until the surface sank far below the sea-level, and above the volcanic series beds of limestone and shale were formed. The Coniston Limestone is a hard compact greyish blue, grey, or nearly black rock; it will not burn into lime, and is of little use for building, for it rapidly decomposes when exposed to the air, when the numerous fossils which it contains stand out in relief, although the unweathered rock shows hardly a trace of organic remains; associated with the limestone are beds of shale, which are highly fossiliferous, containing many corals, crinoids, and brachiopods as *Orthis vespertilio, Heliolites*, &c.; by the similarity of the fossils the Coniston Limestone is known to be of the same age as the Bala Limestone of North Wales.

Commencing at Sunny Brow, 2 miles south-west of Ambleside, we follow these Coniston beds eastwards to the north side of Town Head, Troutbeck, where a north and south fault displaces the beds 1 mile to the southward; thence they run past Stile End and north of Stockdale to High House Fell; from this point the country is very obscure, being covered by drift and peat, but we see the limestone once more at Shap Wells, though here it is much broken up and altered. East of Kentmere an interbedded mass of light flesh-coloured or pink felspathic rock (an old lava) is seen between two limestone beds; it is 700 feet thick in Stockdale Beck and Long Sleddale. The Coniston Limestone is seen again at Keisley, on the east of the Pennine Fault, and it is also brought up on the east of the Lune, in Helm Gill, just outside the county boundary.

This well-marked calcareous stratum rests unconformably on the Borrowdale Beds beneath, and like them has a general southerly dip; the thickness varies from 300 to 600 feet, increasing as we follow it westwards.

UPPER SILURIAN FORMATION. The base of this series is marked by a conglomerate bed, above which come pale and black shales and mudstones, called the *Stockdale Shales*, which are the equivalent of the Tarannon Shales of Wales; they form a narrow band at the surface, not a quarter of a mile wide, immediately to the south of and resting upon the Coniston Limestone, to which they are unconformable. No representatives of the Llandovery Rocks of Wales occur in the Lake district.

Coniston Flags and Grits. "The flags consist of dark blue sandy mudstone, showing fine lines of lamination and splitting into good flags when the cleavage and bedding coincide or nearly so, and into rough slabs when the cleavage is across the bedding. The grits consist of thick beds of tough sandstone, or grit with interstratified slates or flags." These beds are from 6,000 to 7,000 feet in thickness between Windermere (north end) and Troutbeck; from this point they form a very regular band about 2 miles in width eastwards by Applethwaite Common, Hugill Fell, Skeggles Water, Long Crag, and Lord's Seat, to Birkbeck Fells, south of Shap, the strata dipping steadily south-east at from 60 to 80 degrees.

After passing under newer rocks these beds rise up again and form a large tract round Whinfell Beacon and Grayrigg Forest, extending eastwards between Bridge Inn and Howgill Station to the Langdale Fells and Cantley Crags. The Coniston Grits are also brought up on the east side of the Lune Valley, forming Holme Fell, Middleton Fell, and Barbon Fell. Fossils are not very common or well preserved in this great series of hardened sandy mud-beds; *Cardiola interrupta* is a well known shell, orthoceratites and graptolites are not unfrequent.

Bannisdale Slates.—These are correlated with the Wenlock and Lower Ludlow Shales. They may be described as coarse sandy slates, "often much cleaved and jointed, never making good slates, but often large rough slabs, quarried for paving or building stones." The only tolerable slates were formerly worked in the higher part of the valley called Bannisdale, at the very base of the series.

They occupy a very wide stretch of country from Windermere Station, Bowness, and Winster on the west, to Underbarrow, Staveley, Sleddale Forest, Potter Fell, Bretherdale, and Tebay Station, undulating rapidly, with a very irregular surface, due to the frequent alternation of hard and soft beds; "between Kendal and Windermere, the Bannisdale Slates may be said to be crumpled or puckered, so numerous are the rolls." The total thickness of this division is about 5,000 feet; fossils are rare.

Hay Fell and Kirkby Moor Flags.—From the north of Staveley a band of hard, thick-bedded sandstones can be traced running north-east, and with an outcrop about half-a-mile wide to High Borrow Bridge; these are the Kirkby Moor Flags, brought in by a sharp synclinal curve, and bounded on the south side by a broken line of fault. Passing southwards, we find the great mass of this division lies between the main road from Kendal to Kirkby Lonsdale on the west and south, and the Lune on the east, forming all the country round Old Hutton, Audland, and Mansergh. In this high and bleak district Benson's Knot (1,035 feet) and Lambrigg Fell (1,109 feet) are conspicuous points. Fossils are numerous, especially the shells *Holopella gregaria*, *Pterinea retroflexa*, and *Chonetes lata*, occurring generally in layers, which as they decompose form lines of soft brown earthy rock in the hard grey sandstone. The total thickness of this series is above 2,000 feet; it is of the same age as the Ludlow Beds of Shropshire, &c.

Thus the Silurian rocks of the Lake district attain the enormous total thickness of 36,000 feet, of which the lower half is found in Cumberland, and the upper half in Westmorland. The middle of this period was marked by volcanic operations on a grand scale, but the lower and upper series of beds were tranquilly deposited as mud and sand on ancient sea-floors.

INTRUSIVE IGNEOUS ROCKS,—With the exception of the basaltic rock called the Whin Sill, which is described further on, in connection with the Carboniferous Limestone of the Pennine Chain, all the intrusive igneous rocks of Westmorland are found in the Silurian strata.

The well-known "Shap Granite" occupies an area of 4 or 5 square miles on Shap Fells, rising at Wasdale Crag to a height of nearly 1,500 feet; it is a beautiful rock formed of crystalline quartz, plates of black mica, and small crystals of red and white felspar; but its distinguishing feature is the presence of numerous very large oblong crystals of pink felspar, often from 1 to 2 inches long, so that boulders of this rock look "like a lump of pudding studded with large raisins." This granite is largely quarried for building stone, and is polished for many ornamental purposes. It is clearly intrusive, having burst through the aqueous rocks just at the junction of the Borrowdale Series with the Coniston Limestone; the surrounding rocks are greatly altered, the Coniston Limestone being converted into a crystalline marble, while the ash-beds are changed into a foliated felspathic rock.

Many *trap dykes*, or narrow more or less vertical sheets of igneous rock, also traverse the Silurian strata; some of these are *granitic*, like those south-east of Shap, which are doubtless offshoots from the main mass of Wasdale Crag. *Felstone dykes*, usually compact and of a red tint, occur near High Borrowdale, on Potter Fell, east of Staveley, &c.; dykes of *mica-trap* or *minette*, usually compact and dark grey in colour, are seen at Barley Bridge and Gill Bank near Staveley, Docker Fell, Uldale Hend, &c.

CARBONIFEROUS FORMATION.—We now know that no strata belonging to the Old Red Sandstone age occur in Westmorland. Since elsewhere we find rocks more than 10,000 feet thick belonging to this period, we see that there is a marked gap or hiatus in the geological formations of Westmorland.

Carboniferous Conglomerate.—Resting very irregularly on all or any of the Silurian beds, we note thick, coarse conglomerates made up of rounded and angular fragments of Upper Silurian rocks, and between 200 and 300 feet in thickness. These beds are seen south of Gaisgill Station and along the Birkbeck to Bampton and the east end of Ulleswater; detached patches are found at Greyrigg and Skelsmergh Hall north of Kendal. Whether these beds represent old glacial deposits (as Professor Ramsay thinks), or are simply a beach or shore accumulation, is not certainly decided.

Carboniferous or Mountain Limestone.—This is a greyish or greyish-blue rock weathering to a lighter tint. Its main outcrop forms a band 2 or 3 miles wide from Lowther Castle by Shap, to Orton and Crosby Garret; dipping eastward, the beds pass under the Permians of the Eden Valley, on the east side of which they are again brought up by the Pennine Fault to form the escarpment of the Pennine Chain; thick beds of reddish sandstone occur near the base; the thickness of the mountain limestone, with its intercalated bands of sandstone and shale, is not much under 3,000 feet.

In the south of Westmorland similar beds extend from Kirkby Lonsdale westwards by Hutton Roof and Farleton Fell to Burton-in-Kendal and Milnthorpe, and thence northwards to Kendal and across the river Kent to Whitbarrow; in this part the junction with the Silurian beds is everywhere a line of fault. Two outliers occur north-east of Kendal.

The Mountain Limestone forms picturesque escarpments called " scars ; " it is traversed by many fissures, on whose sides ferns grow luxuriantly, and " sink-holes " or " swallow-holes " abound, in which the streams disappear, so that the surface of the rock is usually dry. The loose surface blocks of limestone weather into most fantastic shapes, and are often to be seen on rockeries. Fossils are numerous, especially near the bottom and top of the formation, corals, encrinites, and large brachiopod shells abounding in the shaly bands; in the central compact grey limestone they are rarer, and very difficult to extract.

The Carboniferous Limestone is extensively quarried for building purposes, road-mending, and burning into lime; near Kendal it is cut and polished for marble.

Veins of lead-ore (galena) are worked at Hartley, near Kirkby Stephen; 150 tons, value £1,800, were obtained in 1877. Of hematite (iron-ore), 8,000 tons, value £4,800, were obtained.

In the Carboniferous Limestone of the Pennine Chain there is a fine example of an intrusive sheet of igneous rock, called the *Great Whin Sill*. It is a greyish black basalt, and is finely exposed in Teesdale at the waterfalls called High Force and Caldron Snout; it is seen along the face of the Pennine escarpment, one of the finest sections being at High-Cup Nick, about 4 miles east of Appleby, where the basalt is 73 feet thick, and has baked or altered the shale beds above and below it; it thins in a southerly direction, and disappears east of Brough.

It is difficult to define exactly the upper boundary of the Carboniferous Limestone in the east of Westmorland. The Whin Sill is often taken as the base line of the *Yoredale Beds*, which here consist of alternating limestones, sandstones, and shales, forming the extreme east of the county at Milburn Forest, Stainmoor Forest, Mallerstang Common, Ravenstonedale Fells, &c. There are two or three thin coal-seams in this series, which are worked at three collieries, 1,791 tons of coal being raised in 1877.

In this same region the *Millstone Grit* just enters Westmorland.

PERMIAN FORMATION. Red rocks of Permian age lie between Brough and Kirkby Stephen, and extend thence in a north-westerly direction towards Penrith and Long Marton; they are bounded on the east by the Pennine Fault, and are also faulted on the west against the Carboniferous Limestone. They are well exposed between Appleby and Hilton; at Burrels, near the former town, we see the lower beds to be limestone breccias, called " brockram," 600 feet thick, succeeded by red sandstones and yellow breccias called " rotten brockram," 1,500 feet thick; the Middle Permians or Hilton Shales (200 feet thick) are seen in Hilton Beck to be thin-bedded sandstones and marly shales, containing numerous plant-remains; here, too, is a bed of magnesian limestone, 6 feet thick: the Upper Permians are dark red sandstones, 700 feet thick. The breccias present a strong likeness to glacial drift, and they may possibly mark an old glacial epoch.

SURFACE DEPOSITS.—Under this head we include all the irregular accumulations of clay, gravel, and sand which lie upon the older rocks, often masking them completely and rendering their study difficult.

Glacial Period. -To the educated eye nothing can be clearer than the ice-marks borne by the hard and lofty rocks of Westmorland. When we examine the valleys between Ambleside and Ulleswater, we frequently find striations or scratches on the rocks, the grooves pointing down the valleys, as in Great Langdale and Easdale ; these have been made by stones embedded in ice-masses (glaciers), which occupied all the high central region of the Lake district, and passed outwards in all directions to lower ground. The stony clay, till, or pinel, found in patches on the high grounds, and in larger spreads on the plains, is formed of matter pushed under or before these glaciers. During this intensely cold period there was at least one great depression of the land, to perhaps 2,000 feet below its present level; only the higher peaks would then stand above the sea, and from these icebergs would be detached, carrying blocks of rock (boulders or erratics) to great distances. The granite boulders from Wasdale Crag near Shap must somewhere about this time have been so carried over the Stainmoor Pass into Yorkshire, where we now find them along the east coast ; many are also found in Westmorland on the east and south of Shap.

Most of the lake-basins, as Ulleswater, Windermere, &c., are now believed to have been ploughed out by glaciers ; they lie in lines of weakness (caused by faults), and in the direction along which the glaciers must have advanced ; some, too, occupy the point where two or more glaciers must have met, and where the eroding action of the ice would be great.

The lines or mounds of sand and gravel called "eskers" were probably formed by currents washing and re-assorting the materials of the boulder-clay, during the period of depression.

The Post-Glacial Deposits include the gravels and muds of existing rivers ; the filled up tarns, of which there are many ; the peat-beds, sometimes 15 feet in thickness ; and the deposit of sand and loam made by the sea, which forms the low plain of Beetham and the flat land near the river Gilpin.

PREHISTORIC MAN.—Few traces of those early inhabitants of our isles, whose implements were mainly formed of stone, have occurred in Westmorland. Mr. Evans records a large perforated stone axe-head, 11 inches long, found in a turf moss near Haversham ; a large oval stone, with a groove round the centre, found at Burns near Ambleside, is believed to have been used to sink nets.

No. 39.

GEOLOGY OF WILTSHIRE.

NATURAL HISTORY AND SCIENTIFIC SOCIETIES.

Wiltshire Archæological and Natural History Society; Devizes. Magazine.
Marlborough School Natural History Society. Report.

MUSEUMS.

Literary and Scientific Institution Museum, Devizes.
Museum of the Wiltshire Archæological and Natural History Society, Devizes.
Salisbury and South Wilts Museum.
The Blackmore Museum, Salisbury.

PUBLICATIONS OF THE GEOLOGICAL SURVEY.

Maps. Sheets: 12, Newbury, Hungerford, Andover, Basingstoke, Odiham; 13, Oxford, Reading, Wantage; 14, Marlborough, Amesbury, Westbury; 15, Salisbury, Blandford, Ringwood, Shaftesbury, Cranborne, Wimborne Minster; 18, Wincanton, Sherborne, Beaminster, Langport, Yeovil; 19, Bath, Frome, Axbridge; 34, Stroud, Fairford, Swindon, Chippenham; 35, Bristol Coal-field.

Books. The Geology of Parts of Wiltshire and Gloucestershire, by Prof. Ramsay, &c., 8d. Geology of the London Basin, by W. Whitaker, 13s.

IMPORTANT WORKS OR PAPERS ON LOCAL GEOLOGY.

List (by W. Whitaker) of 169 works, by 89 authors, in Wiltshire Archæological and Natural History Magazine, vol. xiv. p. 107.

1862. Whitaker, W. On the Western End of the London Basin, and on the Grey Wethers of Wiltshire. Journ. Geol. Soc., vol. xviii. p. 258.
1865. Blackmore, Dr. H. P. Discovery of Flint Implements at Milford Hill, Salisbury. Journ. Geol. Soc., vol. xxi. p. 250.
1866. Woodward, H. On the Oldest British Crab, from the Forest Marble near Malmesbury. Journ. Geol. Soc., vol. xxii. p. 493.
1870. Stevens, E. T.—Flint Chips: a Guide to Prehistoric Archæology. 8vo. London and Salisbury.
1871. Whitaker, W. On the Occurrence of the "Chalk Rock" near Salisbury. Geol. Mag., vol. ix. p. 427.
1881. Etheridge, R. and Andrews, Rev. W. R. Purbeck Beds of Vale of Wardour. Q. Journ. Geol. Soc. vol. xxxvii. p. 246.

See also General Lists, p. xxv.

In the magazine of the Wiltshire Archæological and Natural History Society (vol. xiv.) we have a list compiled by Mr. W. Whitaker of 169 papers, &c., by 89 authors which relate to the geology of Wilts. The whole of the surface has been mapped by the officers of the Geological Survey (Messrs. Bristow, Aveline, Hull, and Whitaker), and the northern portion of the county is described in a "Survey Memoir," by Professor Ramsay. Of local workers we must name Messrs. Cunnington, C. Moore, G. P. Scrope, C. J. A. Meyer, T. Codrington, R. N. Mantell, J. C. Pearce, the Rev. P. B. Brodie, Professors Buckman and Rupert-Jones; whilst for our knowledge of the fossils we are largely indebted to Drs. Wright, Lycett, and Woodward, Professors Morris, E. Forbes, Milne-Edwards, Bell, &c. Unfortunately these works are scattered over numerous and expensive scientific publications, as the volumes of the Palæontographical Society, the Quarterly Journal of the Geological Society, the "Geological Magazine," the "Wiltshire Magazine" above referred to, &c. Recently Mr. Hudleston and the Rev. J. F. Blake have written a valuable account of the Coralline rocks, and the latter gentleman has also described the Kimmeridge Clay.

The recent deposits near Salisbury which contain such interesting traces of the early existence of man have been diligently worked by Dr. Blackmore and others, and their "finds" are to be seen in the Blackmore Museum at Salisbury, which contains a splendid series of the tools and weapons both of prehistoric man and of modern savage tribes for comparison. The excellent work entitled "Flint Chips," written by Mr. E. T. Stevens, gives a full account of this admirable collection. In the adjoining rooms of the Salisbury and South Wilts Museum there are some fine specimens of local fossils. At Devizes, geological collections may be seen in the museum of the Literary and Scientific Institution and in the museum of the Wiltshire Natural History Society. At Marlborough College, the School Natural History Society is also doing good work.

The rocks which form the county of Wilts belong mainly to the Secondary or Mesozoic formation: they run across the county in broad irregular bands, which have a general line of direction or *strike* from north-east to south-west: they incline or slant or *dip* to the south-east, not lying now horizontally as they did when they were originally deposited on some sea-floor, but tilted up towards the north-west, so that the edges of a number of successive rock-beds are visible, forming the irregular bands we have mentioned above. As the lowest strata are certainly the oldest or first formed, it follows that those rocks which we find in the north-west of the county, about Bradford, Malmesbury &c. are of the most remote geological age, for as we try to follow these across the strike in a south-easterly direction, we find they pass under newer beds, which in turn pass under the chalk, while the chalk is itself overlain by the Tertiary strata in the south-east corner on the edge of the New Forest.

In describing the strata we shall commence in the north-west with the oldest beds, and then take the others one by one, in ascending order, finally treating of the alluvial and other surface deposits which occur irregularly over the whole area.

But first looking broadly over the entire county we note that it may be divided into two regions, whose wide difference in appearance, agriculture, and population is entirely owing to geological causes. The first of these lies in the north-west, and includes about one-third of the county; it has a level or undulating surface and a south-easterly slope; it is a district of oolitic limestones and clays, and the chalk escarpment forms its boundary. The remainder of Wilts—the southern and eastern parts—is mainly formed of chalk, where breezy downs and long level treeless plateaux have a distinctiveness which instantly reveals to the discerning eye the nature of the rock of which they are composed.

The heights above sea-level of a few points determined by the Ordnance Survey are, Highworth Church, 442 feet ; Cricklade Church, 297 feet ; Wootton Bassett Church, 424 feet ; Christian Malford Church, 187 feet ; Chippenham Church, 184 feet ; Corsham Church, 303 feet ; Warminster Town Hall, 388 feet ; Tower on Downton Hill, 485 feet ; Salisbury Cathedral, 153 feet.

THE LIAS.—The liassic strata only enter Wiltshire in the extreme west, running up the valley of the Box Brook from Bathford by Ditcheridge to a little beyond Colerne. Ditcheridge stands on the *Middle Lias*; but the hard band called the "Rock-bed" is here very thin, while a few feet of blue clay which lie upon it belong to the *Upper Lias*, the beds thus thinning out greatly in this direction. The junction of the Lias with the sands of the Oolite above is marked by wet ground, caused by the springs which are thrown out along the top of the impervious clays.

THE OOLITE.—This great assemblage of strata has received its name from the fact that some of the beds of limestone which it contains have a structure like the roe of a fish: it consists of an alternation of beds of sandstone, limestone, and clay, which, as a whole, attain their maximum development in Gloucestershire, where they form the Cotteswold range. As the oolitic strata are followed in a south-easterly direction, they become greatly attenuated, as has been shown by Professor Hull; still in Wiltshire they are of a very respectable thickness.

LOWER OOLITE.—*Midford Sands.*—These are transition or passage beds connecting the liassic and oolitic formations. They contain liassic ammonites, as *A. bifrons*, *A. opalinus*, &c., and oolitic bivalves, as *Hinnites abjectus*, *Trigonia striata*, &c.; as Professor Phillips remarks, "before the liassic life had come to an end, the oolitic life had begun." The Midford Sands are of a yellow tint, and contain concretionary masses of limestone called "sand-bats;" they occur resting upon the Lias along the sides of the valley of the Box Brook, and are here about 40 feet in thickness.

Inferior Oolite.—This is a buff-coloured sandy limestone, under which lies a rubbly bed full of corals. "It occurs at Elmhurst and Upper Shockerwick, and runs up the sides of the Box valley nearly to Slaughterford: from near Box it stretches through Ashley Green and Bathford to near Bradford." It is well exposed in the bank of the canal opposite Limpley Stoke, and in a gully near Charlcombe church, where it is about 25 feet thick. At Limpley Stoke and Beechen Cliff the Inferior Oolite is almost entirely composed of corals, with various species of such shells as *Trigonia*, *Ostrea*, &c. Thus the small portion of this stratum which enters Wilts, possesses neither the thickness nor the qualities which elsewhere (Cotteswolds, Ham Hill, &c.) make it such a valuable building stone.

The Fullers' Earth.—This is a bed of blue and yellow clay, from 60 to 100 feet thick; in the lower portion it contains bands of limestone, the "Fullers' Earth Rock." It can be traced by Bonner Down and Ashley Wood to beyond Slaughterford, and then returns on the opposite side of the valley past Box. The characteristic fossil is a small oyster named *Ostrea acuminata.* The position of the bed is rendered visible at a distance by the way in which the superincumbent beds of Great Oolite have in places slipped down over the clay; the water which issues plentifully at the junction has lessened the adhesion, and produced the catastrophes. The Fullers' Earth proper is a hydrous silicate of alumina; it is "an unctuous clay, usually of a greenish-brown or greenish-grey colour, sometimes blue: it is opaque, soft, dull, with a greasy feel and an earthy fracture: it yields to the nail and affords a shining streak: it scarcely adheres to the tongue, and falls into a pulpy, impalpable powder when placed in water, without forming a paste with it." (Bristow.) It occurs in veins from 18 inches to 3 feet in thickness, and was formerly largely dug for fulling at the cloth mills, but is not now worked.

The Great Oolite. This division, which is also known as the *Bath Oolite*, furnishes perhaps the most famous building stone in England. It does not occupy much of the surface of Wiltshire; from near Limpley Stoke and Corsham we can trace it to Yatton Keynell and Castle Combe: it is again exposed in the valley which runs east and west by Great Sherston, and we again find it (this time brought up by a fault) on the east of Tetbury. The thickness of the Great Oolite varies from 100 to 200 feet; the upper beds are blue limestones, weathering white when exposed to the atmosphere, and so hard as to have been largely used for mending the roads; below these come the fine oolitic freestones which furnish such valuable building material. At Corsham the stone is worked underneath higher strata (forest marble) by means of shafts (sometimes 70 feet deep) and tunnels; the underground quarries here are perhaps the largest and best worked in the kingdom. Box Hill yields a stone of a very superior quality as to fineness of texture, called Scallet. "The absence of fossils renders the stone more valuable as a freestone, as it then yields more readily to the saw and to the chisel; when fossils are abundant the stone is best adapted for rough building purposes, such as walls and the foundations of houses." (H. B. Woodward.) When first removed from the quarry the stone is soft and moist, and easily carved or sawn, but as it dries, it hardens and becomes an excellent hard white stone. Of fossils, univalve shells, such as *Nerinæa Voltzii*, *Purpuroidea Morrisii*, &c., and many bivalves, as *Trigonia costata*, *Tancredia brevis*, &c., are abundant, especially in the upper limestones.

The Forest Marble (with Bradford Clay).—The strata which own this name occupy a considerable surface area in the north and west of Wiltshire. From Bradford-on-Avon and Corsham (where they are well exposed in the railway cutting), they run northward to Tetbury, and then turning north-east form a tract some 5 or 6 miles wide, on which stand the villages of Ashley, Crudwell, and Kemble. The strata are a variable series of shelly limestones, sands and clays, about (Fig. 91) 60 feet thick near Tetbury, but thickening as they are followed southwards. Near Bradford a thick mass of clay, the *Bradford Clay*, is locally developed in the lower portion of the Forest Marble. It is a pale grey clay about 40 feet thick, containing a little carbonate of lime and enclosing thin slabs of brownish limestone and sandstone; a fossil called the Pear Encrinite (*Apiocrinites rotundus*) is abundant in the Bradford Clay; its fragments are called "coach wheels" by the quarrymen. The large slabs of coarse limestone furnished by the Forest Marble proper, are, according to Professor Buckman "of great value for forming the sides of piggeries and cattle sheds. The smaller pieces broken up form a very durable material for road-making, and some of the thicker blue-centred slabs are used for building purposes. As a soil the Forest Marble is usually poor, but capable of great improvement by draining and cultivation." The remains of the oldest known British Crab (*Palæinachus longipes*) were found by Mr. William Bury, in the Forest Marble near Malmesbury.

The Cornbrash.—This is a rubbly pale-coloured earthy limestone, about 20 feet thick. We can trace it between Trowbridge and Semington, and round Great Chalfield and Atford; from Corsham and Chippenham it runs by Kington, Hullavington and Malmesbury to Charlton and Pool Keynes. At Rodborne it is quarried for rough building stone. The fossils are numerous, and include *Ammonites Herveyi*, *Avicula echinata*, &c. Numerous villages occur along the outcrop of the Cornbrash, as it is a water-yielding stratum; it furnishes a good corn-growing soil.

MIDDLE OOLITE.—*The Oxford Clay (with Kelloway's Rock).*—This formation consists of blue clay, weathering yellow near the surface, and here attains a thickness of about 500 feet. Near its base is a fossiliferous bed of calcareous sandstone called Kelloway or Kellaway's Rock, either after the village of the latter name near Chippenham (Woodward), or because it occurred in pits belonging to a man named Kelloway (Ramsay). The Oxford Clay extends from Bradley and Melksham to Christian Malford; then between Wootton Bassett and Malmesbury the outcrop is 7 miles in width, and so it continues by Purton Station, South Cerney, and Cricklade, towards Lechlade; sections are exposed in the cuttings of the Great Western Railway, and in brickyards near Malmesbury and at Minety station. The Oxford Clay forms a stiff, heavy soil, difficult to cultivate and mostly in permanent pasture; the old Forest of Braydon stood on it, and there are still many woods. Of fossils the large flattish oyster-like shell (*Gryphæa dilatata*) is abundant, ammonites and belemnites are also plentiful. In making the railway cutting near Christian

Malford, about 1840, many fossils were found, including some new forms of cephalopods, which were described by Mr. Pearce under the name of *Belemnoteuthis*, even the impressions of their soft parts were wonderfully preserved. Remarkable branches of some coniferous trees from the Oxford Clay of this locality have been figured by Mr. Carruthers in the "Geological Magazine" (vol. vii.); he there remarks that it would be very desirable if foliage or fruit could also be found.

The Coral Rag. The thick beds of clay we have just described were probably deposited in a moderately deep and tranquil sea. After the accumulation of a great thickness of mud, an elevation of the sea-floor accompanied by a clearing of the waters made the conditions fit for the growth of coral reefs. In the Coral Rag we have a bed of rubbly oolitic limestone, full of the remains of corals and about 20 feet thick; it is accompanied by beds of sand and calcareous sandstones, termed the *Upper* and the *Lower Calcareous Grit*. Near Westbury a bed of iron-ore (hydrated peroxide) has been largely worked; it occupies a narrow strip of ground running north and south near the railway station, and the bed of ore is from 11 to 14 feet in thickness. In 1877 there were raised 79,176 tons of iron-ore, valued at £19,784, but in

Fig. 91.—Quarry at Yatton Keynell, near Corsham.

a Forest Marble. Fissile shelly oolite, resting obliquely on the Great Oolite; 4 feet.
b Great Oolite (upper zone). Regularly bedded massive shelly limestone; 7 feet.
c Great Oolite (Lower zone). Shelly oolite, full of false bedding. The upper part coarse; the lower affording very fine building stone, which is followed underground; 16 feet.

1879 the amount had fallen to 47,623 tons, valued at £9,525. *Ostrea deltoidea* is a very common fossil here. At Steeple Ashton, many fine corals, such as *Thecosmilia annularis*, are found, mostly on the surface of the ploughed fields; at Seend the *Lower Calcareous Grit* is well exposed where the furnaces for the iron-ore stand. From this point we can follow the beds by Westbrook to Calne, and thence by Hillmarton and Purton to Highworth. At the latter place the total thickness is about 100 feet, and fine sections are exhibited where the beds are quarried for building stone. Spines and plates of a sea-urchin (*Cidaris florigemma*) are very characteristic of the Coral Rag. Physically this bed forms a ridge or escarpment rising about 100 feet above the plain of Oxford Clay; by its decomposition it produces a light sandy arable soil. At Calne specimens of a new genus of sea-urchin (*Pelanechinus*) which possessed a flexible test have been found by Mr. Keeping.

UPPER OOLITE. *Kimmeridge Clay.*—Between Semley, Sedghill, and Knoyle Common, we find a bed of bluish shaly clay, about 65 feet thick; we can trace it again on the north-east, between Westbury Station, Worton, and Bulkington; it is then overlapped and concealed by newer strata, but we recover it again at Olney Marsh, and find it thickens northwards to 500 feet at Swindon, where its outcrop is also broad, occupying all the ground between

Swindon on the east and Wootton Bassett and Stratton St. Margaret's on the west. There are large brick-pits in the Kimmeridge Clay at Swindon; here *Ammonites biplex* is common in a bed of sandstone, and limestone nodules occur; *Ostrea deltoidea* is another common fossil. In these pits in 1874 bones of enormous extinct reptiles (*Omosaurus*, &c.) were found, which were extracted and removed with great skill by Mr. W. Davies, of the British Museum, and have since been described by Professor Owen.

Portland Beds.— Strata of this age are exposed in three separate areas in Wiltshire. At Chilmark and Tisbury, in the valley of the Nadder, about 12 miles west of Salisbury, there are large quarries in a siliceous limestone, of which Salisbury Cathedral and Wilton Abbey were built. At Tisbury about 61 feet of stone are seen in the quarries, and the beds here contain beautiful yellow crystals of sulphate of barytes (sugar-candy stone). A coral (*Isastræa oblonga*) which has been converted into flint and chert, is also plentiful at Tisbury. In the valley east of Worton and south of Pottern sandy beds of the same age occur. Lastly, in the stone quarries at Swindon we see about 8 feet of yellowish Portland stone, underlaid by sands 25 feet thick; there is an outlying mass round the village of Bourton.

The Portlandian Beds are a marine shallow-water series, indicating an upheaval of the oolitic sea-bottom.

Purbeck Beds.— Strata of this age occur at two points only: at Teffont Evias, Chicksgrove, and Chilmark Common slabs of a thin compact limestone are raised for tiling; at Swindon we see in the stone quarries about 10 feet of fresh-water marls and limestones of Purbeck age, which have yielded to the diligent researches of Mr. C. Moore, of Bath, more than eighty species of fossils, including teeth of mammals, bones of the oldest true frog, together with fruits and seeds, and such fresh-water shells as *Paludina* and *Bithynia*. They thus show a continuation of that upheaval of the oolitic sea-floor which we have already noted when describing the Portland Beds, and which finally resulted in the formation of a continent occupying part of what is now the Atlantic Ocean.

NEOCOMIAN (OR LOWER CRETACEOUS) FORMATION.- Over the continent just referred to a great river flowed eastwards, the deposits in whose estuary constitute the Hastings Sands and Weald Clay of Kent, Surrey, and Sussex: of the *Hastings Sands* a small outcrop has been mapped by the Geological Survey at Catherine Ford on the Nadder.

Lower Greensand.—The sandy beds of this age are from 25 to 30 feet thick; they commence on the south at Poulshot Green, near to which is an outlier at Seend, where a patch of Lower Greensand rests on Kimmeridge Clay; here it contains a rich bed of iron ore, which has been largely worked. The outcrop broadens between Rowde, Bromham, and Bowden Hill, but soon narrows again to a band about a quarter of a mile wide, which can be traced by the dryness of the ground through Blackland, Corton, and Swindon Reservoir. At Sands Farm, east of Calne, a bed of very pure and white quartz sand has long been worked, and is carried to great distances for domestic purposes. Fossils are scanty in the Lower Greensand, but the curious shell *Diceras Lonsdalii* has been found. The beds are evidently a marine littoral deposit, and indicate subsidence.

THE CRETACEOUS FORMATION.—The lowest division of the cretaceous beds proper is termed the *Gault*. It is a stiff blue clay, full of small spangles of mica, and occasionally it contains calcareous concretions, septaria, and phosphatic nodules; it varies in thickness from 80 to 140 feet, and can be traced continuously, as a band of low ground at the base of the chalk escarpment, from Dilton by Westbury, Coulston, Heddington, Cliffe Typard, Wroughton, to Wanborough Marsh and Hinton Mill. It is worked for brick-making at several points along this line. The Gault is found along both sides of the Nadder Valley, at Donhead and Sutton Mandeville on the south side, and at Dinton on the north, but is absent at East Knoyle.

The *Upper Greensand* is in thickness equal to the Gault. It is a grey or brown sand, speckled with green grains, and containing irregular beds of sandstone and chert. From Shaftesbury it passes eastwards by Charlton and Berwick St. John to Barford St. Martin; then turns west by Knoyle, skirting the chalk by Warminster and Westbury, and then running like a bay into the chalk past Pewsey and Milston to Burbage. From Devizes the outcrop narrows greatly through Cherhill, Compton Bassett, Wanborough, and Little Hinton. Many echinoderms (sea-urchins) have been found in this formation in the neighbourhood of Warminster.

The Chalk.—The well-known pure white earthy limestone which we call "Chalk" occupies a larger area in Wiltshire than all the other strata put together. From Cranborne Chase in the south it stretches over Salisbury Plain, and then passes north-east by Marlborough: its total thickness is here about 800 feet, in which the following subdivisions can be traced. At the base we have a bed called *Chloritic Marl*, about 6 feet thick; it is of a pale yellow colour, with green grains of glauconite and phosphatic nodules. Next comes the *Chalk Marl* (locally called Malm), or Grey Chalk, about 50 feet thick; springs issue from the top, but in summer these become so dried up as to be without water for months together; sharks' teeth are common, with *Ammonites varians* and other fossils. Above this we have the *Lower Chalk*, or Chalk-without-flints, about 300 feet thick; it is softer than the Upper Chalk, and of milk-white or creamy tint. The *Upper Chalk*, or Chalk-with-flints, is about 400 feet thick; the layers of flint which mark the planes of bedding form a conspicuous feature in pits in the Upper Chalk, and enable us at once to distinguish it from the beds below. Fossils are common, mostly rounded sea-urchins, such as *Ananchytes ovatus, Cidaris, Micraster,* and *Galerites*, with remains of sponges and *Ventriculites*.

Between the Upper and Lower Chalk is a hard bed which Mr. Whitaker has named the *Chalk Rock:* it is a hard cream-coloured chalk with layers of green nodules, is well jointed, breaks with an even fracture, and rings under the hammer; sections are exposed near the top of the road-cutting on the northern side of the Pewsey Valley, about 4 miles south-south-west of Marlborough, also at Leigh Hill, and on the top of Whitesheet Hill, where it is about 3 feet thick.

The Chalk forms a bold escarpment facing west and north; in North Wilts "the unequal denudation of the hard and soft chalk forms a striking feature, especially when viewed from the north-west and west. The hard chalk rising abruptly out of the high and broad plateau of the soft chalk, the bare greasy sides and tops of the former contrast strongly with the cultivated plains of the latter; and while the plateau of the Lower Chalk rises and falls in gentle undulations, interspersed with small streams, the Upper Chalk is cut up into numerous ridges and valleys, and is nearly destitute of water." (Areline.)

The "balks," or "linchets," or terraces which run along the slopes of the chalk escarpment, are due to the downwash of soil by rain.

The Upper Chalk forms a white land, with a scanty covering of soil and a short sweet turf; the beech and box grow well upon it, and it affords pasturage for great flocks of sheep; much of it, however, has of late years been brought under cultivation. The great expanse of this rock, which is about equally divided between Hants and Wilts, forms what Pennant calls the "great central Patria of the chalk," the centre whence all the ranges of this rock traversing our island diverge.

There are many pits or quarries where the chalk is worked to burn into lime. "At Bishopston Down, near Warminster, enormous blocks of crystalline carbonate of lime, one of which weighed 50 cwt., were cut into slabs for mantel-pieces. Calcareous spar also occurs in a chalk-quarry at Nook, near Heytesbury, in blocks less both in number and size." (Conybeare.)

At Cherhill, east of Calne, a "White Horse" has been formed by cutting away the turf on the chalk escarpment.

Microscopical examination shows chalk to be composed of the minute shells of animals called foraminifera; it was formed by their slow deposition on the bed of a deep ocean.

TERTIARY PERIOD.—EOCENE FORMATION.—Though the beds we are now about to describe rest upon the chalk, there intervenes a period of time to be reckoned perhaps in millions of years. For during this interval the whole life of the globe had had time to change; not one species of animal or plant is common to the chalk and to these Eocene beds which rest upon it, and such a change could only have been accomplished in a vast period of time. The Eocene strata of England lie in two *basins*, viz., the London Basin and the Hampshire Basin; outliers only of the former occur in Wilts, but of the latter the main outcrop enters the south-eastern extremity of the county.

Hampshire Basin.—On each side of the chalk ridge called Dean Hill, about 5 miles south-east of Salisbury, we find a narrow band of the *Woolwich and Reading Beds*—unfossiliferous plastic clays and mottled sands. On these rests the *London Clay*, forming the Earldom Woods, Bentley Wood, &c. Then we get the *Bagshot Sands*, forming Hamptworth, Landford, and West Wellow Commons, with an outlying patch at Alderbury and West Grinstead.

London Basin.—Capping the hills round Great and Little Bedwin, outliers of the divisions of the Eocene strata occur. The *Reading Beds* and *London Clay* are here very thin, not more than 12 to 15 feet each in thickness, so that we are close to their original western termination.

THE DRIFT OR SURFACE DEPOSITS.—In the counties south of the Thames, no clear evidence of glacial action has yet been made out, and it seems probable that this district was above water when icebergs came sailing over the submerged more central portions of our island. The brick-earths of the Marlborough Downs, the clay-with-flints often found upon the chalk, and the flint gravel and quartzose gravels which are seen lying on the Oxford Clay are of very dubious and uncertain age; the gravels and alluvial deposits which border the present streams are, however, of much more recent formation, and form the last or newest of the geological series.

The Druid Stones, Sarsens, or Grey Wethers are huge blocks of a siliceous sandstone which are strewn in great numbers over the surface of the Chalk. The circles of Stonehenge and Avebury are mainly formed of these stones; the inner circles at Stonehenge, however, are of greenstone, similar, Professor Ramsay says, to some of the Silurian rocks of North Wales. The Grey Wethers are probably the relics of a bed of sandstone contained in the *Bagshot Series* (*Eocene*), which formerly stretched much further to the west than they do at present.

PREHISTORIC MAN.—Wilts is rich indeed in the remains of races of men of whom no written records exist. The earliest traces of man's presence in this district consist of certain roughly-shaped flint implements found in gravels on the sides of the valleys of the Avon, Wiley, &c., at Bemerton, Fisherton, Milford Hill, Lake, Ashford, Britford, Downton, &c. These gravel-beds are from 40 to 100 feet above the present level of the rivers, a fact which alone marks their great age, since the river has, since their formation, deepened its valley to that extent. These rough tools belong to Sir John Lubbock's *Palæolithic Stone Age*. Of much later date, and showing marks of great progress, are the elegantly fashioned and skilfully chipped stone implements of the *Neolithic Age*—the celts, chisels, scrapers, arrow-heads, knives, &c., which have been found scattered on the surface in so many localities in Wiltshire, but more especially at Avebury, Durrington, Everley, Roundway near Devizes, Stonehenge, Upton Lovel, West Kennet, Wilsford, and Winterbourn Stoke. In the mounds or barrows—prehistoric burial-places—which stud the Chalk Downs in such numbers, Sir R. C. Hoare found numbers of such objects, which are noted in his magnificent work on "Ancient Wilts." All or most of these stone objects were the work of men who lived in this country long

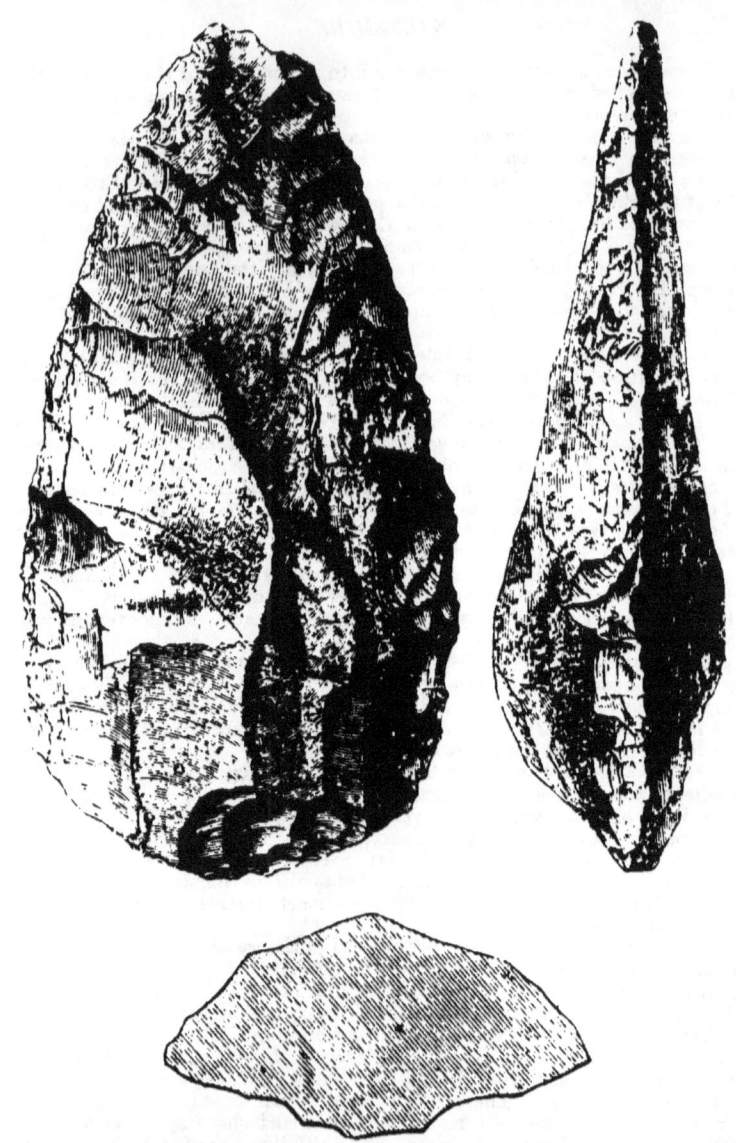

Fig. 92.—Palæolithic Flint Implement (one-half natural size), found in gravel at Elm Grove, Milford Hill, Salisbury. The lower figure shows the cross section.

before the time of the Romans —of men who were ignorant of the properties of metals.

Then we have the grand megalithic structures of Stonehenge and Avebury, and Silbury Hill, a miniature mountain which covers about 5 acres of ground and rises to a height of 130 feet; these structures, too, have usually been assigned to Druidic (*i.e.* pre-Roman) times, but Mr. Fergusson has of late years claimed for them a date about 520 A.D. and believes that the purpose was to commemorate the twelve great battles in which King Arthur overthrew for a time the Saxons, but these are questions which belong rather to Archæology than to Geology.

The composition and origin of the great blocks of rock which compose Stonehenge have been lately investigated by Prof. Maskelyne;. the outer circle is composed of "Sarsen stones"— compact quartzose rocks from the Tertiary strata: the "altar-stone" is a grey sandstone or micaceous grit, resembling the Old Red Sandstone of the Mendip Hills near Frome; 29 of the stones which form the inner circle may be referred to the igneous rock called *diabase*; three others are hornstones, and one is a siliceous schist.

No. 40.

GEOLOGY OF WORCESTERSHIRE.

NATURAL HISTORY AND SCIENTIFIC SOCIETIES.
Evesham Field Naturalists' Club.
Severn Valley Naturalists' Field Club.
Worcestershire Natural History Society.
Worcestershire Naturalists' Field Club.
Malvern Field Club. Transactions and Proceedings.
Dudley and Midland Geological and Scientific Society and Field Club. Proceedings.

MUSEUMS.
Dudley Museum.
Messrs. Burrow's Museum, Great Malvern.
Museum at Dr. Grindrod's, Malvern.
Natural History Society's Museum, Worcester.

PUBLICATIONS OF THE GEOLOGICAL SURVEY.
Maps. Sheet: 44, Evesham, Worcester, Cheltenham. Quarter Sheets: 43 N.E. Woolhope to Malverns; 54 N.E. Henley-in-Arden, Solihull; 54 S.E. Stratford-on-Avon; 54 N.W. Kidderminster, Bromsgrove; 54 S.W. Droitwich, Worcester; 55 N.E. Cleobury Mortimer, Bewdley, Stourport; 55 N.W. Clee Hills, Ludlow; 55 S.E. Bromyard; 55 S.W. Leominster, Weobley; 62 S.W. Wolverhampton, Walsall, Dudley; 62 S.E. Sutton Coldfield, Birmingham, Coleshill; 61 S.E. Much Wenlock, Bridgenorth.

IMPORTANT WORKS OR PAPERS ON LOCAL GEOLOGY.
1856. Symonds, Rev. W. S. Fossils from Keuper Sandstone of Pendock. Journ. Geol. Soc., vol. xi. p. 450.
1858. Morris, Prof. J. Fern from the Coal-measures near Bewdley. Journ. Geol. Soc., vol. xv. p. 80.
1860. Roberts, G. E. Rocks of Worcestershire. London.
1861. Symonds, Rev. W. S. Sections of Malvern and Ledbury Tunnels. Journ. Geol. Soc., vol. xvii. p. 152.
1864. Holl, Dr. H. B. Geological Structure of the Malvern Hills. Journ. Geol. Soc., vol. xx. p. 413.
1877. Harrison, W. J.—Rhætic Section at Dunhampstead, near Droitwich. Proc. Dudley Geol. Soc., vol. iii. p. 115.

See also General Lists, p. xxv.

THIS county lies just to the east of that great region of old rocks which, stretching westwards, forms the whole of Herefordshire and Wales. A line running north and south through the Malvern Hills and the Forest of Wyre Coal-field would separate the hard and ancient *Palæozoic* strata from the softer, later-formed *Mesozoic* beds which form the Valley of the Severn, and extend to the eastern edge of the county.

The Malvern Hills.—The age of the rocks which form the main axis of this well-known ridge has only of late years been correctly ascertained. From Keys End Hill on the south, by Raggedston Hill, Midsummer Hill, Swinyards Hill, the Herefordshire Beacon, Wind's Point, and thence to Worcester Beacon and North Hill, we find masses of crystalline rocks, mostly of a syenitic and gneissic character, but extremely varied in composition, which, though greatly altered, are probably of metamorphic origin, and are certainly much older than any of the surrounding strata.

Running all along the eastern edge of the chain, and extending for many miles in a line nearly due north and south, we have an enormous fault, or dislocation; so that the rocks lying to the east, in the Vale of the Severn, are altogether unlike and of a different age to the harder beds composing the hills, these indeed owing their superior elevation to their greater hardness, which has enabled them to resist denudation better.

The metamorphic rocks of the Malverns have been assigned by Dr. Holl to the LAURENTIAN PERIOD, which comprises the earliest of the stratified rocks; it is safer, however, to style them PRE-CAMBRIAN, as they evidently agree with the rocks described under that name by Dr. Hicks at St. David's, in South Wales, and by Dr. Callaway in the Wrekin and near Lilleshall, in Shropshire. Dr. Holl's views are ably set forth in a paper by him in vol. xxi. of the Geological Society's Journal, and in vol. xxiii. the Rev. J. H. Timins gives the results of a large number of chemical analyses of rocks from different parts of the chain.

Very fine sections were exposed during the construction of the Malvern and Ledbury tunnels of the Worcester and Hereford Railway, and these have been described by the Rev. W. S. Symonds. In the Malvern tunnel bands of chlorite-schist with graphite were found, and fissures which had been filled up with Silurian limestone.

There are about forty trap-dykes to be seen in different parts of the syenitic ridge. These are composed of igneous rock, which has been forced in a liquid state along fissures, or lines of least resistance, in the older beds. Small faults occur, usually running across the ridge. They are marked by lines of fragmentary or brecciated rock, and have been produced during the different movements to which the axis has been subjected.

CAMBRIAN FORMATION.— Resting on the western flanks of Raggedston and Midsummer Hills, there are certain beds known as the *Hollybush Sandstone.* They are of an olive and brownish-green colour, and contain small tongue-shaped shells, known as *Lingula squamosa*, a broader shell *Obolella Phillipsi*, together with tubes which represent the ancient homes of boring worms, such as may now be seen on the sea-shore. Where these sandstones rest against the metamorphic axis before described they are seen to be quite unaltered, and at the junction, moreover, there is a band of breccia composed of fragments of the crystalline rocks, which must accordingly have been already altered and upheaved when the Cambrian beds, themselves of such great antiquity, were deposited upon them.

Black Shales.—These are above 1,000 feet thick, and rest unconformably upon the Hollybush Sandstone, occupying the region about Bransill Castle. Like the sandstones, they contain numerous intrusive masses of volcanic rocks, in the neighbourhood of which they are considerably altered. They contain small trilobites of the genera *Olenus* and *Agnostus*, which are most numerous in a small band of rock in the Valley of White-leaved Oak, between Raggedston and Chase End Hills. A good plan is to take a quantity of the shale home in a basket, dry it, and then split it open, and carefully examine it with a lens.

In the south-west, near Hayes Copse, the black shales are seen to be overlaid by some greenish shales, containing a remarkable net-like fossil, called *Dictyonema socialis*. All these beds dip westwards at a considerable angle. On the east of the Herefordshire Beacon there is a small area occupied by rocks, which are probably of the same age as the Cambrian sandstones and shales above mentioned. They have, however, been so altered by intrusive trap-rocks that all traces of fossils have been obliterated.

SILURIAN FORMATION.—West of the Malverns, and stretching northwards in a long but narrow band as far as Abberley, there occur sandstones, shales, and limestones, of Upper Silurian age. The Lower Silurian beds are missing, having probably not been deposited in this area, which may then have been above the sea-level.

The Llandovery (or May Hill) Sandstone rises to the surface north of the Permian conglomerate in Bromesberrow Park. Thence it sweeps round by Rowick, Howler's Heath, and the Obelisk, to the Herefordshire Beacon, where it is cut out by a fault. Reappearing again on the other side of the hill, we can trace its outcrop as a narrow band hardly a quarter-mile wide up to West Malvern, Cowleigh Park, and Old Storrage, where it widens out, forming an anticlinal axis. Fossils are much more frequent than in the underlying strata. Large shells of the genera *Pentamerus, Atrypa*, &c., occur, with screw-like casts of an annelid, *Tentaculites annulatus*.

This May Hill Sandstone, in a greatly altered condition—turned into quartzite, in fact—also forms the Lickey Hills in the extreme north-east of the county; but this locality has been described in connection with the South Staffordshire Coal-field.

Going westwards we find the Upper Silurian rocks much folded, forming a series of synclinals and anticlinals. First we get some thin bands belonging to the *Woolhope Limestone*, then shales with the *Wenlock Limestone*, and then a great thickness of shale passing up into a nodular band, which represents the *Aymestry Limestone*. The general inclination of all the beds is to the westward, and they pass up gradually into the Old Red Sandstone, which then stretches continuously over a great part of Herefordshire.

Fossils are very numerous in these Silurian beds, and very fine collections may be seen in the Museum of the Malvern Field Club. Dr. Grindrod and other local collectors have also remarkable and extensive suites of specimens.

THE OLD RED SANDSTONE occupies so little of Worcestershire that it may be very briefly considered. It contains remains of land plants, and scales and other remains of large fishes. Traces of giant crustaceans also occur. These beds form the valley of the Teme, between Tenbury and Martley, and are also exposed over a limited area north of Abberley, and in a still smaller patch near Shatterford.

THE CARBONIFEROUS FORMATION. In Worcestershire the *Coal-measures* repose directly on the Old Red Marls, Sandstones, and Cornstones (concretionary earthy limestones), just mentioned. There is thus, as elsewhere in Central England, an absence of the thick beds of limestone and coarse sandstone (Millstone Grit) which constitute elsewhere the base of the Carboniferous Formation; and this is usually explained by supposing the existence in those times of a *land barrier* stretching across England from east to west, upon which, in consequence, these lower beds would not be deposited, but which sank altogether or in part beneath the sea before the deposition of the measures containing workable beds of coal.

The *Forest of Wyre Coal-field* extends from the Abberley Hills northwards in a broad strip, stretching from Bewdley on the east to within a mile of Cleobury Mortimer on the west. From Billingsley it may be traced northwards in a long narrow band, barely 1 mile wide, on the right bank of the Severn: to the north-east and east it is overlaid by Permian strata: its surface area is about 35 square miles. The beds here are the *upper* Coal-measures, which seldom contain coal-seams of much value or thickness. There are two or three beds of workable coal, one of which, about 5 feet thick, has been traced over a large region, but, owing to the inferior quality generally, this coal-field has not yet been thoroughly examined. In the band of limestone which is so remarkably constant in the higher beds of the Midland Coal-fields, the late

Mr. G. E. Roberts found many fossils of interest, such as tiny coiled-up *serpulæ*, with fish teeth and scales, and the small bivalve shell known as *Posidonia*. In shaly sandstones near Kidderminster the same geologist found fronds of ferns showing distinctly netted veins, a very rare character. These were described by Professor Morris in vol. xv. of the Geological Society's Journal. A bed of shale which has been cut through near Dowles Brook is very full of the impressions of ferns and leaves. Thick woods still extend over much of the coal district, in some of which one comes across the remains of old sinkings; and as the ground is undulating, or even hilly, the present collieries present a somewhat more picturesque appearance than is usually the case.

To the north-east the Coal-measures pass under the Permians, and there is a possibility of the future extension of coal-workings in this direction, by sinking through the newer strata. In several places intrusive sheets of basalt have been met with. At Arley Colliery, near Bewdley, a boring showed basaltic rock at a depth of 454 yards; a fair seam of coal occurred 176 yards from the surface. At Shatterford there is a rather remarkable basaltic dyke, locally called "dewstone," which appears connected with a fault running north and south, the downthrow being to the east. There is also a mass of igneous or trap rock near Kinlet.

THE PERMIAN FORMATION.—Strata of this age only occupy a few detached patches in the county, but they present geological phenomena of high interest. Just on the south of the Malverns, in Bromesberrow Park, Permian beds are described as "conglomerates of a dark red colour, filled with small fragments of the adjacent Silurian rocks, to which they are wholly unconformable, and on which they lie as an ancient sea beach." Similar beds occur near Martley, Abberley, Bewdley, Shatterford, on Church Hill, Warshill near Kidderminster, and the Clent Hills near Hagley. Here they are *breccias*, that is, composed of large angular fragments of rock, embedded in a paste of red marl. The rocks are similar to those of the Longmynd, distant some 20 miles to the north-west, and Professor Ramsay has suggested that they were transported from that point by ice in some remote glacial period. No fossils have yet been found in the Permian beds of Worcestershire.

THE TRIAS.—Red marls and sandstones of this period occupy at least three-fourths of the county. In the south they stretch across from the Malverns to Tewkesbury, and thence northwards they form the fertile Severn Valley, with an average width of 8 or 10 miles. Sweeping round by Droitwich and Bromsgrove, they again extend south to Feckenham and Alcester, whilst to the north and east they pass right over into Staffordshire and Warwickshire.

The Bunter Sandstones and Pebble Beds.—These are the lowest Triassic beds, and are only seen north of Stourport and Kidderminster, and on the southern flank of the Lickey Hills. They are well displayed along Kinver Edge, where, as at Nottingham, "Rock Houses," or caverns, have been excavated in the thick-bedded pebbly sandstones. They are seen along the crest of the ridge, and in a detached mass called "Holy Austen Rock." All the pebbles are waterworn, and consist of quartz, syenite, Silurian sandstone and limestone, &c., but no granites and no rocks less ancient than Millstone Grit. Every hard pebble is marked with spots, caused by pressure of pebble against pebble, which are very characteristic, and many are traversed by cracks produced by the same agency.

Keuper Marls and Sandstones.— At the close of the Bunter epoch it would appear that the English area was elevated, and became for a time dry land; for we find here no trace of a fossiliferous limestone—the *Muschelkalk* which intervenes between the Bunter and the Keuper on the Continent. The general dip of the Keuper beds is at a slight angle to the south-east or east. The intercalated sandstone beds generally form escarpments running nearly north and south, the valleys and the plains being composed of the marls. In the east of Worcestershire, however, the beds gently rise, so as to form a trough or synclinal axis, although they roll over and again dip to the eastward as we follow them in that direction. The lower beds are mostly sandy, reaching a thickness of 400 or 500 feet. At Ombersley and Hadley there are quarries which yield building-stones of delicate red and yellowish tints, which have been employed with good effect in the restoration of Worcester Cathedral. The quarries here are from 30 to 40 feet in depth, and contain plant-remains. Near Bromsgrove a remarkable fossil fish (*Dipteronotus cyphus*) was found in the same beds.

The New Red Marls come next in the series. They are finely exposed in the railway cutting near Worcester Station. Brine springs and beds of rock-salt occur at Droitwich and Stoke: at the former place salt has been extracted from the brine for upwards of 1,000 years, as it was one of the sources of revenue granted to Worcester Cathedral by Kenulph, King of the Mercians, in the year 816. The manner of working has been fully described by Mr. Horner. The brine occurs in a sort of subterranean reservoir, proved in several borings to be about 22 inches in depth, with a floor of rock-salt and a roof of gypsum from 40 to 100 feet in thickness: when this is pierced by boring the brine ascends to the surface with force. In 1879 there were obtained in Staffordshire and Worcestershire 230,500 tons of salt.

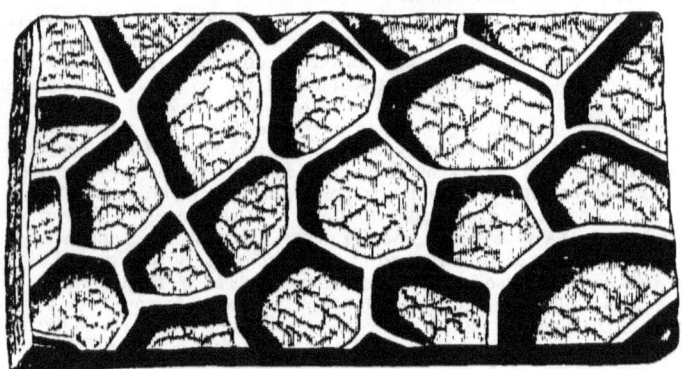

Fig. 93.—Suncracks in Lower Keuper Sandstone.

All these "red rocks" were probably deposited in vast inland salt lakes; as these dried up, the excess of salt was deposited on the lake-bottom. Beds of sandstone occur in the marls, and both are frequently ripple-marked; rain-pittings and sun-cracks often occur, with tracks of extinct reptiles, such as *Rhynchosaurus*, showing that the lake margin was subject to great fluctuations, according as the seasons were wet or dry. In the Upper Keuper Sandstones, at Pendock, the little crustacean *Estheria minuta* occurs in a good state of preservation; in appearance it resembles a small bivalve shell. The thickness of this upper member (Keuper) of the Trias in Worcestershire is probably not less than 1,000 feet.

THE RILETIC BEDS. This name is applied to a series of black and light-coloured shales, with nodular bands of limestone in the upper part, altogether not more than from 50 to 100 feet thick, which everywhere rest upon the Red Marls, and are in turn overlaid by the Lias. Hardly any good sections are known in Worcestershire, but they are pretty well exposed at Dunhampstead, on the Gloucester and Birmingham Railway, near Droitwich, and have been noted by Mr. Kirshaw, at Cracombe, north-west of Evesham, and at Hob-Lench.

THE LIAS. The limestones and bluish clays, which are so well known under this name, enter the county at Tewkesbury, pass thence northwards to Droitwich, and bend round to the Vale of Evesham, thus forming a spur or promontory, on which Pershore is the only town of any importance. The edge is usually marked by a low escarpment overlooking the plains of Red Marl. At Brockeridge and Defford Commons, near Tewkesbury, many bones of great extinct reptiles have been obtained from the *Lower Lias*, together with the *Ammonites* and bivalve shells which everywhere occur in abundance in these beds. At Strensham, beds called the "Insect Limestones" are well seen. There are three or four small outliers of the Lower Lias between Tewkesbury and the Malverns, as at Berrow Hill, Ham Court, Gray Hill, &c.

The *Marlstone*, a hard, ferruginous limestone, is only seen on the northern side of Bredon Hill. Here, too, we have the *Upper Lias Clays*, about 100 feet thick, forming wet and marshy ground, capped by INFERIOR OOLITE freestones, the whole rising to the height of 979 feet, and affording a most commanding view of the Vale of the Severn. This outlier is due to a fault which runs east and west along the southern foot of the hill.

THE DRIFT. By the action of subaërial and submarine causes the surface of the land has been made to assume its present varied aspect. At a comparatively recent geological period there is good evidence that an arm of the sea extended up the Severn Valley, forming the "Straits of Malvern;" indeed, the surface everywhere bears signs of complete submergence beneath the sea. In the Valley of the Avon we find thick beds of purple boulder-clay, full of chalk and limestone pebbles; over this clay, beds of gravel with flints are frequently found. In the centre and north of the county the drift is composed of reddish sand and clay, frequently containing pebbles of a greenish slaty-coloured rock. Fragments of marine shells sometimes occur. In these beds we probably have traces of the last *Glacial Period*, when glaciers and icebergs descended from the north of England bearing stones and mud, which they deposited in mingled heterogeneous masses where they entered the sea or melted. At Harborne near Birmingham there is a fine section showing two boulder-clays separated by a sandy deposit.

Gravels of later age have been deposited by the Severn and the Avon at considerable heights above their present beds, showing how the rivers have deepened the valleys in recent times. In these gravels remains of extinct mammals have been found, but unaccompanied as yet by any of the roughly-chipped flint implements which are at present the earliest known records of the presence of man.

PREHISTORIC MAN. -Of stone implements of the later or Neolithic period, Mr. John Evans found a fine double axe-head of basalt, about 5 inches long by 2 inches broad, near Grimley, together with another, pointed at one end only. One of rather different shape was found in the bed of the Severn at Ribbesford.

An oblong piece of chlorite slate, $5\frac{3}{8}$ inches long, $1\frac{3}{4}$ inches broad, and $\frac{1}{4}$ inch thick, rounded on one face and hollowed on the other, was found in a gravel-pit at Aldington: it has a hole at each corner, and was probably used as a bracer or guard for the left arm in shooting with the bow. A similar specimen was found in another gravel-pit at Lindridge.

No. 41.
GEOLOGY OF YORKSHIRE.
EAST RIDING.

NATURAL HISTORY AND SCIENTIFIC SOCIETY.
Driffield Literary and Philosophical Society.
MUSEUM.
Museum of the Literary and Philosophical Society, Hull.
PUBLICATIONS OF THE GEOLOGICAL SURVEY.
Maps. -Quarter sheets: 95 S.E. Bridlington; 95 S.W. Driffield. (Survey not completed.)
Books. Geology of the Oolitic and Cretaceous Rocks South of Scarborough, by C. Fox-Strangways, 1s.
IMPORTANT WORKS OR PAPERS ON LOCAL GEOLOGY.
List of Works on the Geology of Yorkshire, by W. Whitaker, in Phillips' Geology of Yorkshire Coast, 3rd edition; and in the Guide to Fossils in the Leeds Museum (1d.).
1826. Vernon, Rev. W. Account of the Strata at Bielbecks, near Cave. Annals of Philosophy, 2nd series, vol. xi.
1875. Mortimer R. Well Section in Chalk at Driffield. Journ. Geol. Soc., vol. xxxi. p. 111.
1879. Lamplugh, G. W. Fresh-water Remains in the Boulder Clay at Bridlington. Geol. Mag., p. 393.
1881. Fox-Strangways, C. Geology of York. Science Gossip, p. 169.
1881. Lamplugh, G. W. Shell Bed at Base of Drift at Speeton. Geol. Mag., p. 174.
See also General Lists, p. xxv.

THE physical structure of the East Riding is simple and easily understood, and is directly dependent on the nature of the rocks which constitute its surface. There is a central axis or backbone of chalk, which runs from the Humber northwards to Settrington, and then curves eastwards to Flamborough Head. Older rocks than the chalk form the Vale of York on the west and the Vale of Pickering on the north, whilst newer rocks constitute the Plain of Holderness on the east.

In the study of this part of England we have not the full advantage of the admirable maps which have been executed of other counties by the officers of the Geological Survey, showing the exact position and extent of every bed of rock; but the whole strength of the Survey is now being concentrated in the eastern and north-eastern counties, and doubtless within the next half-dozen years we shall have from this Government Department very full and minute particulars respecting the rocky structure of this region; in fact the map of the north-east portion of this Riding is already completed, but from the time necessary to engrave and colour it, it will probably not be ready for issue for some months. But much has been done by amateur workers, so much indeed that future students will only have details left to fill in.

Passing over the work of Lister (1674), Dr. Alderson (1799, on Hull and Beverley), Storer and Winch (1815-17, on the ebbing and flowing stream in Bridlington Harbour, in " Philosophical Magazine "), we come to William Smith, who in 1821 published the first geological map of Yorkshire (in four sheets). In 1829 Professor Phillips (Smith's nephew) published his " Geology of the Yorkshire Coast," of which a second edition appeared in 1835, and a third (revised by Mr. R. Etheridge) in 1875. The fresh-water deposit at Bielbecks was described by the Rev. W.V. Vernon in 1829-30 (" Philosophical Magazine "). Mr. H. C. Sorby has written several notes on the drift of Holderness, but this deposit has been most thoroughly examined by Mr. Searles V. Wood, jun., and the Rev. J. L. Rome (Quarterly Journal Geological Society, vols. xxiv., xxv.). Professor Judd has admirably described the Speeton Clay in the same publication (vols. xxiv., xxvi.). The Lias has been treated of in a separate work (Blake and Tate on the Yorkshire Lias), and the Chalk has been described by the Rev. J. F. Blake (Proc. Geolog. Association, vol. v.) and Messrs. Meyer, Mortimer, Rev. T. Wiltshire, and others.

In commencing a geological account of the East Riding we shall begin on the west of the district, where the river Ouse forms its boundary. Here the oldest rocks occur, and as we pass from this part eastwards towards the coast we shall meet one after another with beds of newer (later-formed) rocks. In fact all the rock-beds stretch or extend (or *strike* as we say geologically) from north to south, while they slant or dip towards the east, resting one upon another in this direction. The consequence is that a deep boring anywhere on the coast, say at Bridlington or Hornsea, would, after passing through all the intervening rock-beds, reach at last those which we are about to commence with, and which form part of the Vale of York.

THE TRIAS OR NEW RED SANDSTONE. The rocks of this age are in the East Riding entirely of a clayey or marly nature. Their colour is red, a fact which is due to their impregnation with a small quantity of peroxide of iron, and they are entirely destitute of fossils. They stretch across the Ouse westward to Tadcaster, &c., and run north to the Tees, forming in fact the greater part of what is known as the Vale of York; but with their extension in these directions we have at present nothing to do. Northwards they also run beyond the boundary of the East Riding between York and Stamford Bridge, and southwards they pass under and are concealed by the alluvial deposits forming Walling Fen, &c., and bordering on the Ouse. The beds of this division of the Trias are known as the *Keuper Marls*, and may be 600 feet in thickness; a boring made at York reached this depth and was still in the " red rocks." The marls contain thin bands of gypsum and (elsewhere) much rock-salt. Their mode of formation is believed to have been in an inland sea or lake, whose waters were so charged with mineral substances as to be inimical to life. The top of the Keuper marls is generally marked by bands of a grey or blue tint alternating with red, but the whole of the Trias is so covered over with beds of sand, gravel, clay, &c. (drift), of much later age, that good exposures are rare. The height of this region is low, not exceeding some 70 feet above the level of the sea. This fact is due to the soft nature of the rocks, which have been worn away by rain, rivers, glaciers, &c., at a more rapid rate than the harder masses of sandstone and limestone which form hills on the east and on the west of the Vale of York.

THE RHÆTIC BEDS. This is the name given to beds which, though not more than 30 or 40 feet thick in Yorkshire, attain to about one hundred times this thickness in the Rhætian Alps of Lombardy. They form a link between the Triassic beds we have just described and those of the Lias, to which we shall next come, and may be termed " passage beds," connecting the one formation with the other. They constitute the boundary of the red marls on the east, and from their thinness, moderate hardness, and the fact that they furnish no product of economic value, are very difficult to trace on the surface. In fact, there is no good exposure of the Rhætic beds known in the East Riding, and local geologists would do well to search diligently for them.

Professor Phillips records a section on the railway cutting between Barton Station and Howsham. Here were red Keuper Marls (with one greenish-white band near the top) surmounted by shales containing eighteen thin bands of limestone, altogether 100 feet thick, and containing the well-known fossil shell *Gryphæa incurva* in the upper part. In Howsham Wood the Rev. J. F. Blake found fragments of limestone containing characteristic Rhætic fossils, such as *Avicula contorta*, &c., and black shales also occur near. In Bugthorpe Beck are many fragments of sandstone, in one of which Mr. Blake found a saurian bone, and these may represent the famous Rhætic bone-bed. The beds called *White Lias*, which form the top of the Rhætic series, appear also to occur near Pocklington, Garrowby, Acklam Wood, &c. Thus it is pretty certain that the Rhætic beds in the East Riding are composed, as elsewhere, of a lower series of black shales from 10 to 20 feet thick, resting on the Red Keuper Marls, and containing one or two sandy layers (bone-beds), and an upper series of light shales and white earthy limestones (the White Lias), containing such shells as *Pleuromya* and *Modiola*. All along the line we have indicated, from Hotham in the south, by Owsthorpe and Garrowby, to Howsham, a diligent look-out should be kept for landslips, excavations, &c., which may reveal the wished-for sections. In the "Geologist Magazine" for 1858 there is an account of the neighbourhood of Hotham by the Rev. T. W. Norwood, in which it is stated that the "Lowest Lias may be seen in the marl-pits passing conformably into the New Red Sandstone:" but these pits are now closed.

THE LIAS. This is the name given to a formation consisting of beds of bluish shales and clays, interstratified with thin beds of limestone. Its thickness in the East Riding does not exceed 600 feet, and it has a gentle easterly dip of about $2\frac{1}{2}$ degrees, which is the same as that of the beds of Red Marl on which it rests: the Lias can be traced through the whole of England, from the cliffs at Lyme Regis, in Dorsetshire, through Gloucester, Warwick, Leicester, and Lincoln shires, until it crosses the Humber south of Brough. From this point we can trace it as a low escarpment northwards by North Cave, North Cliff, Market Weighton, and Burnby to Kirby Underdale and Howsham, where it crosses the Derwent.

The Lias is everywhere fossiliferous, and the shells, &c., which occur in it are most fortunately not scattered indiscriminately throughout, but especially the more highly organized kinds, such as the ammonites are restricted on the whole to certain levels or horizons, so that when we find for example a certain species of ammonite, we can tell almost exactly in what particular part of the whole thickness of the Lias the quarry is situated from which we obtained the fossil. In the lowest beds appears the first British ammonite *Ammonites planorbis*. These beds are well seen in the marl-pits open in the low escarpment which runs from North Cliff towards the Humber. Then in higher beds occur *Ammonites angulatus, A. Bucklandi*, and *A. oxynotus*. The beds so far may be classed together and called the Lower Lias, and in this district they constitute the greater thickness of the formation.

The *Middle Lias* is represented by the zones of *Ammonites armatus* and *A. spinatus*. The latter shell is the characteristic fossil of the famous ironstone beds of Cleveland, and these are present in the East Riding, although greatly diminished in thickness, and no longer rich in metal. A blue oolitic ironstone crops out near Millington and on the South Dalton road near Market Weighton. At Caldwell, near Sandon, a magnificent spring gushes from a bed of this rock, and thence the Middle Lias extends southwards by Houghton Hall and Hotham.

Indications of the *Upper Lias* are seen at Sancton and Houghton Hall, where *Ammonites communis* and *A. bifrons* occur in soft clays in such numbers that they have been used for mending the roads. The jet rock is absent. Thus the Lias of the East Riding agrees in all its main features with that of Lincolnshire and the rest of England, and differs strongly from that of the North Riding, the change being apparent directly we cross the valley of the Derwent. As a point of interest we may add that the limestone lands of the Yorkshire Lias are never used for lime-burning, being too impure. The breadth of surface occupied by the Liassic beds is on an average 1 mile from east to west; but in the neighbourhood of Kirby Underdale it exceeds 4 miles.

THE OOLITE.—Rocks of this age are exposed in two distinct localities — firstly, in the north-west corner of the Riding, between Kirkham and Norton; and secondly, in a narrow band running southwards from near Sancton by South Cave to the Humber. Between Sancton and South Grimston the Oolitic strata are hidden from view by the chalk of the wolds, which overlaps them so as to rest directly on the Lias. This is unfortunate, as we are thereby prevented from tracing the manner of junction of the two sets of Oolitic beds, which differ in many points.

Taking the southern area first, we find several exposures of the Inferior Oolite near South Cave. There are sandy beds and also layers of good limestone. These represent the *Northampton Sand* and *Lincolnshire Limestone* of the midland counties. Between Drewton and North Newbald they are overlaid by sandy clays containing fossils which prove them to belong to the *Oxford Clay*.

In the northern area the Inferior Oolite extends from Kirkham by Westow and Acklam to South Grimston. The order of succession can be made out fairly well on the sides of the valley of the Derwent. The lowest beds here are the Dogger ironstone of Kirkham and Castle Howard, above which come beds of shale and then the Main or Whitwell limestone, which is quarried at Westow and several other places. This bed is considered to represent the Lincolnshire Limestone on the one hand and the "Millepore Bed" of the Yorkshire coast on the other. Higher up the Derwent Valley, between Mennithorpe and Norton, we find beds of the age of the Oxford Clay, Coralline Oolite, and Kimmeridge Clay. The latter runs eastward across and indeed forms the basis of the Vale of Pickering, although it is so covered over by *drift* as to be hardly ever visible at the surface: it forms the northern or lower part of Filey Bay, while the harder Coralline Oolite runs out to sea as the headland called Filey Brigg. All along their course in this region the Oolitic beds form a belt of varied picturesque scenery, rising to heights of from 300 to 400 feet above the sea, and contrasting well with the more monotonous tracts of the Vale of York on the west, and the chalk wolds on the east. Both the Liassic and the Oolitic strata were, we think, deposited in seas of no great depth, the coast line of which lay to the north and west.

THE NEOCOMIAN OR LOWER CRETACEOUS FORMATION.—This term includes certain strata known as the *Speeton Clay*, from their great development on the coast at Speeton Cliffs, about half-way between Flamborough Head and Filey Brigg. These Speeton beds may be as much as 500 feet in thickness. They contain numerous fossils, such as *Belemnites jaculum, Exogyra sinuata*, &c., with several zones of *Ammonites*. Professor Judd has shown them to be the equivalent of the Neocomian strata of the Continent, and (in their upper part) of the Lower Greensand of the south of England. The cliffs are much obscured by landslips, whilst a great fault or dislocation of the strata probably runs through Speeton Gap; hence their study is one of great difficulty: they have a south-westerly dip of about 10 degrees, passing under the chalk, to which they are unconformable: the Speeton beds are exposed at only one or two places inland, as near West Heslerton and Knapton.

UPPER CRETACEOUS FORMATION.—*The Chalk.* Where the Chalk forms the coast-line it rises in height from 159 feet at Flamborough Head to 450 feet near Speeton. Striking to the west it passes inland by Hunmanby to Thorpe and Settrington. Here the direction changes, and it runs southwards past

Garrowby, Warter, Londesborough, and Goodmanham to North Ferriby on the Humber. All along this curved line the chalk forms a bold escarpment whose extreme height is 808 feet at Wilton Beacon, near Kirby Underdale: Acklam Wold is 751 feet and Hunsley Beacon 531 feet. From this escarpment the surface gradually slopes to the south and east until near Bridlington, Driffield, Beverley, and Hessle the chalk passes under that great accumulation of clay, sand, &c., which forms the surface of the Plain of Holderness. The crescent-shaped mass of chalk thus exposed at the surface includes about 376 square miles, and is 15 miles wide in its central part, but tapers off to about one-third of this amount at its eastern and southern terminations: its total thickness is 800 feet.

At the base we find an irregular bed of *red chalk* well seen at Speeton and from 6 to 20 feet thick. *Belemnites minimus* is the commonest fossil. Over this comes the *grey chalk* about 80 feet thick: up to this point no flints occur. Then we get a mass of flint-bearing chalk 380 feet thick, containing *Inoceramus* as a common shell in its lower part, and sea-urchins (*Micraster*) in the upper. Lastly, the highest and most easterly beds (320 feet thick) are destitute of flints and contain many fossilized sponges, with *Marsupites*, &c. Mr.

Fig. 94.—Thin Slice or Section of hard White Chalk, examined by transmitted light and highly magnified. Entire shells and fragments of *Globigerina* and other foraminifera are seen. (After Nicholson.)

Fig. 94A.—Atlantic Mud or Ooze, composed chiefly of Foraminifera (*Globigerina*, &c.).

Mortimer of Fimber has a fine collection of fossils from the Yorkshire chalk. Danes Dyke, on the coast, is a good spot for the collection of upper chalk fossils. In its scenery the Yorkshire chalk much resembles that of the south of England. The smooth and rounded contour of its hills, intersected by dry winding valleys and almost bare of trees, enables the skilled eye to recognise the nature of the rocks which compose it, with great readiness. When William Smith, "the father of English Geology," visited York, in 1794, he wrote, "From the top of York Minster I could see that the wolds contained chalk by their contour." It may be noticed that the Yorkshire chalk is much harder than that of the south of England: it is quarried at many points for lime-burning, while the flints are used to mend the roads. As to its mode of formation, the chalk appears to be an organically-formed rock, *i.e.*, it is composed mainly of the tiny shells of various species of foraminifera (microscopic creatures of low organization) such as now inhabit the surface of the North Atlantic. As these die their shells fall to the bottom, where they constitute a whitish ooze or mud, which when dried and hardened by heat and pressure forms a mass much like chalk, which rock we therefore think was formed in much the same way at the bottom of a deep ocean but some millions of years ago at the very latest. (See figs. 94 and 94A.)

QUATERNARY PERIOD. —We now pass over a great period of time, to represent which no rocks occur in Yorkshire, and in the surface accumulations of Holderness we come at once upon deposits which were formed when this country had for a time an extremely cold climate, so that great ice-masses (glaciers) covered all the north of England. The stones (boulders) and mud, rubbed by these glaciers off the land form the *Great Chalky Boulder Clay*, which is exposed at the foot of the cliff at Dimlington, and also between Ringborough and Mappleton. On top of this leaden-coloured chalky clay, we find in Dimlington Cliff and in Bridlington Harbour, a shelly deposit which may mark an interval when the climate for a time was warmer. *Nucula Cobbddiæ* is a characteristic shell. Above this, we get a *Purple Boulder Clay* along the coast between Bridlington, Flamborough Head, and Speeton. It is unstratified, tough, and of a purplish-brown colour, containing angular and striated stones. In its lower part it is very chalky, but this material diminishes in amount upwards. The *Hessle Sands and Gravels* occur near Hessle on the Humber, and are overlaid by the Hessle Boulder Clay, which stretches over most of the plain of Holderness.

Lastly, we get fresh-water deposits at one or two points, as at Bielbecks, 2 miles south of Market Weighton. These contain marl-beds full of river and land shells of still-living species, together with bones of large quadrupeds, such as the mammoth and rhinoceros, of species now extinct.

PREHISTORIC MAN.—The East Riding is extremely rich in those flint implements, which belonged to people who inhabited our islands before the use of metals was found out. This Stone Age, as it is termed, is divided into two parts—an older or *Palæolithic*, when the stone implements were of the roughest and rudest character; and a later or *Neolithic*, by which time man had attained considerable skill in the treatment of flint, which was the material almost invariably selected for the manufacture of the various axe-heads, knives, scrapers, arrow-heads, &c., which we find.

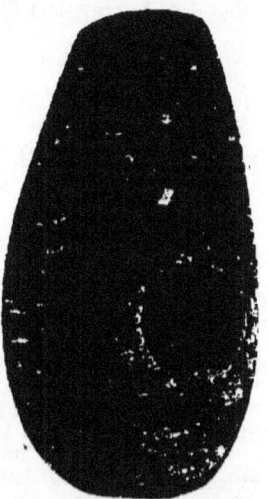

Fig. 95.—Perforated Axe-head of Micaceous Grit, found in a burial-mound (with a skeleton, a bronze knife, &c.) at Rudstone, near Bridlington, Yorkshire.

All the Yorkshire specimens belong to the Neolithic period. Scattered over the Wolds at Bridlington, Sherburn, Acklam Wold, Cowlam, Fimber, Gauton Wold, Rudstone, &c., very many specimens of these ancient stone tools and weapons have been found: indeed, at some places, they are so numerous that it seems probable they mark the site of ancient manufactories, spots on the Wolds where, the flint being specially suited for the purpose, large numbers were made to exchange with neighbouring tribes for other commodities. Those who wish to be able to recognise these prehistoric stone objects, should experiment on lumps of flint with a hammer, holding the flint in one hand and striking it smartly on the edge with the hammer. The shape of the flakes detached will be found to be very different from the naturally fractured flints, such as may be picked up in the bed of a river or on the sea-shore.

No. 42.

GEOLOGY OF YORKSHIRE.

NORTH RIDING.

NATURAL HISTORY AND SCIENTIFIC SOCIETIES.
Whitby Literary and Philosophical Society.
Scarborough Philosophical Society.
Richmond and North Riding Naturalists' Field Club.
Cleveland Literary and Philosophical Society.

MUSEUMS.
Kirkleatham Museum.
Redcar Museum.
Scarborough Museum.
Whitby Museum.

PUBLICATIONS OF THE GEOLOGICAL SURVEY.
Maps.—Quarter-sheets: 95 N.W. Scarborough; 95 S.W. New Malton; 97 S.E. R. Ure; 97 S.W. Hawes; 97 N.E. Richmond; 97 N.W. Swaledale; 104 S.E. Whitby. (Survey not completed.)
Books. Geology of Coverdale, Colsterdale, and parts of Wensleydale and Nidderdale.

IMPORTANT WORKS OR PAPERS ON LOCAL GEOLOGY.
1836. Phillips, Prof. J. The Mountain Limestone.
1855. Phillips, Prof. J. Rivers, Mountains, and Sea Coast of Yorkshire, 2nd edition.
1863. Baker, J. G. North Yorkshire. Lond.
1875. Phillips, Prof. J. The Yorkshire Coast, 3rd edition.
1876. Tate, R., & Blake, J. F. The Yorkshire Lias. Van Voorst. 28s.
1877. Barrow, G. A New Marine Bed in the Lower Oolites of East Yorkshire. Geol. Mag. p. 552.
1880. Hudleston, W. H. Palæontology of Yorkshire Oolites. Geol. Mag., p. 241.

See also General Lists, p. xxv.

THE officers of the Government Geological Survey have for several years been actively engaged in the North Riding; maps of the eastern and western parts will probably be published during the present year (1881), but for the central part (Vale of York) we shall have to wait some time longer. We are not ignorant, however, of the general geological features of the district. The arrangement of its rock-masses has been well described by several authors, some of whom, indeed, have laboured at their elucidation with special care for the sake of the county which gave them birth. First we must place Professor John Phillips; his works on the "Mountain Limestone" and the "Yorkshire Coast," together with his geological map of the entire county (published in 1853) are of the highest interest and accuracy. Mr. J. G. Baker's "North Yorkshire" (1863) is a very useful compendium. The "Yorkshire Lias," by Tate and Blake, and two important papers by Mr. W.

H. Hudleston on the Oolites, lately published in the Proceedings of the Geologists' Association of London, give minute details respecting the beds of the eastern division. Much has also been done by Messrs. Tiddeman, Dakyns, De Rance, L. C. Miall, Binney, M. Simpson, J. Rofe and others. Mr. Wood may be named as a great collector of the fossils of the mountain limestone, and Messrs. Williamson, Bean, and Leckenby for the coast. The museums at York, Scarborough, Whitby, and Richmond, contain local collections of rocks and fossils, which may be freely inspected.

The great extension of the North Riding from west to east causes it to include a large variety of rocks, for it is well known that in England the strata, or beds of rock, run across the country from north-east to south-west, resting one upon the other, the oldest or first formed lying most to the west. This order of succession holds good in North Yorkshire. The beds of the Pennine chain in the west are the oldest, and those which form the south-east corner of the coast between Scarborough and Filey are the newest of all the stratified rocks which enter into the "solid" geology of the district. The glacial deposits of clay and gravel, with travelled blocks of stone of all sizes, rest indiscriminately on the edges of the rock-beds of which we have been speaking, and must be considered separately.

The Carboniferous Formation.— The rocks which bear this name form all the west of the county, from the Pennine ridge as far east as a line drawn from Pierce Bridge, on the Tees, to near Tanfield, on the Ure. Here we can distinguish three main divisions, viz.: (1) the Carboniferous Limestone, also called the Mountain Limestone and Great Scar Limestone; (2) the Yoredale Rocks; (3) the Millstone Grit. The total thickness of these beds is about 3,000 feet, but it is not easy to say exactly how much of this belongs to each division. The reason for this is that as we go northwards we are approaching the shore line of the ocean in which these Carboniferous beds were deposited. Consequently, the Limestone, which to the south, in the West Riding and Derbyshire, forms an undivided mass some thousands of feet in thickness, becomes much reduced as we go northwards, and beds of shale and sandstone are interpolated in it; it is well exposed along Yoredale (or Wensleydale) between Hawes and Middleham.

The Yoredale Rocks, which succeed, were so named by Professor Phillips, from their occurrence higher up on the sides of the same valley; they are composed of sandstones (flagstones) and shales, with some limestone bands; the latter are quarried at Gilling. They contain good veins of lead ore, which are worked at several points in Nidderdale, Swaledale, Wensleydale, and Wharfedale. The total amount of lead ore raised in Yorkshire in 1875 was 4,050 tons; this yielded 2,946 tons of lead and 7,438 ounces of silver. In 1879 there were raised 5,132 tons of ore, yielding 3,713 tons of lead and 6,475 ounces of silver. Arkendale and Old Gang were the most productive mines. The veins in which the ores occur were, doubtless, once fissures or cracks in the rocks, and the minerals they now contain have been deposited from water holding them in solution as it circulated through these fissures.

The Millstone Grit owes its name to the hard coarse beds of sandstone which mainly compose it; altogether it is from 500 to 600 feet thick. Between Masham, Bedale, and Melsonby the Millstone Grit forms comparatively low, flat ground, an undulating moorland gritstone country, whose surface is about 300 feet above the level of the sea. But as we follow it westwards it rises and forms a capping to the highest hills, which indeed owe their existence to the weather-resisting powers of the hard sandstones which compose it; it forms lofty ridges separating Teesdale from Swaledale, and the latter from Wensleydale, rising to heights of over 2,000 feet, and, still rising westwards, crowns Shunnor Fell (2,329 feet), Water Crag (2,186 feet), and Micklefell (2,600 feet).

The scenery of the Carboniferous tract is very varied. On the Pennine escarpment the fine perpendicular "scars" formed by the limestones are conspicuous, and it is to the alternation of hard limestones with softer beds of shale and sandstone that the numerous "forces," or waterfalls of the same region and the dales are due. Subterranean caverns and "swallow-holes" abound in the limestone rocks, and although the surface vegetation is scanty, yet ferns of great beauty line the frequent fissures which traverse the rock. The Millstone Grit forms a somewhat dreary moorland surface.

The fossils of these rocks confirm the ideas already stated concerning their origin. In the limestones we get remains of *crinoids*, and marine shells, such as *Productus giganteus*, *Orthoceratites*, and *Bellerophon*. Messrs. Wood and Rofe both formed fine collections of *Crinoidea* from this district, and of a new genus, named *Woodocrinus* after its discoverer, many very beautiful specimens have been found in Swaledale, near Richmond, showing the long stem, cup-like body, and branching tentacles of this curious "sea-lily." Higher in the series the shales and grits contain scanty plant remains, and even thin coal-seams. Chert is common; it is an impure variety of flint. The splendid collection of fossils formed by the late Mr. E. Wood has been purchased for £720 by Mr. Reed and presented by him to the York Museum.

THE PERMIAN FORMATION. No Coal-measures are present in the North Riding, so the Permian beds rest directly upon the Millstone Grit. Doubtless, the coal-field of Durham was once continuous with that of the West Riding, but before the deposition of the Permian strata, a great upheaval took place in an east and west direction, and from the district so raised between the Tees and Wharfe the Coal-measures were denuded or swept away.

The Permian or Magnesian limestone crosses the Ure half-a-mile west of Tanfield Bridge: thence it runs N.N.W. to Thornton Watlas, forming an escarpment 300 feet high, with an easterly slope. We then only see it in irregular patches near Little Crakehall and Catterick, until it again appears on the south side of the Tees at Pierce Bridge, where it constitutes the lower part of the cliffs.

THE TRIAS, or New Red Sandstone, forms the vale of York, a comparatively low level tract of land which runs from Middlesbrough and Croft on the north, southwards past Northallerton and Myton, to and beyond York. The Red Sandstones and Marls are scarcely ever exposed to view in this district, being deeply covered up by glacial deposits of clay, gravel, &c. The lower Triassic beds are called the *Bunter* (or variegated) *Sandstone*. They run along the west side of the vale, whilst the *Keuper Marls* occupy the eastern side. In a boring made at Middlesbrough, by Messrs. Bolckow and Vaughan, in 1868, after passing through 58 feet of surface deposits they penetrated 81 feet of red marls and clays, containing two beds of gypsum, 2 feet and 6 feet thick respectively. Next came no less than 1,067 feet of white and red sandstone with gypsum veins. At a depth of 1,206 feet, the borer entered a bed of rock-salt, and continued in it to a total depth of 1,306 feet, thus giving a thickness of 100 feet of salt. Here the boring stopped, the total thickness of the salt bed not having been proved. It is in the Trias of Worcestershire and Cheshire that our great stores of salt are found, but they are there attained at one quarter of the above named depth.

THE LIAS.—This is the geological name for a great series of bluish shales and clays which have harder beds of ironstone in their central portion. The whole series is of great thickness in north-east Yorkshire. Mr. Blake gives Lower Lias 378 feet, Middle Lias 470 feet, and Upper Lias 280 feet, making a total of 1,128 feet. These beds form the easterly part of the vales of York and Cleveland. They form, too, the slopes of the moorland hills which are capped by the Oolitic beds, and they are exposed along the sides of many inland valleys, as in Eskdale, Rosedale, Bransdale, &c. The whole series extends from the Derwent Valley between Kirkham and Howsham in a north-westerly direction, past Craike to Dalton and Thormanby, then it runs due north by Osmotherly, but turning to the north-east passes through Stokesley and Guisborough to the coast between Redcar and Staithes. The beds have a gentle south-easterly dip of about 3 degrees, and can be traced southwards along the coast, forming the lower portion of the cliffs as far as Robin Hood's Bay, where, owing to an east and west fault, they extend inland for some distance.

The *Lower Lias* is composed of blue clays with thin bands of impure shelly limestone. In the *Middle Lias* we get a great development of iron ore. One bed known as the "Bottom Seam" is characterized by the fossil *Ammonites margaritatus*, but the famous "Cleveland Main Seam" is situated from 7 to 20 feet above this, and is characterized by the presence of *Ammonites spinatus*. These beds are the equivalents of the marlstone series of the middle and south of England. In the year 1875 the quantity of ore (argillaceous carbonate) raised in north Yorkshire (Cleveland district) amounted to 6,124,794 tons, valued at four shillings per ton net at the mines; in 1879 only 4,750,000 tons were raised, and the net price at the mines was estimated at three shillings per ton only. (See Government publication, "Mineral Statistics of the United Kingdom," by R. Hunt.) The main seam is thickest and most valuable at its northerly outcrop at Eston Nab, where the workable part is from 10 to 12 feet thick, and it decreases in value as we follow it south-east to Whitby and south-west to Thirsk. On an average the rock contains 30 per cent. of metallic iron. This seam may be said to have been first worked by Messrs. Roseby at Skinningrove in 1848, but it is to the firm of Bolckow and Vaughan, who opened the main seam at Eston in 1850, that the chief share of credit is due for the immense energy and vigour with which they laid the foundations of the vast industry now centred at Middlesbrough.

The *Upper Lias* is shaly and contains the jet-rock series near its base. The substance known as jet appears to be formed by the segregation of the bitumen with which the shales are charged. It is not fossil wood, though pieces of wood are often found which have been converted into jet. The manufacture of jet ornaments gives employment to about 1,500 persons, and is chiefly carried on in Whitby: the value of the trade in 1872 amounted to £88,000.

Above these jet-beds we get the *alum-shales*, marked by the shell called *Leda ovum*. They are about 110 feet thick, and in former times the shale of Whitby was our only natural source of potash-alum. The shale having been dug was calcined by means of burning brushwood: it was then washed in large cisterns to obtain the aluminium sulphate, and after evaporation either sulphate or chloride of potash was added, which precipitated the alum in the form of flour. About 25 years ago a Mr. Spence, of Manchester, invented an improved process by which alum could be more cheaply obtained from the shales of the Coal-measures, and the Yorkshire product has, in consequence, been driven out of the field.

The Yorkshire Lias furnishes a rich collecting ground to the fossil-hunter. No fewer than fourteen distinct zones, characterized by different species of ammonites, have been made out, whilst of other shells there is also a great profusion. Bones of large reptiles (the *Ichthyosaurus* and *Plesiosaurus*) characterize the upper portion, where (especially in the jet-beds) remains of fishes are also common.

THE OOLITE.—The oolites of north-eastern Yorkshire present many striking differences when compared with those of the typical region of the Cotteswold Hills. Instead of the thick beds of limestone which there constitute the main feature of the formation, we find in Yorkshire a great development of sandstones and shales full of plant-remains, and containing many shells resembling those which now inhabit estuarine or brackish waters. Even thin seams of coal occur, and altogether it is easy to see that the Yorkshire oolites were mainly formed in a large estuary and in shallow seas near to the coast line, while those of the south-western counties were accumulated in a deeper and more tranquil sea.

In the cliffs of the Yorkshire coast between Staithes and Filey, we get a magnificent section of these beds cutting right across their strike, so that after examining the lowest and oldest beds in the northern part, say between Staithes and Robin Hood's Bay, we find as we walk southwards along the beach higher and higher beds continually coming on, for the beds have a steady southerly dip or slant until at last they pass underneath the chalk at Speeton Cliffs. At Blea Wyke (or Blue Wick), near the Peak, about 60 feet of sandy beds occur above the Lias. These are true passage beds connecting the Lias with the Oolite: they represent the Midford Sands; *Lingula Beanii* is a characteristic fossil. When we attempt to follow the different oolitic beds inland, we find them change very rapidly, becoming much thinner and losing the arenaceous and argillaceous character as we follow them towards the south-west. Three main divisions are clearly traceable.

The *Lower Oolites* form what Professor Phillips called the Moorland Range of hills, which extends from Scarborough and Whitby westward to the Hamilton Hills; the chief height attained is at Burton Head, 1,489 feet, but Hambleton End is 1,300 feet, and the cliffs at the Peak 605 feet above sea-level. Rosebury Topping (1,057 feet) and Eston Nab (800 feet) are detached hills capped by the Lower Oolites. The Howardian Hills are also formed of the Lower Oolite. The bottom bed is called the "Dogger," and contains a seam of ironstone: it has been largely worked in Rosedale, where it swells out so as to form an enormous boss of iron-ore. Mr. Hudleston writes: "On the testimony of the foreman of the miners it exists as a sort of hump or saddle, 600 yards long by 150 yards wide and 80 feet high. It has been 'drifted' in all directions during the last fourteen years for the purpose of extracting the ore, and is now pretty well riddled. There is nothing like it in all this district, and it may fairly be described as the richest ore in North Yorkshire, a perfect nugget of ironstone on a bare hill-side, 700 feet above the level of the sea. Before it was so much quarried and drifted, it used to stand out as a conspicuous cliff, and men hunted foxes into its holes and recesses. There is a tradition that during thunderstorms the lightning frequently struck that cliff of iron; and the country folk said that treasure must be concealed there, which was true enough, whilst others, more superstitious, thought that the devil lay buried beneath. It is a fact, however, that the magnetic iron ore was used as a road-stone for years by people who took it to be a kind of whin-stone or basalt." It contains nearly one-half its weight of pure iron. These Moorland Hills are heathy and barren, and the scenery much resembles that of the Millstone Grit region.

In Gristhorpe Bay there is a seam of coal 9 inches thick; a shaly bed in the Lower Oolite here has yielded 14 species of cycads, ferns, &c.

The *Middle Oolites* form the hills which Professor Phillips called the Tabular Range, because of their flat tops. They extend from the coast between Scarborough and Filey westward across Troutsdale and Newtondale to Kirby Moorside, and the Hambleton Hills west of Helmsley. Here they curve round and stretch by Hovingham to Malton. The beds dip south and south-east: they have a steep but short escarpment on the north, but slope more gently southwards until they pass under the clays of the Vale of Pickering. At their base we can distinguish the *Kelloway Rock*, 80 feet thick at Scarborough, but thinning southwards to 5 feet in Gristhorpe Bay. Then comes the *Oxford Clay* 130 feet, capped by the *Coral Rag* 200 feet thick, which from its hardness forms the upper stratum of the flat-topped hills. All these beds are well seen in Newtondale and in the steep sides of the other dales, formed by streams running from the Moorland Range on the north into the Vale of Pickering, and cutting through the Tabular Hills on their way.

The *Upper Oolite* is represented by the *Kimmeridge Clay*, which forms the greater part of the Vale of Pickering, whose surface is from 70 to 100 feet above the sea-level. This clay is of considerable but unknown thickness, and is only exposed at a few points on the north side of the vale near Helmsley, Kirby Moorside, &c.: it yields *Ostrea deltoidea*, everywhere common in beds of this age.

BASALTIC DYKES.—In Teesdale, from Caldron Snout to near Middleton, a great intrusive sheet of basalt forms the boundary line between the Scar limestone and the Tynebottom limestone (the lowest member of the Yoredale series). From this region the famous dyke of igneous rock known as the Whin Sill runs northward through Durham and Northumberland. Two dykes also run eastward: one of these, passing by Cockfield in Durham, crosses the Tees midway between Yarm and Stockton, and can be traced by Stainton, Langbargh, and Castleton to near the Peak on the coast.

THE DRIFT.—The surface accumulations of the North Riding do not offer so complete a history of the Glacial Period as may be obtained in Holderness, Lincolnshire, &c. The reason of this is that, lying farther north and on higher ground, it was repeatedly swept over by ice-sheets descending from the north, each one pushing before it and destroying the accumulations left by the preceding glacier. Excavations made almost anywhere in the Vale of York reveal large boulders of northern rocks—granite from Shapfell in Westmorland, basalt, limestone &c. Some fine specimens, which were uncovered in excavating the foundations for the new railway station at York, have been placed in the grounds of the Museum there. Many of these blocks were probably brought by a glacier which crossed the Pennine ridge, at the depression known as Stainmoor. With an alteration in the climate the glaciers disappeared, and since that time the only additional deposits have been formed by the rivers, which in the low ground frequently rise above their banks and flood the surrounding country.

PREHISTORIC MAN.—The limestone caverns of the North Riding have not yet yielded evidence of so early an occupation by man as the Victoria Cave near Settle, Kent's Cavern near Torquay, &c. In the cave at Kirkdale near Kirby Moorside, which was discovered in 1821, and examined by Dr. Buckland, there were found embedded in stalagmite the remains of about 300 hyænas, together with the gnawed and broken bones of a large number of other animals, such as the ox, elephant, rhinoceros, &c. The cave is in the Coralline Oolite, and was evidently once a hyæna's den. As to the date of its occupation by them there is little evidence, but it was probably post-glacial. Of the stone implements used by the prehistoric inhabitants of the North Riding many specimens have been found, such as axe-heads (celts), arrowheads, knives, scrapers, &c. These are usually fashioned out of flint, and belong to the later or Neolithic Stone Period. On the oolitic hill-ranges in the east many of the rounded mounds or barrows, which are the burial-places of chiefs, have been opened by Mr. Bateman (see "Ten Years' Diggings"), Canon Greenwell, and others, and found to contain stone weapons, &c., which had been interred with the corpse. Full particulars of all these "finds" are given in Mr. John Evans' great book "The Ancient Stone Implements of Great Britain."

No. 43.

GEOLOGY OF YORKSHIRE.
WEST RIDING.

NATURAL HISTORY AND SCIENTIFIC SOCIETIES.

Yorkshire Philosophical Society, York; Annual Report.
West Riding Union of Natural History Societies, including the Scientific, &c., Societies at Barnsley, Huddersfield, Wakefield, Ovenden, Bradford, Leeds, Goole, York, Selby, Stainland, Liversedge, Heckmondwike, Clayton West, Holmfirth, Mirfield, Honley, Batley, Ripponden, Rastrick. Journal.—"The Naturalist," Huddersfield (4d. monthly).
Geological and Polytechnic Society of the West Riding of Yorkshire. Leeds. Proceedings.
Ripon Scientific Society.
Philosophical and Literary Society of Leeds. Transactions.
Natural History Society, the Friends' School, Bootham, York; organ, "The Natural History Journal," monthly (4d. York).

MUSEUMS.

The "Museum Isurianum," Aldborough.
Bradford Museum.
Giggleswick Museum.
Halifax Museum.
Museum of the Literary and Scientific Society, Huddersfield.
Museum of the Mechanics' Institute, Leeds.
Museum of the Philosophical and Literary Society, Leeds.
The Museum, Weston Park, Sheffield.
The Yorkshire Philosophical Society's Museum, York.

PUBLICATIONS OF THE GEOLOGICAL SURVEY.

Maps.—Quarter Sheet: 92 S.W. Clitheroe; 92 S.E. Ilkley, Denton, Calverley; 93 S.W. Leeds, Tadcaster; 93 N.W. Knaresborough, Boroughbridge; 87 N.E. Snaith, Knottingley; 87 N.W. Wakefield, Pontefract; 87 S.E. Doncaster; 87 S.W. Barnsley; 88 S.E. Holmfirth, Penistone; 88 N.E. Huddersfield, Dewsbury, Halifax; 88 N.W. Todmorden; 88 S.W. Manchester, Oldham. (Survey not completed.)

Books.—Memoir, Geology of the Yorkshire Coal-field, by Prof. A. H. Green, &c., 42s. Geology of Dewsbury, Huddersfield, and Halifax, by A. H. Green, 6d., Geology of Part of the Yorkshire Coal-field (west of Penistone), A. H. Green, 1s. Geology of Leeds and Tadcaster, by Aveline, 6d. Geology of Harrogate, by C. Fox-Strangways, 6d. Geology of Sedbergh, Tebay, &c., by Aveline and Hughes, 6d. Geology of Barnsley, A. H. Green, 9d. Geology of Wakefield and Pontefract, by Green and Russell, 6d. Geology of the Country between Bradford and Skipton, by J. R. Dakyns, 6d.

IMPORTANT WORKS OR PAPERS ON LOCAL GEOLOGY.

List of 96 works on Geology of Yorkshire in Leeds Museum Guide (by Prof. Miall and Mr. Whitaker); full list in Phillips' Geology of Yorkshire Coast, 3rd edition.

1836. Phillips, Prof. J.—The Mountain Limestone.
1837. George, E. S.—The Yorkshire Coal-field. Trans. Leeds Lit. and Phil. Soc., vol. i.
1855. Phillips, Prof. J.—The Rivers, Mountains, and Sea-coast of Yorkshire. Second edition.
1866. Binney, E. W.—On the so-called Lower New Red Sandstones of Central Yorkshire. Geol. Mag., vol. iii. pp. 49, 473.
1867. Hughes, Prof. T. McK.—Break between the Upper and Lower Silurian Rocks between Kirkby Lonsdale and Malham. Geol. Mag.,vol.iv.p. 346.
1870. Spencer, T.—Sketch of Geology of Halifax. Trans. Manchester Geol. Soc., vol. ix.
1873-80. Tiddeman, R. H.—Reports on the Exploration of Settle Cave. Reports of British Association.
1875. Sorby, H. C.—Fossil Forest at Wadsley, Sheffield. Journ. Geol. Soc., vol. xxxi. p. 458.
1878. Davis, J. W., and Lees, F. A.—West Yorkshire: Its Geology, Physical Geography, Climatology, and Botany. London, L. Reeve & Co.
1878. Miall, Prof. L. C.—Geology, in 3rd edition of Whitaker's History of Craven. Cassell, Petter, & Galpin.
1878. Davis, J. W.—The Valley of the Calder. Geol. Mag., p. 500.
1879. Holmes, T. V.—Geology of Sheffield. Science Gossip, p. 169.
1881. Bird, C.- Sketch of the Geology of Yorkshire, 5s.
See also General Lists, p. xxv.

BEFORE commencing to describe what the West Riding is *made of*, that is to say, before beginning to give an account of the rocks which in that region form the outer portion of the earth's crust, it seems very desirable that we should explain clearly what we propose to do.

The surface or soil upon which we walk is made up of broken disintegrated fragments of solid rock masses which lie below, or perchance we may walk upon the solid rock itself. Now, suppose we start from Leeds, or Wakefield, or Sheffield, and walk to the west, we shall soon find ourselves standing on a different kind of rock, first on massive sandstones, then on limestone. If on the contrary we walk eastward, we get, after passing over coal-seams and thick beds of clay, on to a different kind of limestone, and then to a rich soil of red marl forming the Vale of York.

But on the contrary, if we proceed either to the north or to the south, we shall find ourselves walking continuously upon the same or nearly the same bed of rock as that from which we started, until we get north of Leeds, when there is a change in the direction of the beds or strata.

If we turn to a geologist for an explanation of this, we shall learn that the earth's crust in our district is made up of several successive beds, or layers, or strata of rock.

We may note here that every substance which enters largely into the composition of the earth's exterior is technically termed a *rock*, whether it be hard like granite or sandstone, or soft and yielding like shale or clay.

The different beds of rock lie one upon the other, and were all originally formed at the sea-bottom out of mud, sand, &c., brought down by rivers, or washed off coasts. Thus, when they were made, each layer in its turn was quite horizontal or flat.

But, owing to the action of deep-seated volcanic forces, it is certain that no portion of our globe remains long in a state of rest ; it is either elevated or depressed. We know that what is now dry land has been many times below the sea, and *vice versâ*, or as Tennyson finely puts it—

" There rolls the deep where grew the tree.
O earth, what changes hast thou seen !
There, where the long street roars, hath been
The stillness of the central sea."

Now, when England was upheaved long, long ago above the waters, not once only, but several times, with intervals of rest and intervals of depression, the upheaving forces acted at one time most strongly along a line ranging from Morecambe Bay to the mouth of the Tees, and at another time along an axis extending right down from the Cheviots to the river Trent, forming, in fact, the Pennine Chain.

This is the main reason why we find the greatest elevations on the north and on the west, and why Mickle Fell, the culminating point of the Yorkshire hills, rising to a height of 2,580 feet, is found in the extreme north-west corner of the county. Thus, the different layers or beds of rock have, on the whole, a slope or *dip* towards the east and south, while they come to the surface or *crop out*, one after the other, as we walk over their upturned edges from east to west, the lowest and consequently the oldest rocks coming up last in the west. On the contrary, when we pass from south to north, we walk along the outcrop or line of extension of the rocks, and consequently do not see the same changes.

LITERATURE OF THE SUBJECT.- In the third edition of Phillips' "Geology of the Yorkshire Coast" there is a full list of works on Yorkshire geology, compiled by that chief of geological bibliographists, Mr. W. Whitaker. There is also a capital list by Mr. L. C. Miall in the "Guide to the Fossil Collection in the Leeds Museum." So numerous are these books and articles that it is impossible to do more here than to indicate a few of the principal ones. Professor Phillips' books, "The Rivers, Mountains, and Sea-Coast of Yorkshire" and his "Mountain Limestone" are classical works. In the Proceedings of the Geological and Polytechnic Society of the West Riding of Yorkshire there is a rich store of information, whilst the "Geological Magazine," the "Colliery Guardian" and the Quarterly Journal of the Geological Society contain very numerous and valuable papers on the subject. A concise account of the coal-field, with a very clear outline map, will be found in Hull's "Coal-Fields of Great Britain," in connection with which the Report of the Royal Coal Commission of 1871 should be studied. The Geological Survey has accurately laid down on coloured maps, every bed of rock, and these maps, together with the "Memoir on the Geology of the Yorkshire Coal-field," lately issued by the Survey, constitute a most complete and thorough account of the geological structure of the whole region.

THE SILURIAN FORMATION. Beds of this age are the oldest rocks which occur in Yorkshire, and they occupy a limited area only, in the north-west corner. The lowest bed of rock is called the *Coniston Limestone* and is of the same age as the *Bala Limestone* of North Wales; it occurs in Helm Gill, west of Dent: it is a bluish-grey limestone and is overlaid in the Sedbergh district by a considerable thickness of shaly beds, containing a trilobite (*Trinucleus concentricus*) and a bivalve shell (*Strophomena depressa*). All these beds are of Lower Silurian age, and a great interval elapsed before the succeeding beds were laid down upon them, during which the Lower Silurians were hardened, upheaved, and folded. Upon their upturned edges the *Coniston Flags and Grits* were deposited, and these have been shown by Professor Hughes to belong to the Upper Silurian Period, and upon them we find the *Bannisdale or Ireleth Slates*, the equivalents of the Ludlow and Wenlock beds of Shropshire. All these beds are exposed in the valley of the River Rawthey, a tributary of the Lune, and they also run eastwards, along the northern edge of the Craven fault, past Ingleton and Settle, to Gordale Beck. Thus they are found at the bottom of the deep valleys at Horton-in-Ribblesdale, Crummack Beck, Clapham, Chapel-le-Dale, and Kingsdale. At Gordale the slates contain beds of tough blue crystalline sandstone. The dip here is to the south-south-west. The number of boulders of Silurian grits which appear on the surface near Kilnsey-in-Wharfedale probably indicates the presence of similar beds beneath the covering of drift and alluvium which there conceals the valley-bottom.

THE CARBONIFEROUS FORMATION.— It has been considered that the Old Red

Sandstone, so well developed in Scotland and in Herefordshire, is possibly represented by scattered patches in the north of England. East of Sedbergh in the Valley of the Rawthey we find a mass of coarse red conglomerate, 2 or 3 miles in length and resting unconformably on the Coniston Grits : it is between 200 and 300 feet thick, and contains scratched pebbles. Professor Ramsay considers the conglomerate may, perhaps, have been formed in some remote glacial period, but of late the idea has been gaining ground that it is a shore deposit and really belongs to the Carboniferous Limestone, of which it forms the base. These red conglomeratic beds contain no fossils.

In the great series of Carboniferous rocks there are four clear and, on the whole, well-marked subdivisions.

1. *The Carboniferous or Mountain Limestone.* Ranging east and west from Clitheroe to Skipton, and again from near Ingleborough, by Settle, to Kilnsey and Hartlington, we have a region where the rocks which form the surface are chiefly of a tough pale grey, rather crystalline and nearly pure limestone : its base is often a limestone matrix full of pebbles, often of Silurian rocks, but varying according to the nature of the underlying strata. Sometimes, too, we find beds of sandstone at the base, often of a reddish tint, which some geologists have assigned to the old red sandstone, but which are shown by the fossils they contain to be truly of Carboniferous age. This *Mountain Limestone* is a deposit of great thickness, and, as already noted, we owe its appearance at the surface to two great systems of upheavals, running nearly at right angles to each other. The mountain limestone of Derbyshire, which forms the well-known tract between Buxton and Wirksworth, is, notwithstanding its great thickness, a surprisingly compact and homogeneous mass ; but this mass, dipping to the north-east, passes beneath the later-formed beds, and is hidden from view until we get it again brought up to light north of Clitheroe, Colne, and Skipton. Here we find it to contain several beds of shale, but the limestone still preponderates; it is the *Scar Limestone* of Phillips. Between Malham and Gargrave we find the beds of shale thickening ; then, still going northwards, beds of sandstone begin to split up the higher part of the limestone. Under Ingleborough there is a nearly undivided calcareous mass, 400 or 500 feet thick ; but at Alston Moor there are no less than twenty different limestones, amounting altogether to 470 feet in thickness, obscured by the interposition of no less than 1,686 feet of sedimentary strata. Further still to the north these mechanical admixtures increase in amount, while the calcareous strata diminish, and at length, in the northern parts of Northumberland, the limestone district has become a valuable coal-field ; Phillips estimates the thickness of Scar Limestone under Penyghent at 400 feet : south of this point its base is only seen where it reposes on Silurian rocks, near Malham Tarn, Ingleton, &c. The Geological Survey estimates its thickness near Clitheroe at above 3,250 feet.

In the great upheavals and depressions to which the beds have been subjected, numerous dislocations or slips have taken place, by which the beds have been cracked, and one side lifted higher than the other. Such dislocations are geologically termed *faults*. The North and South Craven faults, which run from north-east to south-west, are good examples. In these faults rich deposits of lead ore are often met with : it appears to have been deposited from water trickling through the crevices until they were filled up. In 1875 there were raised from the Yorkshire mines 4,050 tons of lead ore, which produced 2,950 tons of lead, and 7,438 oz. of silver ; in 1879 there were raised 5,132 tons of lead ore, valued at £50,220 ; the greater portion of the ore, however, came from the North Riding (see " Mineral Statistics," by R. Hunt, for 1875).

As limestone is slightly soluble in rain-water, we get little depth of soil in parts where this rock forms the surface, but a short thick herbage grows, which fattens sheep, and to this fact is due the fame of the pasturage of Craven. The water, moreover, of the limestone tracts is always hard from the presence of the dissolved carbonate of lime. Caverns are numerous, and are invariably due to rain-water. Sinking through surface cracks and then coursing along subterraneous passages, the rain-water enlarges the crevices or joints which traverse the rock in all directions. In this way the river Aire springs forth at once a full-sized stream from under Malham Cove, a huge cliff of limestone 285 feet high, the water having chiefly come, no doubt, by an underground route from Malham Tarn, a picturesque little lake, situated on high ground to the north. Sometimes a portion of the top of these underground river-courses falls in, and then we can see the stream forcing its way and creating a great noise by dashing together the stones which have fallen in; such are the *Pots*, which occur in many of the dales, as Gingle Pot and Hurtle Pot, near Ingleton. Of the caverns, that in Ingleborough has been explored by Mr. James Farrer for a distance of nearly half a mile from the mouth, thus piercing the very heart of the mountain.

We may add that the limestone has a high reputation both for burning into lime and as road-metal, for both of which purposes it is largely employed. In several places the rock exhibits very remarkable foldings and contortions, as at Draughton, Skipton, &c. Of the region between Malham and Bolton Abbey, Mr. W. H. Dalton says:—"The beds are often vertical and look as if they were tied in knots. If a book were put through a turnip-cutting machine and subsequently daubed with mud, the task of determining the page and line of every visible and invisible fragment would be little more perplexing than has been the deciphering of the contorted and broken limestones of South Craven." But the task has been done and done well, and we have the results in the admirable maps, sections, and memoirs of the Geological Survey.

2. *The Yoredale Rocks.*—These beds were so termed by the late Professor Phillips from their development in Yoredale (or Wensleydale), the Valley of the Ure: the name, however, was an unfortunate one, for the more detailed and minute work of the Geological Survey has shown these beds to be wanting, or to be doubtfully present, in the Yore Valley. The Yoredale beds, where present, consist of alternations of flagstones, shales, limestones, and thin seams of coal. From the moors about the upper part of the river Derwent in the south-west, these beds run up west of Halifax to Bowland Forest, curving round thence to the east by Harrogate. The thickness of the strata is very variable, amounting to as much as 2,300 feet in the Ribble Valley, but only 600 feet at Hebden Bridge, near Halifax. This, and the occurrence of thin seams of coal, would seem to show that in this region there were then great alternations in the state of the surface, for the Carboniferous Limestone was undoubtedly formed in a deep and pure sea, whilst to form the Yoredale Rocks much muddy sediment and sand would be necessary; and lastly, the coal-seams show that the beds were elevated more than once above the sea-level.

The black bituminous shales which occur in the Bowland district have led to numerous trials for coal there, which, it need scarcely be said, could have no chance of success.

3. *The Millstone Grit.*—The strata which bear this name would seem to indicate a still more persistent shallowing of the carboniferous sea, for they consist in the main of coarse sandstones, with beds of shale between, and some thin, yet occasionally workable, seams of coal. The beds of grit or coarse sandstone are four in number: the lowest is known as the Kinder Scout Grit, because it caps the Peak in Derbyshire, also forms the hard rocky tops or plateaux of the outlying masses of Ingleborough, Whernside, &c., and the mighty escarpment of Gragreth. The highest grit-bed is called the "Rough Rock" and is well seen in the bold ridge known as Otley Chevin. Near Skipton it is so coarse as to be a conglomerate containing pebbles 2 or 3 inches long. Fossils are few, but plant-remains are not uncommon, all of the same species as in the coal-beds above. The officers of the Geological Survey

estimate the thickness of the Millstone Grit near Skipton at 3,333 feet, with eight seams of coal.

All along Western Yorkshire, from the Peak country by Halifax, Sowerby Bridge, and Keighley, the Millstone Grit extends, forming a hilly, somewhat barren country, often moorland, and with an easterly inclination or dip: it then turns eastward, and the Rough Rock caps all the hills between the Wharfe and the Aire, and forms outliers above Keighley on both sides of the Worth: it dips here to the south-east under the Lower Coal-measures. At Pateley Bridge, near Harrogate, are the well-known Brimham Rocks, huge irregular crags which rise in fantastic shapes from the moorland: they are due to local variations in the hardness of the rocks and are the result of long exposure to the action of the weather.

At Horsforth, near Leeds, flaggy beds in the Millstone Grit are largely worked, and the "Yorkshire Stone," so much used for building and other purposes, comes from here. Addingham Crag and the Ilkley Crags, near Bradford, are also formed of Millstone Grit. In this district the sandstones are very irony; thousands of chalybeate springs issue from them, and when ploughed they are seen to form a red soil. The coarse red sandstones of Bramham Moor, Plumpton, and Knaresborough were believed by Sedgwick and Phillips to be of Permian age, but they have been conclusively shown by Mr. Binney and others to belong to the Rough Rock of the Millstone Grit.

4. *The Coal-measures.*—The Yorkshire coal-field forms the northern portion of a great mass of coal-bearing strata, which extends southwards in Derbyshire and Notts, as far at least as the River Trent: it is bounded on the west and north by the moors of Millstone Grit, but on the east by the escarpment of the Magnesian Limestone, underneath which it dips. Thus it forms the half of a tilted basin, of which the eastern portion lies buried under newer, later-formed rocks. In Notts, the coal-seams have been reached by boring through these newer beds at Shireoaks Colliery and elsewhere, and the coal-seams will probably be followed in a similar way in Yorkshire.

The Yorkshire coal-seams are believed, with good reason, to have once been continuous with those of Lancashire. But the upheaval of the Pennine Chain severed the connection: and another, probably earlier, upheaval in the north, acting along an east and west line, separated the Yorkshire from the Newcastle Coal-field.

Thus, in the Carboniferous age, we must imagine that Yorkshire formed part of the sea-floor of a deep and clear sea in which the Mountain Limestone was formed, probably from the remains of countless numbers of minute organisms and shells. Then this sea shallowed, and we get the Millstone Grit formed; lastly, during the formation of the Coal-measures, even terrestrial conditions prevailed.

The mode of accumulation of the thick beds of carbonaceous matter which constitute our coal-seams was long a vexed and difficult question. It is now, however, pretty clear that they are the compressed remains of dense forests, which grew on clayey flats but little above the level of the sea. Under every coal-seam we find a bed of under clay (or *spavin* as it is termed in Yorkshire). This was the soil in which the ancient forest grew, and it is usually full of the rootlets (*stigmaria*) of a plant known as the *Sigillaria*, whose stems and branches are among the commonest of our coal fossils. In a piece of coal it is not easy to discern the vegetable tissues by the naked eye, but this may be done by grinding down a slice until it is so thin as to be semi-transparent, and then examining it with a microscope. In the beds of sandstone and shale which lie between the coal-seams, the fossils are much more easily recognisable. Here we find the fronds of delicate ferns, and the remains of gigantic clubmosses (*Lepidodendron*); reed-like *Calamites*, too, occur, which are allied to the horse-tails of our marshes and ditches. At Wadsley, near Sheffield, some stumps of Coal-measure trees were uncovered in excavating the site of the Lunatic Asylum; they are sigillariæ, and Mr. Sorby has shown from the character and extension of their roots, that the prevailing winds were then, as now, from the west.

Animal remains are not so common. The shells are chiefly fresh or brackish water kinds, and the fish and reptiles such as would be likely to occur in inland waters or estuaries.

We may divide the Coal-measures of the West Riding into three divisions Lower, Middle, and Upper.

Lower Coal-measures. These are about 1,300 feet thick, and are also known as the Gannister beds; they include all the strata between the Rough Rock (top of Millstone Grit), and the Blocking or Silkstone coal.

At Elland, Bradford, &c., there are some well-known beds of flagstone, and from 100 to 130 yards below these are the lowest workable coal-seams. The upper one is known as the Halifax Hard bed, and the Halifax Soft bed which is 30 yards lower, is itself only 40 yards above the Rough Rock. Owing to the upheaval on the west, these beds *crop out*, or come to the surface, along a line passing west of Bradfield, Penistone, and Huddersfield. The roof or top of the Hard Bed is a black shale full of shells, such as *Goniatites*, *Orthoceras*, and *Aviculopecten*, which is very characteristic of this particular level. A hard siliceous stone, known as calliard, or gannister, is also found above and below the Hard Bed; it is ground up to line furnaces with, and also used as road-metal and for rough walls. The Hard-bed coal is from $1\frac{1}{2}$ to 3 feet thick, but the Soft bed not more than 2 feet.

The *Better Bed Coal* is about 30 feet above the Elland Sandstone; it is extensively worked by the Low Moor Iron Company, being used by them for smelting the clay ironstone of the district; it is very free from sulphur, and the excellence of the iron made by this firm is in a great measure due to this seam. A thin stratum of shale, less than 1 inch thick, associated with the Better Bed Coal, is so full of fish-remains as to be named by Mr. J. W. Davis a *bone-bed*. The clay underlying this coal is of very good quality, and is made into fire-bricks, tubes, terra-cotta work, &c. The total quantity of fire-clay raised in Yorkshire in 1875 was 183,961 tons, but in 1879 the amount had diminished to 155,938 tons.

Middle Coal-measures. These contain the most valuable coal-seams. The Silkstone, or Sheffield, or Black Shale coal of South Yorkshire, is a soft and bituminous bed of great purity, producing an excellent house coal, and is also well adapted for coking; it is without doubt the same bed as the "Arley Mine" of Lancashire, and thus was originally spread over not less than 10,000 square miles. It is rarely above 5 feet in thickness, and can be traced northwards with certainty as far as Cawthorne, where it consists of two beds, each $2\frac{1}{2}$ feet thick, with a dirt-band between; here it breaks up into several beds, and dies away. Proceeding northwards we soon meet with its equivalent, however, in the *Blocking Coal*, which is about 2 feet thick, and which can be traced up to Pudsey.

Another important seam is the Barnsley, which is 4 feet thick near Sheffield, 7 or 8 feet at Rotherham, and 10 feet at Barnsley. Passing by Darton and Crigglestone, we find it so broken up by dirt-seams as to be worthless, but it recovers itself, and is then known as the Warren House Coal, and further north still as the Gawthorpe seam. Professor Green has ably explained these variations by showing that the beds probably grew in various swamps, all on about the same level, but separated by muddy channels. Above the Barnsley Coal we get the Wath Wood seam, which is known north of Barnsley as the Woodmoor Coal. Still higher in the series, and cropping out further east, we have the Shafton Coal, probably the same as the Nostell seam, and which lies about 430 yards above the Barnsley Coal. Altogether the Middle Coal-measures are about 900 feet in thickness near Bradford, but nearly 3,000 feet in South Yorkshire. Of the intervening beds of sandstone, some, by their superior hardness, have been enabled to resist the denuding forces which have worn away the softer shales. They now stand out, forming bold escarpments and lines of hills, contributing much to the beauty of the otherwise tame and flat Coal-measure tracts; among these hard sandstones we may name the Oak's Rock, the Cudworth Rock, and the Woolley Edge Rock. We may here call attention to the aid which has been rendered by intelligent workers in collieries, who have saved any remarkable fossils they met with. Aided in this way, Professor L. C. Miall, of the Leeds Museum, has done excellent work in elucidating the nature of the carboniferous fishes. Of *Megalichthys* the most perfect known skull is in the Leeds Museum, and also the palatal plate of *Ctenodus* upon which the genus was founded by Agassiz. Mr. Miall has done even more important work among the *Labyrinthodonts*, an extinct order of reptiles, allied to the frogs, but attaining a length of 8 or 10 feet; their bones often occur in the roof of the coal-seams, and they should be diligently looked for and preserved. The total amount of coal raised in the West Riding in the year 1875 was 15,425,278 tons, this being the produce of 523 collieries; in 1879 the amount was 16,024,249 tons from 525 collieries. The Coal-measures also yield a great quantity of *ironstone*, which usually occurs in rounded lumps or nodules, forming more or less regular layers. We get the iron in two states of combination: first with sulphur, when it has a yellow appearance, and is known as iron pyrites or coal brasses; of this about 2,500 tons were raised in 1875, valued at £1,250; and 1,650 tons, valued at £825 in 1879; secondly, the metal occurs united with carbonic acid, when it is called carbonate of iron. As this has always much clayey matter associated with it, it is commonly known as argillaceous carbonate or clay ironstone. Black Band is the most valuable kind of ironstone, and is distinguished by the large amount of carbonaceous matter which it contains. Of the carbonate, 159,089 tons were raised in the West Riding in 1875. At the same time there were 37 furnaces in blast, which produced 267,153 tons of pig iron. This shows that a very large quantity of iron ore is imported from the Cleveland and Lincolnshire iron districts. In the smelting of the ore it is estimated that 695,557 tons of coal were used. In 1879 there were 31 furnaces in blast, and 218,805 tons of pig iron were made.

Upper Coal-measures.—We assign to this division certain beds of coarse reddish grit, known as the Red Rock of Rotherham, which occur east of that town and which are supposed to be not far from the true centre of the basin.

Resources and Future Extension. As the most valuable beds of coal—the Silkstone and the Barnsley seams—become exhausted towards their outcrop, they will be and in fact are being followed eastwards along their dip. The country east of Barnsley and Wakefield is as yet almost untouched, and although the depth to the top hard coal is here not much under 2,000 feet, while to the Silkstone seam it is another 1,000 feet, yet in the Lancashire coal-field collieries are now being worked at depths quite as great as these, and there can be little doubt that working at a depth of even 4,000 feet is possible under the improved systems of ventilation. It would even seem probable that coal may be got at Selby, Doncaster, Tickhill, &c.; for if the coal-beds rise up to the east as they do to the west, then the depth there may not be so great as to prevent profitable working. In fact, the existence of beds belonging to the Coal-measures has already been proved as far east as the river Trent, by borings at Scarle in Lincolnshire, which reached them at about 1,900 feet. Although plant-remains were brought up, the exact position of the strata in the carboniferous series could not, unfortunately, be determined; most probably, however, they belonged to the highest division.

THE PERMIAN FORMATION.—A line drawn from Nottingham to Tynemouth will run along a narrow band of limestones, reddish sandstones, and marls, known as the Permian beds. They were so named by Sir R. Murchison from their great development near Perm, in Russia.

In Yorkshire the Permian strata usually rest upon sandy red beds of Carboniferous age, these latter owing their colour to the infiltration from above of water containing salts of iron. For a long time these red sandstones (including the Pontefract Rock and other beds, really belonging to the Upper Coal-measures) were classed as of Permian age. The lowest true Permian beds are (1) false-bedded quicksands in South Yorkshire, then come (2) beds of yellowish dolomitic limestone, surmounted by (3) marls with sandstones and gypsum. Above these come (4) other beds of limestone, and on the top we again get (5) marls and sandstones with gypsum. The following towns indicate the position of the beds from south to north. They enter Yorkshire near Adston and pass thence by Tickhill, Conisborough, Pontefract, Castleford, Tadcaster, Knaresborough, and Ripon. The width of the outcrop is from 3 to 6 miles and the beds have a uniform dip to the eastward. One consequence of this is that in the centre of the coal-basin, say east of Rotherham, or near Pontefract, these Permian deposits appear to rest *conformably* upon the Carboniferous rocks, *i.e.* dipping in the same direction and at the same angle. Passing northwards, however, we find that north-east of Leeds, the Permians rest on the Millstone Grit, whilst near Richmond they even repose upon the Yoredale rocks. This proves that in reality they are altogether *unconformable* to the Carboniferous rocks in this district, for on

Fig. 96.—Section showing an unconformity between the Pontefract Rock and the Magnesian Limestone, near Pontefract.

1 Thickly-bedded Sandstone, Pontefract Rock.
2 Shaly Sandstone and Shale.
3 Magnesian Limestone.

The Pontefract Rock is situated near the top of the Middle Coal-measures. It readily disintegrates into a loose porous soil on which the liquorice plant grows well. Pontefract is supplied with water by wells in this sandstone; its thickness varies from 65 to 140 feet.

the north the latter were upheaved and some thousands of feet of strata worn away, before the deposition of the later-formed marls, sandstones, and limestones of the Permian system, which run in a straight line over the contorted and denuded edges of the Carboniferous strata. (See fig. 96.)

Professor Ramsay has shown that the Permian rocks were probably deposited in a large inland salt lake, something like the Caspian Sea of our own times. The fossils are few and dwarfed, a shell called *Axinus obscurus* having the widest range. The limestone is generally a *Dolomite*, that is, composed of carbonate of lime plus carbonate of magnesia. The waters of the inland sea to which we referred just now must have been so highly charged with these salts, as to have repeatedly deposited the excess upon the sea-floor, thus leading to the formation of solid beds of dolomite. The total thickness of the Permian rocks is here from 300 to 400 feet, but the different members of the formation vary much in thickness. Thus in the railway cutting at Tadcaster, the middle marls (No. 3) thin away to a mere seam, so that the upper limestone (No. 4) rests almost directly upon the lower, and at the base of the former there is a thin bed of gravel formed of lower limestone pebbles. The magnesian limestone yields some excellent building stone usually of a yellowish tint, compact and fine-grained, but variable in quality and requiring careful selection in the quarry. The Anston quarries furnished the stone for the front of the Museum of Practical Geology, in Jermyn Street, London, in which there is said to be not a single bad block. There are other important quarries at Brodsworth, Cadeby, and Park Nook, near Doncaster, Huddlestone near Sherburne, and Smawse near Tadcaster. Other buildings which may be mentioned as proving the excellence of the stone, are the keep of Conisborough Castle, Tickhill Church (15th century), Huddlestone Church (15th century), Huddlestone Hall (16th century), Roche Abbey (13th century). Westminster Hall was built of stone from Huddlestone, York Minster of Jackdaw Craig stone, while that of Smawse was used for Beverley Minster. The smoky atmosphere of towns does not, however, appear to suit the magnesian limestone. The grand new church of St. George at Doncaster showed blocks on the exterior in a state of decay within one year after its erection.

THE TRIAS.—Passing still eastwards, we now come to the rocks which form the western part of the Vale of York. The strata here are often covered over and obscured by deposits of *Drift* or *Warp*, presently to be described, but where visible we see them to be mainly soft, thick-bedded reddish sandstones, with inter-stratified pebble-beds. They belong to the lower division of the Trias, which is known as the *Bunter or Variegated Sandstone*. In the south, Tickhill is built on an outlying patch of the lower sandstone, while the High Common east of the town shows the pebble-beds. From this point northwards, we get few opportunities of seeing the Bunter beds: they pass by Thorne and Snaith, and form the plain west of York. In a boring at Walmgate, York, 161 yards of soft brick-red sandstone were passed through. At Aldborough and Ripon the Bunter sandstone was formerly used as a building stone, but the quarries are not now worked. Near the latter town there are several curious natural pits or shafts 60 or 70 feet deep and from 50 to 100 feet across. They have been well described by the Rev. J. S. Tute, and appear to be due to the dissolving away by water of certain beds of red marl and gypsum contained in the underlying Permian beds. The overlying beds of Bunter Sandstone, then having nothing to support them, fell in. At Sharrow, one fell in during the night, about 1850, and alarmed the inhabitants of a neighbouring house, who in the morning found little more than the breadth of the road between themselves and the pit. About 1830, some men at Bishop Monckton were making a stack near the Old Hall, and had left it for a while; on their return they found that the ground had given way beneath the stack, which had disappeared. The men hastened in great consternation to their master, exclaiming, "It's gone! it's gone!" The hole still remains a receptacle for rubbish.

The upper division of the Trias or *Keuper Marls and Sandstones* barely enters the West Riding near Aldborough, and does not call for notice here.

POST-PLIOCENE FORMATION.—*The Glacial Period.*—We must now pass over an immense interval of time, to be numbered in millions of years, during which such deposits as the Liassic clays and the chalk of the East Riding were being formed. These rocks probably once stretched much farther to the west than they do now, running up probably to the foot of the Pennine Chain. But during the course of unnumbered ages they have gradually been worn away, until now they lie altogether on the east of the Ouse.

So far as the researches of men have gone at present, it seems that about a quarter of a million years ago, this earth was so situated with regard to the sun, that the northern hemisphere was in large part covered by a great ice-cap, which, descending southwards, ground over our rocks, tearing them and smoothing them, and carrying on the débris to be deposited as a thick coating on surfaces far distant from their native home. Thus it is that in Yorkshire we find irregular beds of tough blue, brown, and yellow clay, full of pebbles of rocks which occur further north, and alternating with beds of gravel and sand. Proofs of this ice action are also seen in the grooves and striæ left on the bare hill-sides in Wharfedale, along Rombald's Moor, Ingleborough, &c.: the low rounded hills of Craven are mainly hummocks of boulder-clay. Great glaciers, coming down from the Howgill fells and the Cumbrian mountains, appear to have been divided by the Pennine Chain, the western ice sweeping down Lancashire, and the eastern ice proceeding in an east and south-east direction through Craven, and reaching as far as Airedale at least. Owing to the gradual sinking of the land, the sea advanced over the Vale of York and up the Valley of the Aire, washing and re-distributing much of the Boulder-clay, which is consequently now scarcely recognisable as such. The eastern slopes of the Pennine Hills, however, south of the Aire basin, do not show any signs of glaciation.

Post-Glacial Deposits.—On the extensive low and flat lands which stretch westwards from the Humber and run up either side of the Ouse, nearly to York, we find a deposit of fine clay, which is termed *Warp*. The greater part of this great clay flat is now about 50 feet above the level of the sea, but it was deposited at a time when the land was slightly lower than at present, and when the tide rolled daily over the area now occupied by fertile corn-fields: the warp appears to consist of sediment washed off the land or brought down by rivers, and is in fact a true tidal alluvium.

Of the terraces of *River-Gravels* which occur at varying heights on the sides of our river-valleys, and of the peat deposits of the moors, we have not space left to speak. These things belong to times comparatively recent, and they pass on into the alluvial deposits which are forming at the present day along our river-courses.

Traces of Prehistoric Man. We have now obtained evidence of the existence of the human race at a period long anterior to anything referred to in written history, or even in tradition.

The relics which furnish the proof are the tools and weapons *made of stone*, which were used by these early inhabitants of our country, and their great antiquity is shown by the circumstances under which they are found, and by the absence of any historical account of the period when they were in use, either in England or any other part of the Old World. We find savage nations, however, still retaining the knowledge of the manufacture and use of such stone objects, and from them much has been learnt respecting the stone implements which an English workman would simply regard with amazement.

Antiquarians and geologists have recognised two periods in the stone age. In the earlier called the *Palæolithic* or older stone age the implements are roughly-chipped flint nodules, of large size, and either pointed or oval in form. These are found at considerable depths in *river drift*, as in the brick-pits near the Thames, &c.

None of this type have yet been found north of Peterborough, and it has been suggested that the north of England was then perhaps so covered with glaciers as to be unfit for habitation. A diligent look-out should be kept in all gravel-pits, especially those 40 or 50 feet above the level of the present rivers, for there we are most likely to meet with such objects.

The second and later stone period is called the *Neolithic*. Now we find the implements much lighter, more elegantly shaped, intended for a greater number of purposes, and moreover very frequently polished or rubbed. The people of this time as well as the preceding were quite ignorant of the use of metals. Later still the art of smelting copper ore seems to have been found out, and by the admixture of a little tin, *bronze* was produced, which for a long time remained in use. Finally *iron* was discovered.

Stone objects of the Neolithic period are not so common in the West Riding as on the Wolds of East Yorkshire, or the Moors of the North Riding. One reason may be that they have not been so diligently looked for, but we must also remember that the principal material used was *flint*, and this would be plentiful on the chalk wolds, while it does not occur west of the Ouse. From Adel, near Leeds, Thoresby recorded as long ago as 1715, two flint arrow-heads, and in 1868, the boys of the Reformatory there, found three beautifully finished arrow-heads, and several javelins, knives, flakes, &c.; scrapers also intended for cleaning skins, and *cores* or masses of flint from which pieces had been struck off. The spot seems in fact to have been a kind of manufactory, to which flint was brought from the chalk-wolds further east.

Several polished celts, or axe-heads of flint, together with one ground at the edge only, were found 6 or 7 feet below the surface, in a sand-bed near York, and nearly a quarter of a mile from the river which is thought to have deposited the sand. In Leeds a perforated hammer-head, made of greenstone, was found 12 feet deep in gravel, while sinking for foundations for the works of the North Eastern Railway in Neville street.

A link uniting the ages of bronze and stone was found at Broughton-in-Craven in 1675, and is recorded by Thoresby. A stone axe-head was found there in an urn, together with a small bronze dagger and a hone.

The most important evidence, however, as to the early existence of man in the West Riding has been furnished by some caves of the limestone district. In Cave Ha, near Giggleswick, Professor T. McK. Hughes found broken pottery, flint-flakes, a stone bead and bones—especially one molar tooth of a bear. At the more important Victoria Cave, near Settle, very important discoveries have been made by Mr. R. H. Tiddeman and Mr. Jackson, aided by private subscriptions and an annual grant from the British Association. This cave is in the Carboniferous Limestone and is about 1,400 feet above the sea. The floor was covered with stones fallen from the roof: under these was a blackish layer containing charcoal, with enamelled ornaments of metal, showing that the cave had probably been inhabited by the Romano-Britons in the troubled times which followed the withdrawal of the Romans about 450 A.D. Next came a layer, about 6 feet thick, with many flint implements, bone harpoons &c.; this was evidently of Neolithic age. Bones of the badger, reindeer, horse, &c., were found in and under this layer, but then the explorers came on a lower bed of laminated clay, containing scratched boulders of Silurian limestone; lastly, at the bottom of all, came numerous bones of the hyæna, mammoth, rhinoceros, &c., and a single bone of a shape much resembling a human fibula (one of the leg-bones). Prof. Busk, however, now assigns this bone to the skeleton of a bear. Now if the laminated clay was deposited during the Glacial period (and the ice-scratched boulders in it would seem to indicate this), then the bones found *beneath* it prove the existence of these various animals in the West Riding *prior to the glacial epoch*, which would be a fact of high interest and importance. It has, however, been argued that the clay and boulders may have been washed into the cave by floods, &c., and the evidence is hardly complete. The exploration is still being pursued, and search is being made for the bed of the ancient stream which originally hollowed out the cavern. The lowest (and therefore presumably the oldest) bones yet found are those of a small species of wolf.

No. 44.

GEOLOGY OF NORTH WALES.

Natural History and Scientific Society.
Oswestry and Welshpool Naturalists' Field Club.

Museum.

Carnarvon Museum.

Publications of the Geological Survey.

Maps.—Anglesey, Sheets 77 N., 78; Carnarvonshire, Sheets 74 N.W., 75, 76, 77 N., 78, 79 N.W., 79 S.W.; Denbigh, Sheets 73 N.W., 74, 75 N.E., 78 N.E., 78 S.E., 79 N.W., 79 S.W., 79 S.E., 80 S.W.; Flintshire, Sheets 74 N.E., 79; Merionethshire, Sheets 59 N.E., 59 S.E., 60 N.W., 74, 75 N.E., 75 S.E.; Montgomeryshire, Sheets 56 N.W., 59 N.E., 59 S.E., 60, 74 S.W., 74 S.E.

Books.—Memoirs, vol. iii. On the Geology of North Wales, by Professor Ramsay and J. W. Salter, 1866, price 13s. (New Edition preparing.)

Important Works or Papers on Local Geology.

1858. Ramsay, Prof.—Physical Structure of Merioneth and Carnarvonshire. The Geologist, vol. i. p. 169.
1867. Geikie, Prof.—On the Volcanic Rocks of the British Isles. Rept. British Assoc., Dundee, p. 49.
1872. Hicks, Dr. H.—On Menevian Fossils. Q. Journ. Geol. Soc., vol. xxviii. p. 173.
1872, &c. Stanley, Hon. W. O.—Stone Circles, Weapons, &c., found in Anglesea and North Wales. Archæological Journal.
1872. Symonds, Rev. W. S.—Records of the Rocks. Murray, 12s. 6d.
1874. Mackintosh, D.—Glacial Traces in the Lake District and North Wales. Q. Journ. Geol. Soc., vol. xxx. p. 179.
1875. Davies, D. C.—Phosphorite Deposits of North Wales. Q. Journ. Geol. Soc., vol. xxxi. p. 357.
1876. Ramsay, Prof.—Physical History of the Dee. Q. Journ. Geol. Soc., vol. xxxii. p. 219.
1876. Mackintosh, D.—Beds of Drifted Coal near Corwen. Q. Journ. Geol. Soc., vol. xxxii. p. 451.
1876. Davies, D. C.—On some of the Causes which have helped to shape the Land on the North Wales Border. Proc. Geol. Assoc., vol. iv. p. 340.
1876. Davies, D. C.—Drift of the North Wales Border. Proc. Geol. Assoc., vol. v. p. 423.
1876. Langley, A. A., and Bellamy, J. C.—Slate Quarrying in the Festiniog District. Proc. Inst. Civil Eng., vol. xlvi. p. 211.
1877. Evans, J. F.—Mines of the Parys Mountain, Anglesea. Trans. Manch. Geol. Soc., vol. xiv. p. 357.
1877. Davies, D. C.—Relation of the Carboniferous Strata to the Permian. Q. Journ. Geol. Soc., vol. xxxiii. p. 10.

1877. Hughes, Prof. McK.—Silurian Grits of Corwen. Q. Journ. Geol. Soc., vol. xxxiii. p. 207.
1877. Phillips, J. A. —On certain Eruptive Rocks of North Wales. Q. Journ. Geol. Soc., vol. xxxiii. p. 423.
1878. Hicks, Dr. H. On some Pre-Cambrian Areas in Wales. Geol. Mag., p. 460.
1878. Hughes, Prof. McK. Pre-Cambrian Rocks of Bangor. Q. Journ. Geol. Soc., vol. xxxiv. p. 137.
1879. Hicks, Dr. H.—Classification of Pre-Cambrian Rocks. Geol. Mag., p. 433.
1879. Ruddy, Thos. On the Upper Part of the Cambrian (Sedgwick) and Base of Silurians of North Wales. Q. Journ. Geol. Soc., vol. xxxv. p. 200.
1879. Hicks, Dr. H. Pre-Cambrian Rocks of Carnarvon and Anglesey. Q. Journ. Geol. Soc., vol. xxxv. p. 295.
1879. Bonney, Prof. Quartz-Felsite, &c., at Base of Cambrians in North-West Carnarvon. Q. Journ. Geol. Soc., vol. xxxv. p. 309.
1879. Mackintosh, D. - Erratic Blocks of ... East of Wales. Q. Journ. Geol. Soc., vol. xxxv. p. 439.
1879. Hughes, Prof. McK.—Further Observations on Pre-Cambrian Rocks of Carnarvon. Q. Journ. Geol. Soc., vol. xxxv. p. 682.
1880. Hughes, Prof. McK.- Geology of Anglesey. Q. Journ. Geol. Soc., vol. xxxvi. p. 237.
1880. Callaway, Dr. C. Pre-Cambrian Geology of Anglesey. Geol. Mag., p. 117.
1880. Hicks, Dr. H. Pre-Cambrians in Harlech Mts. Geol. Mag., p. 519.
1881. Bonney, Prof. T. G.—Serpentine, &c., of Anglesey. Q. Journ. Geol. Soc., vol. xxxvii. p. 40.
1881. Callaway, Dr. C.- Archæan Geology of Anglesey. Q. Journ. Geol. Soc., vol. xxxvii. p. 210.

See also General Lists, p. xxv.

UNTIL half-a-century ago, it may be said that no scientific investigation had been made of the nature and relations of the rocks of North Wales. With all the other old rocks, both of this country and the Continent, they were termed "Transition" or "Grauwacke," and it was considered impossible to trace in them any separate series, or to determine their order of succession.

In the year 1831, however, one of the best geologists that ever buckled on a hammer—the late Adam Sedgwick, the Woodwardian Professor of Cambridge addressed himself to the arduous undertaking of examining and describing this region. The task occupied him many years, and indeed he may be said to have worked at it until his death in 1873; he was the first to distinguish the main subdivisions of the strata, and so he prepared and cleared the way for the subsequent minute mapping of the rocks by the Government Geological Survey. To the assemblage of beds of rock, which he found to form the crust of the earth in North Wales, Professor Sedgwick gave the name of CAMBRIAN, from Cambria, the ancient name for Wales.

While this ardent geologist was attacking Wales from the north, where, as we now know, the arrangement of the rocks is most complicated and very difficult to decipher, his friend Mr. (afterwards Sir) R. I. Murchison, had begun to work on the rocks of South Wales and Shropshire, where the beds are much less disturbed, and the whole problem of their arrangement and succession easier: to the strata seen there and in the adjacent counties, Murchison gave the name of SILURIAN.

Subsequently, when the Geological Survey came to *minutely* examine the *whole* region of Wales (1842-46), it was found that the *upper* beds of Sedgwick's Cambrian system were identical with the *lower* layers of Murchison's Silurian; the latter geologist at once appropriated this extension of his system, and on the geological maps published by the Survey, only a small area is shown as Cambrian, the dividing line being drawn at the base of the Lingula Flags; of late years, however, fresh discoveries have led British geologists to reconsider the question, and in this work the Cambrian system will be carried up to the base of the Arenig Group of rocks, a classification which is in accordance with the views of Lyell, Phillips, and Geikie.

Further discoveries made within the last few years — mainly by Dr. Hicks have shown that in Wales there exist rocks older than either the Silurian or Cambrian strata, comparable perhaps with the Laurentian rocks of Canada, which are the oldest known on the face of the globe.

Every one who has visited North Wales must (if he has thought upon the subject at all) have wished to know something of the rocks which form all the beautiful scenery of that region, and to which the climate, the soil, and the number, character, and occupations of the people are so largely due. We shall endeavour to describe the various strata in succession, commencing with those which have been determined to be of greatest age.

PRE-CAMBRIAN FORMATION.—In the Geological Survey maps, there are several parts of North Wales where rocks are coloured either as igneous, or

Fig. 97.—Contorted Pre-Cambrian strata on the west coast of Holyhead Island, Anglesea.

as "altered Cambrian," or "altered Silurian," which later work has shown to belong to a series of rocks older even than the Cambrian beds, and which are now known as "Pre-Cambrian." In these old rocks Dr. Hicks distinguishes three divisions: (1) the lowest or oldest he styles DIMETIAN, consisting of granitoid and gneissic rocks, which are overlaid by (2) compact felspathic and quartzose beds (some of which are old lavas)—this is the ARVONIAN group; lastly, (3) we have the PEBIDIANS, schistose slaty rocks, usually of a green colour, with volcanic agglomerates and breccias.

These Pre-Cambrian beds form more than one-half of Anglesey; they occur between Bardsey Island and Nevin on the west side of the Lleyn peninsula; between Bangor and Carnarvon; the so-called syenite north of Festiniog; near Dolgelly, &c. The thickness of these Pre-Cambrian rocks is estimated at 18,000 feet; they are entirely unfossiliferous. At Rhoscolyn, near Holyhead, a beautiful green rock is quarried as "Mona marble;" it is a variety of serpentine.

CAMBRIAN FORMATION.—The Cambrian rocks of Wales occur in Merioneth and Carnarvonshire. Their fossils show them all to be of marine origin, though Professor Ramsay has suggested that the reddish sandy beds, in which no fossils occur, may have been deposited in great lakes.

Harlech and Barmouth Beds.—These form a broad mountainous tract of country, of which the towns of Barmouth and Harlech mark the northern and southern limits respectively; on the west is the sea, while Craig-y-Penmaen marks the eastern limit; the ridge of Llawlech runs north and south, nearly in the centre of this Merionethshire anticlinal of Cambrian strata. The rocks are mostly greenish-grey grits, with some bands of slate, traversed by many dykes of igneous rock; except worm-tracks and worm-borings, no fossils are known from this district; the total thickness of the strata exposed may be 6,000 feet.

Llanberis Beds.—The Cambrian strata of Merioneth, dipping under newer rocks, again rise to the surface in a long narrow strip running from Bangor and Bethesda on the north-east, past Nant Francon, Penrhyn, and Llanberis, to the coast at Dinas Dindle in Carnarvon Bay. At the base is a bed of conglomerate, above which lie purple and green slates, and green and grey grits, altogether 3,000 feet thick. The slates are largely quarried, and the quarry at Penrhyn yields about 120,000 tons per annum; it is 500 feet in depth, and is worked in steps or terraces each about 40 feet in height. Fossils are still wanting, with the exception of the traces of burrows of marine worms (*Chondrites*), which occur in the fine-grained sandstone of Moel-y-ci, near Bangor. (See fig. 103, p. 336.)

At St. David's in South Wales strata of about the same age as those at Harlech, Barmouth, and Llanberis have yielded numerous fossils, and it is not impossible that a minute and untiring search, conducted by geologists who knew how to recognise obscure traces of life, might bring to light animal remains from the more northern Cambrian region.

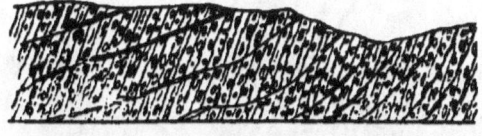

Fig. 98.—Section on the north side of Llyn Padarn, near Llanberis. This shows a conglomerate, consisting of slaty pebbles in a slaty matrix. When deposited, the pebbles doubtless lay with their longer axes parallel to the planes of bedding. Immense lateral pressure has, however, since that time acted on the strata, and has elongated the pebbles in the lines of cleavage.

Menevian Beds.—Menevia was the name of a Roman station at St. David's in Pembrokeshire, where these beds were first examined by Messrs. Salter and Hicks. In North Wales they circle round the Cambrians of the Harlech and Barmouth tract, being composed of black and grey flags and slates, with some beds of sandstone, altogether 600 feet thick; at Maentwrog (in the Waterfall valley), and near Dolgelly, several species of trilobites have been found.

Lingula Flags.—These consist of black and grey micaceous slates and sandstones, which rest upon the Menevian beds and are about 5,000 feet in thickness; they are chiefly developed in Merioneth, where they can be traced from Traeth Bach north of Harlech, north-east to Maentwrog, then southwards to Dolgelly and along the estuary of the Mawddach. The Lingula Flags were named after a shell (*Lingulella Davisii*), discovered in them in 1846, by Mr. L. Davis; worm-tracks abound, and several trilobites occur, many of which have been obtained from the slate quarries, near Festiniog. Small quantities of gold occur in quartz lodes, which traverse these beds at Dol-y-frwynog and Cwm Eisen, on the Mawddach. In 1877 there were raised 139 ozs. of gold, valued at £627, from the mines of Clogan and Cefn Coch. The Lingula Flags are traversed by many intrusive dykes of greenstone; the great mass of Rhobell Fawr, lying north of Dolgelly, has always been considered to be of this nature.

Tremadoc Slates.—These are dark earthy slates about 2,000 feet thick, named after the little town of Tremadoc, in Carnarvonshire, where they were first examined by Sedgwick in 1846; they contain a bed of pisolitic iron ore by which they can be traced to Dolgelly. Their fossils have been diligently collected by Messrs. Ash and Homfray, and comprise sixty-eight species, of which no fewer than sixty are peculiar to this division; they include the first crinoids, star-fishes, and cephalopods.

On the whole, the Cambrian rocks of North Wales are not geologically so interesting as the later-formed Silurian rocks; they consist of a great mass of beds from 12,000 to 15,000 feet in thickness, which seems to have slowly accumulated as sediment upon the floors of shallow seas; the life which at first existed in the waters must have been very scanty, and the consequent absence of fossils deprives the rocks of one great source of interest to the geologist, whilst neither by remarkable scenery, nor by valuable economic products are these shortcomings compensated for.

SILURIAN FORMATION. —The rocks we have next to describe will ever be connected with the name of that veteran geologist, the late Sir Roderick Murchison. He named them Silurian because the district in which the best types of the rocks occurred (Shropshire and the neighbouring Welsh counties) was formerly inhabited by a British tribe called the *Silures*.

Arenig Beds (or Lower Llandeilo Series). These are beds of grit and black slates 800 feet thick, which crop out on the west side of the Arenig range in Merioneth; they lie unconformably on the beds beneath, for, tracing them northwards, we find near Bangor and Carnarvon, the Arenig slates lying directly on purple slates and conglomerates of Lower Cambrian age, the entire Upper Cambrian series being absent.

Fig. 99.—Plaited surface of a bed of Lingula Sandstone, with the tracks of worms (*Helminthites*) upon it. South of Maentwrog, riverside. From the Lower Lingula Flags. (After Salter, Geol. Survey.)

The remarkable fossils called *graptolites* first occur in these beds, forty-eight species having been found; forty-seven species of trilobites also occur.

Llandeilo Flags. —Above the Arenig slates we find more slates and flaggy sandstones 3,000 feet in thickness, but sometimes greatly exceeding this amount, by reason of thick interbedded masses of volcanic ash and lava. These are the strata which mainly form the mountains of Cader Idris, Aran Mowddwy, the Arenigs, and the Moelwyns, a great crescentic chain, whose peaks, rising to a height of about 3,000 feet, are the highest points in Merionethshire. Dipping to the east under newer rocks, the Llandeilo Flags rise again to the surface on the east of the Berwyns at Craig-y-Glyn; still further east they form the Breidden Hills. Among the fossils *Orthoceras* (a long straight shell) is characteristic, as well as the trilobites *Ogygia Buchii* and *Asaphus tyrannus*.

The hill of Penmaenmawr, 6 miles west of Conway, consists of a greenish-grey felspathic rock, probably an altered dolerite, which bursts through black slates. It is largely quarried, being easily cut into "setts," which are much used for paving roads in Liverpool, Manchester, &c.

Bala and Caradoc Beds. —After the remarkable display of volcanic activity which took place in North Wales during the deposition of the Llandeilo Flags, there appears to have been an interval of comparative repose, during which the material of the black and blue slates and sandy beds, which form the lower Bala beds, were deposited on a shallow sea-bottom; then fresh vents, probably submarine, were found, and another great series of ash-beds and lavas bears witness to the renewed action of the igneous agencies on a grand scale. If we include all these volcanic deposits, the extreme thickness of the Bala beds is not less than 10,000 feet. We can trace this division from Towyn and Aberdovey north-east by Dinas Mowddwy to Bala; then it curves round to Bettws-y-coed and Criccieth. Two thin beds of limestone occur, called the Rhiwlas or Bala Limestone, and the Hirnant Limestone; from these limestone bands great numbers of fossils have been obtained, including brachiopod shells, star-fishes, crinoids, and more than one hundred species of trilobites.

The eastward inclination or dip of the beds south of Bala Lake, carries them under a long strip of Upper Silurian rocks, presently to be described, on the east of which they rise again, and form the range of the Berwyns, which for 30 miles marks the boundary between Montgomery and Merioneth; in this region the Bala Limestone is present, but the Hirnant band has thinned out.

Phosphorite.—In 1863 a miner discovered a nodular bed of phosphate of lime, at the *top* of the Bala Limestone at Cwmgwynen, 5 miles west of Llanfyllin; it has since been found and worked at Craig Rhiwarth in the Berwyns; in 1877 six tons were raised, valued at £7 10s.; it is used in the manufacture of artificial manures.

It was in the Snowdon district that the most copious outpouring of volcanic products during this epoch took place; these old ashes and lavas now form the wildest and grandest region of Wales, including Snowdon (3,571 feet), Moel Siabod, Carnedd Llewelyn, Carnedd Dafydd, Y-Glyder-Fawr, and Moel Hebog.

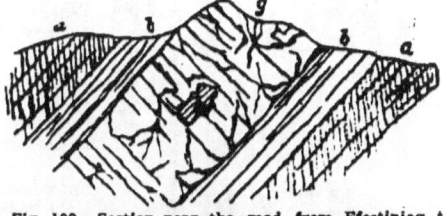

Fig. 100.—Section near the road from Ffestiniog to Dolwyddelan, showing the effect of intrusive igneous rocks on the cleavage of slate. A bed of greenstone has here baked or porcellanized the slates in its immediate neighbourhood and these are not cleaved, but have a speckled appearance like "snake-stone."

a Cleaved Slate.   *b* Altered Slate, uncleaved.
g Greenstone, with included fragments of slate.

Geological Structure of Snowdon. -- The highest mountain in England or Wales owes its elevation to several circumstances; among these may be named the positive hardness of the beds of rock of which it is formed, and the position in which they lie, bent into a synclinal curve, so that the effect of rain and frost and ice has been reduced to a minimum.

The base of the mountain is formed of slate and fossiliferous grits of the Bala beds; on these lie three great beds of felspathic porphyry; still higher are beds of felspathic, sandy, and calcareous ashes; and highest of all are the relics of a sheet of felspathic lava, here and there perched on the ashes in outlying fragments. (See figg. 101.)

Caradoc and Bala beds also occur in Anglesea, running from Parys Mountain on the north to Cymmerau Bay on the south. The famous Parys Mountain is composed of compact felspathic rocks, usually considered volcanic, and black shales containing graptolites.

The immense mass of copper pyrites, which was long worked here, has now been almost entirely quarried away, with the exception of certain minor branching veins. The amount of copper ore raised in 1877 was 3,613 tons, valued at £9,253, and yielding 136 tons of copper.

These mines were first opened in 1768, since which time they have yielded copper ore to the value of more than one million sterling.

Ochre and Umber are also worked in the Parys mountain and Mona mines;

these are earthy iron ores (*Limonite*) used in the manufacture of paints; in 1877 there were raised 2,440 tons, worth about £1 per ton.

Lower Llandovery Beds.—These may be called beds of passage between the Lower and Upper Silurian Formations, for of the 128 species of fossils which they have yielded (mostly obtained near Builth and Llandovery) 93 also occur in the Bala Beds below, while 83 are common to the rocks above: in North Wales, however, these beds appear unfossiliferous. We first see them near Bwlch-y-groes, about 5 miles south-east of Bala Lake, and they run southwards in a narrow strip past Malwyd to Dolfriog and Bryn Crugog; they consist of beds of grey grit intermingled with shales, altogether 1,000 ft. thick.

Thus the Lower Silurian period was one of great volcanic activity in North Wales, and in the rocks which now exist we have evidence of two principal epochs of eruption; the first of these occurred during the deposition of the Llandeilo Flags, and is indicated by the igneous rocks of Aran Mowddy, Cader Idris, Arenig, and Moelwyn; then the other is marked by those of the Snowdon district, which lie among the Bala beds. These volcanic rocks consist partly of massive sheets of felstone, varying in colour and texture, and partly of thick accumulations of tuff or ash. The former are true lava-flows, the latter point to frequent showers of volcanic dust and to the settling of such dust and stones on the sea-bottom, where they mingled with the ordinary sediment, and with shells, corals and other organisms. Some of these ashy deposits attain a great thickness: thus at Cader Idris they are about 2,500 ft. thick, the accumulated result of many eruptions: northwards this mass thins entirely away, and the ordinary sedimentary strata take its place. Equally local are the massive beds of felstone, which represent the submarine lava-flows of the time; sometimes they still preserve the slaggy vesicular character which marked their surface, when the melted rock was in a state of motion along the sea-bottom, an evidence of the existence and position of true submarine volcanoes during the Lower Silurian period in Wales.

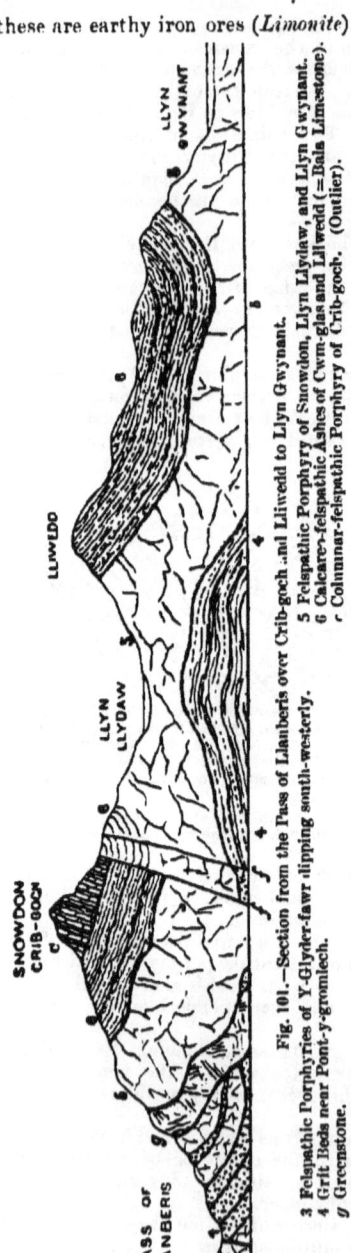

Fig. 101.—Section from the Pass of Llanberis over Crib-goch and Lliwedd to Llyn Gwynant.

3 Felspathic Porphyries of Y-Glyder-fawr dipping south-westerly.
4 Grit Beds near Pont-y-gromlech.
5 Felspathic Porphyry of Snowdon, Llyn Llydaw, and Llyn Gwynant.
6 Calcareo-felspathic Ashes of Cwm-glas and Lliwedd (=Bala Limestone).
c Columnar-felspathic Porphyry of Crib-goch. (Outlier).
g Greenstone.

No trace of the original craters or cones, however, is now left; these have long since been swept away.

The Lower Silurian rocks of Montgomeryshire are traversed by numerous metallic lodes, which chiefly occupy lines of fault or dislocation, which have fissured and cracked the solid rocks; heated waters rising from below have deposited ores of various metals on the sides of these cracks until they have become quite filled up. The produce of this county in 1877 was: lead ore, 8,066 tons, containing 60,182 ozs. of silver, value £143,288; zinc ore, 2,714 tons, value £10,444; copper ore, 45 tons, value £314. The Van mine near Llangurig is by far the most productive.

UPPER SILURIAN FORMATION.—The *Upper Llandovery Beds* being absent in North Wales, we next see strata called the *Tarannon Shales*, lying unconformably on all the beds beneath: they are fine, smooth, grey, or light-blue slaty beds, called "paste rock" by Sedgwick; their thickness is about 1,000 feet. They can be traced all the way from Conway in a narrow band by Cerrig-i-druidion, to the valley of the Tarannon river in Montgomeryshire; they have yielded as yet no fossils.

Wenlock Series.—The base of this division consists in North Wales of grits and shales 3,000 feet thick, named by Sedgwick the *Denbighshire Grits*, and extending from Conway to Llanrwst, Pentre Voelas, Derwen, Nant-yr-Eira, and ending off at Llandewi-ystrad-enny, in Radnorshire. On the whole, then these *Denbighshire Grits* may be considered as the equivalents of the *Woolhope Shale and Limestone*, which in Hereford and Shropshire form the lower part of the Wenlock Series; indeed, the grits pass into shales as we trace them to the south-east. They form a very sterile tract of country.

The Wenlock Shales proper, appear resting on the lower gritty beds over a large area in Denbigh, forming the country drained by the Elwy and its tributaries, and extending south-east past Bryn Eglwys to Corwen, Llangollen and Glyn Ceiriog: dipping under the newer rocks of the Vale of Clwyd, they rise up again on its eastern side, to form the fine range of Moel Fammau (1,845 feet) and Moel Fenlli (1,600 feet).

Southwards the Wenlock Shales occupy much ground in the centre and south-east of Montgomeryshire, round Llanfair, Llanlugan, and Montgomery; no limestones occur in the Wenlock Beds of Wales.

No traces of the highest Silurian strata—*the Ludlow Beds*—have yet been discovered in North Wales; they lie further to the south-east.

Thus in Lower Silurian times a group of volcanic islands comparable to the Lipari Isles, or to the neighbourhood of Santorin in the Grecian Archipelago, stood in North Wales, and from both subaërial and submarine openings great quantities of lava and ashes poured out; this seems to have been followed by consolidation and elevation of the strata, and the formation of the lowest Upper Silurian (Llandovery) beds as a beach deposit; then, depression taking place, the whole region sank beneath the waters, and the remaining Upper Silurian rocks were laid down as sediment, on the floor of a moderately deep and tranquil sea.

OLD RED SANDSTONE.—Of this well-known formation only a thin representative of the upper beds occurs in North Wales. In Anglesea we can trace red and grey sandstones and cornstones, with a bed of quartz conglomerate at the base, running in a long narrow strip north and south from Dulas Bay to Llangefni; in Denbigh similar beds crop out from Llysfaen to Ruthin, and again 3 miles north of Llangollen; no fossils have been found. Possibly these red beds should be rather classed with the Carboniferous Formation, than with the Old Red Sandstone; further enquiry is necessary to settle this point.

CARBONIFEROUS FORMATION.—In Anglesea we find a complete series of Carboniferous rocks, which owe their preservation, to their situation on the downthrow (north-west) side of a great fault, which ranges across the island from north-east to south-west, displacing the strata to the amount of 2,000 feet.

The *Carboniferous Limestone*, the *Millstone Grit*, and the *Coal-measures* rise up in succession, and are overlaid unconformably by Permian Beds; the land of country they form extends from Hirdre-faig and Red Wharf Bay to Malldraeth Bay. There are 2 small collieries, which in 1877 raised 1,330 tons of coal; the quantity of coal remaining unworked is estimated at five millions of tons.

The same succession of Carboniferous strata can be traced on the other side of the Menai Straits.

Carboniferous Limestone. Commencing with the grand promontory of Great Orme's Head, the "Mountain Limestone" runs southwards, underlying the red rocks of the Vale of Clwyd, resting here, in fact, in a hollow or trough formed by Silurian strata. Between Llanedilan and Cyrn-y-brair, the Limestone is shifted to the east by the great Bala fault, one of the longest in Britain, which has been traced from the coast of Merioneth through Bala Lake into Cheshire; from this point the limestone runs south to Llan-y-blodwell, in Shropshire; an outlying patch occurs west of Corwen. The Carboniferous Limestone forms a range of noble hills, whose escarpment faces the west; its thickness is here from 1,000 to 1,500 feet; it contains many mineral veins, one of the best known of which is the "Great Minera Vein," which occupies a line of fault running N.W. and S.E., crossing the Bala Fault at a right angle. Of lead ore there were raised in 1877, in Denbigh, 2,898 tons, valued at £38,000, and yielding 2,338 tons of lead and 12,900 ozs. of silver; in Flintshire, 3,184 tons of lead ore were obtained, value £42,000, yielding 2,560 tons of lead, and 17,689 ozs. of silver; Denbigh also yielded 2.238 tons of zinc ore (value £10,000), and Flint, 1,873 tons (value £7,500). The Mountain Limestone is also largely quarried for building and for burning into lime.

The upper part of the Carboniferous Limestone passes into thick shales, which probably represent the *Yoredale Beds* of Yorkshire, and these are overlaid by coarse sandstones—the *Millstone Grit;* the two divisions are together about 1,000 feet in thickness.

Coal-measures. In the Denbighshire Coal-field we find shales and sandstones 3,000 feet thick, containing 7 workable seams of coal; the area of this coalfield is 47 square miles; Wrexham and Oswestry mark its eastern boundary, Ruabon stands near the centre. In 1877 there were 59 collieries at work, which raised 1,622,500 tons of coal.

The Flintshire Coal-field extends from Mold along the west bank of the Dee, having an area of 35 square miles; from 51 collieries 855,750 tons of coal were raised in 1877.

In the same year about 50,000 tons of fire-clay were also obtained from the Coal-measures of North Wales, and 42,000 tons of iron ore, valued at £25,000.

Although we now only meet with Carboniferous rocks on the eastern borders of Wales (and in Anglesea), yet at the time of their deposit the coal-containing beds probably stretched far to the west, covering all the lower country, but lapping round the mountains of Snowdonia, which then stood as islands or low hills; all this mass of strata and much more has since been worn away by the agents of denudation.

PERMIAN red marls and sandstones overlie the Coal-measures east of Wrexham; to their occurrence in Anglesea we have already alluded.

TRIAS. The Vale of Clwyd is formed of red sandstones belonging to the *Bunter* division of the New Red Sandstone; the low country along the west bank of the Dee at Eccleston, Pulford, Holt, &c., is also formed of similar rocks: these are the newest stratified rocks which occur in North Wales.

SUMMARY. Considering now all that we have said of the rocks of North Wales, we find them to consist of a great thickness of old and hard rocks, which have been raised by those slow and long-continued movements of the earth's crust which are continually taking place, and then exposed by the removal of the newer rocks which once more or less covered them up. If, as some of the early geologists believed, the various strata were laid one above the other as regularly as the coats of an onion, then these Welsh rocks would for ever be invisible and unknown, for they would be buried beneath all the layers of rock which form the centre and east of England. But owing to a tilting up of the west side of our country, and to the forces of denudation - the sea, rain and rivers, frost and ice—by which higher rocks have been worn away, the Cambrian and Silurian rocks of North Wales contribute to form the wonderfully varied surface of this little island, this "epitome of geology," as Britain has well been called.

The present surface features of North Wales appear to be of high antiquity. Looking back through an enormous interval of time, we see that the Cambrian and Silurian rocks were greatly changed and denuded before the deposit of the Old Red Sandstone and Carboniferous Limestone on their upturned edges; then, after the formation of the Millstone Grit, there came changes in the physical geography of the region, which converted it into a vast delta a swamp, in which grew the plants whose compressed remains form the seams of coal. Elevation followed after this, and all through the Oolitic, Cretaceous and Tertiary epochs it is probable that North Wales remained above the level of the sea, having, indeed, a higher altitude than at present. During all this time the forces of the atmosphere rain and rivers, frost, snow and ice—would be acting upon the surface and wearing it away unequally, the softer and more exposed strata suffering most. In this manner thousands of feet of solid rock have been worn away, so that the surface on which we now walk, is far below the old surface (or surfaces) which it had in past times. And this wear and tear is still going on, every stream, every rivulet is carrying to the sea, sediment composed of particles of the rocks which constitute the land, so that the surface of the country is being gradually lowered; reckoning in this way it is only a question of time as to how long it will take before North Wales is entirely swept into the sea. The rivers would perhaps take six millions of years to do it; but, in saying this, we do not take into account the upheaving forces below the surface, these may neutralize or more than neutralize the levelling effects of denudation. However, the time is distant! So that looking to the future, the geologist can contemplate with equal serenity the idea that waves may roll where Snowdon now towers, or that "the highest mountain in England or Wales" may be elevated to a height in excess of Mont Blanc both these things have happened in the past to parts of what is now Wales, and they may happen again in the future.

THE GLACIAL PERIOD.— The commencement of the Glacial Period found Wales carved out into hills and valleys, pretty much as it is now. What was the cause of the intense cold which prevailed in this country, at a comparatively recent geological epoch (that is to say about a quarter-of-a-million years ago) it does not now concern us to enquire, neither is it thoroughly understood. That such a cold period did prevail was not imagined until the year 1845, when Agassiz, coming fresh from the study of glacial conditions in Switzerland, declared that he recognized in the mountainous parts of Great Britain, phenomena which could only have been produced by the former presence of great masses of ice.

The signs of old glaciers in North Wales have been admirably described by Professor Ramsay. About the commencement of the Glacial Period it would appear that the country generally stood at a higher level than at present, and Britain was joined to the Continent; as the cold increased, sheets of ice gathered on the mountains of Scotland and the Lake District, and pressed southwards over what is now the bed of the Irish Sea, scraping the coast-hills of North Wales and passing over Anglesea and the Lleyn peninsula, on whose rocks, when bared, we see still the grooves made by the stones embedded in the ice. These grooves point to the north-east, showing the direction whence this great glacier advanced. At the same time all the Snowdonian mountains and the heights south of them, were overlaid by ice to the depth of several hundreds of feet; this ice passing downwards along the valleys, smoothed the rocks into rounded flowing outlines (*roches moutonnées*) and carried great blocks of rock (*blocs perchés*) far from their parent homes. When we walk up the passes of Llanberis, Nant-ffrancon, or indeed any mountainous valley of North Wales, we are treading on the bed of an old glacier, and if our eyes have been trained to the observance of ice-work, as it is now being done in Switzerland, Greenland, &c., we shall note all the proofs of its former presence in this region.

After this continental period, a great depression of the surface of the land appears to have taken place, of which we find traces in beds of gravel and sand, resting on the hill-sides at various levels until, at last, we find them at a height of 1,170 feet, near the summit of Moel Tryfan, where they are well seen in the Alexandra slate-quarries; 20 or 30 species of shells have been found here, mostly of cold-water or Arctic kinds. In beds of sand and gravel overlying boulder-clay, near Corwen, much drifted coal has been found, and there is much coal debris in the drift round Ruabon and Wrexham; this may have been brought by floating ice of no great thickness.

Fig. 102.—Saddle Quern, for grinding corn, &c.; found in a hut-circle at Ty Mawr, Anglesea.

Then elevation again took place and the land rose above the waters; again glaciers began to form on the hill-tops, though not of such great size as before. Icebergs, detached from the glaciers, carried blocks of felspathic rocks from the Arenigs, eastwards over the submerged plain of Cheshire, until they dropped their burdens near Wolverhampton and Birmingham, where great numbers of boulders are now to be found. By the grinding action of the ice, hollows were scooped in the softer beds of rock, and these now form lake-basins; almost all the lakes of North Wales, such as Bala Lake, the Lakes of Llanberis, Llyn-llydaw, Arenig, &c., owe in this way their existence to the work of glaciers. But the climate began to ameliorate, slowly the glaciers retreated up the valleys, leaving behind them streams and semi-circles of stones (*moraines*) which they had borne onward; finally all the ice disappeared, the climate became temperate, and the Geological record merges into the Historical.

PREHISTORIC MAN.—Perhaps the earliest traces of man yet found in North Wales, are the human bones and flint implements found in the caves at Cefn, on the western side of the Vale of Clwyd; these caves are in the western escarpment of the Carboniferous Limestone, that rises from under the New Red Sandstone which fills the lower part of the valley; bones of extinct species of elephant, rhinoceros, hippopotamus and bear, with the spotted hyena and reindeer, were also met with, embedded in the floor of these caves; Prof. Ramsay considers that some of these cave deposits may date back to pre-glacial times.

Of the *Neolithic Stone Age*, when man had attained considerable skill in the manufacture of stone tools, having learnt to grind and polish them, we meet with frequent relics. Round Ty Mawr, in Anglesea, the Hon. Owen Stanley has traced the remains of many hut-circles, and in the debris he has found hammer-stones, flint celts or axe-heads, scrapers, whetstones, querns for grinding corn, &c. (See fig. 102.) The cromlechs, of which 29 are known in Anglesea, are immense stones poised on other or supporting blocks; they are probably the remains of tumuli or burying-places, the mound of earth by which they were once surrounded having been removed.

In the Amlwch Parys Mine a stone hammer (of basalt), nearly a foot long, was found in some ancient workings.

Similar tools, weapons, such as arrow-heads, spear-heads, &c., have been found in Carnarvonshire, at Bangor, Aber, Dwygyfylchi, Llandudno, Nantlle, Penmaenmawr and Tomen-y-mur; in Denbighshire, at Brynbugeilen, Moel Fenlli, Pentrefoelas and Tynewydd; in Merioneth, at Carno, Harlech, Llanaber, and Maesmore near Corwen; and in Montgomeryshire, at Carno, Llanbrynmair and the Snow Brook Lead Mine, Plinlimmon.

Tin and copper appear to have been the first metals discovered, for the implements which succeeded the stone tools, and which are made after their patterns, are formed of *bronze;* but this, too, had been superseded by iron. at the time of the Roman invasion, B.C. 55.

No. 45.

GEOLOGY OF SOUTH WALES.

Natural History and Scientific Societies.
South Wales Institute of Engineers; Swansea. Proceedings.
Cardiff Naturalists' Society; Report and Transactions.
Royal Institution of South Wales; Swansea.

Museums.
Cardiff Free Library and Museum.
Museum of the Royal Institution of South Wales, Swansea.
University for Wales Museum, Aberystwith.

Publications of the Geological Survey.
Maps. - Brecknockshire, Sheets 36, 41, 42, 56 N.W., 56 S.W., 57 N.E., 57 S.E.; Carmarthenshire, Sheets 37, 38, 40, 41, 42 N.W., 42 S.W., 56 S.W., 57 S.W., 57 S.E.; Cardiganshire, Sheets 40, 41, 56 N.W., 57, 58, 59 S.E., 60 S.W.; Glamorganshire, Sheets 20, 36, 37, 41, 42 S.E., 42 S.W.; Pembrokeshire, Sheets 38, 39, 40, 41, 58; Radnorshire, Sheets 42 N.W., 42 N.E., 56, 60 S.W., 60 S.E.

Books. Memoirs, vol. i., including De la Beche on the Formation of the Rocks of South Wales; Ramsay on the Denudation of South Wales; Smyth on the Gogofau Mine, &c., 1846. price 21s. Vol. ii. part 2, Hunt and Smyth on the Mines of Cardigan and Montgomery, &c., 1848, price 21s. Iron Ores of South Wales, 1861, price 1s. 3d.

Important Works or Papers on Local Geology.
List by W. Whitaker of 667 books, papers, &c., on the Geology of Wales, in British Association Report for 1880, p. 397.

1867. Moore, C. Abnormal Conditions of Secondary Deposits in Somerset and South Wales. Q. Journ. Geol. Soc., vol. xxiii. p. 449.
1871. Harkness and Hicks. St. David's Promontory. Q. Journ. Geol. Soc., vol. xxvi. p. 384.
1872. Hicks, H. Tremadoc Rocks at St. David's. Q. Journ. Geol. Soc., vol. xxix. p. 39.
1872. Ramsay, Prof. A. C. River Courses of England and Wales. Q. Journ. Geol. Soc., vol. xxviii. p. 148.
1872. Symonds, Rev. W. S. Records of the Rocks. Murray, 12s. 6d.
1875. Brown, T. F. South Wales Coal-field. Proc. S. Wales Inst. Eng. vol. ix. p. 59.
1875. Hicks, H. —Arenig and Llandeilo Rocks of St. David's. Q. Journ. Geol. Soc., vol. xxxi. p. 167.
1875. Hopkinson, J., and Lapworth, C. Arenig and Llandeilo Graptolites. Q. Journ. Geol. Soc., vol. xxxi. p. 631.
1876. Hicks, H. On some Areas where the Cambrian and Silurian Rocks occur as a Conformable Series. Rep. Brit. Assoc. for 1875, p. 69.

1877. Moore, Chas.—The Liassic and other Secondary Deposits of the Southerndown Series. Trans. Cardiff Nat. Soc., vol. viii. p. 53.
1877. Taylor, Dr. W.—The Gower and Doward Bone Caves. Trans. Cardiff Nat. Soc., vol. viii. p. 79.
1877. Hicks, H.—Pre-Cambrian (Dimetian and Pebidian) Rocks of St. David's. Q. Journ. Geol. Soc., vol. xxxiii. p. 229.
1878. Keeping, Prof. W.—Notes on Geology of Aberystwith. Geol. Mag., p. 532.
1879. Hicks, Dr. H.—Pre-Cambrian (Arvonian) Rocks of Pembrokeshire. Q. Journ. Geol. Soc., vol. xxxv. p. 285.
1879. Harris, W. H.—Geology of Cardiff. Science Gossip, p. 99.
1880. Perkins, C. H.—The Anthracite Coal and Coal-field of South Wales. British Association Report, p. 220.
1880. Woodward, H. B.—Geology of Swansea. Science Gossip, pp. 171, 198.
1881. Keeping, W.—Geology of Central Wales. Q. Journ. Geol. Soc., vol. xxxvii. p. 141.

See also Trans. Woolhope Club (numerous references to, and papers on South Wales); and Trans. of South Wales Institute of Mining Engineers.

See also General Lists, p. xxv.

THE rocks which form South Wales are in the main a prolongation of the same beds which constitute the northern division of Cambria; one great difference is due to the fact, that volcanic action was not nearly so rife in this southern area during the deposition of the Silurian strata; as a consequence of this, the rocks are of a more uniform degree of hardness, so that the scenery is less grand and striking.

On the other hand, owing to the great development of the Carboniferous rocks, the mineral wealth of South Wales far surpasses that of the more beautiful and tourist-attracting region further north.

The rocks of Wales may be said to form the foundations of the British Isles. Although these old rocks only appear *at the surface* in the west, yet they extend eastwards, undulating beneath newer rocks, right under London and the east of England. Lying thus at the bottom, it is evident they must be the oldest or first-formed of the various layers of aqueous or sedimentary rocks, which constitute our islands. Though geology does not profess to estimate in years the age of any system of rocks, yet it is certain that all the great leaders in that science, would concur in affirming that, in all probability, the old rocks which we see forming the peninsula of St. David's were deposited as mud and sand on a sea-bottom more than one hundred millions of years ago. Yet we know well that they were not the first rocks which formed a solid crust on the surface of our planet.

In studying South Wales, then, let us commence on the extreme west, where the oldest rocks crop out or come to the surface, and then describe the beds as they rest one upon another, layer upon layer; a course which will lead us to the south-east, where the newest rocks of the region form the cliffs near Cardiff; in this task we shall find the maps published by the Geological Survey to be of the greatest assistance.

PRE-CAMBRIAN FORMATION.—The Government geological survey of South Wales was made between 1840 and 1855, at a time when much less was known respecting rock-masses, especially igneous rocks, than at present; the use of the microscope has, in fact, within the last ten years made a great difference in our ideas respecting rocks, and many which were formerly thought to be truly *igneous, i.e.*, to have once been in a completely melted state, from which they have cooled down and crystallized out, are now known to be metamorphic aqueous rocks. Thus, on the survey map of Pembrokeshire, a ridge of syenite is shown at St. David's, piercing Cambrian rocks, and flanked on each side by altered Cambrian rocks. But thanks to the minute researches of Dr. Henry Hicks, F.R.S., continued during the last fteen years, we know now that the so-called syenite belongs to an extremely old series of rocks, the oldest in the British Isles, which we now term Pre-Cambrian, and in which Dr. Hicks has even been able to make out several divisions or sets of beds. The oldest group he styles DIMETIAN. It is composed of "quartzose rocks, granitoid gneiss, and compact granitoid rocks, with bands of crystalline limestone;" above these come the ARVONIANS, "breccias, halleflintas and quartz felsites;" and lastly the PEBIDIANS, including volcanic ashes, breccias, and agglomerates with micaceous and chloritic schists.

These old rocks form a ridge running east-north-east and west-south-west from Llanhowell, past the city of St. David's to the coast at Porth-lisky; they are unfossiliferous.

CAMBRIAN FORMATION. —Resting on the flanks of the old "Pre-Cambrian Island" formed by the rocks we have described above, we find conglomerates composed of rounded masses of quartz embedded in a purple matrix; these form the base of the whole Cambrian Formation; they are overlaid by green, purple, and grey flaggy sandstones, 3,000 feet thick, with intercalated red shales, which contain the earliest undeniable traces of fossils yet known on the surface of the globe, including small brachiopod shells (*Lingulella ferruginea* and *L. primæva*) annelids, phyllopods, polyzoa, pteropods, a crustacean, two sponges, and several trilobites; Porth-Clais Harbour is the best place to find fossils; the sandstones have been used in the restoration of the cathedral at St. David's.

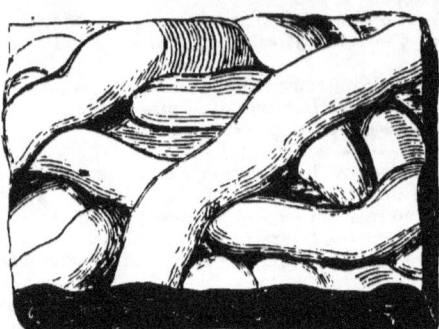

Fig. 103.—Filled up Burrows of Marine Worms (*annelides*) in Cambrian Sandstone.

Menevian Beds (so named from Menapia, the old Roman station at St. David's).—These are grey and black flagstones and slates 600 feet thick. The characteristic fossils are trilobites, as *Paradoxides Davidis*, nearly 2 feet long. *Erinnys venulosa*, the trilobite with the largest number of rings, occurs here, in conjunction with *Agnostus* and *Microdiscus*, the two genera with the smallest number. Blind trilobites are also found, as well as those which have the largest eyes, such as *Microdiscus* on the one hand and *Anopolenus* on the other.

Porth-y-rhaw, Caerbuddy, and Craclli are the best points at which to examine the Menevian beds.

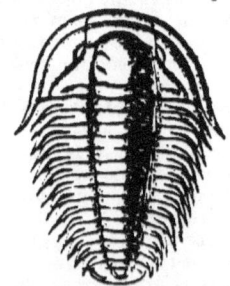

Fig. 104.—*Agnostus rex*, Lower Cambrian.

Fig. 105.—*Olenus micrurus*, from the Lingula Flags, Lower Cambrian.

Lingula Flags.—Grey and bluish sandstones and slates 2,000 feet thick; these rest conformably on the lower rocks, as may be seen in Whitesand Bay, Solva Harbour, near Caerfegga, &c.: they also form the north-west corner of Ramsey Island, where the fossil shell, *Lingulella Davisii*, after which these beds are named, occurs abundantly.

The Tremadoc Slates include dark earthy flags and sandstones with some iron-stained slates on the top, and are about 1,000 feet in thickness; they are seen in the north-east of Ramsey Island, at the north end of Whitesand Bay, and round Tremanhire and Paran. The fossils are numerous and interesting, and include the first known crinoids, star fishes, lamellibranchs and cephalopods. The same pressure which has produced the phenomenon of cleavage has also distorted the fossils, so that they are often difficult to recognise. Fig. 106 shows a well-known trilobite, *Angelina Sedgwickii*, in its natural shape (*A*), and (*B*) as we often meet with it in slaty rocks.

SILURIAN FORMATION.—Why do we draw a great line of demarcation at the top of the Tremadoc slates? Mainly because, in the succeeding strata of which we shall now have to speak, we find a great change in the forms of life—the fossils; "of seventy species of fossils found in the Tremadoc, only four pass up into the Arenig beds" (Etheridge.) To effect such an extinction of old, and introduction of new species, must, as we know, have required a great interval of time.

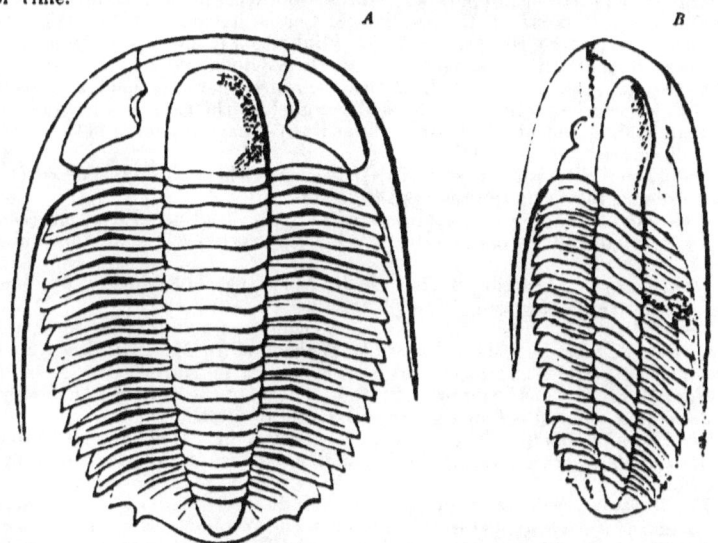

Fig. 106.—*Angelina Sedgwickii*, in its natural condition (*A*), and distorted by cleavage (*B*), from the Upper Tremadoc Beds, Cambrian Formation.

The Arenig Beds; black slates, 4,000 feet thick, traceable from the north side of Whitesand Bay to beyond Llanveran; numerous fossils have been found in a quarry under Llanvirn, in the cliffs north of Trwyn-hwrddyn in Whitesand Bay, &c., *graptolites* appearing for the first time, and in great numbers.

Llandeilo Flags.—In Abereiddy Bay we see black slates, flags, and sandstones, with some bands of limestone, and also beds of indurated volcanic ash, the result of eruptions of submarine volcanoes during the deposition of the strata; a considerable, but uncertain thickness of the beds is shown, perhaps 3,000 feet. These Llandeilo Flags extend to Fishguard and the Precelly Hills (1,754 feet high); curving eastward, we trace them past St. Clare and Carmarthen to Llandeilo, Builth, and Llandrindrod. At Llandewi Felfry, Lampeter Felfry, and Llandeilo, there are good beds of limestone—the oldest in Wales—which are quarried for lime, and which have yielded numerous fossils, as the trilobites *Ogygia Buchii*, *Asaphus tyrannus*, with many double graptolites, &c.

Bala or Caradoc Beds.—No sharp line of demarcation has been drawn between these and the Llandeilo Flags, nor in South Wales is there any rigid line between the Bala beds and the *overlying* newer *Lower Llandovery Rocks;* together these occupy a wide area, including the whole of Cardiganshire, and extending thence into the surrounding counties, forming, in fact, the desolate, almost desert-like region of "Mid-Wales." The strata undulate so that their real thickness can scarcely be determined, but it must be considerable, perhaps above 10,000 feet; there is an absence of the lava flows and ash-beds which distinguish the Bala series in the Snowdon district, instead of which we find clay-slates, with grits and hard sandstones, and conglomerates or pudding-stones in the upper part. Fossils are nowhere abundant, though the surfaces of the sandstones are often covered with ancient worm-tracks; in slates at the lead-mines, 8 miles east of Aberystwith, the Rev. J. F. Blake noticed many small red depressions on the rocks, which proved to be casts of Dentalian Foraminifera. The cliffs round Aberystwith show the beds to be wonderfully contorted, and to be dipping westwards towards the sea. Further east workable slates occur at Devil's Bridge, Corris, Tregaron, Strata Florida, Goginan, &c.; there the great mass of Plinlimmon (2,481 ft.) is formed of another group of grit-beds, with some quartz conglomerates.

At Gogofau, near Pumpsaint, in Carmarthenshire, there are quartz veins in hard Caradoc grits, which were worked for gold in the time of the Romans; several gold ornaments, including a beautifully-wrought gold necklace, have been discovered here.

The Lower Silurian rocks of Mid-Wales are rich in the ores of several humbler minerals; the returns for 1877 show:

 Cardigan. Copper ore, 127 tons, value £583; lead ore, 5,850 tons (containing 47,284 ozs. of silver), value £79,244; zinc ore, 588 tons, value £2,204.

 Carmarthen. —Lead ore (from Rhandir Abbot, near Llandovery) 687 tons (silver, 2,064 ozs.), value £8,216.

UPPER SILURIAN. The strata of this division rest unconformably on the lower rocks. First we find the *Upper Llandovery Beds* (or May Hill Sandstone), grey and yellowish sandstones, 800 feet thick, seen in Marlows Bay (west of Milford Haven); a long, narrow strip crops out east of the town of Llandovery, and a similar outcrop is found 5 or 6 miles west of Builth.

The Tarannon Shales.— These are pale-coloured, or purple unfossiliferous shales and slates, which rest on the May Hill Sandstone in the localities above named.

The Woolhope Beds are well developed in Radnorshire; at Corton near Presteign, at Nash Scar, and near Old Radnor, the Woolhope Limestone is seen as a thick and massive, sometimes subcrystalline rock; the eruptive trap rocks which form the hills of Hanter, Worsel, Stanner, and Old Radnor, have metamorphosed the limestone. The latter rock is largely quarried, for there is no other calcareous bed between this region and the coast-line of Cardigan. Among fossils, the Barr trilobite (*Illænus Barriensis*), and shells, crinoids, and corals, are frequently met with.

Wenlock Beds. These are thick beds of shale, traceable from south of Llandeilo, in a long narrow strip (1 to 4 miles wide), past Myddfai, Builth, and thence to Newtown: they dip to the south-east.

Ludlow Beds. —These are also shales, with a band of concretionary limestone (Aymestry limestone), passing up into fine-grained sandstones. They form the surface of the eastern half of Radnorshire.

OLD RED SANDSTONE. The passage from the Silurian beds into the rocks above, is marked by a change in the colour of the rocks from grey to red. The strata we have now to describe consist of red, yellow, and chocolate-coloured sandstones, red marls, and irregular layers of sandy concretionary limestone called cornstones, the total thickness being above 10,000 feet. The "Old Red" forms the greater part of the counties of Hereford and Monmouth; it enters Brecknockshire from the east, forming all the surface from Crickhowel to Hay, Talgarth, Brecon, and Mynydd Epynt, rising in the Brecknock Beacons to a height of 2,862 feet; passing out of Brecknock the outcrop narrows rapidly, and runs as a band 3 or 4 miles wide by Detwydd to the mouth of the Towy, and thence to Narberth. All along this line from east to west—viz., from Crickhowel to Narberth—the Old Red Sandstone dips to the south and disappears under Carboniferous rocks; passing underneath the entire coal-field of South Wales it rises up again on the south, but here its continuity is broken by the sea; we find it running east and west of Milford Haven, forming the south half of Caldy Island; passing across the centre of the peninsula of Gower (from Burry Holmes to Oxwich Bay), and then patches of it crop out from near Bridgend to Cardiff, whence it passes to the main mass in Monmouthshire.

At Skrinkle Haven and West Angle on the coast of Pembroke, the passage upwards from the Old Red Sandstone to the Carboniferous Limestone is excellently displayed; the two formations are evidently quite conformable, shales, sandstones, and limestones alternating in turn at the junction.

In the great thickness of red beds, fossils are far from common; perhaps the presence of oxide of iron (which tinges the rocks red) was inimical to life; scales, plates, and spines of ganoid fishes, with obscure plant-markings, are all the remains of life which the Welsh Old Red has yielded, and they favour the idea generally held, that the beds are of fresh-water origin, having been deposited in great lakes or estuaries.

CARBONIFEROUS FORMATION.—The famous coal-field of South Wales occupies 1,000 square miles in the south of the principality. It has a length from west to east of 85 miles, between St. Bride's Bay in Pembroke and Pontypool in Monmouth, and the extreme breadth from north to south is 20 miles. In form it is somewhat like a pear, with the narrow end to the west; a pear, too, out of which the sea has taken great bites, as witness the large openings of Carmarthen Bay and Swansea Bay.

Carboniferous Limestone.—This well-known blue or grey, tough, compact, semi-crystalline rock, forms at the surface a rim round the entire coal-field, except at the extreme western end in Pembroke, where both Limestone and Millstone Grit are absent (or have parted with their usual characters), and where consequently the Coal-measures rest directly upon Silurian rocks; below the surface the limestone underlies the whole coal-field; all the colliery pits would if continued downwards at last enter it. It is finely seen in the bold coast cliffs near Stackpole, where it is much contorted. The thickness of the Carboniferous or Mountain Limestone varies from 500 to 1,000 feet; it is largely quarried to use as a flux in the iron-furnaces, and for building and burning into lime; it also contains valuable deposits of iron-ore (hematite) and lead-ore (galena). The amount of brown hematite raised in 1877 was 77,320 tons, valued at £39,471; of this amount 49,084 tons were from the Mwyndy mine: of ochre 295 tons were raised at the Garth iron mine.

Two or three *outliers* mark the former extension of the Carboniferous rocks far to the north of the coal-basin; chief of these is Pen Cerrig Calch, north of Crickhowel, a grand isolated hill of Mountain Limestone capped with Millstone Grit.

The Millstone Grit is a coarse sandstone—the "Farewell Rock" of the miners, who know that it marks the base of the workable seams of coal. Accompanying the Carboniferous Limestone, it forms, like that rock, a rim or edge all round the coal-field. It is best seen on the north crop from Kidwelly, eastwards by Cross Inn, to Penderyn: here it forms a table-land in which rivers rise and run southwards to the sea, cutting deep north and south valleys across the coal-field, their original direction having been determined by numerous faults which cross the district in the same direction. After sweeping round the Blorenge and passing east of Pontypool the Millstone Grit is seen only at intervals on the south crop, as south of Caerphilly and north of Bridgend: its thickness varies from 250 to above 1,000 feet.

B B

The Coal-Measures.—These are of enormous thickness—perhaps above 10,000 feet. They are divided into an upper and a lower series, separated by a great thickness of sandstone, called the Pennant Grit. Owing to the basin-shaped arrangement of the beds, which distinguishes more or less all our coal-fields, the Upper Coal-measures occur in the centre of the district, while the lower beds crop out both to the north and south, between these upper central strata and the Millstone Grit. Two great undulations of the strata (anticlinal axes) have been traced, one running from Risca to Aberavon, and the other from Cwm Neath to Kidwelly; the effect of these folds is to bring up the Lower Coal-measures within a workable distance of the surface. The lower Coal-measures are rich in coal and ironstone, and contain numerous marine shells and entomostraca, which have been diligently collected by Dr. Bevan, and by Mr. W. Adams, F.G.S., the eminent mining engineer of Cardiff.

In the Upper Coal-measures, whose base is marked by the Mynyddislwyn coal, only plant-remains are found, together with the shell *Anthracosia*, which resembles a small fresh-water mussel.

There are 25 coal-seams above 2 feet in thickness, giving a total thickness of 84 feet of workable coal; and the quantity of coal which it is possible to get, may be estimated at 30,000 millions of tons, which at the present annual rate of consumption, will last for about 2,500 years.

The change in the character of the coal-seams in passing from east to west is well known. At the eastern end of the basin, near Pontypool, &c., the coals are chiefly bituminous and make a very superior coke, continuing so to Rhymney, between which and Dowlais a slight change takes place, becoming more marked at Cyfarthfa, where the seams become anthracitic; this quality continues through Hirwaun Common, the head of the Neath Valley, across to Ynyscedwin in the Swansea Valley, the Twrch Valley, and thence to Mynydd Mawr and the Gwendraeth; at the two latter places anthracite of a very pure quality is worked for hop and malt drying, distillery purposes, &c.; further west in Pembroke all the coal is anthracitic. Now ordinary or bituminous coal, becomes converted into anthracite or stone coal, by the loss of its gaseous constituents; whose removal may be effected by heat from below, together with disturbance of the strata, by which fissures or faults allowing the gases to escape are produced. Dislocations of this nature are numerous in Pembrokeshire, and it is possible that masses of igneous rocks may exist beneath this western area, of whose presence the alteration of the coal-seams is an indication.

The following table shows the amount of coal raised in the South Wales Coal-field in 1877 :—

No. of Collieries.	County.	Tons of Coal.
268	Glamorganshire	11,889,600
3	Brecknockshire	141,885
10	Pembrokeshire	76,400
44	Carmarthenshire	526,450
116	Monmouthshire	4,350,785

Of iron-ore from the beds of clay iron-stone (carbonate of iron), which are numerous in the Lower Coal-measures, the amount raised in 1877 was 367,316 tons, valued at £220,390; a considerable quantity of fire-clay was also raised.

All the great iron-works as Risca, Pontypool, Blaenavon, Ebbw Vale, Merthyr Tydvil, and Aberdare, stand upon the Lower Coal-measures, situated most favourably with the coal and iron around and beneath them, and the Mountain Limestone within a very short distance.

THE TRIAS.—There is very little of this formation in South Wales. Patches of *Dolomitic Conglomerate* lie west of Cardiff near Newton Nottage, Coychurch, St. Fagans, and Llandaff; these are overlaid by *Keuper Red Marls*, which are also seen at the base of the cliffs near Penarth.

In the conglomerate at Newton Nottage, Mr. T. H. Thomas detected in 1878, and Mr. Sollas has described under the name of *Brontozoum Thomasi*, some gigantic three-toed footprints made either by some great bird like the emu, or by some strange and now extinct reptile ; the slab has been removed to the Cardiff Museum.

THE RHÆTIC FORMATION.—The cliffs between Penarth and Lavernock expose the finest British section of the grey marls and black shales, which constitute the Rhætic beds. At the base, at Penarth Head, are the strikingly coloured red marls of the Trias ; then come the grey or "tea-green" Rhætic Marls, 28 feet thick, surmounted by black shales 24 feet thick, which contain the famous bone-bed, a hard sandy layer full of teeth, scales, bones, &c., of fishes and reptiles. Above these as we walk towards Lavernock we see the *White Lias Series*, 18 feet thick, composed of grey and brown sandy shales, with irregular beds of limestone. Fossils are very numerous, including the shells *Avicula contorta*, *Cardium Rhæticum*, &c., and the pretty star-fish, *Ophiolepis Damesii* (rare).

THE LIAS.—Lower Lias shales and limestones lie upon the Rhætic beds on the south coast of Glamorgan, from Penarth, westwards past Barry Island, Aberthaw, and Dunraven Castle, to the mouth of the river Ogmore, and run inland as far as Cowbridge, where they rest upon the upturned edge of the Carboniferous Limestone. The Lias is, in this district, close to its ancient shore-line, and at Sutton and Southerndown it becomes a conglomerate formed of limestone pebbles, representing an ancient beach. From quarries at Brocastle, Llanbethian, Laleston, Bridgend, Ewenny, &c., Mr. C. Moore has obtained more than fifty species of corals, with many shells, *Ammonites planorbis*, *A. angulatus*, *A. Bucklandi*, &c. The Lias limestones are worked in this district to make cement, &c.

SURFACE DEPOSITS.—*The Glacial Period.*—The traces of the former presence of glaciers are not so clear and abundant in South as in North Wales; for this the nature of the rock-masses, the minor elevation of the land, and its position further south are mainly accountable. "There are well-marked grooves or striations of a glacial character on the rocks of the coast of Pembroke at Trwyn-hwrddyn, Porth-clais, and Pel-dal-deryn. Boulder-clay with ice-scratched stones is common, and chalk-flints mingled with stones native to the district are not uncommon on the mainland and in Ramsey Island. Flints are also found in quantity, mingled with ice-scratched erratics, all along the low ground of Glamorganshire north of the Bristol Channel between Cardiff and Bridgend, and boulder-clay is indeed common, here and there, all over South Wales." Large boulders of Millstone Grit (known as "Arthur's Stone") and of Old Red conglomerate are numerous on the peninsula of Gower. The Raised Beaches and Submerged Forests which occur at several points on the coast-line mark oscillations of the land since the Glacial Epoch.

PREHISTORIC MAN.—Probably the earliest traces of human habitation of this region are the stone implements found below the present floors of the caves in the Carboniferous Limestone of Gower and Pembroke.

The numerous caves of Gower were explored by Col. Wood and Dr. Falconer in 1860 : here, beneath layers of stalagmite, great numbers of bones of extinct species of mammals—as the mammoth, long-haired rhinoceros, cave lion, cave bear, hyæna, &c., were found, together with flint flakes and a very fine flint arrow-head.

The caves near Tenby and on Caldy Island gave almost precisely the same evidence ; they were explored in 1865 by Mr. Smith, of Gumfreston, and the Rev. H. H. Winwood.

Of stone tools found elsewhere, we may note a polished celt, or axe-head, found at Cardiff, and now in the British Museum; it is 4½ inches long, but appears to have been worn down by much use, and to have been re-ground : a very similar specimen was found at Melyn Works, Neath. A remarkable stone axe-head, made of some felspathic rock, and perforated for a handle, occurred at Llanmadock in Gower, and is now in the Museum of the Royal Institution at Swansea.

The method by which holes were bored in those days is illustrated by a highly-polished siliceous stone, pointed at each end, which was found in the ruins of St. Botolph's Priory, Pembrokeshire: close to it lay a water-worn pebble, in which a hole had been bored to the depth of half-an-inch, apparently by friction with the pointed end of the other stone: lastly, a flint flake or knife was found in a barrow (burial-mound) near Hay, Breconshire.

To go back to the period when metals were unknown in this country, and when the only tools used by the savage tribes who then inhabited South Wales were made of stone, bone, or wood, we should have to pass over an interval of time of great duration. Of the later or "Bronze Age," to which the earliest metal objects belong, there are scanty traces in this district; while of antique iron objects few or none have been found; and, indeed, this is what we should expect, for while stone, if undisturbed, is practically indestructible. iron rapidly rusts away under the influence of air and moisture.

APPENDIX.

GLOSSARY OF GEOLOGICAL TERMS.

ABNORMAL.—Not regular or usual.
AERIAL.—Relating to the atmosphere.
AGGLOMERATE.—An accumulation of rock-fragments which have been ejected from a volcano.
ALGÆ.—Seaweeds.
ALLUVIUM.—Matter washed down by rain and rivers: as river-gravels, lake-deposits, the material which forms deltas, &c.
ANEROID.—A peculiar kind of barometer (usually of about the size and form of a watch), consisting of a metallic box partially exhausted of air and provided with a thin flexible lid. This lid rises and falls (and so moves an index hand) according as the pressure or weight of the air diminishes or increases.
ANTHRACITE.—A hard, compact and stony coal; almost pure carbon.
ANTICLINAL.—Dipping in opposite directions, like the two sides of an ordinary house roof.
AQUEOUS.—Formed by or in water.
ARENACEOUS.—Sandy; composed of grains of sand.
ARCHIPELAGO.—A cluster of islands.
ARGILLACEOUS.—Clayey; composed of clay.
ATMOSPHERIC.—Belonging to the atmosphere.
ATTRITION.—Wearing away by rubbing or friction.
AURIFEROUS.—Containing gold.
AXIS.—A central line.
BASIN.—A set of beds resting one upon the other and all dipping to a common centre.
BASSET.—Or out-crop.
BED.—A layer of any kind of rock.
BIND.—A tough clayey shale.

BITUMEN.—Mineral pitch or tar.
BITUMINOUS.—Containing or yielding bitumen.
BLACK-BAND.—A kind of clay ironstone containing sufficient carbon to calcine itself.
BOSS.—Any rounded or projecting mass of rock.
BOULDER.—Any large, usually rounded, mass of stone which has been transported from a distance by the agency of ice.
BRACHIOPODA.—Bivalve shells; one valve usually larger than the other, and perforated for attachment to the rocks: in the interior are two loops or spiral processes to which fleshy arms were attached.
BOG.—A wet spongy morass composed of peat or other decaying vegetable matter.
BRACKISH.—Water which is only slightly salt.
BREAK.—An interval of time unrepresented by any deposit.
BRECCIA.—A rock composed of *angular* fragments cemented together.
CAINOZOIC.—The third, and latest formed, of the three great series of stratified rocks; includes all beds in which fossils of still existing species are found.
CALCAREOUS.—Composed of, or containing lime.
CALC-SPAR, CALCITE.—Crystallised carbonate of lime.
CARBONIFEROUS.—Coal bearing; yielding coal.
CEPHALOPODA.—The highest division of the mollusca, including the nautilus, ammonite, &c.

CHALYBEATE.—A term applied to water holding iron in solution.

CHLORITIC.—Containing green grains.

CHERT.—An impure variety of flint.

CLAY IRONSTONE.—An impure carbonate of iron.

CLEAVAGE.—A structure produced in some rocks, especially slates, in consequence of which they will split into thin plates independent of the planes of bedding; it is due to great pressure.

CLUNCH.—A tough coarse clay.

CONCHOIDAL.—Shell-like: applied to the curved surfaces which some rocks exhibit when broken.

CONCRETIONS.—Rounded lumps or nodules of any mineral matter.

CONFORMABLE.—When successive beds of rock lie parallel one to another.

CONGLOMERATE. A rock formed of rounded pebbles cemented together.

CONTEMPORANEOUS.—Formed at the same time.

CONTORTIONS.—When rocks are abruptly bent or folded they are said to be contorted.

COPROLITES.—Properly, the petrified dung of any animal.

CROP.—Where a bed of rock comes to the surface.

CRUSTACEA.—Animals such as the crab and lobster which have a hard external covering divided into segments.

CULM.—Impure shaly coal or anthracitic shale.

DEBRIS.—Any loose material resulting from the waste of rocks.

DECOMPOSITION.—A breaking up or separating.

DELTA.—The alluvial land formed at the mouth of many rivers.

DENUDATION.—The wearing away of rocks so as to lay bare others that lie beneath them.

DEPOSIT.—Any matter such as mud, sand, &c., which has settled down from suspension in water.

DETRITUS.—Matter worn off or rubbed from rocks, as gravel, mud, &c.

DIP. The amount (measured in degrees) by which any stratum deviates from a horizontal position.

DISINTEGRATION.—Breaking up of rocks by frost, rain, &c.

DRIFT.—Matter which has been carried along; usually applied to deposits which have been transported by ice, as the glacial drift.

DYKE.—The wall-like intrusions of igneous rocks which are frequently seen cutting across other strata.

EPOCH. A period of time; age or era.

EROSION.—A wearing away.

ERRATICS OR ERRATIC BLOCKS.—Rocks which have been transported from a distance, generally by ice.

ESCARPMENT.—The steep slope of a line of hills formed by the outcrop of some hard rock bed.

ESTUARY.—The wide tidal mouth of a river.

FAULT.—Any break in the continuity of a bed of rock. On one side of the crack, fissure, or fault, the beds have been elevated (upthrow side), and on the other depressed (downthrow side). The inclination or slope of a fault is almost always towards the downthrow side. Faults are the result of movements of the earth's crust whereby strata have been fractured and displaced.

FAUNA. The entire group of animals of any region or epoch.

FELSPATHIC.—Containing much felspar.

FERRUGINOUS.—Containing iron; usually red or rusty in appearance.

FIRE-CLAY.—Clay which contains no alkalies, and which can, therefore, resist great heat without melting.

FISSILE.—Splitting into thin plates, as slate.

FLORA.—The plants of any region or epoch.

FLUVIATILE.—Formed or deposited by a river.

FLUVIO-MARINE.—Deposits formed in or near estuaries, and, therefore, containing both fresh water and marine fossils.

FORMATION.—Any set of strata which have their principal characters, as shown by age, fossils, &c., in common.

FOSSIL.—The remains or traces of any animals or plants which by natural processes have become embedded in rocks.

FOSSILIFEROUS.—Containing fossils.

FRIABLE.—Easily crumbled or broken into powder.

APPENDIX.

GENERA.—Plural of genus.

GENUS.—A kind; any set of animals or plants which closely resemble one another.

GEODE.—A hollow nodule, usually containing crystals.

GORGE.—A narrow rocky valley with steep sides.

GRIT.—A coarse or sharp sandstone.

HONE.—A whetstone; stone for sharpening tools.

HORIZON.—The line where sky and sea (or earth) appear to meet: also applied to the particular level or position in a formation at which any given fossil usually occurs

HYDRAULIC CEMENT.—A cement which sets or becomes hard under water.

ICHTHYODUROLITE.—Fossil fin-spines of fishes.

IGNEOUS.—Fire-formed.

IN SITU.—In its natural position.

INTERCALATED.—Placed between or interposed.

INTERSTRATIFIED.—Occurring between other strata.

INTRUSIVE.—Thrust or forced in.

JOINTS.—The natural cracks or fissures by which all rocks are more or less divided into blocks.

LACUSTRINE.—Belonging to or formed in a lake.

LAMINATED.—In very thin layers.

LENTICULAR.—Shaped like a double convex lens.

LOAM.—A mixture of sand and clay.

MARL.—A limy clay.

MATRIX.—The rock in which anything, as a fossil or a crystal, is embedded.

MESOZOIC.—"Middle Life;" the name of the second great division of the stratified rocks.

METAMORPHIC.—Altered or changed in form or structure.

MICACEOUS.—Containing much mica.

MORAINES.—The stony material or detritus brought down by glaciers.

NEVE.—A mass of snow partly converted into ice; forms the upper part of glaciers.

NODULE.—Any lump or concretion of mineral matter.

ORE.—Mineral matter from which some metal can be obtained.

ORGANIC.—Produced by or having organs; includes the vegetable and animal kingdoms.

OUTCROP.—The line where any stratum rises to the surface; called "basset" by miners.

OUTLIER.—A fragment of any formation which has been separated from the main mass by denudation.

OVERLAP.—The extension of upper beds over and beyond lower ones.

PARTING.—A thin bed not above a few inches in thickness.

PERCHED BLOCKS.—See Erratics.

PERCOLATE.—To pass or filter through a porous substance.

PETRIFY.—To convert into stone.

PETROLOGY.—The science which studies rocks.

PLUTONIC.—Igneous rocks which have cooled at great depths and under great pressure, as granite, syenite, &c.

PORPHYRY.—Any igneous rocks which contain large crystals embedded in a compact matrix.

RAVINE. A deep narrow valley with precipitous sides.

ROCK. Used geologically for any materials which form part of the crust of the earth, whether hard or soft.

RUBBLE.—A mass of loose angular stones.

SCARP.—A steep precipitous face of rock.

SCHIST.—A metamorphic foliated rock more or less fissile.

SECTION.—Any exposure of rocks, whether natural or artificial.

SEDIMENT.—Matter which has settled down from suspension in water.

SEPTARIA.—Nodules of impure limestone or ironstone crossed by cracks, filled (usually) with calcite.

SHALE.—Hardened laminated clay.

SILICA.—The oxide of silicon; forms quartz, flint, sand, glass, &c.

SLICKENSIDES.—The polished striated appearance often presented by the sides of faults, caused by friction.

SPECIES.—A set of individuals (animals or plants) so nearly alike that they all may have sprung from one parent.

STRATUM (PL. STRATA).—A layer or bed of rock.

STRATIFIED.—Arranged in layers.

STRIATED.—Finely grooved, or scratched.

STRIKE.—A line on the surface of any bed drawn at right angles to the dip.

SUB-AERIAL.—Under the air; in the atmosphere.

SUB-AQUEOUS.—Under the water.

SUBMARINE.—Under the sea.

SUBMERGENCE.—A sinking under water.

SUBSIDENCE.—A sinking or settling down.

SUPERFICIAL.—Near or on the surface.

SUPERPOSITION.—Resting one above the other; in regular order.

SYNCLINAL.—Applied to beds of rock which dip from opposite directions inwards to a common centre.

SYNCHRONOUS.—Occurring at the same time.

SYSTEM.—See Formation.

TILL.—A Scotch term for boulder clay.

UNCONFORMABLE.—Not parallel one to another; the lower and older rocks having been disturbed, tilted or eroded before the upper strata were deposited upon them.

WATERSHED.—The line which separates two river-basins or drainage areas.

WEATHERING.—The wasting away of rocks by the action of the weather.

WHEAL.—A corruption of "Huel," the Cornish word for "mine."